Plant Virus and Viroid Diseases in the Tropics

K. Subramanya Sastry · Thomas A. Zitter

Plant Virus and Viroid Diseases in the Tropics

Volume 2: Epidemiology and Management

Springer

K. Subramanya Sastry
Department of Virology
SV University
Tirupathi, Andhra Pradesh
India

Thomas A. Zitter
Department of Plant Pathology
and Plant-Microbe Biology
Cornell University
Ithaca, NY
USA

ISBN 978-94-007-7819-1 ISBN 978-94-007-7820-7 (eBook)
DOI 10.1007/978-94-007-7820-7
Springer Dordrecht Heidelberg New York London

Library of Congress Control Number: 2013933113

Printed on acid-free paper

Springer is part of Springer Science+Business Media (www.springer.com)

Foreword

This textbook entitled "Plant Virus and Viroid Diseases in the Tropics-Volume 2: Epidemiology and Management" is an ideal introduction to the subject written for students, practitioners and researchers by Drs. K. Subramanya Sastry and T. A. Zitter. The authors focused on physical and biological factors that favor epiphytotics, including weeds that can constitute virus reservoirs as well as breeding foci for vectors. Virus properties, survival and spread by invertebrate vectors, nematodes and fungi, are followed by the description of spatial dynamics, disease gradients, forecasting, and mathematical modeling techniques. Seven important tropical diseases their ecology and epidemiological aspects are described in detail.

In the Chap. 1, the well organized overview of virus survival and spread, disease forecasting, disease gradients and progress curves, systems analysis and simulation models are covered.

The Chap. 2, the authors have presented the outlines of the management of tropical plant virus and viroid diseases, the selection of virus-free seed and propagules, techniques of production of virus-free plant materials by biotechnological strategies, barrier cropping, time of planting, eliminating of weeds, use of insecticides and use of transgenic plants-the most effective control measure.

The important role of plant quarantine is stressed. Certification schemes for seed-borne viruses in lettuce, barley, pea, beans, soybean, cowpea and peanuts, as well as production of virus-free plants and certification schemes through tissue culture techniques are discussed for cassava, citrus, potato, sugarcane and other plants. Cultural practices and phytosanitation in virus and viroid diseases, as well as cross-protection in fourteen virus diseases are described in detail.

Control of vectors by oil and insecticides are discussed in great detail, as is also the use of aluminum mulches for vector control. Sources of resistance and transgenic approaches to viral and viroid diseases, benefits and risks, are enumerated. Technical guidelines for exchange of germplasm and breeding lines, with detailed descriptions of methods of plant importation are described.

This authoritative review will provide a unique education platform to the readers, so that they can keep in touch with the latest developments in the field of tropical plant virus diseases. Drs. K. Subramanya Sastry and T. A. Zitter have substantial practice in tropical plant viruses and this invaluable book is meticulously researched. It will appeal to all those with an interest in tropical plant virus diseases and their control.

New Brunswick, USA Karl Maramorosch

Preface

Many of the world's most important food crops are grown in the tropics and major crops like rice, maize, wheat, sorghum, barley, tomato, chillies, okra, peas, peanut, sunflower, cucurbits, pigeonpea, etc., are raised through true seed, whereas cassava, potato, sweet potato, sugarcane, cocoa, avocado, apples, banana, and other fruit crops are grown through vegetative propagated materials like tubers, sets, rhizomes, cuttings, budwood, etc. Almost all these crops are affected by important virus and viroid diseases besides fungal and bacterial diseases. However, emergence of new viruses and virus strains of existing viruses, along with changing contexts due to agricultural intensification and climate change have been creating new challenges and demanding an even greater effort to overcome hurdles to increase agricultural productivity, food availability, and economic development. These diseases are responsible for heavy yield losses. We have definite chemical measures against fungal and bacterial diseases, whereas until now no promising viricides have been developed to control virus spread.

Disease-free crops and plants are of great economic and social importance in feeding the world population. The thrust of the book Volume-2 is on virus and viroid disease in the tropics in order to provide the latest information on ecology, epidemiology, and management of virus and viroid diseases in southeastern Asian countries, the African, and South American continents, which fall within the tropical zone. Plant viruses are a matter of great concern globally, but effective management measures against plant viruses requires a clear understanding of their ecology and epidemiology.

Environmental factors like rainfall, wind velocity, soil conditions, temperature, and moisture play a major role in crop production. Among the major virus diseases that are encountered in tropical zones are tungro, yellow mottle and hoja blanca in rice, mosaic in sugarcane, mosaic in cassava, tristeza in citrus, swollen shoot in cacao, sterility mosaic in pigeonpea, rosette and bud necrosis in peanut, necrosis in sunflower and legumes (vegetables and ornamental crops), leaf curl in cotton and tomato, and ringspot in papaya. Key factors for emergence of new plant virus and virus-like diseases include the intensification of agricultural trade (globalization), changes in cropping systems (crop diversification), and climate change.

In this second volume, the list of plant virus genus and species according to 9th ICTV classification and the latest techniques of plant virus diagnosis are included. In the Chap. 1 along with information on various aspects of ecology and epidemiology of plant viruses of tropics, we examine the physical and biological factors which are favorable for epiphytotics to develop. Various aspects related to survival and spread of virus and viroids are also presented as well. For an easier understanding of epidemiology, aspects of disease progress curves, mathematical modeling techniques, and systems analysis and simulation models are discussed.

In the Chap. 2, comprehensive information on plant virus management are included. The ultimate goal of plant pathologists is to effectively manage the virus and viroid diseases of tropical crops. This topic is quite extensively covered on various relevant aspects including integrated disease management practices. In this chapter, various aspects of disease management like the production of virus-free planting materials through certification schemes for crops like cassava, sweet potato, potato, citrus, banana, grapes, strawberry, pome, stone fruits, ornamental bulbous crops that helps in production of virus-free planting materials are discussed. Similarly, new steps on true seed certification schemes for certain legumes are provided. Cultural practices including rouging, border cropping, plant density, elimination of the virus sources, etc., are discussed. Vector control through the application of insecticides and oils or both are found to be effective in certain virus–host combinations are presented. Available success stories of different horticultural crops with cross-protection techniques are included. Development of pathogen-resistant transgenics for the management of virus and viroid diseases are also added. In the present world globalization, plant quarantines play a major role in almost all countries to exclude the entry of new diseases while importing the germplasm from other countries for research and agricultural purposes.

Information on the key factors of virus epidemiology in certain tropical countries is an important step towards the development of management measures against virus and viroid diseases. Identification of risk factors that contribute to virus outbreaks need to be intensified and integrated disease management (IDM) strategy in reducing the impact of these virus diseases needs to be continued throughout the tropical countries. Nevertheless, integrated control measures have evident benefits and should be fostered and promoted as a means of enhancing crop productivity to meet the increasing demands of burgeoning human population. Originally, the authors have planned to confine to the aspects of virus epidemiology and management of tropical zone only. But to provide more information and clarity of the subject, it was inevitable for us to include the research results of temperate crops also, since some of the crops are grown in both zones.

It is hoped that the information provided in this volume on various aspects of virus and viroid diseases of tropical crops would be useful to research scientists, seed companies, quarantine personnel and institutions of both research and teaching.

Tirupathi, India — K. Subramanya Sastry
Ithaca, USA — Thomas A. Zitter

Dr. K. Subramanya Sastry
Emeritus Professor
Department of Virology
SV University
Tirupathi, AP, India
Email: kssastry1944@yahoo.com

Dr. Thomas A. Zitter
Professor of Plant Pathology
Cornell University
Ithaca, NY 14853
Email: taz1@cornell.edu

Acknowledgments

In order to develop Volume-2 on Plant Virus and Viroid Diseases in the Tropics: "Epidemiology and Management", a large number of plant virologists of both National and International presence have immensely helped. We have been benefited by the critical suggestions and comments made by Prof. J. M. Thresh; Prof. F. J. Morales; Dr. G. P. Rao; Dr. N. Vinod Kumar; Dr. K. Bikas Mandal; Prof. H. R. Pappu; Prof. K. V. S. Sarma; Dr. R. A. Naidu; Dr. S. J. Singh; Dr. D. B. Parakh; Dr. S. V. Rao; Prof. V. Muniyappa; Dr. K Muralidharan; Dr. B Viswanath; Dr. A. M. Anthony Johnson; Dr. M. Vinayak Hegde; Dr. S. Edison; Dr. R. Selvarajan; Dr. G. Nagaraja, B. N. K. Kumari and others. The single person who has taken all the brunt and hardship in finalizing the manuscript is Prof. P. Sreenivasulu, Former Head of the Department of Plant Virology, SV University, Tirupathi (India) and we record appreciation for his sincere hard work and devotion to the plant virology subject. We are highly indebted to Dr. P. Lavakumar and IITA for allowing us to use the literature cited in the text. We profusely thank Prof. D. V. R. Saigopal, Head of the Department of Plant Virology, SV University, Tirupathi, for timely guidance, and critically going through all the chapters. Virologists helpful during my (Prof. T. A. Zitter) career include Dr. D. E. Purcifull, Dr. F. W. Zettler, Dr. E. Hiebert, Dr. J. R. Edwardson, Dr. R. Provvidenti, Dr. D. Gonsalves, Dr. J. A. Tomlinson and Dr. D. G. A. Walkey. We wish to thank our entomological colleagues including Dr. J. N. Simons and Dr. J. H. Tsai, and horticultural support from H. Y. Ozaki and V. L. Guzman. Special thanks are extended to the technicians Geisha Echenique, Louis Hsu and Jessica Drennan.

We thank Elsevier, CABI, Springer, Academic Press and other publishers for providing permission for using the illustrations and photographs from their earlier publications. Throughout preparation of the Volume-2, Mr. C. Nagaraja has devoted maximum time in computerizing the book for which the authors are highly indebted. Both the authors acknowledge the help provided by their family members during the preparation of this volume. The meticulous care taken by staff of Springer publishers in bringing out this publication at an early date with a nice display of the book is gratefully appreciated and acknowledged.

K. Subramanya Sastry
Thomas A. Zitter

Contents

Acronyms

ACMV	*African cassava mosaic virus*
AMV	*Alfalfa mosaic virus*
ACLSV	*Apple chlorotic leaf spot virus*
ApMV	*Apple mosaic virus*
ASSVd	*Apple scar skin viroid*
ASGV	*Apple stem grooving virus*
ASPV	*Apple stem pitting virus*
ArMV	*Arabis mosaic virus*
AGVd	*Australian grapevine viroid*
ASBVd	*Avocado sunblotch viroid*
BaMV	*Bamboo mosaic virus*
BBMV	*Banana bract mosaic virus*
BBTV	*Banana bunchy top virus*
BSV	*Banana streak virus*
BaMMV	*Barley mild mosaic virus*
BSMV	*Barley stripe mosaic virus*
BYDV	*Barley yellow dwarf virus*
BaYMV	Barley yellow mosaic virus
BCMV	*Bean common mosaic virus*
BDMV	*Bean dwarf mosaic virus*
BGMV	*Bean golden mosaic virus*
BGYMV	*Bean golden yellow mosaic virus*
BLRV	*Bean leaf roll virus*
BPMV	*Bean pod mottle virus*
BYMV	*Bean yellow mosaic virus*
BCTV	*Beet curly top virus*
BMYV	*Beet mild yellowing virus*
BtMV	*Beet mosaic virus*
BWYV	*Beet western yellows virus*
BYNV	*Beet yellow net virus*
BYV	*Beet yellows virus*
BYVMV	*Bhendi yellow vein mosaic virus*
BVY	*Blackberry virus Y*
BlCMV	*Black eye cowpea mosaic virus*

BMoV	*Blackgram mottle virus*
BlScV	*Blueberry scorch carlavirus*
BBSV	*Broad bean strain virus*
BBWV	*Broad bean wilt virus*
BMV	*Brome mosaic virus*
CabLCV	*Cabbage leaf curl virus*
CaCV	*Capsicum chlorosis virus*
CdMV	*Cardamom mosaic virus*
CLV	*Carnation latent virus*
CarMV	*Carnation mottle virus*
CNFV	*Carnation necrotic fleck virus*
CRSV	*Carnation ringspot virus*
CMoV	*Carrot mottle virus*
CMDV	*Carrot mottley dwarf virus*
CTLV	*Carrot thin leaf virus*
CBSD	*Cassava brown streak disease*
CBSUV	*Cassava brown streak Uganda virus*
CBSV	*Cassava brown streak virus*
CsCMV	*Cassava common mosaic virus*
CMD	*Cassava mosaic disease*
CVMV	*Cassava vein mosaic virus*
CaMV	*Cauliflower mosaic virus*
CeMV	*Celery mosaic virus*
CLRV	*Cherry leaf roll virus*
CMLV	*Cherry mottle leaf virus*
CRLV	*Cherry rasp leaf virus*
ChiLCV	*Chilli leaf curl virus*
CSNV	*Chrysanthemum stem necrosis virus*
CSVd	*Chrysanthemum stunt viroid*
CVB	*Chrysanthemum virus B*
CBLVd	*Citrus bent leaf viroid*
CDVd	*Citrus dwarfing viroid*
CEVd	*Citrus exocortis viroid*
CLBV	*Citrus leaf blotch virus*
CiLV	*Citrus leprosis virus*
CMBV	*Citrus mosaic badna virus*
CiMV	*Citrus mosaic virus*
CPsV	*Citrus psorosis virus*
CRSV	*Citrus ring spot virus*
CTV	*Citrus tristeza virus*
CVV	*Citrus variegation virus*
CYMV	*Citrus yellow mosaic virus*
ClYVV	*Clover yellow vein virus*
CNV	*Cocao necrosis virus*
CSSV	*Cocao swollen shoot virus*

CCCVd *Coconut cadang-cadang viroid*
CFDV *Coconut foliar decay virus*
CoRSV *Coffee ringspot virus*
CoYMV *Commelina yellow mottle virus*
CLCuV *Cotton leaf curl virus*
CABMV *Cowpea aphid borne mosaic virus*
CpBMV Cowpea banding mosaic virus
CCMV *Cowpea chlorotic mottle virus*
CpGMV *Cowpea golden mosaic virus*
CpMV *Cowpea mosaic virus*
CPMoV *Cowpea mottle virus*
CPSMV *Cowpea severe mosaic virus*
CGMMV *Cucumber green mottle mosaic virus*
CMV *Cucumber mosaic virus*
CuNV *Cucumber necrosis virus*
CPFVd *Cucumber pale fruit viroid*
CVYV *Cucumber vein yellowing virus*
CABYV *Cucurbit aphid-borne yellows virus*
CYSDV *Cucurbit yellow stunt disorder virus*
CymMV Cymbidium mosaic virus
DMV Dahlia mosaic virus
DsMV *Dasheen mosaic virus*
DBV *Dioscorea bacilliform virus*
DLV *Dioscorea latent virus*
EACMV *East African cassava mosaic virus*
EMV *Eggplant mosaic virus*
EMDV *Eggplant mottled dwarf virus*
EMARaV *European mountain ash ringspot-associated virus*
FBNYV *Faba bean necrotic yellows virus*
GarMbLV *Garlic mite-borne latent virus*
GFkV *Grapevine fleck virus*
GLRaV *Grapevine leafroll associated virus*
GLRV *Grapevine leafroll virus*
GVA *Grapevine virus A*
GBNV *Groundnut bud necrosis virus*
GRSV *Groundnut ring spot virus*
GRV *Groundnut rosette virus*
HPV *High plains virus*
INSV *Impatiens necrotic spot virus*
ICMV *Indian cassava mosaic virus*
ICRSV *Indian citrus ringspot virus*
IPCV *Indian peanut clump virus*
IYSV *Iris yellow spot virus*
JYMV *Japanese yam mosaic virus*
LYSV *Leek yellow stripe virus*

LBVV	*Lettuce big vein virus*
LiYV	*Lettuce infectious yellows virus*
LMV	*Lettuce mosaic virus*
LNYV	*Lettuce necrotic yellows virus*
LSV	*Lily symptomless virus*
LLV	*Lolium latent virus*
MacMV	*Maclura mosaic virus*
MCMV	*Maize chlorotic mottle virus*
MDMV	*Maize dwarf mosaic virus*
MMV	*Maize mosaic virus*
MRFV	*Maize rayado fino virus*
MRDV	*Maize rough dwarf virus*
MSV	*Maize streak virus*
MSpV	*Maize stripe virus*
MNSV	*Melon necrotic spot virus*
MYSV	*Melon yellow spot virus*
MeYVMV	*Mesta yellow vein mosaic virus*
MLBVV	*Mirafiori lettuce big-vein virus*
MiLV	*Mirafiori lettuce virus*
MYMV	*Mungbean yellow mosaic virus*
OCSV	*Oat chlorotic stunt virus*
OSDV	*Oat sterile dwarf fijivirus*
ORSV	*Odontoglossum ringspot virus*
OLCV	*Okra leaf curl virus*
OkMV	*Okra mosaic virus*
OYVMV	*Okra yellow vein mosaic virus*
OLV-2	*Olive latent virus 2*
OYDV	*Onion yellow dwarf virus*
OrMV	*Ornithogalum mosaic virus*
OuMV	*Ourmia melon virus*
PMV	*Panicum mosaic virus*
PaLCuV	*Papaya leaf curl virus*
PRSV	*Papaya ring spot virus*
PYFV	*Parsnip yellow fleck virus*
PWV	*Passion fruit woodiness virus*
PEBV	*Pea early browning virus*
PEMV	*Pea enation mosaic virus*
PMV	*Pea mosaic virus*
PSbMV	*Pea seed-borne mosaic virus*
PLMVd	*Peach latent mosaic viroid*
PRMV	*Peach rosette mosaic virus*
PBND	*Peanut bud necrosis disease*
PBNV	*Peanut bud necrosis virus*
PCV	*Peanut clump virus*
PeMoV	*Peanut mottle virus*

PStV	*Peanut stripe virus*
PZSV	*Pelargonium zonate spot virus*
PepMV	*Pepino mosaic virus*
PMMV	*Pepper mild mosaic virus*
PMMoV	*Pepper mild mottle virus*
PeMV	*Pepper mottle virus*
PepRSV	*Pepper ringspot virus*
PVBV	*Pepper vein banding mosaic virus*
PVMV	*Pepper veinal mottle virus*
PYMV	*Pepper yellow mottle virus*
PVCV	*Petunia vein clearing virus*
PPSMV	*Pigeon pea sterility mosaic virus*
PMWaV	*Pineapple mealybug wilt associated virus*
PYMoV	*Piper yellow mottle virus*
PPV	*Plum pox virus*
PnLV	*Poinsettia latent virus*
PopMV	*Poplar mosaic virus*
PAMV	*Potato aucuba mosaic virus*
PLRV	*Potato leaf roll virus*
PMTV	*Potato mop-top virus*
PSTVd	*Potato spindle tuber viroid*
PVA	*Potato virus A*
PVC	*Potato virus C*
PVS	*Potato virus S*
PVT	*Potato virus T*
PVX	*Potato virus X*
PVY	*Potato virus Y*
PYDV	*Potato yellow dwarf virus*
PYMV	*Potato yellow mosaic virus*
PYVV	*Potato yellow vein virus*
PYV	*Potato yellowing virus*
PoLV	*Pothos latent virus*
PDV	*Prune dwarf virus*
PNRSV	*Prunus necrotic ringspot virus*
RBDV	*Raspberry bushy dwarf virus*
RpRSV	*Raspberry ringspot virus*
RBSDV	*Rice black streaked dwarf virus*
RDV	*Rice dwarf virus*
RGSV	*Rice grassy stunt virus*
RHBV	*Rice hoja blanca virus*
RNMV	*Rice necrosis mosaic virus*
RRSV	*Rice ragged stunt virus*
RSV	*Rice stripe virus*
RTBV	*Rice tungro bacilliform virus*
RTSV	*Rice tungro spherical virus*

RTV	*Rice tungro virus*
RWSV	*Rice wilted stunt virus*
RYMV	*Rice yellow mottle virus*
RGMV	*Ryegrass mosaic virus*
SDV	*Satsuma dwarf virus*
SLV	Shallot latent virus
ShVX	*Shallot virus X*
SBWMV	*Soil-borne wheat mosaic virus*
SrMV	Sorghum mosaic virus
SACMV	*South African cassava mosaic virus*
SBMV	*Southern bean mosaic virus*
SRBSDV	*Southern rice black streaked dwarf virus*
SbBMV	*Soybean blistering mosaic virus*
SbCMV	*Soybean chlorotic mottle virus*
SbDV	*Soybean dwarf virus*
SMV	*Soybean mosaic virus*
SSSV	*Soybean severe stunt virus*
SLCV	*Squash leaf curl virus*
SqMV	*Squash mosaic virus*
SqVYV	*Squash vein yellowing virus*
SLCMV	*Srilankan cassava mosaic virus*
SCV	*Strawberry crinkle virus*
SLRSV	*Strawberry latent ringspot virus*
SMoV	*Strawberry mottle virus*
SCRLV	*Subterranean clover red leaf virus*
SCSV	*Subterranean clover stunt virus*
SBYV	*Sugar beet yellows virus*
SCFDV	*Sugarcane Fiji disease virus*
SCMV	*Sugarcane mosaic virus*
SCSMV	*Sugarcane streak mosaic virus*
SCYLV	*Sugarcane yellow leaf virus*
SuCMoV	*Sunflower chlorotic mottle virus*
SNV	*Sunflower necrosis virus*
SHMV	*Sunn-hemp mosaic virus*
SPCSV	*Sweet potato chlorotic stunt virus*
SPFMV	*Sweet potato feathery mottle virus*
SPLV	*Sweet potato latent virus*
SPLCV	*Sweet potato leafcurl virus*
SPMMV	*Sweet potato mild mottle virus*
SPMSV	*Sweet potato mild speckling virus*
SPSVV	*Sweet potato sunken vein virus*
SPV	*Sweet potato virus*
SPVD	*Sweet potato virus disease*
TaBV	*Taro bacilliform virus*
TEV	*Tobacco etch virus*

TLCV	*Tobacco leaf curl virus*
TMGMV	*Tobacco Mild Green Mosaic Virus*
TMV	*Tobacco mosaic virus*
TNV	*Tobacco necrosis virus*
TRV	*Tobacco rattle virus*
TRSV	*Tobacco ring spot virus*
TSV	*Tobacco streak virus*
TStV	*Tobacco stunt virus*
TVCV	*Tobacco vein clearing virus*
TVMV	Tobacco vein mottling virus
TASVd	*Tomato apical stunt viroid*
TAV	*Tomato aspermy virus*
TBRV	*Tomato black ring virus*
TBSV	*Tomato bushy stunt virus*
ToCV	*Tomato chlorosis virus*
TCSV	*Tomato chlorotic spot virus*
TGMV	*Tomato golden mosaic virus*
TICV	*Tomato infectious chlorosis virus*
ToLCV	*Tomato leaf curl virus*
ToMV	*Tomato mosaic virus*
ToMoV	*Tomato mottle virus*
TPCTV	*Tomato pseudo-curly top virus*
ToRSV	*Tomato ring spot virus*
TSWV	*Tomato spotted wilt virus*
ToTV	*Tomato torrado virus*
TYLCV	*Tomato yellow leaf curl virus*
TriMV	*Triticum mosaic virus*
TBV	*Tulip breaking virus*
TCV	*Turnip crinkle virus*
TuMV	*Turnip mosaic virus*
TYMV	*Turnip yellow mosaic virus*
ULCV	*Urdbean leaf crinkle virus*
WBNV	*Watermelon bud necrosis virus*
WMV	*Watermelon mosaic virus*
WSMoV	*Watermelon silver mottle virus*
WSSMV	*Wheat spindle streak mosaic virus*
WSMV	*Wheat streak mosaic virus*
WCCV	*White clover cryptic virus*
WClMV	*White clover mottle virus*
WTO	World Trade Organization
WTV	*Wound tumor virus*
YMMV	*Yam mild mosaic virus*
YMV	*Yam mosaic virus*
ZYMV	*Zucchini yellow mosaic virus*

Chapter 1
Ecology and Epidemiology of Virus and Viroid Diseases of Tropical Crops

1.1 Introduction

Generally for distinguishing the differences between the terms ecology and epidemiology, there are no clear cut differences. The ecology describes the factors influencing the behavior of a virus in a given physical situation. These factors include host-range, tissue tropism, pathogenesis and host responses. It is a fundamental concept based on the relational properties of the virus and is not a property of the environment. Epidemiology is the study of the determinants, dynamics and distribution of disease caused by virus and virus-like pathogens in host populations over time and space. It includes a dimensional aspect of the factors determining the spread of a virus into a given situation, as it is influenced by environmental factors, rate of pathogen (virus), reproduction, mode of viral dispersal, efficiency of virus survival, level of aggressiveness of the virus, level of host plant resistance, and others. The epidemics is the progress of disease in time and space. The magnitude of disease intensity depends on the initial level of disease and the rate of increase, the disease intensity either low or high at a given time during the epidemics. A directory of plant pathology defines an epidemics as a "wide spread increase, usually limited in time, in the incidence of infectious disease". When the epidemics occur over large areas, with high levels of disease, it is known as pandemic. The epidemics may cause minor to moderate losses and occasionally epidemics may also get out of control and become extremely wide spread. For example, in sub-Saharan Africa, since 1900 there were more than 15 epidemics due to rosette virus in groundnut with losses of up to £ 200 million for epidemics. Similarly, *African cassava mosaic virus* epidemics have caused £ 40 million loss annually affecting 60,000 ha of cassava. The two terms, ecology and epidemiology are often interlink but each has a specific meaning (Hull 2002).

Kranz (1990) defines epidemiology subject as "the science of populations in populations of host plants, and the diseases resulting there from under the influence of the environment and human interferences". The foundations of the modern plant epidemiology were laid down by Gregory (1968). He examined in great detail how different forms of disease gradients in space arose and how these

K. S. Sastry and T. A. Zitter, *Plant Virus and Viroid Diseases in the Tropics*,
DOI: 10.1007/978-94-007-7820-7_1,

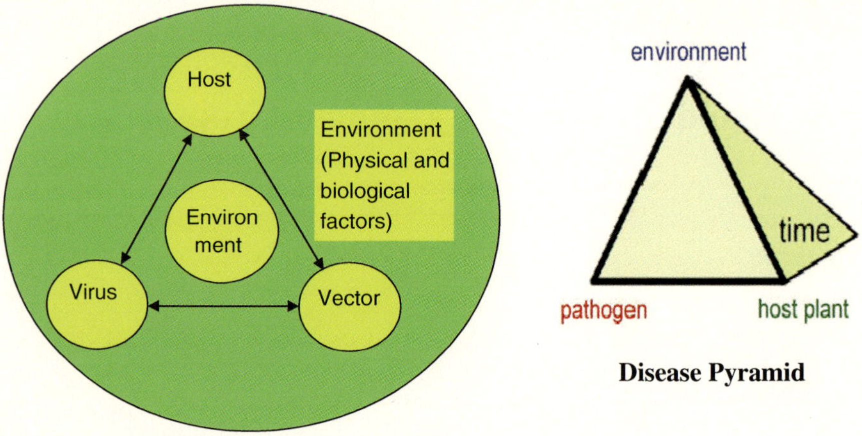

Fig. 1.1 Diagrammatic representation of interactions between plant-virus-vector as influenced by environmental factors

developed over time. For the past two decades molecular techniques are extensively used in ecological and epidemiological studies and the word molecular ecology is coined. The molecular ecology is most useful for the management of tropical plant virus diseases. Fargette et al. (2006) have extensively studied molecular ecological aspects of *Cassava mosaic virus, Rice yellow mottle virus,* and *Banana streak virus*. Concepts and techniques of molecular ecology are described in text books (Brown 2002; Beebee and Rowe 2004; Freeland 2005).

Understanding plant virus disease epidemiology is vital to formulating viable disease management practices in a given agro-ecosystem. Various factors that influence the development and progress of plant disease epidemics are variable from the region to region and hence researchers working in different regions have to understand disease epidemics before attempting to control the virus and virus-like diseases. Thus different groups of researchers are involved in studying the epidemiology of a particular crop-virus-vector-environment system (Fig. 1.1).

The factors responsible for virus disease epidemics are (1) The host should be susceptible. (2) The virus should be virulent to cause the disease. (3) The presence of a potential vector. (4) When the above cited factors are present, then several environmental factors play an important role in the disease development. For pedagogical purposes, the triangle is sometimes expanded to a square, pyramid (Fig. 1.1) or multi-dimensional structure to either emphasize the effects of time and / or space on disease, to reflect the influence of humans on diseases (Francl 2001).

Some pathologists represent the specific conditions required for disease development in the form of "Disease Pyramid". Specific conditions must be present for biotic disease to develop. There must be a susceptible host plant, the pathogen and environmental conditions conducive for disease development and must come together in a given point in time. These conditions make up what is called the "Plant Disease Pyramid". Biotic disease cannot occur if one of these pieces is missing (Fig. 1.1).

Due to the hectic work involved in the research of plant virus epidemiology, only a few research groups are engaged in studying the epidemiology of virus and viroid diseases. In addition, a very long time is required to generate meaningful data about particular crop-virus-vector system. Further, this study requires constant involvement and periodical visits to the fields to achieve a better understanding of the various factors that contribute to the virus disease epidemics. Even with all these limitations, a meaningful group of disease forecasting systems were developed against a number of virus diseases.

1.2 Epidemiological Concepts

Epidemiology is the science of disease in populations, and is a quantitative discipline with strong conceptual foundations and practical applications. Epidemics which is also known as epiphytotics, of plant diseases will have major impacts on agricultural and horticultural crops with major socio-economic and political consequences. The epidemics are defined as the phenomenon in which a pathogen spreads severely and affects many plant individuals within a population over relatively large area and within a relatively short time frame. An epidemiological tool-kit includes methodologies for analyzing disease progress in time and space, the derivation of key parameters determining the rate of epidemic progress and the extent of control required for disease management. The complexity of biotic interactions affecting plant epidemics is shown by introducing genetic variation in host and pathogen. Epidemic development of virus diseases is influenced by (1) Environmental factors, (2) Rate of virus/viroid replication, (3) Mode of virus dispersal, (4) Efficacy of virus survival, (5) Level of aggressiveness of the virus, and (6) Level of host plant resistance.

For more details on the epidemiological concepts, the reader may refer to representative books/review articles (Thresh 1974a, 1980; Harrison 1981; McLean et al. 1986; Thresh 1991, 2006a, b; Anderson 2005; Jeger 2009; Waggoner and Aylor 2000; Jones et al. 2010; Jones and Barbetti 2012).

To understand the intricacies of various factors that contribute to disease epidemics, involvement of plant virologists/plant pathologists, entomologists, crop specialists and statisticians are needed to facilitate the fruitful epidemiological investigations. Having this in foresight as early as 1978, a scientific society "International plant virus epidemiology" (IPVE) was formed under the chairmanship of Prof. Mike Thresh at the third "International Society of Plant Pathology" (ISPP) congress held at Munich, Germany. The first plant virus epidemiology symposium was held at Oxford, UK in 1981 and IPVE has conducted 11 workshops at 2 years intervals on this topic and the latest 12th Conference was held at Arusha, Tanzania in 2013. The proceedings of the IPVE were published (Plumb and Thresh 1983; McLean et al. 1986; Fereres et al. 2000; Thresh et al. 2004).

Basic information on plant virus ecology and epidemiology is available from different sources (Zitter 1977; Maramorosch and Harris 1981; Plumb and Thresh 1983; Jeger and Chan 1995; Waggoner and Aylor 2000; Zadoks 2001; Fereres et al. 2000; Thresh 2003; Thresh and Fargette 2003; Jeger et al. 2004; Anderson and Morales 2005; Fargette et al. 2006; Madden et al. 2007; Jones 2009; Makkouk and Kumari 2009; Pappu et al. 2009; Jones et al. 2010; Patil and Fauquet 2011; Jones and Barbetti 2012).

In addition, researchers have worked out the information related to plant virus ecology and epidemiology in different crop-virus-pathosystems, for example, exhaustive information is available on *Tospoviruses* (Reddy et al. 1983a; Van Os et al. 1993; Gitaitis et al. 1998; Culbreath et al. 2003; Wells et al. 2003; Coutts et al. 2004a; Jones 2004; Ranganath et al. 2006; Pappu et al. 2009; Naidu 2013); *Begomoviruses* (Muniyappa 1980; Cohen and Antignus 1994; Thresh et al. 1994; Sserubombwe et al. 2001; Fondong et al. 2002; Morales 2004; Morales and Jones 2004; Anderson and Morales 2005; Patil and Fauquet 2011); *Tomato mottle Geminivirus* (Polston et al. 1996); *Rice tungro virus* disease complex (Yadav and Mishra 1990; Anjaneyulu et al. 1994; Hibino 1996; Varma et al. 1999; Osmat 2000; Azzam and Chancellor 2002; Muralidharan et al. 2003; Chancellor et al. 2006; Krishnaveni et al. 2011); *Rice yellow mottle virus* (Reckhaus and Andriamasintscheno 1997; Fargette et al. 2006); *Rice hoja blanca virus* in rice (Morales and Jennings 2010); *Maize streak virus* (Rose 1978; Asanzi et al. 1994; Bosque-Perez and Buddenhagen 1999; Smith et al. 2000; Magenya et al. 2008); *Maize chlorotic dwarf virus* in maize (Madden et al. 1990a); PVY, PLRV in potato (Jones 1981; Shahid Ali et al. 2013); *African cassava mosaic virus* (Bock 1983; Fauquet et al. 1988; Fauquet and Fargette 1990; Thresh 1991; Fargette et al. 1993, 1994a, b; Holt et al. 1997; Legg et al. 1997; Calvert and Thresh 2002; Jeger et al. 2004) and *Cassava brown streak virus* in cassava (Jeremiah et al. 2013); *Cucumber mosaic virus* in chilli, tomato and eggplant (Kiranmai et al. 1998) and in lupine (Thackray et al. 2004); *Barley yellow dwarf virus* in barley (Irwin and Thresh 1990; Burges et al. 1999); *Yellow vein mosaic virus* in okra (Chellaiah and Murugesan 1976; Khan and Mukhopadhyay 1985; Pun and Doraisamy 2000; Kalita et al. 2005); *Tobacco leaf curl virus* in tobacco (Valand and Muniyappa 1992); *Tomato leaf curl virus* in tomato (Sastry et al. 1978; Saikia and Muniyappa 1989; Ramappa et al. 1998; Jeger et al. 2004); *Tomato yellow leaf curl virus* (Mazyad et al. 1979; Ioannau and Iordanou 1985; Cohen et al. 1988; Aboul-Ata et al. 2000; Czosnek 2007); *Yellow mosaic disease* of horsegram in *Macrotyla uniflorum* (Horsegram) (Muniyappa 1983); *Yellow mosaic disease* of soybean (Gupta and Keshwal 2003); *Soybean mosaic virus* in soybean (Irwin and Goodman 1981; Almeida et al. 1994; Irwin et al. 2000); *Pigeonpea-sterility mosaic virus* in pigeonpea (Singh and Rathi 1997; Kumar et al. 2008a); *Bean yellow mosaic virus* in lupin (Thackray et al. 2002); *Bean golden mosaic virus* (Anderson and Morales 2005); *Bean leaf roll virus* and *Chickpea chlorotic stunt virus* in legumes (Makkouk and Kumari 2009); *Groundnut rosette virus* (Naidu et al. 1998); *Indian peanut clump virus* in groundnut (Delfosse 2000); *Iris yellow spot virus* in onion (Pappu et al. 2007); *Tomato spotted wilt virus* in lettuce and pepper (Coutts et al. 2004a);

Tobacco streak virus in sunflower (Nagaraju et al. 2003; Shivasharanayya and Nagaraju 2003; Upendhar et al. 2006; Chandra Mohan et al. 2006; Lakhmod et al. 2007; Lokesh et al. 2008; Kumar et al. 2008a); and virus diseases in sweet potato (Alicai et al. 1999; Byamukama et al. 2004). Epidemiological studies were also carried out against major virus diseases of certain fruit crops viz., tristeza in citrus species (Gottwald et al. 2002); plumpox virus in stone fruits (Dallot et al. 2003; Budzanivska et al. 2011); *Blackberry yellow vein associated virus* in blackberry (Poudel et al. 2013); bunchy top virus in banana (Allen 1978; Smith et al. 1998).

All these cited examples are concerned to epidemiological studies of different virus diseases which have various types of insects as vectors. Epidemiological studies were also carried out on viroid diseases which have no insect-vectors and generally spread through true seed and vegetative plant material. Even weed and collateral hosts helps in viroid disease spread as seen in PSTVd in tomato and *Chrysanthemum mottle viriod* in chrysanthemum (Yamamoto and Sano 2006; Verhoeven et al. 2010).

1.3 Conditions Favorable for Epiphytotics

Virus and viroid pathogens will not always be able to cause a disease unless environmental conditions and suitability of the host are also favourable for survival, multiplication, and entry of the pathogen into the plant and further development of the disease. For causing (epidemics) epiphytotics, the pathogen has to be virulent in the first place; the host has to be susceptible, and the environmental conditions like temperature, relative humidity, soil moisture, soil pH, and soil type, need to be favourable (Nayudu 2008; Jones 2013). Disease will not develop if anyone of the three conditions is not fulfilled.

If all the criteria are not fulfilled, such as susceptible host and pathogen are present but the environment is not conductive to the pathogen infecting and causing disease, disease cannot occur. Likewise, if the host is susceptible and the environment favors the development of disease but the pathogen is not present, there is no disease (Fig. 1.2). Sometimes a fourth factor of time is added as the time at which a particular infection occurs, and the length of time conditions remain viable for that infection, can also play an important role in epidemics.

1.3.1 Physical Factors

(a) **Air temperature**

The changes in air temperature will alter the distribution and prevalence of viruses and vectors. However, one cannot anticipate the emergence of entirely new problems as a consequence of global warming.

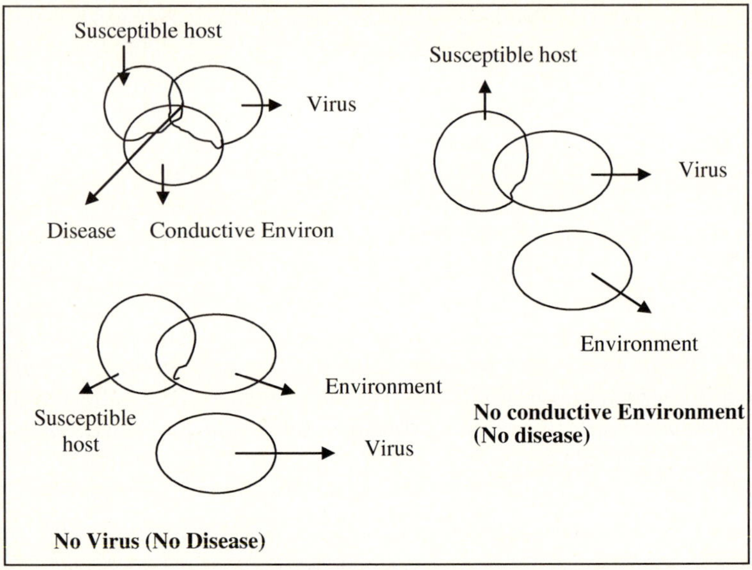

Fig. 1.2 Criteria for virus disease development

There is a close correlation between the air temperature and the insect vector population primarily the whitefly (*Bemisia tabaci*). Whitefly transmitted geminiviruses are major problems in almost all tropical countries. Among the major geminiviruses are: *African cassava mosaic virus* in Africa; *Bean golden mosaic virus* in Brazil, Puerto Rico and Guatemala; *Cotton leaf curl virus* in Pakistan and India; *Tomato leaf curl virus* and *Rice tungro virus* in southeast Asian countries and *Tomato yellow leaf curl virus* (TYLCV) in Israel, Spain, Italy and some other tropical countries. In recent years, *Torradoviruses* which includes *Tomato torrado virus,* has been associated with the green-house whitefly, *Trialeurodes vaporariorum* in Poland as well as *B. tabaci* elsewhere.

Weather data collected at UAS, Bangalore where experiments involving the incidence of TLCV and whitefly populations were conducted, indicated high temperature, low humidity and low rain fall from January to May, coincided with the increase in the whitefly vector population. The low whitefly populations during July–November were perhaps due to low temperatures, high rainfall, and high humidity (see Fig. 1.3).

The impact of the virus diseases on yield is presented in the Chapter-3 of Volume I. Temperature plays a key role in the population dynamics and activity of whiteflies.

Temperatures of 20–30 °C favour large populations of whitefly and are associated with high fecundity, and greater longevity (Cock 1986; Fargette et al. 1994a, b). In Latin America, *Bean golden mosaic virus* (BGMV) was transmitted by *B. tabaci*, and the optimum vector population was at the warmer temperatures of 26.7 °C (26.5 %) or above. Generally, the inciting virus is most important at

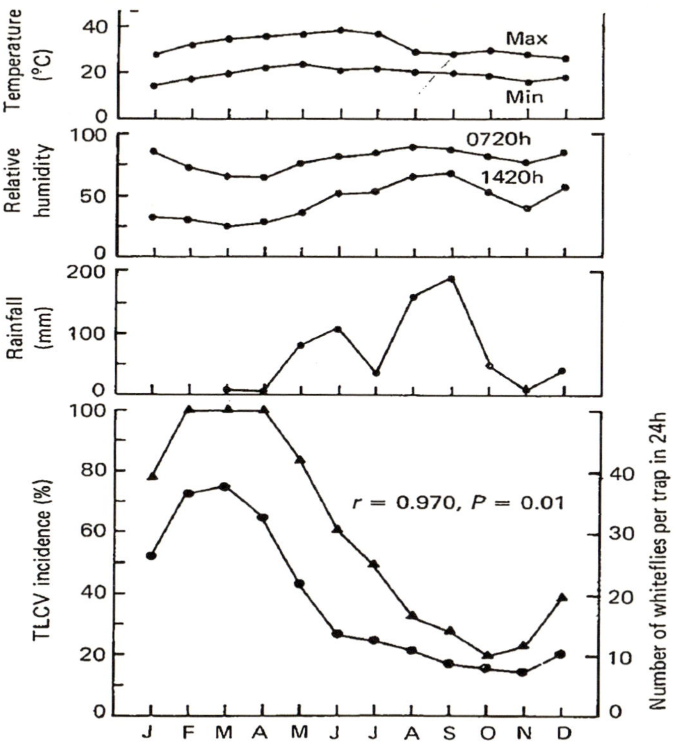

Fig. 1.3 TLCV incidence in relation to whitefly numbers and weather conditions during 1983; ●, whitefly density; ▲, TLCV level. *Source* Saikia and Muniyappa (1989)

elevations below 2,000 m, where whitefly populations, inoculum sources and warm temperatures are more common (Anderson and Morales 2005; Morales 2006).

In India epidemiological studies were conducted against *Tomato leaf curl virus* (TLCV) transmitted by *B. tabaci* by planting tomato crop in March, July and November months by Saikia and Muniyappa (1989). In the March (summer) planted tomato crop the TLCV incidence appeared 2 weeks after planting; spread was initially slow but from 5 weeks onwards the incidence increase rapidly, reaching 100 % by 11 weeks. In the July planted crops, symptoms of TLCV were first observed 3 weeks after planting, and then increased slowly in incidence, reaching 59 % at 14 weeks after planting. In the November planted tomato crop symptoms first appeared 4 weeks after planting and a maximum of 66 % TLCV incidence was observed 14 weeks after planting (Fig. 1.4).

The air temperatures will also have marked effects on the rate of multiplication and movement of airborne virus vectors. For example, aphid vector populations will be reduced at higher temperatures and with reasonably warm temperatures the winged alates tend to fly.

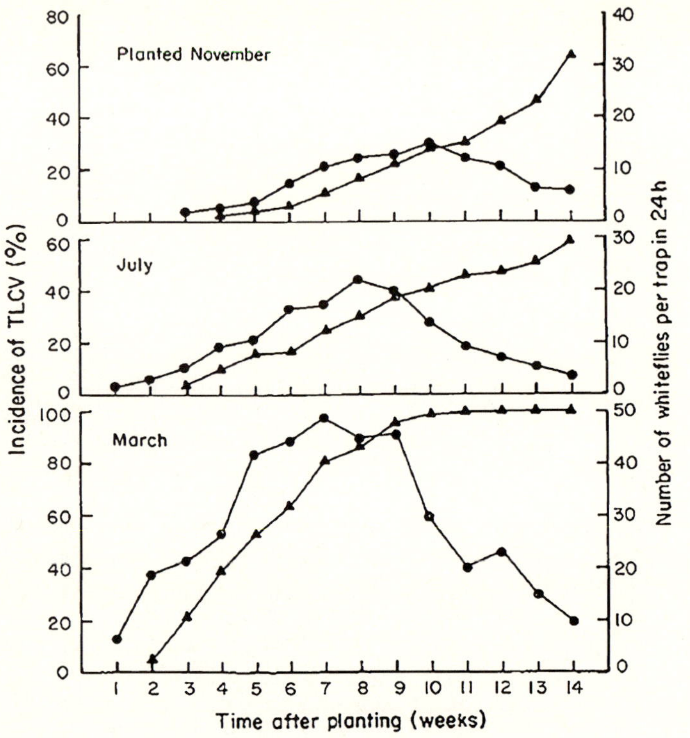

Fig. 1.4 Spread of TLCV in relation to whitefly numbers in different seasons during 1983; ●, whitefly, density; ▲, TLCV level. *Source* Saikia and Muniyappa (1989)

In central India growth of the potato crop under high temperatures (30–35 °C) and dry weather during September–October months favors thrips activity to the maximum extent and thus potentially cause higher incidence of *Groundnut bud necrosis virus* (GBNV). In the crops planted after October 15th, low activity of thrips and thus low GBNV incidence occurs. Therefore, potato planting after the end of October is helpful in reducing GBNV disease incidence by avoiding crop exposure to thrips vectors (Somani et al. 2007); similar studies have been reported for GBNV (also called *Peanut bud necrosis virus*) in greengram (Sreekanth et al. 2002). Similarly in India, the high incidence of *Tomato leaf curl virus* in tomato (Saikia and Muniyappa 1989) and *Indian peanut clump virus* (IPCV) in peanut (Delfosse 2000) are related to humidity and temperature.

The air temperature will also have marked effects on the rate of multiplication and movement of the air-borne virus vectors. For example aphid populations at higher temperatures will be reduced, while at reasonably warm temperatures the winged alates tend to fly.

An economically important virus of rice is *Rice tungro virus* which is caused by two viruses, *Rice tungro bacilliform* and *Rice tungro spherical viruses* which are transmitted by the leafhopper vector *Nephotettix* species and present in almost all Southeast Asian countries. Epidemiological studies on rice viruses revealed that there is no uniformity in time of appearance of the vector populations.

In west Bengal (India) the insects appear towards the end of dry season (March–April), gradually increasing during the wet season and reaching a peak during September–October. The population declines slowly during winter season (November–February) (Mukhopadhyay 1986). In Bangladesh (Alam and Islam 1959) *N. virescens* multiplies in aus crop (April–August) and reaches a peak towards the end of aus crop and beginning of aman crop (July–August). In double cropping areas of Malaysia, *N. virescens* increases rapidly during May–June, reaches a peak during July, and declines sharply thereafter (Lim 1969). Large numbers of *N. virescens* also appear in the seedbeds and migrate to late planted crops (Bottenberg et al. 1990).

The spread of soil-borne fungal vectors primarily depend on the temperature of the soil. High ambient temperature and comparatively low relative humidity apparently accelerates the growth and development of the fungal vector. In West Africa, during December–January, when average relative humidity falls from a normal 80–90 % to about 40 %, the *Polymyxa* vector development is most rapid (Thouvenel and Fauquet 1980). In India, at temperatures less than 25 °C, the incidence of IPCV is negligible, but the high temperatures that prevail during the monsoon growing season, are conducive to natural transmission. High temperatures followed by outbreaks of rain appear to break dormancy of the resting sporangia of *Polymyxa* vector (Reddy 1991).

(b) **Rainfall and humidity**

The spread of certain viruses like *Cucumber green mottle mosaic virus* in river waters, and *Tobacco mosaic virus* in leachates of glass house running water is a regular feature in virus spread (Koenig 1986). The role of rain water in the spread of IPCV transmission by fungal vector *Polymyxa graminis* has been determined by Delfosse et al. (2002). In India, IPCV occurs mainly in the rainy season (July–November) in peanut crops and the peanut crop escapes the disease in post rainy season (December–March). Even in West Africa when the peanut crops are suggested to be grown during the rainy season, heavy losses have been sustained (Delfosse 2000; Delfosse et al. 2002). In almost all tropical countries *Bemisia tabaci* transmitted *Geminiviruses* are widespread and heavy rains reduce virus spread by decreasing whitefly populations (Fargette et al. 1994a, b). More details on role of water in virus spread can be obtained from review articles of Koenig (1986) and Kontzog et al. (1988).

From Nigeria, Fajinmi et al. (2011) while working with *Pepper veinal mottle virus* (PVMV) on Bell pepper have observed increased populations of aphid vectors in all agro-ecological zones from March, onset of rainy season, and reached a peak in August and then declined from September, onset of the dry season. The Humid forest and derived Savanna agro-ecological zones had the highest populations of aphids and types of aphid species. The ability to transmit PVMV to a healthy pepper plant varied between aphid species, and *R. maidis* failed to transmit PVMV from infected to healthy plants. The most probable significant PVMV vector was identified as *Myzus persicae* (Sulzer) and *Aphis gossipii* (Glover).

Early studies in South Africa by Storey and Bottomley (1928) established a positive relationship between the final incidence of *Groundnut rosette virus* and the total rainfall during the preceding dry season. This influenced the survival of volunteer self-grown groundnut seedlings and crop debris and hence the abundance of initial source of inoculum. Other positive relationships between the incidence and pre-season rainfall that influences the prevalence of vectors and inoculum sources have been established with *Cotton leaf curl virus* and *Maize streak virus* (Thresh et al. 2003).

(c) **Wind**

Through wind currents vectors like whiteflies, thrips, and mealybugs are blown to distant places and hence viruses like *Begomoviruses*, TSWV, TSV and *Cocoa swollen shoot viruses* can spread to new areas (Strickland 1950). Primarily the wind helps in the flight of air-borne vectors and also in determining the direction of virus spread. Freak wind conditions may cause annual epidemics of virus diseases. The wind direction and speed will influence the flights of whiteflies, leafhoppers, aphids, and other insect vectors. For example, aphids carrying *Barley yellow dwarf virus* and *Maize dwarf mosaic virus,* across America when blown by northerly jet winds. Similarly, aphid vectors of *Lettuce necrotic yellows virus* (LYNV) were introduced into New Zealand through air streams (Close and Tomlinson 1975). Other vectors like brown leafhoppers (*Nilaparvata lugens*) transmitting *Rice ragged stunt virus* (RRSV) and *Rice grassy stunt virus* (RGSV) in rice are dispersed annually over long distances from tropical to temperate regions of Asia (Thresh 1983). The leafhopper *Macrosteles fascifrons* is probably blown across the U.S. and into the Canadian plains (Chiykowski and Chapman 1965). Regarding the spread of mealybugs, at very young stages they are carried some distances by wind and are responsible for the spread of Swollen shoot disease of cocoa (Thresh 1958). *Rice yellow mottle virus* which is mechanically transmitted in rice, the field spread also takes place due to wind-currents (Sarra et al. 2004).

(d) **Light**

The activities of insects that are vectors of plant virus diseases are influenced by light. The whitefly (*Bemisia tabaci*) shuns light and is invariably seen on the lower side of the leaf. The flight of *Myzus persicae* is inhibited at night by low light intensity. The dramatic increase in flight activity in most insect vectors around sunset is due to an intrinsic biological urge to take-off at dusk or twilight.

In West Africa, Atiri et al. (2000) have extensively studied all the above cited climatic factors in relation to epidemiology of economically important virus diseases and their spread. Case studies of *Maize streak mono-gemini virus*, *Okra mosaic tymovirus* and *African cassava mosaic begomovirus* demonstrate that the most important factor that influences the incidence and spread of virus diseases is climate.

That is because climate influences: (1) virus disease outbreaks; (2) the rate of development and activity of virus vectors and also of their migration; and (3) the phenology of crops, weeds and wild hosts that harbour plant viruses. Rainfall, temperature and wind are identified as key weather components in virus pathosystems involving maize (cereal), okra (vegetable) and cassava (tuber crop), and are therefore important to determine the most suitable period in which to undertake crop protection measures.

1.3.2 Biological Factors

1.3.2.1 Sources of Inoculum

Initial sources of infection must be present for disease outbreaks to occur. Sources that are numerous and present within or alongside crops are particularly dangerous. Under field conditions the initial virus inoculum will be infected stocks of seed or vegetative planting material. However, there are also other additional sources like weed hosts volunteer plants, adjacent crops, viruliferous vectors etc., and discussed in detail in this chapter. For example in India, farmers grow groundnut, sunflower, vegetables like okra, cotton in adjacent fields. For viruses like TSV, they can act as a virus source. Infected sunflower crops can serve as a virus source for groundnut (Vemana and Jain 2010).

Infected trees, shrubs or herbaceous plants growing around crops may provide a reservoir of virus infection (Duffus 1971; Bos 1981). Such plants may be the only foci of infection for the crop, or they may be the primary source, with secondary spread occurring from plants that become infected within the crop.

1.3.2.2 Infected Seed Stocks

Virus transmission through seed is important for virus dissemination as discussed previously and nearly 231 viruses are seed-borne (Sastry 2013a). The infected plants from the seed act as foci for virus inoculum in the field and tend to be randomly distributed within crops are particularly important. This facilitates spread by contact (e.g., BSMV) or by vectors (e.g., LMV). A basic approach to disease control is to develop and release virus-free stocks of seed for use by farmers. There have been few attempts to produce virus-free seeds in groundnut, lettuce, beans etc., and there is obvious scope for improving the current unsatisfactory situation of non-availability of disease-free seed in large quantities. One possibility is to develop cultivars in which seed transmission occurs infrequently or not at all. Another method is to select seed exclusively from healthy plantings raised in isolated areas or outside the main periods of crop production (Sastry 2013a).

Under field conditions the diseased plants developed from the virus-infected seeds will serve as the initial foci for the virus spread with in the field.

The percentage of seed transmission depends on age of the plant and also the virus strain involved. For example *Pea seed-borne mosaic virus*- P-1 and P-2 isolates are transmitted through seed at 24 % and 3 %, respectively (Johansen et al. 1996). Similarly three isolates of *Peanut mottle virus* (PMV) the seed transmission rate is 0.3 %, 0 and 8.5 % for M1, M2, M3 isolates in Starr peanut (Adams and Kuhn 1977). Even the survival of the virus in the seed for the long periods also plays an integral role for virus perpetuation in the off seasons.

1.3.2.3 Infected Vegetative Propagules

The virus/viroid infected vegetative propagules cause serious problems to infected stocks of planting materials of cassava, yam, sugarcane, potato, sweet potato, banana, citrus, grapevine and many other important vegetatively propagated crops, when once they are infected. The wide-spread use of vegetative propagation for the multiplication of many horticultural crops results in the spread of the diseases through propagules such as cuttings, tubers, runners and bulbs. Since infection by most viruses is completely systemic, then any propagule is likely to be infected. Thus vegetative propagation presents a very efficient method of vertical virus spread without the virus having the difficulty of entering and establishing infection in a new healthy plant.

Moreover, there is a similar need for virus-free planting material for use in field by farmers, as the progeny derived from the virus infected vegetative planting material generally are infected. Importation of infected-budwood that often remains symptomless is a serious concern in virus epidemiology as it spreads virus and viroid diseases from one country to another, eluding routine diagnostics followed for quarantine certification. Selection, chemotherapy, thermotherapy, cryotherapy, electrotherapy and meristem-tip culture *in vitro* can all be used to develop virus-free stocks. These can then be multiplied on a large scale for general distribution. There has been lot of progress on these lines in different crops in the tropics and details are provided in Volume-I (Sastry 2013b). The situation is likely to change as methods of production of virus-free plants of different horticultural crops through meristem-tipculture become more commercially orientated, and attempts are made to meet the requirement for increased number of virus-free planting materials in number of crops (NCS-TCP 2008). More details about meristem-tipculture and micropropagation techniques for virus-free crop production in various crops are provided in the Chap. 2 of this book.

1.3.2.4 Debris, Volunteers and Stubbles as Virus Sources

When crops are harvested there is an opportunity for debris to remain and persist at the site until further plantings are made. Infected plants and debris that survive in this way will be important foci of infection for *Tobacco mosaic virus* and *Potato virus X*. The importance of crop debris and 'volunteer' seedlings will depend on the harvesting methods used, the efficiency of harvesting, the effectiveness of land preparation and other cropping practices. Plant virus survival is also influenced by rainfall and temperature regimes and the interval between plantings. The risks will increase as cropping is intensified and becomes increasingly specialised. This is apparent from experience with rice, sugarcane and other crops and is likely to occur even more frequently in the future. The neglected fields with heavy virus infections (e.g., leaf curl disease of tomato, yellow mosaic virus in legumes) can also serve as a source of pathogen and vector to the newly planted crops of some *Solanaceous* and *Leguminosae* (*Fabaceae*) families. In Australia, in self generating clover pastures, *Bean yellow mosaic virus* (BYMV) which is nonpersistently aphid-transmitted is a major crop production problem.

Another source of infection within a crop is infected volunteer plants regenerating from a previous season's crop. These *volunteer* plants are common in potato and sugar beet crops, and may frequently be virus-infected. Aphid spread of *Potato virus* Y and *Potato leafroll virus* has been shown to result from volunteer tubers in England (Doncaster and Gregory 1948), and regenerated beet debris is a source of *Beet mosaic* and *Beet western yellow viruses* in Washington State (Howell and Mink 1971). Volunteer peanut plants could potentially serve as a bridge for both TSWV and thrips vectors (Chamberlin et al. 1992). In the Columbia Basin region of Washington, volunteer carrot plants were found to be the over wintering source for both *Carrot thin leaf* and *Carrot motley dwarf viruses* (Howell and Mink 1977), and *Onion yellow dwarf virus* has been shown to overwinter in volunteer onions, before infecting the new season's crop (Louie 1968). More information is available on self-sown 'volunteer' seed of crops including groundnut, rice, onion and garlic which when infected has acted as 'bridging hosts' for viruses and their vector between successive crops. Similarly, *Cotton leaf curl virus,* which is transmitted by whiteflies in the Sudan, spreads from infected cotton plants that regenerate from stumps of the previous year's crop (Tarr 1951).

The infected rice stubbles with tungro viruses in both boro and kharif (rainy season) crop plays a major role in virus perpetuation. Similar observations have also been made by John and Rao (1971). Prasada Rao and John (1974) further observed that leafhoppers when collected from infected stubbles carry the virus and were able to transmit to the new crop. Anjaneyulu (1975) established that rice stubbles are the reservoirs of the virus between crops. Kondaiah and John (1977) found the presence of the virus in dried rice leaves even after 40 days of storage. Detailed studies on the retention of RTV in rice stubbles and the efficiency of the stubbles for acting as the source of the virus under field conditions were made by Tarafder and Mukhopadhayay (1979, 1980). They observed that the infected stubbles of both boro and kharif crop could retain the virus for 75 days, but the virus content was found to decline with age. The virus content of the stubbles of kharif crop was found to be higher than that of the boro crop. Mukhopadhyay (1980) reported that the infected stubbles when chopped off and mixed or puddled in soil either in the laboratory or under field conditions failed to act as the source of the virus. Ling and Palomar (1966) from Philippines have also reported that root debris and stubble of infected plants carry RTV in them.

1.3.2.5 Weed Hosts

(a) **Weeds as virus reservoirs**

Virus spread from weeds within a crop may occur through highly mobile vectors such as aphids or other winged insects or through less mobile vectors such as wingless beetles or soil-borne nematodes or fungi. Both persistent and non-persistently transmitted viruses can be spread by number of different insect vectors and more details are found in the Chapter-4 of Volume I.

Weeds within crops are also important as sources of infection of various nematode-transmitted viruses (Murant 1981). Many common weeds of arable land are the natural hosts of the nematode vectors (McNamara and Flegg 1981), and the nematode-transmitted viruses are also frequently seed-transmitted to a high level in these weeds (Murant and Lister 1967). Consequently the ecology of these viruses is closely related to that of their vectors, and the longevity of nematode-transmitted viruses in weeds and are the major factors in the epidemiology of nematode transmitted viruses.

The spread of the majority of the aphid-borne virus diseases depends heavily on weed hosts, which serve as inoculum sources. Some viruses have a limited host range and are confined to weed hosts of one plant family. Several viruses infecting economic plants perpetuate on many weeds belonging to *Solanaceae, Leguminosae* (*Fabaceae*) and *Malvaceae.* For example, PVY (Conover and Fulton 1953; Simons 1956; Anderson 1959; Badami and Kassanis 1959; Pontis and Feldman 1963; Sastry et al. 1974; Gooding Jr. et al. 1975; Prasada Rao and Yaraguntaiah 1976; Chagas et al. 1977; Mariappan and Narayanaswamy 1977a, b; Vicente et al. 1979); *Eggplant mosaic virus* (EMV) (Debrot et al. 1977); *Pepper mottle virus* (PeMV) (Nelson and Wheeler 1972; Purcifull et al. 1975) and *Tobacco etch virus* (TEV) (Anderson 1959); perpetuate on the weed hosts belonging to *Solanaceae.* The above mentioned viruses are all aphid- borne and are transmitted in a nonpersistent manner. Semi-persistently transmitted viruses like *Beet yellows* and *Beet yellow stunt* also have some weed hosts (Duffus 1973). Among the aphid-transmitted persistent viruses which survive on weed hosts are *Beet western yellows* (Duffus 1964; Wallis 1967), *Lettuce necrotic yellows* (LNY) (Stubbs and Grogan 1963) and *Barley yellow dwarf viruses* (Bruehl 1961).

Occasionally, a single virus can perpetuate on many weed hosts. *Tomato spotted wilt virus* has 925 host plant species representing 70 dicots and monocots families (Pappu et al. 2009); *Alfalfa mosaic virus* (AMV) occurs on more than 50 plant species belonging 14 families (Hull 1969; Feldman and Gracia 1970; Kviscala 1975). Similarly, *Maize streak virus* has been isolated from at least 22 species of grasses from various parts of Africa (Storey and McClean 1930). In the tropics and sub-tropics, many grasses are found to be hosts for aphid-transmitted *Sugarcane mosaic virus*. Bhargava (1975) reported *Pennisetum purpureum, Eleusine indica* and *Echinochloa* spp. as natural hosts for this virus. In Queensland (New South Wales), Teakle and Grylls (1973), isolated *Sugarcane mosaic virus* (SCMV) strains from 23 naturally-infected species of *Poaceae* (*Gramineae*) and each strain was named after an important perennial host, viz. (1) Johnson grass (*Sorghum halepense*), (2) sugarcane, (3) sabigrass (*Urochloa mosambicensis*), and (4) Queensland blue couch grass (*Digitaria didactyla*).

There are also a large number of reports from different countries which indicate that the most commonly occurring virus diseases of economic crops perpetuate on a large number of weed and other hosts as listed in the following table.

African cassava mosaic virus and *East African cassava mosaic Cameroon virus*	Ogbe et al.(2006), Alabi et al. (2008)
Alfalfa mosaic virus	Kviscala (1975)
Bean mosaic virus	Robertson and Klostermeyer (1962)
Bean pod mottle virus	Moore et al. (1969)
Beet curly top virus	Magyarosy and Duffus (1977)
Beet mild yellowing virus	Hartleb and Bauer (1977)
Cabbage black ring virus	Polak (1967), Schmelzer and Schmelzer (1974)
Cassava mosaic virus	Ndunguru et al. (2013)
Cowpea mosaic virus	Alconero and Santiago (1973), Lima and Nelson (1977), Joshi et al. (1977)
Cucumber mosaic virus	Valenta (1960), Stubbs and Grogan (1963), Khurana (1970), Tomlinson et al. (1970), Gracia and Feldman (1972), Bhargava et al. (1975), Joshi and Dubey (1975), Bruckart and Lorbeer (1976), Komm and Agrios (1978), Suteri et al. (1979), Kiranmai et al. (1998), Sacristan et al. 2004)
Lettuce mosaic virus	Moreno et al. (2007a)
Lettuce necrotic yellows	Randles and Carver (1971)
Maize streak virus	Mesfin et al. (1992)
Okra mosaic virus	Givord (1978), Atiri (1984), Alegbejo et al. (2008)
Pea seed-borne mosaic virus	Stevenson and Hagedorn (1973)
Potato viruses	Cadman and Chambers (1960), Vilasov and Abdukarimova (1965), Kaczmarek (1976)
Soybean mosaic virus	Bhargava et al. (1969), Benda (1970), Singh and Bhargava (1975), Ram and Chatterjee (1977)
Tobacco ring spot virus	Dias (1977)
Tobacco streak virus	Prasada Rao et al. (2003a, b), (2006), Sharman et al. (2009), Vemana and Jain (2010)
Tomato leaf curl virus	Mariappan and Narayanaswamy (1977b), Sastry et al. (1978), Saikia and Muniyappa (1989), Ramappa et al. (1998), Haider et al. (2006), Sivalingam and Anupam Varma (2007), Pramesh et al. (2013), Swarnalatha et al. (2013)
Tomato spotted wilt virus	Floresca et al. (1978), Groves et al. (2002), Chatzivassiliou et al. (2007), Okazaki et al. (2007), Pappu et al. (2009,)
Turnip mosaic virus	Arnold and Bald (1960), Laird and Dickson (1972)

(b) **Weeds as breeding foci for vectors**

The most dangerous weeds are those on which vectors also overwinter. It is an established fact that to protect a crop from the insect pests coming from disease-free weed hosts, the use of only one or two systemic insecticidal sprays for insect management are sufficient. On the other hand, a small number of viruliferous vectors will cause virus infections if they feed on a susceptible crop. The disease incidence mainly depends on the period of the vector activity, which differ greatly from place to place, season to season and also from year to year. Cockbain (1969) reported that *Acyrthosiphon pisum*, the vector of *Pea enation mosaic virus* and *Bean leaf roll virus*, overwinters on hosts like *Medicago* spp. *Trifolium* spp. and *Vicia* spp., and spreads these diseases to many leguminous crops. In Western New York, *Thrips tabaci* multiplies on 25 weed hosts and thrips populations were highest on the brassicaceous weeds, *Barbarea vulgaris*, *Synopsis arvensis* and *Phalspi arvense*. Some of the weed species were hosts of both thrips vector and *Iris yellow spot virus* as seen in onion (Smith et al. 2011).

Crop rotation as a means of *Peanut clump virus* disease management is difficult as the virus and its fungal vector (*Polymyxa graminis*) multiplies in several graminaceous crop species that are grown commonly in groundnut producing areas. Furthermore, the common weeds like *Cyprus rotundus, Cynodon doctylon, Tridax procumbens,* are also hosts of this virus in India (Ratna et al. 1991).

(c) **Weed hosts as symptomless carriers**

The presence of certain viruses, like *Cucumber mosaic virus, Pepper mosaic virus, Tobacco ring spot virus, Lettuce mosaic virus* and *Tobacco streak virus* are carried in certain weeds which are symptomless carriers and make the problem still more complex since they escape detection (Kemper 1962; Adsuar 1964; Tomlinson et al. 1970; Sharman et al. 2009).

The importance of weeds and other non-crop species as sources of virus infection, has been reviewed by various workers (Duffus 1971; Sastry 1984).

1.3.2.6 Wild Hosts

The role and importance of wild hosts is determined by their prevalence and potency as sources of virus inoculum and by their proximity to crop strands. For example, in the Western region of Ghana, forest trees like *Cola chlamydantha, Ceiba pentandra, Adansonia digitata, Cola gigantean* and *Sterculia tragacantha* are the hosts of the *Cocoa swollen shoot virus* (CSSV) from which the spread of the virus occurs to nearby cocoa plants (Posnette et al. 1950; Attafuah and Tinsley 1958; Dzahini-Obiatey et al. 2010)

In the studies conducted by Tugume et al. (2008) in Uganda, 24 wild species of *Convolvulaceae* were found to be infected with *Sweet potato feathery mottle virus* (SPFMV) under field conditions. Only *Ipomoea eriocarpa, I. hederifolia, I. wightii*, and *I. tenuirostris* previously were known as natural hosts for SPFMV in Uganda

(Karyeija et al. 1998). *I. hederifolia* and *I. wightii* are infected with SPFMV also in Louisiana (Clark et al. 1986). More details about wild hosts as virus sources can be obtained from the review articles of Duffus (1971), Sastry (1984), Cooper and Jones (2006).

1.4 Cropping Systems and Practices

From different regions of the world there is abundant evidence that the adopted cropping practices will have a profound influence on virus incidence. Considerable attention has been given to crop husbandry practices and ways in which they affect the spread of pests and pathogens, including viruses and their vectors (Thresh 1981, 1982). The frequency with which damaging epidemics occur in crops compared with natural vegetation has been stressed by various workers (e.g. Browning 1981). This has led to the view that epidemics are largely man-made phenomena, caused by the disruption of the equilibria that develop in nature when pathogens coevolve with their hosts.

For example, some aphid vectors when they invade crops tend to alight and accumulate on the outer rows of plantings and in the lee of hedges or other barriers. With other species the effect is less pronounced (Dean and Luuring 1970) and there is an equally varied response to crop spacing. *Aphis craccivora* and *M. persicae* alight preferentially on widely spaced plants that stand out clearly against a background of bare soil but not all species behave in this manner (Halbert and Irwin 1981). These findings indicate the complex implications of changes in crop spacing, weed control and thinning practices and in field size, shape, disposition and boundary (Thresh 1982). For some crops, sowing dates or cultural practices can be changed to make plantings less vulnerable to incoming vectors and so decrease virus spread as discussed by Plumb and Thresh (1983) in relation to *Barley yellow dwarf virus*. Similarly, growers of sugar beet and groundnut are advised to achieve a continuous canopy of foliage as early as possible to decrease the losses due to viruses with leafhopper, aphid or thrips vectors (Heathcote 1972). Elsewhere growers of early-sown vegetables who use plastic mulches to warm the soil and conserve moisture can use a colour that will repel incoming whitefly or aphid vectors. There is undoubtedly much more that can be achieved by further work along these lines, but the development of a fully integrated approach to plant virus and viroid disease management will require the close collaboration of virologists, entomologists, and agronomists.

1.4.1 Mono- and Mixed Cropping

A multidisciplinary approach is also required for work on mixed cropping systems, using two or more species planted together or in overlapping sequence. Such systems are traditional in many regions including large areas of Africa, South and

Central America. However, they have, until quite recently, been little studied by agricultural scientists who have been mainly concerned with promoting mono-cropping. It is now realized that mixed cropping can be an effective means of land use and gives stable yields that do not necessarily depend on expensive artificial fertilizers and pesticides.

Mono-cropping (Repeated cultivation of a single crop species on the same land) may result in a marked increase in the occurrence of diseases, particularly those caused by soil-borne viruses. *Grapevine fanleaf disease* is difficult to control as vines are usually continually cultivated on the same site. The disease is widespread in Europe, Asia, South America and North America, where it is transmitted by the soil-inhabiting nematode *Xiphinema index.* Even if the vines are removed, the viruliferous nematode vectors can survive for some years on the remnants of old roots (Hewitt et al. 1962), and a fallow period of up to 10 years may be required for successful eradication (Vuittenez 1970). Even in some crops, sole cropping has a particularly deleterious effect in favouring the spread of *Bean common mosaic virus, Soybean mosaic virus, Cucumber mosaic virus* and many other aphid-borne viruses that are transmitted in the nonpersistent manner. In Philippines due to mono culture of rice, both the leafhopper vectors and primary tungro virus inoculum was continuously available which has led to severe tungro epidemics every year (Ling 1972).

1.4.2 Introduction of Crops into New Areas

Virus epidemics may occur when crops are grown in regions remote from their normal endemic growing area. Despite these difficulties, there is abundant evidence from different agro-ecological regions of big changes that have occurred in the distribution and prevalence of several well-known diseases. Rice tungro disease in South-East Asia is a particularly important example as discussed earlier. The emergence of tungro and other diseases was influential in drawing attention to the hazards associated with an intensification of cropping practices. This was associated with the displacement of many traditional locally selected land races of rice by a relatively small number of specially bred cultivars, some of which were vulnerable to infection (Thresh 1989). Rice yellow mottle and maize streak in many countries of sub-Saharan Africa and rice hoja blanca in South/Central America are other diseases that have become prevalent following an intensification of cropping practices.

Spread has been facilitated by the ability of the virus to remain infective for years in the resting spores of the fungus vector *Polymyxa betae*. The spores are carried in soil, on crop debris, and along water courses and drainage channels, and the spores can also be disseminated in contaminated stocks of seed and seed potatoes. This is likely to be the means by which the fungus-borne *Rice stripe mosaic virus* spreads from Africa and became established in South America (Morales et al. 1999).

Cocao (*Theobroma cacao*) was introduced to Nigeria and Ghana in West Africa in the early twentieth century where it rapidly became an important crop (Posnette 1981). Swollen shoot disease was first observed in Ghana in 1936 (Steven 1937) and has since spread throughout the West African growing area. The virus has been shown to be transmitted by a mealybug (*Pseudococcid*) (Posnette 1947), and indigenous native trees and shrubs have been shown to be the source of the virus (Posnette et al. 1950; Dale 1962).

Similarly, the beetle-transmitted *Rice mottle virus* occurs in rice crops, and is only in a remote area of Kenya. This virus is thought to have originated from local, indigenous grasses (Bakker 1970, 1974).

1.5 Host Properties

(a) **Host diversity**

In the search for virus and vector-resistant cultivars, host diversity is required, and substantial collections of genotypes have been assembled at the main International Crop Research Centers and elsewhere (Thurston 1977). The use of host plant resistance has obvious advantages in controlling virus and other diseases compared with the use of pesticides (Buddenhagen 1983). It is particularly important to identify multiple types of resistance for use in developing resistant cultivars that are durable, and this is particularly important as they do not "breakdown" quickly or readily due to the emergence of resistance-breaking virus strains (Harrison 2002).

(b) **Host genetic vulnerability**

An earlier example of genetic vulnerability occurred in Ghana and elsewhere in West Africa where cacao production was based initially on plantings of the introduced Amelonado variety from South/Central America (Thresh et al. 1988). This cultivar was grown successfully for a time until it became apparent in the 1930s and 1940s that it was extremely vulnerable to *Cacao swollen shoot virus*. Millions of Amelonado trees died within 2 years of infection and more than 193 million trees have been removed by the authorities in continuing attempts to restrict spread (Ampofo 1997). More recently, the farmer-selected Ebwanateraka cultivar of cassava was severely affected during the 1990s epidemic of cassava mosaic disease in Uganda and neighboring parts of western Kenya (Otim-Nape et al. 2000). The cultivar was popular with farmers because it was high yielding and had other favorable attributes and so it was being grown almost exclusively at a time in several of the most important cassava-growing districts. The epidemic caused severe food shortages and farmers responded by switching to other crops and to local cultivars of cassava that were less vulnerable than Ebwanateraka. Farmers have also adopted the officially released mosaic-resistant cultivars which now account for a substantial part of the current production.

It is also well established that the host-resistance and prevailing disease pressure play pivotal role in virus disease incidence. For example at Uganda, Katono et al. (2013) have tested five cassava genotypes with varying levels of tolerance to cassava brown streak disease (CBSD) and evaluated the effect of genotype and location (prevailing disease pressure) on CBSD spread. The results indicated that genotype and location significantly affected CBSD incidence because of high whitefly vector population. Due to high disease pressure and vector population even tolerant genotypes were also highly susceptible.

From these and other experiences there are obvious advantages to be gained from a detailed evaluation of cultivars under a wide range of conditions and for several years, before they are recommended and adopted widely by farmers. It is particularly important to avoid undue reliance on a single cultivar or small group of related cultivars of similar genetic background.

(c) **Host longevity**

It is well known that the crops grown from seed are harvested within 12–16 weeks (e.g. groundnut and lettuce) or a few months (e.g. brassicas and cereals) from planting. Thus there is only a restricted opportunity for viruses to become prevalent and the only ones likely to be damaging are those that can spread rapidly or persist very effectively between successive crops. The opportunity for spread is greater in biennial crops and greatest with long-lived woody perennials (Vander plank 1949). These include tree crops such as cacao, cherry, citrus, coconut, grapevine, oil palm, peach, and walnut, for which the time-scale of epidemics is relatively unimportant. This is because they can develop over many years and ultimately cause serious damage, even though annual spread is slow and mainly localized (Thresh 1983).

(d) **Vegetative propagules**

The propagules of the majority of the vegetatively propagated plant materials are infected when once the mother plant is systemically virus infected. When the infected propagule is used for planting, vectors will widely disseminate the virus within or on vegetative propagules into entirely new regions and over distances far greater than those traversed by natural means. Virus-infected plants so introduced into plantings become particularly dangerous foci of infection, as they occur from the outset and tend to be randomly scattered, which facilitates further spread (Thresh 1983). Furthermore, viruses may build up over successive cycles of vegetative propagation, leading to the progressive degeneration of stocks. This has long been familiar to potato and temperate fruit growers, and also occurs with such tropical crops as avocado, banana, cassava, citrus, pineapple, sweet potato, taro, and yams. Viruses can ultimately become prevalent in vegetatively propagated crops even if the growth cycle is of short duration and virus spread is slow.

(e) **Host mobility**

Many viruses can be disseminated efficiently and far when in or on the seed of crops, on weeds or wild plants, and this is particularly important in the epidemiology of viruses with slow-moving ectoparasitic nematode vectors (Murant 1981). Furthermore, there is an extensive and expanding traffic in crop seed, seedlings, or vegetative propagules due to the activities of commercial growers, nurserymen, international companies, and researchers, and to the use of regions with a favorable climate or cheap labor to produce plants and plant material for export and sale for cultivation elsewhere.

Another aspect of plant mobility is the use of seed-beds or nurseries to raise plants for subsequent transplanting to wider spacing's at cropping sites that are sometimes remote from the source of the planting material. For example, there is an extensive traffic in plants and plant material between Africa and Europe and also from South/Central America to North America and Europe. Virus spread can occur before or after transit and new or extended opportunities occur as the plants are handled, or when they become mixed during collection, handling, grading, distribution, and transplanting. This facilitates mechanical transmission and also disrupts any groups of infected plants so that new "contacts" are established with healthy neighbors and further spread occurs. Transplants are used widely with rice, tobacco, tomato, and many other crops to make the most effective use of the land, irrigation water, and growing seasons available. This may permit sequential cropping or involve raising plants in glasshouses or plastic structures before being transfer outside. Such practices are becoming increasingly widespread as methods of production become specialized and this further exacerbates disease problems (Thresh 1982; Bos 1992).

1.6 Pathogen Properties

The basic properties of viruses and sub-viral agents are furnished in Chapter-2 of Volume I (Sastry 2013b).

(a) **Viruses and viroids as plant pathogens**

The classification of plant viruses provides understanding of the causal agent about its composition and mode of spread, which will help to formulate management measures depending on the mode of spread. Attempts to classify plant viruses began as early as 1935, but Holmes (1939) proposed the first internationally accepted classification. Subsequently several systems of classification were proposed but the proposal of classification of Lwoff et al. (1962) has gained more acceptance. About 45 years earlier, the International Committee for the Nomenclature of Viruses (ICNV), whose name is subsequently changed to the present International Committee on Taxonomy of Viruses (ICTV) and it has developed internationally agreed taxonomy of viruses including the names of virus taxa including species and sub-viral agents. Subsequently 9 ICTV reports were finalized and the latest is the 9th report (King et al. 2012). The list of 91 taxonomic groups of plant viruses including type member and its particle morphology, genome and its vector is presented in the Table 1.1.

Table 1.1 Alphabetical presentation of different taxonomic groups of plant viruses and their characteristic features. *Source* ICTV 9th Report (2012)

Sl. no.	Genus/group	Type member	Number of members	Particle morphology and size (nm)	Genome	Vector	Virus vector relationship
1	*Alfamovirus*	*Alfalfa mosaic virus*	1	Bacilliform 28–58 × 18	ssRNA (+)	Aphid	Nonpersistent
2	*Allexivirus*	*Shallot virus X*	8	Flexuous filaments 800 × 12	ssRNA (+)	Mite	–
3	*Alphacryptovirus*	*White clover cryptic virus 1*	16	Isometric 30	dsRNA	–	–
4	*Ampelovirus*	*Grapevine leafroll-associated virus 3*	8	Flexuous filaments 1800–2200	ssRNA (+)	Mealybug	–
5	*Anulavirus*	*Pelargonium zonate spot virus*	1	Quasi isometric 25–35	ssRNA (+)	–	–
6	*Aureusvirus*	*Pothos latent virus*	4	Isometric 30	ssRNA (+)	–	–
7	*Avenavirus*	*Oat chlorotic stunt virus*	1	Isometric 35	ssRNA (+)	–	–
8	*Babuvirus*	*Banana bunchy top virus*	3	Isometric 18–20	ssDNA	Aphid	Persistent
9	*Badnavirus*	*Commelina yellow mottle virus*	25	Bacilliform 130 × 30	dsDNA (RT)	Mealybugs	Semi-persistent
10	*Begomovirus*	*Bean golden yellow mosaic virus*	192	Bipartite	ssDNA	Whiteflies	Circulative
11	*Benyvirus*	*Beet necrotic yellow vein virus*	3	Straight rods 390 × 20	ssRNA (+)	Fungus	–
12	*Betacryptovirus*	*White clover cryptic virus 2*	4	Isometric 38	dsRNA	–	–
13	*Brambyvirus*	*Blackberry virus Y*	1	Flexuous filaments 800 × 14	ssRNA (+)	?	?
14	*Bromovirus*	*Brome mosaic virus*	6	Isometric 26	ssRNA (+)	Beetle	–
15	*Bymovirus*	*Barley yellow mosaic virus*	6	Flexuous filaments 270–290 × 12–15	ssRNA (+)	Fungus	Persistent
16	*Capillovirus*	*Apple stem grooving virus*	2	Flexuous filaments 640 × 12	ssRNA (+)	–	–
17	*Carlavirus*	*Carnation latent virus*	52	Flexuous filaments 620–700 × 11	ssRNA (+)	Aphid, whitefly	Nonpersistent
18	*Carmovirus*	*Carnation mottle virus*	19	Isometric 28–30	ssRNA (+)	–	–
19	*Caulimovirus*	*Cauliflower mosaic virus*	9	Isometric 50	dsDNA (RT)	Aphid	Nonpersistent

(continued)

Table 1.1 (continued)

Sl. no.	Genus/group	Type member	Number of members	Particle morphology and size (nm)	Genome	Vector	Virus vector relationship
20	*Cavemovirus*	*Cassava vein mosaic virus*	2	Isometric 45–50	dsDNA (RT)	–	–
21	*Cheravirus*	*Cherry rasp leaf virus*	3	Isometric 30	ssRNA (+)	Nematode	–
22	*Cilevirus*	*Citrus leprosis virus C*	1	Bacilliform 50–60 × 110–120	ssRNA (+)	Mite	?
23	*Citrivirus*	*Citrus leaf blotch virus*	1	Flexuous filaments 960	ssRNA	–	–
24	*Closterovirus*	*Beet yellows virus*	11	Flexuous filaments 1250–1450 × 10	ssRNA (+)	Aphid	Semi-persistent
25	*Comovirus*	*Cowpea mosaic virus*	15	Isometric 28	ssRNA (+)	Beetle	Semi-persistent
26	*Crinivirus*	*Lettuce infectious yellows virus*	13	Flexuous 1800–2000 × 13–14	ssRNA (+)	Whitefly	Semi-persistent
27	*Cucumovirus*	*Cucumber mosaic virus*	4	Isometric 28	ssRNA (+)	Aphid	Nonpersistent
28	*Curtovirus*	*Beet curly top virus*	3	Geminate 18–22	ssDNA (monopartite)	Leaf-hopper	Circulative, Persistent
29	*Cytorhabdovirus*	*Lettuce necrotic yellows virus*	9	Bacilliform 227 × 68	ssRNA (-)	Aphid	Persistent
30	*Dianthovirus*	*Carnation ringspot virus*	3	Isometric 31–34	ssRNA (+)	Nematode	–
31	*Emaravirus*	*European mountain ash ringspot-associated virus*	3		ssRNA	–	–
32	*Enamovirus*	*Pea enation mosaic virus-1*	1	Isometric 22–28	ssRNA (+)	Aphids	Circulative, Non-propagative
33	*Endornavirus*	*Vicia faba endornavirus*	7	None reported	dsRNA	Seed	–
34	*Fabavirus*	*Broad bean wilt virus 1*	5	Isometric 25	ssRNA (+)	Aphids	Nonpersistent
35	*Fijivirus*	*Fiji disease virus*	8	Isometric 71	dsRNA	Plant-hopper	Persistent
36	*Foveavirus*	*Apple stem pitting virus*	5	Flexuous 600–800 × 12–15	ssRNA (+)	–	–
37	*Furovirus*	*Soil-borne wheat mosaic virus*	6	Rigid rods 280–330 × 20 92–160 × 20	ssRNA (+)	Fungus	–
38	*Hordeivirus*	*Barley stripe mosaic virus*	4	Rigid rods 100–150 × 20	ssRNA (+)	–	–

(continued)

Table 1.1 (continued)

Sl. no.	Genus/group	Type member	Number of members	Particle morphology and size (nm)	Genome	Vector	Virus vector relationship
39	*Idaeovirus*	*Raspberry bushy dwarf virus*	1	Isometric 33	ssRNA (+)	–	–
40	*Ilarvirus*	*Tobacco streak virus*	19	Isometric 26–35	ssRNA (+)	Thrips	??
41	*Ipomovirus*	*Sweet potato mild mottle virus*	5	Filaments 800–950	ssRNA (+)	Whitefly	Persistent
42	*Lolavirus*	*Lolium latent virus*	1	Flexuous filaments	ssRNA	–	–
43	*Luteovirus*	*Barley yellow dwarf virus-PAV*	6	Isometric 25	ssRNA (+)	Aphid	Circulative
44	*Machlomovirus*	*Maize chlorotic mottle virus*	1	Isometric 30	SsRNA (+)	Beetles	Persistent
45	*Macluravirus*	*Maclura mosaic virus*	6	Flexuous filaments 650–675 × 13–16	ssRNA (+)	Aphid	Nonpersistent
46	*Maculavirus*	*Grapevine fleck virus*	1	Isometric 30	ssRNA (+)	–	–
47	*Mandarivirus*	*Indian citrus ringspot virus*	1	Flexuous 650	ssRNA (+)	Grafting	–
48	*Marafivirus*	*Maize rayado fino virus*	7	Isometric 31	SsRNA (+)	Leaf-hopper	Persistent
49	*Mastrevirus*	*Maize streak virus*	29	Geminate 22 × 38	ssDNA	Leaf-hopper	Persistent
50	*Nanovirus*	*Subterranean clover stunt virus*	6	Isometric 17–19	ssDNA	Aphid	Persistent
51	*Necrovirus, Alpha*	*Tobacco necrosis virus A*	3	Isometric 17	ssRNA (+)	Fungus	–
	Necrovirus, Beta	*Tobacco necrosis virus D*	3	Isometric 17	ssRNA (+)	Fungus	–
52	*Nepovirus*	*Tobacco ringspot virus*	36	Isometric 28	ssRNA (+)	Nematode	–
53	*Nucleorhabdovirus*	*Potato yellow dwarf virus*	10	Rhabdo/Bullet shape 380 × 75	ssRNA (-)	Leafhopper	Persistent
54	*Oleavirus*	*Olive latent virus 2*	1	Bacilliform	ssRNA (+)	Inoculations	–
55	*Ophiovirus*	*Citrus psorosis virus*	6	Filaments flexuous 1500–2500	ssRNA (-)	–	–
56	*Oryzavirus*	*Rice ragged stunt virus*	2	Isometric 65	dsRNA	Planthopper	Persistent
57	*Ourmiavirus*	*Ourmia melon virus*	3	Bacilliform 30 × 18.5	ssRNA (+)	??	??

(continued)

Table 1.1 (continued)

Sl. no.	Genus/group	Type member	Number of members	Particle morphology and size (nm)	Genome	Vector	Virus vector relationship
58	*Panicovirus*	*Panicum mosaic virus*	2	Isometric 25–30	ssRNA (+)	–	–
59	*Pecluvirus*	*Peanut clump virus*	2	Rods 190–245 × 21	ssRNA (+)	Fungus	–
60	*Petuvirus*	*Petunia vein clearing virus*	1	Isometric 43–46	dsDNA (RT)	–	–
61	*Phytoreovirus*	*Wound tumor virus*	3	Isometric 70	dsRNA	Leaf-hopper	
62	*Poacevirus*	*Triticum mosaic virus*	2	–	–	–	–
63	*Polemovirus*	*Poinsettia latent virus*	1	Isometric 34	ssRNA (+)	–	–
64	*Polerovirus*	*Potato leafroll virus*	13	Isometric 24	ssRNA (+)	Aphids	Circulative, Persistent ????
65	*Pomovirus*	*Potato mop-top virus*	4	Rods 250–300 × 18–20	ssRNA (+)	Fungus	–
66	*Potexvirus*	*Potato virus X*	37	Flexuous filaments 470–580 × 13	ssRNA (+)	–	–
67	*Potyvirus*	*Potato virus Y*	146	Flexuous filaments 680–900 × 11	ssRNA (+)	Aphids	Nonpersistent
68	*Pseudovirus*	———	20	Isometric	ssRNA (RT)	–	–
69	*Rymovirus*	*Ryegrass mosaic virus*	3	Flexuous filaments 700 × 15	ssRNA (+)	Mite	–
70	*Sadwavirus*	*Satsuma dwarf virus*	1	Isometric 26	ssRNA (+)	–	–
71	*Sequivirus*	*Parsnip yellow fleck virus*	3	Isometric 30	ssRNA (+)	Aphid	Semi-persistent
72	*Sirevirus*	*Glycine max SIRE 1 virus*	5	Isometric	ssRNA (RT)	–	–
73	*Sobemovirus*	*Southern bean mosaic virus*	14	Isometric 28–30	ssRNA (+)	Beetle	Semi-persistent
74	*Solendovirus*	*Tobacco vein clearing virus*	2	Isometric 50	dsDNA	–	–
75	*Soymovirus*	*Soybean chlorotic mottle virus*	4	Isometric 50	dsDNA (RT)	–	–
76	*Tenuivirus*	*Rice stripe virus*	7	Flexuous > 400 × 8	ssRNA (-)	Plant-hopper	Persistent

(continued)

Table 1.1 (continued)

Sl. no.	Genus/group	Type member	Number of members	Particle morphology and size (nm)	Genome	Vector	Virus vector relationship
77	*Tepovirus*	*Potato virus T*	1	Flexuous filaments 640 × 12	ssRNA (+)	–	–
78	*Tobamovirus*	*Tobacco mosaic virus*	33	Rigid rods 300 × 18	ssRNA (+)	Mechanical Contact	–
79	*Tobravirus*	*Tobacco rattle virus*	3	Short & long rigid rods 160–215 × 22 14–114 × 22	ssRNA (+)	Nematode	–
80	*Tombusvirus*	*Tomato bushy stunt virus*	17	Isometric 30	ssRNA (+)	–	–
81	*Topocuvirus*	*Tomato pseudo-curly top virus*	1	Geminate 18 × 30	ssDNA (monopartite)	Treehopper	Circulative
82	*Torradovirus*	*Tomato torrado virus*	2	Isometric 28	ssRNA	Whitefly	–
83	*Tospovirus*	*Tomato spotted wilt virus*	8	Isometric 80–110	ssRNA (-)	Thrips	Persistent
84	*Trichovirus*	*Apple chlorotic leaf spot virus*	7	Flexuous filaments 680–78 0 × 12	ssRNA (+)	–	–
85	*Tritimovirus*	*Wheat streak mosaic virus*	5	Filaments 700 × 15	ssRNA (+)	Mite	Semipersistent
86	*Tungrovirus*	*Rice tungro bacilliform virus*	1	Bacilliform 110–400	dsDNA (RT)	Leafhopper	Semi-persistent
87	*Tymovirus*	*Turnip yellow mosaic virus*	26	Isometric 29	ssRNA (+)	Beetle	Semi-persistent
88	*Umbravirus*	*Carrot mottle virus*	7	Isometric 52	ssRNA (+)	Aphid	Persistent
89	*Varicosavirus*	*Lettuce big-vein associated virus*	1	Rod 320–360 × 18	ssRNA (-)	Fungus	–
90	*Vitivirus*	*Grapevine virus A*	8	Flexuous filaments 800	ssRNA (+)	Mealybugs	Semi-persistent
91	*Waikavirus*	*Rice tungro spherical virus*	3	Isometric 30	ssRNA (+)	Leafhopper	Semi-persistent

These distinctive features of plant viruses and viroids will help to identify the exact etiological agent which helps in framing suitable management measures.

An appreciation of the risks caused by emerging plant viruses is critical in tropical areas that rely heavily on agriculture for subsistence and rural lively-hood (Anderson et al. 2004). The most important viruses and viroids that cause devastating diseases are *African, Srilankan* and *Indian cassava mosaic viruses, Barley yellow dwarf virus, Sugarcane mosaic virus, Groundnut rosette virus, Cocoa swollen shoot virus, Maize streak virus, Rice yellow mottle virus, Rice tungro viruses, Rice hoja blanca virus, Banana bunchy top virus, Banana bract mosaic virus, Banana streak virus, Groundnut bud necrosis virus, Tobacco streak virus, Tomato spotted wilt virus, Tomato yellow leaf curl viruses, Cotton leaf curl virus, Bean golden mosaic virus, Potato leaf-roll virus, Citrus tristeza virus, Coconut cadang–cadang* and many more.

Some of the virus diseases mentioned above have very wide host range whereas others have narrow host range. For example, CMV has a host range of more than 750 plant species in 90 families. Similarly, TSWV and TSV have more than 800 alternate and weed hosts. On the other hand, viruses like *Onion yellow dwarf virus* is confined to *Liliaceae* family members. Similarly, *Cocao swollen shoot virus* has very narrow host range. The survival of viruses to a larger extent depends on their host range and modes of transmission.

(b) **Stability and concentration of viruses**

Viruses like TMV and ToMV are highly stable and survive for months to years in plant debris, water and soil, and thus these viruses have greater survival chances in the environment. On the other hand, virus like TSWV is very unstable but yet survives in nature because it has very wide host range and active thrips vectors (Pappu et al. 2009). The concentration of the given virus is highly variable depending on temperature, plant nutrition and plant age. In case of vector transmitted viruses, the virus concentration has a vital role on the rate of spread in the fields. Even within the plant, the virus concentration is highly variable depending on the leaf age. Usually young leaves have higher titer of virus compared to the mature and old leaves.

1.6.1 Virus Mutation and Evolution

Plant virus mutation occurs during several different biological processes like replication slippage or it can be induced by irradiation, chemical treatments etc. Incorporation of a non-complementary nucleotide during duplication of DNA or RNA gives rise to point mutations, which alters the genetic information. Usually RNA viruses have high mutation rates due to their dependence on error-prone RNA dependent RNA polymerase for their replication (Jenkins et al. 2002). The mutation rate of RNA viruses was thought to be higher than DNA viruses

(Domingo and Holland 1997). The mutation rate of *Begomoviruses* is considered to be affected by type of virus, host plant, age of the host plant and inoculum homogeneity. Geminiviruses especially *Begomoviruses* and *Mastreviruses*, have high mutation rate in wild as well as cultivated hosts (van der Walt et al. 2008; Arguello-Astorga et al. 2007; Sanz et al. 1999).

The genetic shuffling caused by recombination between viruses (Revers et al. 1996) and between viruses and their plant hosts, including integration into the host genome, contributes to the emergence of "new" virus diseases and more virulent strains of existing viruses. In addition, viruses have a remarkable ability to adapt to extremely diverse and changing ecological niches. The presence of recombinant forms of *Begomoviruses* may play a role in their diversification and adaptation to the new hosts and environment. This might be one of the reasons that geminiviruses have such a diverse nature with more than 200 species described so far.

Two important features of viruses are their ability to mutate and also to produce novel forms through recombination or pseudo-recombination. This behavior accounts for the diversity of viruses and enables them to adapt to new crop species or cultivars and to changes in environmental conditions. Consequently, it has important implications for virus evolution and epidemiologically by leading to new disease problems. Year after year, the number of viruses infecting plants are increasing due to mutation and recombination. Recombination events may occur mainly in plants infected with two or more viruses. The 9th ICTV report has recognized a global total of nearly 2285 species of viruses infecting microbes, plants and animals (King et al. 2012). Large numbers of new plant viruses and new strains were identified in *Begomovirus*, *Tospovirus* and *Potyvirus* groups. Nearly 77 % of the known plant viruses were reported to be infecting cultivated plants, 11 % infecting agricultural weeds, 6 % infecting wild hosts and remaining 5 % infecting either cultivated or wild hosts as shown in the Fig. 1.5 (Wren et al. 2006).

Another example is *Tomato apical stunt viroid* a *Pospoviroid* was suggested to be a recombinant between a *Potato spindle tuber viroid* and a *Citrus exocortis viroid* (Spence and Barbara 2000). Inter-molecular recombination between two components in bipartite *Geminiviruses* has also been reported. For example, pseudo recombination between the DNA-A of *Tomato mottle virus* and DNA-B of *Bean dwarf mosaic virus* (BDMV), both *Begomoviruses* resulted in generation of new BDMV DNA-B by inter-molecular recombination (Gilbertson et al. 1993).

1.6.2 Occurrence of New Virus Strains

The possibility of new virus strains being produced by mutation, or by recombination of viral nucleic acids among the plant viruses has been well established. Even genetic masking may also apparently appear to produce new variants differing in vector transmissibility and/or host range. Plant viruses like *Alfalfa mosaic, Cucumber mosaic, Tobacco mosaic, Tobacco ringspot* and *Turnip mosaic viruses* are highly variable and have developed a large number of strains. Whereas

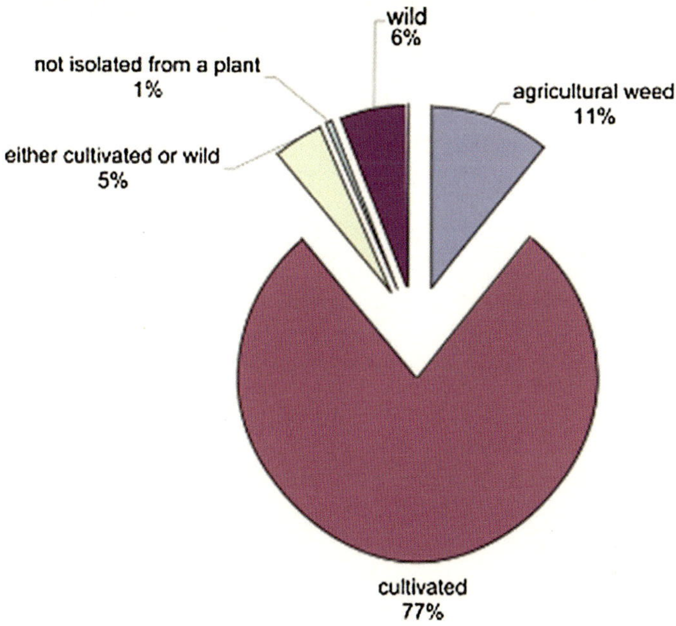

Fig. 1.5 Frequencies of sources of plants from which recognized plant viruses were initially isolated. *Source* Wren et al. (2006); PLOS Biology (UK/US)

other viruses like *Cauliflower mosaic, Maize dwarf mosaic* and *Squash mosaic viruses* have shown less strainal variations. Such new strains may arise quite suddenly and cause serious losses in crops that up to that time have been tolerant to, or unaffected by, the particular virus concerned (Hull 2002; Nayudu 2008). For example, sweet potato genotypes described as resistant to *Sweet potato feathery mottle virus* (SPFMV) in Peru, were found to be susceptible in East Africa, because of diversity in SPFMV (Gibson et al. 1998; Kreuze et al. 2000).

There is a lot of diversity within the *Geminiviruses* due to frequent recombination (Padidam et al. 1999; Garcia-Arenal et al. 2003; Seal et al. 2006). In a similar manner, resistance in cotton against *Cotton leaf curl virus* was broken down in Pakistan (Mansoor et al. 2003). The classification of *Geminiviruses* was made based on the extent of genome variation. The documented evidences are not clearly correlating with the well established host ranges and other properties of these viruses (Harrison and Robinson 1999). *Geminivirus* taxonomy and nomenclature is growing in complexity with the number of genomic sequences deposited in sequence databases. Taxonomic and nomenclatural updates are published at regular intervals (Fauquet et al. 2000, 2003). A system to standardize virus names, and corresponding guidelines, has been proposed (Fauquet et al. 2000). This system is now followed by a large number of virologists working on geminiviruses in the world, making *Geminivirus* nomenclature more transparent and useful. In 2003, due to difficulties inherent in species identification, the ICTV *Geminiviridae*

Study Group proposed new species demarcation criteria, the most important of which being an 89 % nucleotide (nt) identity between full-length DNA-A component nucleotide sequences for *Begomovirus* species. This threshold has been utilized since then with general satisfaction.

Even there is lot of diversity within the *Tospovirus* genus. There are more than 30 distinct tospovirus species affecting a wide range of horticultural crops. The basis for this diversity within the *Tospovirus* genus could be due to several reasons: Expansion of cultivation of new crops in the vicinity of other susceptible crops, expanding host range of thrips vectors, the divided genome of these viruses which make them amenable to genetic reassortment either in the host plant or there thrips vectors. On a regional basis, tropical regions of Asia and South America continue to have the greatest diversity of *Tospoviruses* (Pappu and Sevgan 2013).

1.6.3 Viruses in New Cultivars

Numerous examples exist in which local virus-resistant or tolerant cultivars have been replaced by new cultivars that prove to be highly susceptible to virus infection. The new cultivars are frequently bred for special agronomic characters, such as high yields and crop uniformity, by breeders working in distant horticultural areas or even in different countries. Often, less consideration is given to the local disease situation in the areas where the new cultivars have been developed.

Such a case occurred with new cultivars of maize. Maize was introduced into Europe from South America in the sixteenth century and has been cultivated for more than 400 years in Italy and other Mediterranean countries. In these areas, *Maize rough dwarf virus*, which is transmitted by the planthopper (*Laodelphax striatellus*) occurs in wild grasses. The virus caused no problem in maize, however, until new, high yielding hybrid cultivars were introduced into Europe from America after the Second World War (Harpaz 1972).

1.6.4 Indigenous Viruses

This virus grouping can be illustrated by reference to those causing diseases of three important tropical food crops: rice, maize, and cassava. The main virus diseases of each of the three crops are restricted in distribution to either one or two of the three continents where the crop is widely grown (Table 1.2). The assumption is that the virus or viruses responsible are indigenous and spread into the crop from wild hosts or remain associated with the crop as it was domesticated from wild progenitors. For example, it is known that each of the three main viruses of rice infects wild rice species, or other members of the *Poaceae* from which spread occurred to rice when the crop was domesticated or introduced from elsewhere. Each of the three rice viruses is restricted to one continent viz., *Rice*

Table 1.2 Region-wide distribution of the main virus diseases of rice, maize, and cassava

Virus disease	Africa	Asia	South/Central America
Rice hoja blanca	–	–	+
Rice tungro	–	+	–
Rice yellow mottle	+	–	–
Maize streak	+	–	–
Maize rayado fino	–	–	+
Cassava mosaic	+	+	–
Cassava brown streak	+	–	–
Cassava frog skin	–	–	+

Note: + Present; – Absent. *Source* Thresh (2006b)

hoja blanca in South America, *Rice tungro* in Asia and *Rice yellow mottle* in Africa. The importance of these viruses is likely to have increased during the twentieth century with an intensification of cropping practices. There is a similar situation with cassava which was first introduced to Africa from South America in the sixteenth century and to India in the eighteenth century. Once in the Old World, cassava encountered indigenous cassava mosaic viruses that were already present in Africa or Asia and these now are causing serious problems (Swanson and Harrison 1994). Maize was also transported by the Portuguese from South America to Africa, where it first encountered *Maize streak virus*, which infects a wide range of indigenous African grass species and maize.

In Africa groundnut rosette and several other viruses have a similar evolutionary "background" and this has important implications for those concerned with crop improvement and crop protection projects. Inevitably, problems arise when exotic germplasm is first introduced into a region for crop diversification and plant breeding purposes and it is exposed to viruses not previously encountered and for which there is little or no inherent resistance. This has been an important factor in cassava breeding programs in Africa and India where little use has been made of South American genotypes because of their vulnerability to Old World *Cassava mosaic viruses* (Porto et al. 1994).

Another implication of the limited distribution of many viruses is the importance of quarantine which controls the movement of plant material between and sometimes within countries to prevent or at least decrease the risk of further dissemination of viruses and their vectors. This need for effective quarantine procedures is widely recognized, but is not always easy to implement because of technical problems and the difficulty in controlling the continually increasing movement of people and plant material for trade, tourism and crop improvement purposes. These trends and attempts to eliminate trade barriers and statutory controls partially negate the improvements that have been made in quarantine procedures due to the introduction of new and sensitive methods of pathogen detection that also facilitate a greater ability to intercept the infected materials at the quarantine stations.

The problems that have been encountered in enforcing quarantine controls have led to the suggestion that they are of limited value because pests and pathogens will eventually become established in all the areas where agro-ecological conditions are suitable. However, this "inevitability concept" is unduly pessimistic and there are cogent arguments for enforcing quarantine controls to delay the introduction of pests and pathogens for as long as possible and to provide the opportunity to introduce resistant cultivars and make other contingency arrangements for use should be implemented (Kahn 1989; Hewitt and Chiarappa 1977). These objectives have been promoted by the UN Food and Agriculture Organization which has published guidelines for the quarantine procedures to be used with some of the most important crops (Frison and Putter 1989).

1.6.5 Man-Made Introduced Plant Viruses

Across the world, as air and sea travel has become easier for movement to distant places man has started looking for the diverse seed material for research and agricultural purposes, and thus such man-made dispersal of seed or planting material has made introduction of newer viral diseases to spread inter-continentally. The introduction of virus and viroid diseases has also been introduced either by private seed companies or through international gene banks. Many new crops have been introduced into other countries to increase the diversity of the food crops. During this process of phytosanitary inspection, many virus and viroid diseases causing pathogens were intercepted at quarantine stations although some have entered in the new places. This aspect has been discussed in Chap. 2 of this book.

At present there are about 231 seed-borne plant virus and viroid diseases and seed transmitted viruses are distributed in 24 virus groups. Greater number of seed transmitted viruses are found in *Potyvirus* (35), *Nepovirus* (28), *Cryptovirus* (28), *Ilarvirus* (14), *Tobamovirus* (7), *Potexvirus* (7), *Comovirus* (6), *Carlavirus* (5), *Carmovirus* (5), *Cucumovirus* (5) and in the remaining groups seed transmitted viruses ranged from 1 to 4 (Sastry 2013a). The worldwide distribution of many economical important viruses like *Bean common mosaic virus, Soybean mosaic virus, Pea seed-borne mosaic virus, Peanut mottle virus, Peanut stripe virus* were introduced into new areas through seed exchange (Hampton and Braverman 1979; Goodman and Oard 1980; Demski et al. 1984).

An inevitable consequence of the long history of traffic in seeds, plants and plant material carrying viruses and virus vectors have been introduced to completely new areas. Many have become established and caused severe problems in areas sometimes far from those where they originated. For example *Plum pox virus* (Sharka disease) was introduced into England during 1965 through infected root stocks from Germany (Cropley 1968). The fiji disease of sugarcane is quite a serious disease in Fiji Islands before this disease was introduced to Madagascar and other countries (Kahn and Mathur 1999). *Potato spindle tuber viroid* was introduced from North America into Russia, Poland and other countries.

Citrus tristeza virus causes an even more widespread virus disease as it has been introduced in plants or in bud wood to almost every country wherever Citrus species are grown (Bar-Joseph et al. 1981). Another virus infecting citrus species is *Citrus leprosis virus* (CiLV) of the *Rhabdoviridae,* is transmitted by the mite *Brevipalpus phoenicis,* and existed on sweet oranges in Florida during 1920s which was not noticed in subsequent periods. During the 1930s, CiLV appeared in South America and is now a major disease of citrus in Brazil. Citrus growing areas in Latin America including Panama and Costa Rica are threatened by this disease. Serious losses were also reported in Africa, the Americas, and Spain. Moreover, these losses continue as new virus strains or aphid vectors reach new areas. *Banana bunchy top virus* has also been widely disseminated in vegetative propagules and it has been introduced through infected banana suckers into Pakistan, Hawaii, New Caledonia and other countries.

Many other viruses of vegetatively propagated crops including potato, sweet potato, hop, yam, temperate and tropical fruits (banana, grapes, citrus, stone fruits), and ornamentals have a similarly wide distribution. This is due to the inadvertent traffic in infected propagules in spite of the adoption of quarantine controls. Similar problems arise with seed-borne viruses and this explains the cosmopolitan distribution of several damaging viruses including *Alfalfa mosaic virus*, *Bean common mosaic virus*, *Cucumber mosaic virus*, *Soybean mosaic virus*, and *Tomato mosaic virus*. *Tobacco mosaic virus* also has a very wide distribution due to the traffic in smoking and chewing tobacco. When these products are prepared from infected leaves, the hands or clothing of farm workers can become contaminated and responsible for transfer of inoculum to crop stands during routine cultural operations.

Israel provides a particularly notable example of *Tomato yellow leaf curl virus* that has extended its geographic range in recent years. When introduced to Spain it largely displaced the leaf curl virus of tomato that occurred previously (Sanchez-Campos et al. 1999). The virus also spreads to the Dominican Republic which was attributed to the introduction of infected seedlings from the Mediterranean region (Polston et al. 1994). Infection has since become prevalent in many other parts of Central and North America and has caused serious losses.

To combat the risk of introducing viruses through seed and vegetative propagated materials, countries are required to comply with international obligations as defined by the World Trade Organization (WTO) agreement on the application of sanitary and phytosanitary measures (WTO 1995).

1.6.6 Synergism in Certain Plant Virus Interactions

Synergism is expressed as an increase in the severity of symptoms, virus accumulation, viral movement in plants, and as an effect on yield of the crops. An example of synergism was reported from sub-Saharan Africa (SSA) with the emergence of a new variant as a result of recombination between two distinct viruses, namely *African cassava mosaic virus* (ACMV) and *East African cassava mosaic virus* (EACMV), both of the genus *Begomovirus* (Zhou et al. 1997). The disease caused by the new variant, EACMV-Ug, has caused a serious epidemic in

Uganda (Otim-Nape et al. 1997) and has been identified in Tanzania and Congo. Deployment of resistant cultivars is bringing a measure of control to the pandemic in Uganda, but urgent measures are being worked out in the countries where EACMV-Ug has been identified or that may be at immediate risk.

Another example is the interaction of *Maize chlorotic mottle virus* in combination with either *Maize dwarf mosaic virus* (MDMV), strain a/b or with *Wheat streak mosaic virus* (WSMV) which causes a synergistic reaction and has resulted in *Corn lethal necrosis* disease (Uyemoto 1983). In general when an individual virus infects a susceptible crop, the impact on yield losses will not be to the same extent of a virus disease which has occurred due to interaction of two viruses (Goldberg and Brakke 1987).

In Africa, severe *Sweet potato virus* (SPV) is due to infection of *Sweet potato chlorotic stunt virus* (SPCSV) and *Sweet potato feathery mottle virus* (SPFMV) (Nome et al. 2007). From the International Potato Center (Peru) Untiveros et al. (2007) have reported the co-infection of *Sweet potato chlorotic stunt virus* (SPCSV, genus *Crinivirus*) with *Sweet potato feathery mottle virus* (SPFMV, genus *Potyvirus*) and this resulted in *Sweet potato virus* disease (SPVD), a synergistic disease that is widely distributed in the sweet potato (*Ipomoea batatas*) growing regions of the world. Since both SPCSV and SPFMV are common and often detected as part of multiple co-infections of severely diseased plants, the occurrence of synergistic interactions with other viruses was investigated. The data from their studies showed that SPCSV, but not SPFMV, can cause synergistic diseases in sweet potato with all viruses tested, including members of the genus *Potyvirus* (*Sweet potato latent virus*, *Sweet potato mild speckling virus*), *Ipomovirus* (*Sweet potato mild mottle virus*), *Cucumovirus* (*Cucumber mosaic virus*), and putative members of the genus *Carlavirus* (*Sweet potato chlorotic fleck virus* and C-6 virus). Untiveros et al. (2007) have further reported that the presence of a third different virus in plants affected with SPVD increased the severity of symptoms even further compared with SPVD alone. There was a positive correlation between increase in virus accumulation and symptom expression in double and triple SPCSV-associated co-infections.

Even in sweet potato based on the virus and its strains involved, the synergistic or antagonistic reaction is noticed. For example, *Sweet potato feathery mottle virus* (SPFMV-C) titer was enhanced in sweet potato infected plants in the presence of *Sweet potato chlorotic stunt viruses* (SPCSV), due to mixed infection only mild symptoms were produced, suggesting that enhancing the accumulation of the potyvirus component was alone not sufficient for SPVD development. Both studies of Kokkinos and Clark (2006a) and Mukasa et al. (2006) showed that titers of SPCSV were decreased in the mixed infections as compared to single infections, suggesting an antagonistic effect.

Another example of synergistic effects due to whitefly transmitted viruses is reported in zucchini squash. Gil-Salas et al. (2011) have reported yellowing disease in zucchini squash produced by mixed infections of *Cucurbit yellow stunting disorder virus* and *Cucumber vein yellowing virus*. Even Murphy and Bowen (2006) have demonstrated synergy in bell pepper plants co-infected with *Cucumber mosaic virus* and *Pepper mottle virus*. The mixed virus infection has led to greater disease severity than individual viral components. A synergistic interaction between two viruses was manifested through host growth parameters relative to effects of each virus alone. If each viral treatment results in a characteristic systemic symptom and the mixed infection are not lethal but extremely severe, proof of synergy based on symptom severity will be difficult.

Due to synergistic reaction, increased seed transmission was noticed in certain virus-host combinations. For example, seed transmission of SMV in seeds from soybean plants infected with either SMV alone or in combination with *Bean pod mottle virus* was 6 and 11 %, respectively (Ross 1963). Similarly *Southern bean mosaic virus* (SBMV) was 12 % in cowpea but increased to 20 % in the presence of *Cowpea chlorotic mottle virus* (Kuhn and Dawson 1973). Similarly seed transmission of *Turnip yellow mosaic virus* (TYMV) in *Arabidopsis thaliana* in doubly infected with TYMV and TMV was 70 % as compared to TYMV infection alone which was only 31 % (De Assis Filho and Sherwood 2000).

From the details provided above regarding synergism in plant viruses, it is very clear that some viruses are highly devastating and responsible for yield losses; usually viruses persist throughout the life of systemically-infected plants; due to recombination between viruses, a large number of strains have developed; combined infection of viruses in a host results into synergistic reaction and enhanced yield losses. These factors clearly stress the importance of managing the virus diseases for which quick, low cost, field applicable techniques for early diagnosis of viruses have to be identified.

1.7 Detection and Diagnosis of Plant Viruses and Viroids

1.7.1 Plant Virus Diagnosis

Since agricultural and horticultural crops are being threatened by different virus and viroid diseases besides the wide range of biotic stresses, before framing any management measures against virus diseases, identifying the causal virus species is most important. Quick identification of a plant virus species in the crop is pivotal for monitoring surveillance, the vector association, and consequential ecological and epidemiological significance of the virus vector and disease. Various virological methods over the years have helped in this effort. The identified methods should be simple and least expensive, and have to be used in most situations, while more expensive ones may be essential in the other purposes.

As seen in the last 50 years, plant viruses are becoming more widespread and causing enormous yield losses. Since there are threats of virus epidemics, the movement of viruses and the vectors around the world has been documented regularly and quarantine restrictions are being implemented in places wherever necessary. In any country, before framing any management measures against virus diseases, one has to know the etiological agent of the virus disease. Since the particle morphology, genome, and vector involved are different from virus to virus, quick diagnosis of the causal virus must rely upon molecular tests. The detailed information about virus, genera, and their mode of spread, particle morphology and size is provided in Table 1.1, based on the plant virus species classified according to the 9th ICTV report (King et al. 2012).

1.7.2 Selection of a Diagnosis Method

The four general approaches to virus diagnosis include: (a) biological (b) physical (c) serological (d) molecular technique(s). Selection of the most appropriate diagnostic method should involve some critical appraisals focusing on the objective to be pursued: (i) Certification of mother plants, sanitation and eradication (ii) Large surveys to evaluate incidence, or screening tests for surveillance of the spreading of a disease (iii) For rapid detection of virus and viroid diseases in incoming and outgoing plant materials at quarantine stations (iv) In breeding for disease resistance and transgenic programmes.

Recent developments in molecular detection technologies resulted in more convenient, effective and specific assays, have opened the doors to a greater use of these tests for detecting seed-borne viruses and vegetatively propagated plant material-borne plant viruses. These assays will help the tree growers, crop consultants and plant protection personnel from having to rely excessively on symptomatology and/or time consuming diagnostic procedures and permit early detection of viruses. The advantages of having different methods are obvious as they result not only in increased sensitivity but also expand the range of applications. Each detection procedure is directed towards a special characteristic of the virus transmissibility, serological properties, nucleic acid characteristics and sequences. These detection techniques will increase the reliability of virus-free seed production and also in screening of germplasm.

The latest developments in molecular biology/biotechnology have an impact on seed health testing, especially for seed-borne viruses and also vegetative propagule transmitted plant viruses. These new methods are generally much more rapid and sensitive than the older techniques. But the main disadvantages with these techniques are that they are more complicated and often require the use of sophisticated and expensive laboratory equipments. Nevertheless, seed testing stations having responsibility for indexing a large number of seed lots for economically important viruses may find it advantageous to adopt these new procedures. With the increasing knowledge along with additional adaptation and modification make these methods more suitable for routine seed health testing. However, they should be viewed as management tools in conjunction with other diagnostic procedures, crop acquaintance, biology of the virus and the ecology of the vector as well as the disease.

Global plant clinics in CABI-Bioscience were also established, funded by the DFID, provides expert service from the UK for diagnosis of plant diseases, free of charge (see www.globalplantclinic.org). This expert service was based on the experience of an analyzing hundreds of samples each year from around the world.

The establishment of regional plant disease clinics will extent this concept, especially when it is back stopped by knowledge base resources such as the Global Plant Clinics. Such regional clinics will effectively ease the burden on existing personnel and facilities, and will reduce overlap and duplication. Looking into the importance of plant disease diagnosis, Ausher et al. (1996) have furnished the details of role of plant clinics in plant disease diagnosis and the need for education in developing countries.

In the case of certification, the need to use the most sensitive method should be stressed, accepting the risk of "false positives". For this reason, for evaluation of plant material the selection of a diagnostic test having sensitivity and specificity is required. It should enable the detection of virus/viroid to be eliminated/discarded most effectively because it affords the most accurate diagnosis of healthy plants with high confidence when the target pathogen is not detected. In the case of large-scale surveys or screening tests for surveillance, the selection of one, two or several methods should be on an evidence-based approach, evaluation of cost per analysis, calculation of post-test probability of disease and consideration of different scenarios with different prevalence.

1.7.3 Routine Tests Used in the Early Stages of Plant Virus Research

In the early stages of plant virus research, plant viruses were identified based on symptoms, particle morphology/size and a simple antibody tests like the precipitin test and the double gel diffusion tests. By using differential hosts, researchers would get the clue about tentative identification of the virus, but the length of time required to get results based on differential reaction on indicator hosts was an obstacle. With the advent of molecular diagnostic tests the use of indicator hosts for identification has received lower priority.

1.7.4 Electron Microscopy

According to the 9th ICTV report of plant viruses, there are 991 recognized plant virus species and in majority of the cases, sufficient data is available both on virus structure and genome characteristics. Most plant viruses have distinct virion morphology and size. The virus particles may be spherical, rigid rods, flexuous rods, bacilliform, bullet shaped, geminate etc., Among the currently recognized virus genera, the particles of five viruses are enveloped (*Cytorhabdovirus, Nucleorhabdovirus, Tospovirus, Umbravirus, Cilevirus*), and the rest are non-enveloped. Enveloped particles usually appear as bullet-shaped. Particles of 4 genera (*Mastrevirus, Curtovirus, Begomovirus, Topocuvirus*) are geminate; Particles of 6 genera (*Badnavirus, Tungrovirus, Alfamovirus, Oleavirus, Ourmiavirus,*

Soymovirus) are bacilliform; Particles of 8 genera (*Varicosavirus, Tobamovirus, Tobravirus, Furovirus, Pecluvirus, Pomovirus, Benyvirus, Hordeivirus*) are short or long rods; Particles of 19 genera (*Tenuivirus, Ophiovirus, Potyvirus, Ipomovirus, Macluravirus, Rhymovirus, Tritimovirus, Bymovirus, Closterovirus, Crinivirus, Potexvirus, Carlavirus, Allexivirus, Capillovirus, Foveavirus, Tricovirus, Vitivirus, Mandarivirus, Ampelovirus*) are filamentous; and particles of the remaining genera are isometric. It is therefore apparent that the shape of virus particles can narrow down the process of identification of possible viruses and provide an immediate clue for their short list for further diagnosis.

Electron Microscopy (EM) is being used routinely in advanced virology labs, as a tool for routine diagnosis of the viruses. Negative staining of the viruses with phosphotungstic acid (PTA) or uranyl acetate and following visualization by EM will provide rapid and accurate results. In most cases particle information about shape and size is sufficient for the tentative identification of the virus (Zechmann and Zelling 2009). The various morphological shapes of plant viruses are provided in Fig. 1.6.

1.7.5 Protein-Based Diagnosis of Plant Viruses

The detection of plant viruses depends upon the surface and sequence properties of viral proteins. Based on the specific properties of the coat-proteins, different sero-diagnosis tests are developed. Tests like immunodiffusion, ELISA and its variants are widely accepted for 30–35 years and were highly successful in identifying the virus diseases of a large number of crops.

Based on the availability of the facilities, earlier workers have produced polyclonal or monoclonal antibodies (Sander and Dietzgen 1984). In the case of polyclonal antibodies, they contain antibodies to all available epitopes on the antigen, whereas in the case of monoclonal antibodies they contain antibodies to one epitope. As a result of research carried out with different virus-host combinations, researchers have indicated that monoclonal antisera are more specific than polyclonal antisera and can be used to differentiate strains of many viruses (Dietzgen and Sander 1982; Halk and De Boer 1985).

The application of serology for the detection of plant viruses has been in use for more than half a century. However, the use of advanced immunodiagnostic methods for the identification and detection of viruses has made virus detection easier, more sensitive and at a reasonable cost. Specific recognition between the antibodies and their homologous antigens was recognized in the 1950s. Various diagnostic tests often provide valuable clues to etiological studies, but every *in vitro* test has its own innate limitations.

(a) **Precipitin tests**

Antigenic analysis of isometric viruses is perhaps best done by precipitin tests (Ouchterlony test) in agar gels (Bercks et al. 1972; Ball 1974). Elongated viruses, and those which have a strong tendency to aggregate or diffuse poorly into the

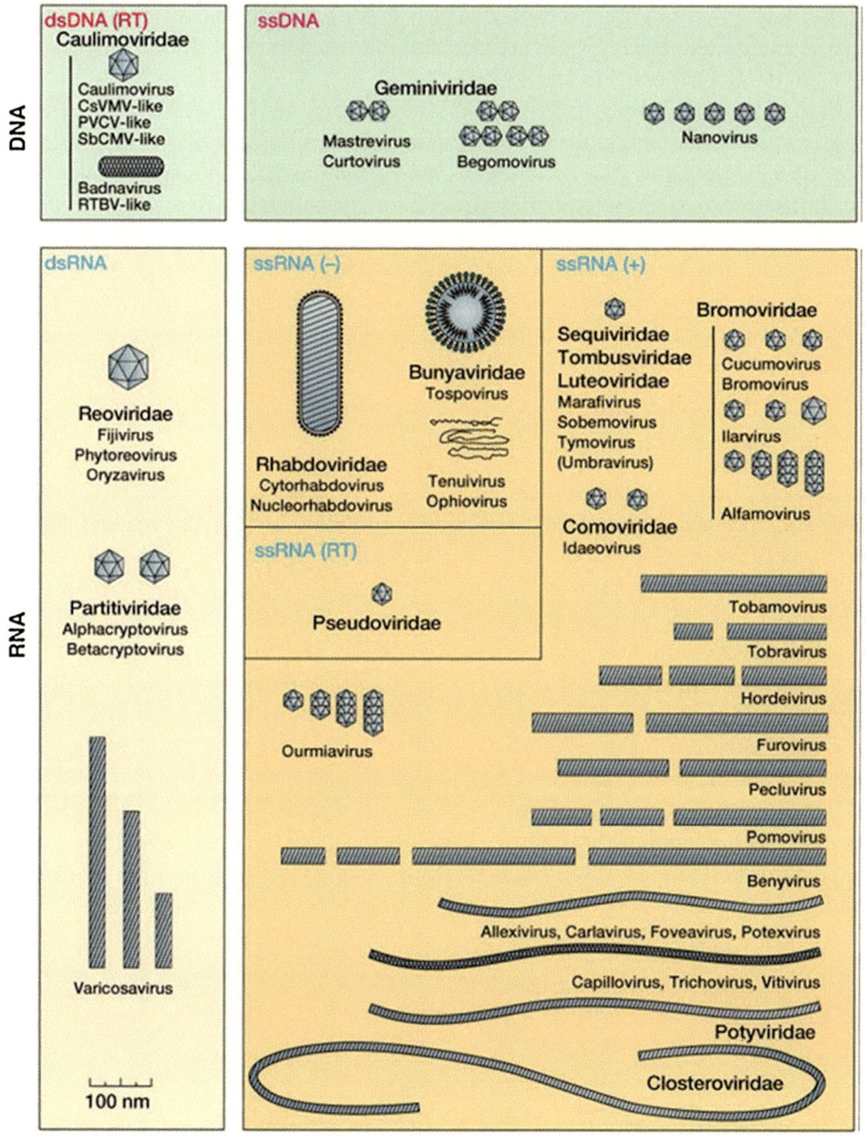

Fig. 1.6 Families and Genera of viruses infecting plants. *Courtesy* MHV van Regenmortel et al. eds., Virus Taxonomy: 7th Report of ICTV (Elsevier)

agar, can be assisted to diffuse effectively in the agar gel double-diffusion test after breakdown of particles with detergents (Purcifull and Batchelor 1977), pyrrolidine (Shepard et al. 1974) or by sonication (Tomlinson et al. 1965). Antisera to the intact viruses do not always react with the degradation products; under these circumstances it may be necessary to use degraded antigens for immunization (van Regenmortel 1967; von Wechmar and van Regenmortel 1968; Moghal and Francki 1976; Chu and Francki 1979).

In the agar gel double-diffusion tests, the comparisons between closely related antigens are easier to establish. Under balanced conditions, a fusion of precipitin lines usually indicates that the antigens are serologically identical, provided that the antiserum used is homologous for one of the antigens (Bercks et al. 1972; Hollings and Stone 1975; Paul et al. 1980). Spur formation under balanced conditions indicates antigenic differences. The resolving power of the agar gel double-diffusion test can be further refined by intragel cross-absorption (van Regenmortel 1967), which may reveal small differences between antigens not seen in the spur formation test. Immunoelectrophoresis may also reveal electrophoretic differences between the antigens.

(b) **Enzyme-linked immunosorbent assay**

Enzyme-linked immunosorbent assay (ELISA) technique utilizes the ability of antibodies produced in animals to recognize proteins, usually the coat-protein of the virus of interest. This technique was initially developed by Clark and Adams (1977). Antibodies are fixed to the surface of a well within a microtitre plate, and sap extract from the plant/partially purified virus is added to the well. If the virus of interest is present in the plant, it will bind to the antibodies fixed on the surface. Any unbound extract is washed-off before a secondary antibody that recognizes the first antibody is added. The secondary antibody allows for indirect detection of the virus because it has a reporter molecule attached to it, usually an enzyme that acts on a substrate that changes colour, which is detected visually by a calibrated microtitre plate spectrophotometer (Clark and Adams 1977). With careful calibration, ELISA can be quantitative as well as qualitative.

Depending on the research needs the following assays of ELISA are frequently used for plant virus detection.

(1) Double antibody sandwich-ELISA (DAS-ELISA)
(2) Direct antigen coating-ELISA (DAC-ELISA)
(3) Plate trapped antigen-ELISA (PTA-ELISA)
(4) Triple antibody sandwich-ELISA (TAS-ELISA)
(5) Protein A-sandwich-ELISA (PAS-ELISA)

Since ELISA techniques are simple, less time consuming and without much cost, they are routinely used universally for virus diagnosis in vegetable, fruits crops, ornamentals and food crops. For more details one can refer to the following review articles: Bar-Joseph and Garnsey (1981), Hampton et al. (1990), Rao et al. (2008), Ahlawat (2010), and Narayanasamy (2011).

This method can be used for testing multiple plants for a single virus by using one well per plant sample, or alternatively a single plant can be simultaneously tested for many viruses on a single plate with different antibodies coated to each well in duplicate or triplicate for reproducibility.

(c) **NCM-ELISA/DIBA**

Nitro cellulose membrane-ELISA (NCM-ELISA) is most popularly called by scientific community as Dot-immuno binding assay (DIBA). The principle of DIBA is almost the same as that of ELISA; except that antigen or antibody is bound to nitrocellulose membrane instead of polystyrene plate and that the product of the enzyme reaction is insoluble. NCM–ELISA has been used with success to detect several plant viruses (Abad and Moyer 1992; Mukasa et al. 2003; Zein et al. 2007; Ahangaran et al. 2009).

DIBA is of two types, direct DIBA and indirect DIBA. The procedure for both methods is the same as that of direct and indirect ELISA. The extract from infected plant material or seed extract is spotted on nitrocellulose membrane (NCM) instead of a microplate. In this technique, a few microliters of sap can be spotted on the NCM which absorbs small amounts. After incubating and washing the antigen, a blocking agent is added to saturate any unoccupied binding sites on the NCM. Then the NCM is incubated in a solution IgG rabbit antibody specific to the antigen to be detected and unbound antibody is removed by washing. Horse radish peroxidase or ALP conjugated anti-rabbit second IgG is added to NCM; incubated and washed to remove unbound antibody. Finally, the enzymes specific substrate is added which is hydrolyzed to form water-insoluble colored spots or dots on the membrane. Horse radish peroxidase, the most frequently used enzyme, will produce purple spots after addition of substrate. Alkaline phosphatase reacts with napthol AS-MX phosphate mixed with 5-chloro-2-toludinediazonium chloride hemizene chloride (fast red TR salt) to produce red dots, or with diazotized 4-benzolamino-2, 5-dimethoxyaniline $ZnCl_2$ (fast blue BBN salt) to produce blue dots on the white background of the NCM.

The main advantages of DIBA are: (1) rapid, simple and economical; (2) highly sensitive and detects as low as 50–100 picogram of antigen; (3) requires only a single crude specific antiserum for each test virus and also a single generally applicable enzyme conjugate; (4) the cost of nitrocellulose is less than that of plastic microplates; (5) while testing for the presence of seed-borne viruses, part of the seed can be used for testing and the remaining can be used for sowing; and (6) DIBA can be directly adopted to field situations.

(d) **Dot-blot hybridization (DBH)**

This hybridization format simply answers the question of whether a plant is infected or not infected by a virus. Dot blotting does not distinguish between the number and size of hybridized molecules, since the hybridization signal is the sum of all sequences recognized by the probe. However, the technique is rapid and versatile in identifying specific nucleic acid sequences in samples ranging from crude plant sap to highly purified preparations. The DBH was conducted by using non-isotopic digoxigenin labeled probes and isotopic labeled probes for identification of virus and viroid diseases of plants.

Dot-blot hybridization (DBH) is extensively used for the detection of plant viruses and viroid diseases of vegetable and fruit crops. Even though this test generally does not distinguish types and sizes of nucleic acids, it can be very useful for qualitative detection since this method can discriminate closely related but different target sequences. Isotope hybridization by using P^{32}, I^{125} and I^{128}, the virus and viroid diseases were tested, wherever lab facilities existed. For example *Plum pox virus* (PPV-D) was detected in infected orchards by using various lengths of radio actively labeled probes (Wetzel et al. 1990). The detection limit was of about 5 pg of purified virus per assay. In Columbia, a strain of *Soybean mosaic virus* infecting *Passiflora* spp. was identified by this method (Benscher et al. 1996). From India, Borah et al. (2008) have used this technique for the detection of *Citrus yellow mosaic badnavirus* from citrus species. The DBH using radio labeled RNA probes were able to detect serotypes of prunus was shown to be as sensitive as DBH using radioactively labeled probes for *Cherry leaf roll virus* (CLRV) (Mas et al. 1993). By following this technique Pesic and Hiruki (1986) have detected AMV in infected Alfalfa pollen by using AMV specific ^{32}p labeled cDNA probe.

The non-isotopic digoxigenin labeled probes are widely used successfully. Presently molecular hybridization is encouraged using non-isotopic digoxigenin labeled probes. For example this technique has been employed for the past 20 years for detection of viruses like *Apple mosaic virus* (ApMV), *Prunus necrotic ringspot virus* (PNRSV), *Prune dwarf virus* (PDV), *Plum pox virus* (PPV), and *Apple chlorotic leaf spotvirus* (ACLSV) (Pallas et al. 1998b); *Citrus psorosis virus* (CPsV), *Citrus variegation virus* (CVV) and *Citrus tristeza virus* in citrus (Loconsole et al. 2009; Barbarossa and Savino 2006); *Cherry mottle leaf virus* in cherry (James et al. 1999); *Lettuce infectious yellows virus*, *Zucchini yellow mosaic potyvirus* and *Beet yellows clostero virus* (Harper and Creamer 1995). Two *Potyviruses* infecting peanut (Dietzgen et al. 1994) and TSV in infected sunflower, gherkin, pumpkin, marigold, globe amaranth (Sarovar and Saigopal 2010). Rodriguez et al. (2011) identified three *Begomoviruses*, viz., *Bean golden mosaic virus* (BGMV); *Soybean blistering mosaic virus* (SbBMV) in bean and soybean respectively by using this technique. DBH was also used in the diagnosis of viroid diseases viz., *Citrus exocortis viroid* (Bagherian et al. 2009); *Potato spindle tuber viroid* (Khan et al. 2009); *Chrysanthemum stunt viroid* (Hooftman et al. 2001). The application of spot hybridization for the detection of DNA and RNA viruses in plant tissues has been reviewed by Maule et al. (1983) and Pallas et al. (1998a).

(e) **Tissue blot immunoassay**

Tissue blotting, like ELISA, utilizes antibodies raised against viruses. The preparation of imprinted membranes (nylon, nitrocellulose, polyvinylidene fluoride etc.) is carried out by briefly pressing a fresh section of virus infected tender shoots, leaves, flowers, or fruits directly on membrane to make a print and the virus is detected by labeled probes, often chemiluminescent. The procedure is less labor-intensive than ELISA, rapid, sensitive, simple (no virus extraction is required), inexpensive (minimal equipment is needed), suitable for testing of 1000–2000 samples per day, and the samples can be taken in the field and processed some time later. By this technique, even viroid diseases have been diagnosed (Hadidi et al. 1991; Hurtt and Podleckis 1995). This technique does not involve disruption of tissue or extraction of the antigen from the targeted plant sample. The fresh infected plant material can be used as imprint tissue on NCM. By following this technique virus disease of cereals, vegetables and fruit crops were identified (Garnsey et al. 1993; Makkouk and Comeau 1994; Makkouk and Kumari 1996; Khatab Eman et al. 2012). Kits are available for a number of viruses, notably from the International Centre for Agriculture Research in the Dry Areas (ICARDA) at Syria, which offers kits for 19 viruses infecting legumes.

(f) **FTA® classic card technology**

Naidu et al. (2013) have evaluated the use of FTA® Classic Cards for plant virus sample collection and shipment for identification. Virus infected leaf samples were collected from India, Bangladesh, Nepal, Cambodia, Tajikistan, Indonesia and pressed gently on FTA® cards, allowed to air dry and brought to a central facility for virus testing. Using a simplified protocol, nucleic acids bound to FTA® card matrix were eluted and used in RT-PCR or PCR for the detection of viruses. The amplified DNA fragments were cloned and nucleotide sequence determined. The derived sequences were compared with corresponding sequences available in Gen Bank to confirm the identity of virus(es) present in individual samples. This technique will be highly useful in epidemiological studies. The FTA® cards do not pose any risk of spreading alien pathogens from one country to the other. Thus, FTA® cards can be stored indefinitely for downstream applications such as virus diversity analysis and molecular characterization of viruses.

(g) **Immuno electron microscopy**

Immuno electron microscopic (IEM) methods have the great advantage of being able to apply to tissue homogenates and requires very low quantities of virus and antiserum (Derrick and Brlansky 1976; Milne and Lesemann 1978; Roberts et al. 1980; Conti et al. 1980; Nicolaieff and van Regenmortel 1980; Lesemann and Paul 1980). Relationships can be studied in different ways: (i) by the differential trapping of virus particles on electron microscope grids coated with antisera to different viruses (ii) by endpoint dilution of an antiserum which effectively coats (decorates) the virus particles (iii) by the observation of a clumping reaction at different serum concentrations. Pretreatment of grids with protein A of *Staphylococcus aureus* may enhance virus adsorption (Shukla and Gough 1979), but meaningful results occur only under certain conditions (Lesemann and Paul 1980; Milne 1980). It may be useful to follow trapping of virus particles on serum-coated grids with decoration (Milne and Luisoni 1977), particularly as trapping may be susceptible to bias due to non-specific adsorption of virus particles to the grids (Lesemann et al. 1980). Decoration of particles by antisera to distantly related viruses may be difficult to distinguish from non-specific decoration, however, especially in crude sap. The visualization of immunological reactions on electron microscopic grids is one of the most sensitive serological techniques. This technique is unique as it combines direct visualization of virus particles with the specificity of a serological reaction. However, heavy capital cost and requirement of skilled man power are the main drawbacks in IEM.

1.7.6 Nucleic Acid Based Methods

1.7.6.1 Polymerase Chain Reaction

Nucleic acid-based methods are more advantageous as any region of viral genome can be targeted to develop the diagnostic tests, which are sensitive, specific and determines the genetic relationships. In plant virology, the most frequently used molecular techniques are first molecular hybridization and followed by the polymerase chain reaction (PCR).

PCR is an *in vitro* method in which DNA sequences are rapidly amplified with very high specificity and fidelity using oligonucleotide primers and thermostable DNA polymerase. The amplified fragments (amplicons) are detected by gel electrophoresis and are characterized by nucleic acid hybridization, restriction enzyme digestion sequencing.

It is generally considered that closely related viruses share a greater nucleotide sequence which is referred as conserve sequences than those that are distantly related. Conserve sequence present in an isolate or strain of virus but absent or different in another strain of virus or species can be used for detection of viruses. PCR allows the amplification of very low amounts of target nucleic acids (Saki et al. 1985). This technique has been used successfully to detect very low amounts of nucleic acids of plant viruses and viroids.

Use of PCR in the diagnosis of plant virus diseases has become very common laboratory practice, as its advantages for speed, sensitivity and specificity are far more important than its drawbacks like risk of contamination, sensitivity to inhibitors, complexity, cost and several modifications to solve these problems have been developed with success. In general, PCR, with all its variants, is currently a basic tool in diagnosis, either alone or preferentially in combination with other techniques.

PCR offers several advantages as the organisms do not need to be cultured before their detection, and as it affords high sensitivity at least theoretically, enabling a single target molecule to be detected in a complex mixture, and it is also rapid and versatile. In fact, the different variants of PCR have increased the accuracy of detection and diagnosis, and opened new insights into our knowledge of the ecology and population dynamics of many pathogens, including plant viruses, thus providing a valuable tool for both basic and applied studies in plant virology.

1.7.6.2 Application of PCR and Its Variants

(a) **Detection of ssDNA plant viruses**

Like ssRNA viruses, ssDNA viruses also require a template with the production of a replicative intermediate. The ssDNA can be amplified by using a complementary DNA and later the second strand synthesis is required.

The PCR technique is extensively used against whitefly transmitted DNA viruses (e.g. viruses of the genera *Geminivirus*, and *Nanovirus*). Among whitefly transmitted viruses *Bean golden mosaic virus* (Gilbertson et al. 1991); *Tomato yellow leafcurl virus* (Navot et al. 1992; Leam Khang et al. 2005; Mason et al. 2007); *East African cassava mosaic Cameron virus* (Alibi et al. 2008); *Indian cassava mosaic virus* (Makesh Kumar et al. 2005); and among *Banana bunchy top virus* (*Nanovirus*) (Selvarajan et al. 2007; Prakash et al. 2010) were identified by PCR. Even Rojas et al. (1993) have identified leafhopper transmitted *Beet curly top geminivirus* by this technique. This aspect has been reviewed by a number of workers (Moriones and Garcia-Andres 2008).

(b) **Detection of dsDNA plant viruses**

The direct PCR is being used to detect the double stranded DNA (dsDNA) viruses. In this technique dsDNA is directly used as template and the primers can be used directly for amplication of the genome. The PCR cycles are performed like other methods of amplification. This direct PCR method is beneficial for detection of dsDNA viruses of family *Caulimoviridae* and the method is simple and within a short period virus can be diagnosed. *Citrus mosaic virus* (Baranwal et al. 2003), *Citrus yellow mosaic virus* (Baranwal et al. 2003; Borah et al. 2009); *Cocoa swollen shoot virus* (Quainoo et al. 2008) are detected by PCR and its variants.

(c) **Reverse Transcription-PCR (RT-PCR) (for RNA viruses and viroids)**

The nucleic acid of about 70 % of plant viruses have RNA genome which is single stranded and they are detected by reverse transcription-PCR. Before PCR can be used for virus diagnosis, to amplify viral RNA, the RNA must be reverse transcribed to DNA.

Reverse Transcriptase (RT) is used to make a complimentary DNA (cDNA) copy of viral RNA prior to PCR. First, a DNA oligo (primer) is annealed to the viral RNA, and then the RT enzyme transcribes the RNA sequence to a cDNA sequence. These DNA copies are then used as the template for PCR. In the case of RNA-viruses, oligonucleotide primers, flanking part of the genome of the virus are extended by a thermostable DNA polymerase in a series of denaturation and extension steps that exponentially increase the target DNA.

Some of the ssRNA which were detected by reverse transcriptase RT-PCR are *Tobacco streak virus* in sunflower (Kumar et al. 2008a; Sharman et al. 2008; Sarovar et al. 2010); Onion yellow dwarf and Leek yellow stripe viruses in garlic (Takaichi et al. 1998, 2001); *Peanut stripe virus* (Gillaspie et al. 2001; Dietzgen et al. 2001) and *Tomato spotted wilt virus* in peanut (Jain et al. 1998); *Cowpea aphid borne mosaic virus* in cowpea (Salem et al. 2010); Sweet potato viruses (Kokkinos and Clark 2006b); *Sugarcane mosaic virus* in sugarcane (Viswanathan et al. 2010; Subba Reddy et al. 2011); *Zucchini yellow mosaic virus* in squash (Hosseini et al. 2007; Simmons et al. 2011; Sharma et al. 2013); *Cucumber mosaic virus* in banana (Singh et al. 1995; Hu et al. 1995); *Potato virus Y* and *Potato leaf roll virus* in dormant potato tubers (Singh and Singh 1996; Awan et al. 2010; Crosslin and Hamlin 2011).

Double stranded RNA viruses of *Reoviridae*, *Betaflexiviridae* and *Partiviridae* are detected with the help of the PCR method. Fang et al. (2001) identified *Rice black streaked dwarf fijivirus* in maize which is dsRNA virus. Even *Southern rice black streaked dwarf virus*, a new proposed *Fijivirus* species in the family *Reoviridae* was also identified by PCR (Zhou et al. 2008). *Cherry mottle leaf virus* (CMLV) of *Prunus avium* belonging to genus *Trichovirus* was identified by RT-PCR, which was a hundred times more sensitive than dot blot hybridization. Even IC/RT-PCR and PC/RT-PCR were effective for the detection of CMLV in herbaceous and woody tissues (James et al. 1999). Hadidi et al. (1995) have also identified virus affecting apples and pears by IC-RT-PCR and molecular hybridization

The PCR techniques briefly described here are being used in diagnosis of virus and virus-like diseases (phytoplasma and viroid) of various crops and diagnosis of the pathogen in the vector which provides useful information in epidemiological studies (Mason et al. 2007; Sastry 2013a, b).

PCR application for viroid disease diagnosis: In recent years diagnostic techniques such as nucleic acid hybridization and reverse transcription coupled with polymerase chain reaction (RT-PCR) have become popular for diagnosis of plant viroid diseases because of their relatively simplicity and high sensitivity (Shamloul et al. 1995; Mumford et al. 2000; Bernard et al. 2006; Hassan et al. 2008). RT-PCR was successfully applied in the detection of viroids from potato (Boonham et al. 2004); pome (Hadidi and Yang 1990) and stone fruits (Shamloul et al. 1995; Hadidi et al. 1992, 1997), citrus (Levy et al. 1992; Yang et al. 1992; Ito et al. 2002), grapevine (Puchta and Sanger 1989; Rezaian et al. 1992; Staub et al. 1995; Wah and Symons 1997), avocado (Schnell et al. 1997), apricot (Pallas et al. 2003). While studying *Apple scar skin viroid* detection by RT-PCR, Sipahioglu et al. (2007) reported that out of the three conventional extraction methods, silica-capture RNA extraction method was found to be superior for total RNA extraction as compared to citric acid buffer and lithium chloride methods.

From Iran, Bagherian et al. (2009) have developed a technique (RT-PCR-DBH) by incorporating RT-PCR and dot blot hybridization (DBH) for the detection of hop stunt and exocortis viroid diseases in citrus. In this method, instead of using nucleic acid extracted directly from the plants, RT-PCR products are subjected to dot-blot hybridization. This method was about 1000-times more sensitive than Southern blot and 100-times more sensitive than PCR in detecting hop stunt viroid in citrus. It avoids the use of ethidium bromide, a carcinogenic dye used in electrophoresis of PCR products. The misinterpretation of results using poly-acrylamide or agarose gel electrophoresis analysis is completely avoided by RT-PCR-DBH. Pallas et al. (2003) in their review article described the application of RT-PCR-DBH in detection of *Hop stunt viroid* in apricot. Earlier Shamloul et al. (2002) have developed a novel multiplex RT-PCR probe capture hybridization (RT-PCR-ELISA) for simultaneous detection of six viroids in four genera: Apscaviroid, Hostuviroid, Pelamoviroid and Pospiviroid. The majority of viroid diseases are seed transmitted and sometimes distributed through the commercial seeds. For example, Singh et al. (2009) have detected the presence of *Citrus exocortis viroid* (CEVd) in the seeds as well as the seedlings that developed from infected seeds of *Impatiens walleriana* and *Verbina x hybridia* by RT-PCR assay.

A multiplex RT-PCR has been designed to simultaneously detect fragments specific to seven citrus viroids or variants, using 6 % PAGE, on the basis of different molecular sizes (Ito et al. 2002). The entire process only takes 2 days and has facilitated the survey of citrus orchards for seven viroids, e.g. CEVd, citrus bent leaf viroid (CBLVd) and its variant CVd-I-LSS, HSVd, citrus viroid III (CVd-III), CVd-IV and citrus viroid OS (CVd-OS). The survey showed a common occurrence of multiple viroid infection by three viroids (Ito et al. 2002). Similarly, a multiplex RT-PCR has been developed to identify six viroids belonging to four genera and in two families, using RT-PCR probe capture hybridization for the identification of specific viroids (Shamloul et al. 2002). In addition to conventional RT-PCR, real-time PCR (using TaqMan® chemistry) has also been developed for viroids (Boonham et al. 2004). Non-radiolabeled hybridization has been successfully used for detection of *Potato spindle tuber viroid* (PSTVd) in true potato seeds (Goldbach et al. 1992; Borkhardt et al. 1994). Array technology has also been used for identification of viroids like Potato spindle tuber viroid (Agindotan and Perry 2008).

Seed-borne viroids can also be identified by R-PAGE as seen in the case of PSTVd infected true potato seeds (TPS) and also viroid infected dormant potato tubers (Singh et al. 1988). Hence, R-PAGE can be used to assess the PSTVd content of various potato seed lots before planting either from seeds or from *In vitro* seedlings, when germplasm is valuable and available in small quantities.

(d) **Duplex PCR**

Ghosh et al. (2008) have simultaneously detected one RNA virus (*Citrus tristeza virus*) and one DNA virus (*Citrus mosaic virus*) from naturally infected plants by using duplex PCR. Duplex RT-PCR technique helps in identification of mixed infection of *Tobamoviruses* in tomato and bell pepper (Vinayarani et al. 2011) and was positive in identification of number of viruses in different host combinations. Even both tristeza and greening diseases in citrus were also detected (Ghosh and Das 2012). Gupta et al. (2007) have simultaneously detected *Citrus mosaic virus* and *Indian citrus ringspot virus* by this technique. Even in peanut two *Potyviruses* and two *Cucumoviruses* were differentiated by using duplex RT-PCR assays. These assays would be useful for testing peanut leaves or seeds for virus identification in epidemiological studies, seed testing or in post entry quarantine (Dietzgen et al. 2001). Simultaneous detection of viruses causing mosaic in sugarcane has been identified by this technique (Viswanathan et al. 2008).

(e) **Multiplex PCR**

This technique allows simultaneous amplification of several viruses or viroids using a mixture of specific primer. It is very useful in plant virus detection because different viruses frequently infect a single crop or alternate hosts. Multiple virus species or strains are detected in a single reaction by combining oligonucleotide primers specific for different viruses. It is important that the amplicons are of different lengths and that there is no cross-reactivity among them. Multiplex PCR is also found to be very useful for the detection of several viruses in a simple reaction (Mumford et al. 1996; Nie and Singh 2001; Lopez et al. 2006; Alabi et al. 2008; Selvarajan et al. 2011; Abarshi et al. 2012; Liu et al. 2012). This method was also used to detect six citrus viroids and one virus in infected citrus species (Ito et al. 2002).

(f) **Real-time PCR**

Real-time PCR is an advanced type of multiplex-PCR. It is a quantitative procedure that helps to detect accumulation of PCR products during the PCR reaction. Instruments for conducting this test are commercially available. Some of the virus and viroid diseases diagnosed in crop plants by Real-time PCR are already cited in this chapter.

(g) **Immunocapture PCR (IC–PCR)**

This technique combines capture of virus particles by antibodies with amplification by PCR. In this method, the virus is adsorbed by the antibody bound to a surface, then removed by heating with a nonionic surfactant such as Triton X-100. The nucleic acids are then amplified using RT–PCR. This method is especially useful in concentrating virus particles from plant species where virus titre is low, or where compounds that inhibit PCR are present; for example, plum tree sap containing *Plum pox virus* (Wetzel et al. 1992) and sugarcane sap containing *Sugarcane streak mosaic virus* (Hema et al. 2003). It has also been used for detection of the episomal *Banana streak virus*, parts of whose genome are naturally present within the banana genome, and therefore there is a high chance of false positives from standard PCR tests (Harper et al. 1999). The application of this method of virus detection in plants is particularly useful in species or tissues containing inhibitory substances. In addition, antibody-mediated virus purification is usually simpler than other methods of isolation. More details about IC-PCR can be had from the review article of Mulholland (2009).

(h) **Nested PCR**

In this method, four primers are used in two PCRs; they are carried out with the first reaction increasing the amount of template for the second. The method is particularly useful where the virus is in very low titre or plant extract with inhibitors of DNA polymerase are present. Low-specificity oligonucleotides, usually degenerate, are used in the first rounds of amplification. Then, an aliquot of the reaction is placed into a fresh tube for a second PCR with primers that anneal within the first amplicon. This, at once, increases the target molecule and dilutes inhibitors. This method has been used successfully to detect members of *Vitivirus* and *Foveavirus* species in grapevines (Dovas et al. 2003). Adkins et al. (2008) have also detected *Squash vein yellowing virus* from *Momordica charantia* by following this technique.

(i) **Nucleic Acid Sequence Based Amplification (NASBA)**

NASBA is an isothermal amplification method that can be used to detect RNA targets. The reaction requires the use of three enzymes, AMV-RT for reverse transcription and to obtain double stranded cDNA, RNase H to hydrolyze the RNA fragment of the hybrid molecule DNA-RNA and T7 RNA polymerase to produce a large amount of anti-sense, single strand RNA transcripts corresponding to the original RNA target. It can be achieved by using two specific primers, one of them including at 5′ end the T7 promoter, NTPs and also dNTPs. The entire NASBA process is performed at 41 °C for 60 min and the typical level of amplification is at least a factor of 10^9. The detection of NASBA products can be assessed by chemiluminescent or colorimetric detection using an internal specific probe digoxigenin labeled or in a real-time assay using molecular beacons (Amplidet RNA) (van Beckhoven et al. 2002; van der Wolf et al. 2004). NASBA-beacon assay yields results in less than 1 h (Robert and Kerst 2001), and offer the advantages that no contaminating DNA is amplified, is performed at 41 °C without the need of a thermal-cycler, and requires only 60 min reaction affording high levels of sensitivity, superior in some cases to real-time PCR (Scuderi et al. 2007).

This technology has been applied for detecting plant viruses such as *Apple stem pitting virus* (Klerks et al. 2001), *Plum pox virus* (Olmos et al. 2007), *Potato virus Y* and *Arabis mosaic virus* (Szemes and Schoen 2003). The sensitivity of this method has proven similar to that obtained by real-time RT-PCR when applied to PPV detection (Olmos et al. 2007).

(j) **Loop-mediated isothermal amplification**

Loop-mediated isothermal amplification (LAMP) is another type of isothermal amplification that is being increasingly used in the field for virus diagnosis and offering sensitivity and economic costs (Notomi et al. 2000). The method requires a set of four specifically designed primers that recognize six distinct sequences of the target and a DNA polymerase with strand displacement activity. The amplification products are stem-loop DNA structures with several inverted repeats of the target and cauliflower-like structures with multiple loops, yielding >500 μg/ml. The LAMP reaction was enhanced by the addition of loop primers (Nagamine et al. 2002) which helps in reducing time and increasing sensitivity. The amplification takes place at 60–65 °C for 60 min. Although it was initially developed for DNA viruses, it can be adapted to amplify RNA (RT-LAMP) (Fukuta et al. 2003a, b). The method has been applied to the detection of some plant viruses such as PPV, with a sensitivity level similar to that obtained by real-time PCR (Varga and James 2006). *Squash leafcurl virus* was also detected by Kuan et al. (2010) by this technique.

(k) **Nucleic acid hybridization tests**

Molecular hybridization-based assays were first utilized in plant pathology to detect viroid diseases like *Potato spindle tuber viroid* (Owens and Diener 1981) and adapted to virus detection (Hull 1993). However, certain problems associated with the use of radioactive probes are relatively low sensitivity and complexity of these techniques and the development of amplification-based assays have minimized new improvements and applications.

Depending upon the nature of the target, the nucleic-acid hybridization technique may be of three different types: Southern blotting, Northern blotting and Western blotting. In the Southern blotting the target and the probe are DNA; in the Northern blotting on the other hand, the probe is DNA, but the target is RNA. In Western blotting the probe is antibody protein and the target is the protein. The essential steps involved in hybridization techniques are:

(a) Isolation of the nucleic acid from the test sample
(b) Denaturation of the nucleic acid
(c) Application of the target nucleic acid on the membrane
(d) Production of radioactive labeled probe
(e) Hybridization of the probe with the target and
(f) Recording the reaction endpoint.

Today the most common molecular hybridization format for the detection of viruses is non-isotopic dot-blot hybridization using digoxigenin-labeled probes (DIG). This technique has been employed for detection of *Apple mosaic virus* (ApMV), *Prunus necrotic ringspot virus* (PNRSV), *Prune dwarf virus* (PDV), PPV, and *Apple chlorotic leaf spot virus* (ACLSV) in important fruit crops (Pallas et al. 1998a). The advantage of DIG probe is that it can be reused 3–4 times without significant loss of sensitivity of the probe. Furthermore, multiple RNA riboprobes or polyprobes have been used to detect different viruses (Ivars et al. 2004; Herranz, et al. 2005) and they can be associated with tissue printed or squashed material in addition to the spotted extracts.

More details on nucleic acid hybridization techniques can be had from book chapters of Hull (1993), Singh and Nie (2002), and Nayudu (2008).

(l) **Fluorescent in situ hybridization**

Fluorescent *in situ* hybridization (FISH) combined with microscopically observation was done in number of bacteria species (Ercolini et al. 2006). FISH was also applied against certain plant viruses like *Tomato golden mosaic virus* (Bass et al. 2000); *Cowpea mosaic virus* in cowpea (Carette et al. 2002); *Banana streak virus* in banana (Heslop-Harrison and Schwarzacher 2002); *Tomato yellow leaf curl virus* in tomato (Ghanim et al. 2009; Skaljac and Ghanim 2010). Ghanim et al. (2009) have reported that FISH technique is simple, rapid, inexpensive method for the localization of virus transcripts in plant and insect vector tissues. This method is based on fluorescent *in situ* hybridization using short DNA oligonucleotides complementary to an RNA segment representing a virus transcript in the infected plant or insect vector. The DNA probe harbors a fluorescent molecule at its 5′ or 3′ ends. The protocol is: simple fixation, hybridization, minimal washing and confocal microscopy, provides a highly specific signal. Ghanim et al. (2009) have tested the reliability of the protocol by localizing two phloem-limited plant virus transcripts in infected plants and insect tissues: *Tomato yellow leaf curl virus* (TYLCV)

(*Begomovirus*: *Geminiviridae*), exclusively transmitted by the whitefly *Bemisia tabaci* (*Gennadius*) in a circulative non-propagative manner, and *Potato leafroll virus* (*Polerovirus*: *Luteoviridae*), transmitted by the aphid *Myzus persicae* (Sulzer). Transcripts for both viruses were localized specifically to the phloem sieve elements of infected plants, while negative controls showed no signal. TYLCV transcripts were also localized to the digestive tract of *B. tabaci*, confirming TYLCV route of transmission. Compared to previous methods for localizing virus transcripts in plant and insect tissues that include complex steps for *in vitro* probe preparation or antibody raising, tissue fixation, block preparation, sectioning and hybridization, the method described here provides very reliable, convincing, background-free results within much less time, effort and cost. Bass et al. (2000) by using FISH technology have observed that *Tomato golden mosaic virus* (TGMV) infection has increased the size of nuclei in some cases altered the nuclear architecture and induced chromatin condensation in the infected host plants.

1.7.7 Microarrays

Arrays, both microarrays and macroarrays, have been used for some years as a tool for visualizing relative changes in global expression levels of mRNAs, as well as single nucleotide polymorphism typing and host-pathogen interactions (Kato-Maeda et al. 2001; Boonham et al. 2002; Lee et al. 2003; Lareu et al. 2003). Microarray technology is based on the well-established highly exploited principle of nucleic acid hybridization. Microarrays are high-density arrays with spot sizes smaller than 150 microns. ssDNA probes are irreversibly fixed as an array of discrete spots to a surface of glass, or membrane or polymer. A typical microarray slide can contain up to 30,000 spots. Since the development of microarray technology for gene expression studies, new approaches have been extending their application to the detection of pathogens. Microarrays are generally composed of thousands of specific probes spotted on to a solid surface (usually nylon or glass). Each probe is complementary to a specific DNA sequence. They are generally membrane-based with spot sizes of greater than 200 microns. Arrays printed with probes corresponding to a large number of virus species or any type of pathogen can be utilized to simultaneously detect all those viruses within the tissue of an infected host. The labeled target molecule is denatured and allowed to hybridize with the arrayed probes. Excess target is washed from the surface and spots where labeled target molecules have bound, become fluorescent under appropriate lighting conditions. The position of a visible spot corresponds to the presence of a particular virus in the plant sample (Boonham et al. 2007).

(a) **Probes for microarrays**

The primary importance in making an array is probe design. Probes determine the sensitivity of the array and the amount of information that they provide. Access to sequence databases and powerful sequence alignment software is therefore essential.

Of importance is knowledge of the genomic strategy of each target species, whether it is single-stranded or double-stranded, RNA or DNA, positive or negative sense.

The probes can be prepared in at least three basic formats: (a) PCR fragments arrayed on nylon membranes, hybridized against cDNA samples radioactively labeled, called macroarrays (Richmond et al. 1999) (b) PCR products spotted onto glass slides and DNA labeled with fluorescent dyes (Richmond et al. 1999; Zimmer et al. 2000; Wei et al. 2001); and (c) oligonucleotides of different length (from 18 to 70 bp) arrayed and hybridized with the same type of labeled DNA material (Lockhart et al. 1996; Loy et al. 2002, 2005; Fessehaie et al. 2003; Peplies et al. 2003). Another type of microarray is called the nanochip (Sosnowski et al. 1997; Nanogen, Inc., San Diego, CA 92121, USA) based on an electronically addressable electrode array that provides direct electric field control over the transport of charged molecules to selected micro locations and concentration over an immobilized substrate. A particular feature of this system is that biotinylated immobilized molecules can be either oligo capture probes or amplified PCR samples. Hybridization is detected and analyzed by fluorescent oligo probes. By regulating the electric-field strength, hybridization stringency can be adjusted for homologous interactions. Nanochips have shown high specificity and accuracy to diagnose plant viral and virus-like pathogens affecting potato, due to their ability to discriminate single nucleotide changes (Ruiz-Garcia et al. 2004). The Potentiality of microarray technology in the detection and diagnosis of plant diseases is very high due to the multiplex capabilities of the system. Moreover, it can be coupled with other systems, i.e. to perform nucleic acid extraction on the chip (Liu et al. 2007), achieve PCR reactions and their detection on the same device (van Doorn et al. 2007) or even mix all the systems in one (Lee et al. 2006), providing the possibility of automation that can be of great importance and utility. This possibility, with the coupling of the previous steps of the analyses (extraction, PCR, detection) promises a wider use in future protocols (Bonants et al. 2005; Lee et al. 2006; Boonham et al. 2007; van Doorn et al. 2007; Liu et al. 2007).

Further advancement in the microarray technology is the membrane (Nylon)–based microarray which is inexpensive and has potential to detect hundreds of pathogens in a single assay. For this, no specialized instrumentation or facilities are required, materials and reagents are relatively inexpensive, probes can be easily spotted onto small pilot arrays and evaluated for effectiveness and membranes can be reused for multiple times. Based on the technique, Agindotan and Perry (2007) have detected PVY and PLRV either singly or in mixed infection. Kokkinos et al. (2006) have also identified viruses infecting sweet potato by this technique. The sensitivity of this described microarray was comparable with that of DAS-ELISA, with conventional RT-PCR and real time RT-PCR and are being 10^3 and 10^5 times more sensitive.

Although there is a great potential for microarray technology in the diagnosis of plant diseases, the practical development of this application is still in progress. For example, following the methodology utilized for genetic analysis (Brown and Botstein 1999), large numbers of DNA probes used in two-dimensional arrays have allowed thousands of hybridization reactions to be analyzed at the same time (Hadidi et al. 2004). Until now, the microarray technology focuses its use in multiplex format of similar or very different pathogens, taking advantage of the number of probes that can be employed in one chip (Bonants et al. 2002; Schoen et al. 2002, 2003; Fessehaie et al. 2003; Franke-Whittle et al. 2005; Bonants et al. 2005; Boonham et al. 2007; van Doorn et al. 2007; Pasquini et al. 2008). With the availability of genomic sequences of plant viruses and the rapid development of microarray technology, as well as a renewed emphasis on detection and characterization of quarantine-plant viruses, there is a rush in the European Union to set up this technology and apply it to virus detection. Several international projects have developed diagnostic microarrays for plant viruses and are being used for detection. For example detection and differentiation of four cucurbit infecting tobamoviruses (Lee et al. 2003) and four potato viruses (PVY, PVX, PVA, PVS) (Boonham et al. 2003) were assayed by microarray methodology and was proved to be potential in viral diagnostics.

Since this method is completely generic, it can be used to detect all viruses whose sequence is currently available, but it is being increasingly used in functional genomics studies. Looking into the wide application of this technique and also having the advantage of screening large number of samples, number of microarray chips were constructed and are widely used. For example Pasquini et al. (2008) have developed a long 70-meroligonucleotide DNA microarray capable of simultaneously detecting and genotyping PPV strains and the thus, developed PPV microarray is capable of detecting and identifying all the strains of PPV infected peach, apricot and *Nicotiana benthamiana* leaves. A microchip detecting potato viruses, PVA, PVS, PVM, PVX, PVY, PLRV, in both single and mixed infections and the chip was also designed to distinguish between the main strains of PVY and PVS, was developed and tested (Bystricka et al. 2003, 2005).

More details on microarray technology one can have from be review articles of Hadidi et al. (2004), Barba and Hadidi (2007), Boonham et al. (2007), Hadidi and Barba (2008).

1.7.8 Recombinant DNA Technology

For plant virus diagnosis, the rDNA technology proved to be very reliable and sensitive for development of quick diagnostic tests and also the generation of transgenic crops with new or improved traits.

rDNA technology has been extensively used for:

(1) Production of polyclonal antibodies based on recombinant protein expressed in heterologous system.
(2) Production of recombinant antibodies by phage display technology.
(3) Single chain fragment variable antibody (scFv).
(4) Plantibody production: both agro-bacterium-mediated transformation and particle bombardment have been used to induce antibody genes into plants.

For more details or rDNA technology and its application in plant virology, one can refer to Pappu et al. (1995), Rybicki (2008), Sreenivasulu et al. (2012).

1.7.9 DNA Barcodes Use as Genetic Markers

DNA barcode is essentially a short stretch of nucleotide sequence that aid in the specific identification of species, strains or sub-strains. They are used to resolve pathogen taxonomy, phylogeny and for identifying pathogens and pests (Kumar and Sharma 2010). DNA barcodes are also known as DNA markers or DNA finger prints.

The plant virus and viroids are characterized by sequencing the specific genes and the data generated is used to interpret origin and spread of the pathogen, taxonomy and phylogeny. For diversity assessment, gene targets are selected based on the pathogen that comprise, ribosomal Internal Transcribed Sequence (ITS), mitochondrial cytochrome oxidase sub unit (COI), histone, virus coat protein etc., The concept is simple and a sample of the specimen is processed to produce the barcode. This is then matched against a library of known barcodes and in this way the specimen is identified. To do this, the barcode data base must first be constructed. The DNA barcode libraries are under construction for the medicinal plants of several nations including South Africa and Nigeria.

The approach has been used at IITA (Nigeria) for assessing the diversity of *Cassava brown streak virus*, *Banana bunchy top virus* and several other pathogens including fungi/oomycetes like *Colletotrichum, Pythium* and *Cercospora* species. Information generated from these studies has provided valuable clues to understanding the origin and drivers of spread, and the identification of previously uncharacterized pathogens (Alabi et al. 2008; Kumar and Sharma 2010).

This technique has been used in the diagnosis of *Alstroemeria yellow spot Tospovirus*, *Prunus necrotic ring spot virus*, *Potato yellow vein virus* etc. The DNA barcoding offers accurate identification of plant virus and virus-like diseases and focuses on strengthening the link between traditional and molecular taxonomy.

1.7.10 Conclusions

Sensitivity is dependent on the specific characteristics of the detection technique and on sampling protocols and sample preparation whereas, the main factors that determine specificity in PCR-based methods are primer selection and amplification conditions. In any event, more in-depth knowledge of the genome of pathogenic viruses and viroids will certainly enable more primers to be developed that target known pathogenicity and virulence genes (Louws et al. 1999). A hypothetical comparison of the 10 molecular tools most frequently used for plant virus diagnosis was compared by Lopez et al. (2006) and presented in the Table 1.3 with details on sensitivity, specificity, feasibility, rapidness and cost.

The 10 techniques listed in the Table 1.3 are routinely used worldwide for plant virus diagnosis. Overall, molecular techniques based on different PCR amplification and especially a real-time PCR are leading to high-through-put, faster and more accurate detection method for virus diseases. Other techniques like microarray, fluorescence *in situ* hybridization (FISH), molecular hybridization are also having great potential in plant virus diagnosis.

Table 1.3 Comparison of sensitivity, specificity, feasibility, rapidness and cost of different techniques in detection of plant viruses

Technique	Sensitivity[a]	Specificity[b]	Feasibility[c]	Rapidness	Cost
Molecular hybridization	+ [d]	++++	++	+	+++
FISH	++	++	+++	+	++
Conventional PCR	+++	++++	+++	+++	+++
Nested PCR in a single tube[e]	++++	++++	+++	++	+++
Multiplex PCR	++++	++++	+++	+++	+++
Multiplex nested PCR	+++	++++	+++	+++	+++++
Real-time PCR[f]	++++	++++	++	+++	++++
NASBA[g]	+++++	+++++	++++	+++++	+++
LAMP	+++++	++++	++++	++++	++
Microarrays	+	+++++	+	++	+

Source MM Lopez et al. 2006

[a] Sensitivity: probability of detecting true positives

[b] Specificity: probability of detecting true negatives

[c] Feasibility: practicability in routine analysis, execution and interpretation

[d] The number of + symbols indicates how methods rate regarding each considered criterion, from acceptable (+) to optimum (+++++)

[e] Coupled with hybridization and colorimetric detection

[f] Using TaqMan probes

[g] Using Molecular Beacons probes

Due to the nature of conventional, nested, or multiplex PCR, practical questions regarding the high level of sensitivity (up to 1 target per reaction) and the amplification of an enormous number of copies of the target sequence should be taken into account (Louws et al. 1999). However for each particular virus, an evaluation of the techniques using likelihood ratios must be performed.

The authors have just provided a brief overall diagnostic technique information on plant virus and viroid detection and recommend that the reader may refer to text books/review articles (Bar-Joseph and Garnsey 1981; Torrance, and Jones 1981; Jones and Torrance 1985; Lange 1986; Singh and Sharma 1989; Hampton et al. 1990; Khurana and Garg 1993; Barbara et al. 1995; Bar-Joseph et al. 1995; Banttari and Khurana 1998; Gallitelli 2004; James et al. 2006; Makkouk and Kumari 2006; Punja and Boer 2007; Baranwal and Ahlawat 2008; Rao and Maneesha Singh 2008; Prakash and Singh 2010; Naidu and Hughes 2003) for detailed information on plant virus and viroid diagnosis.

1.8 Pathways of Virus Spread

Plant viruses must have one or more effective means of dispersal from infected to healthy host plants. This is necessary to maintain the sequence of infection that ensures the long-term survival of viruses and viroids despite fluctuating and sometimes unfavourable environmental conditions. Consequently, the various

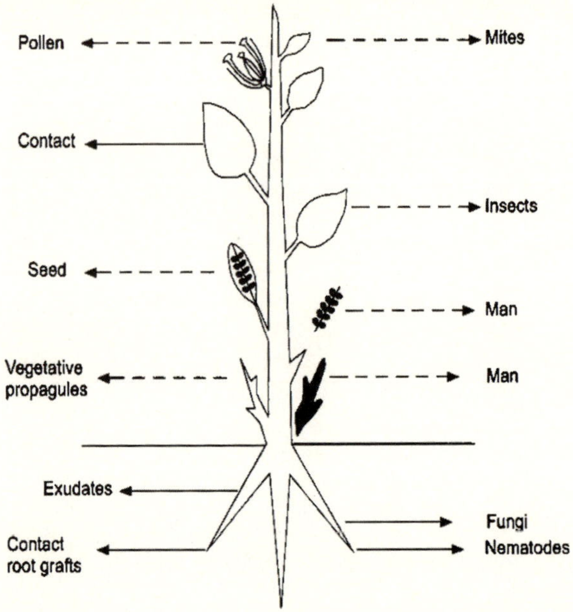

Fig. 1.7 Diagrammatic representation of the principal means of plant virus spread above-(*top*) and below ground (*bottom*) and autonomously (*left*) or by means of vectors (*right*). *Note* ——— = Local spread; ----- = Local and distant spread. *Source* Thresh JM and IITA, Nigeria

means of dispersal are an important feature of the 'epidemiological competence' of viruses and have received much attention from researchers.

Prof. JM Thresh prepared a review entitled "Plant Viruses and Virus Diseases in Africa" as a part of UK Government foresight project "Infectious Diseases: Preparing for the future", and he emphasized the importance of epidemiological studies for effective management of virus diseases to achieve the increased yields. In his article he mentioned the different methods of dispersal which are illustrated diagrammatically in Fig. 1.7, in which a distinction is made between spread above and below ground and between spread by autonomous means and by animal or fungal vectors. It is important to appreciate that any one virus exploits only one or a limited number of infection routes and there is considerable specificity. For example, each vector-borne virus is transmitted by only one or a small number of related species and various methods of virus transmission are depicted in the diagram below.

1.8.1 Contact Transmission

Direct spread by contact obviates the need for vectors but is largely restricted to viruses that are stable and highly infectious and occur at high concentration in their hosts. Another limitation is that spread tends to be localized and dependent on direct contact between plants, or on the activity of man or other vertebrate animals moving through crops, leading to contamination. The opportunity for direct spread in this way is limited, and is less important with plant viruses than with those of humans or livestock. However, contact-borne viruses and viroids of plants are likely to become increasingly important in at least some parts of tropics with increased production of flower and vegetable crops for local use or export. This involves the repeated handling of densely planted crops during transplanting, harvesting and other operations, which facilitates certain virus and viroid spread by contact. Moreover, the increased use of fertilizers, irrigation and protected year-round cropping under glass or plastic increases rates of crop growth and so enhances vulnerability to infection and the opportunity for virus spread to occur. Some of the examples of viruses which are contact transmitted are TMV, ToMV, PVX, *Cucumber green mottle mosaic virus*, and *Rice yellow mottle virus* in crops like tobacco, tomato, potato, cucurbits, rice etc.

1.8.2 Long Distance Dispersal

(a) **Seed transmission**

Germplasm is one of the most requisite factors for any country's research programme and hence the germplasm exchange is taking place. Under WTO, the food grains are also exchanged in higher quantities. Nearly 231 plant viruses and viroids are transmitted through seed.

Approximately 18 % of the described plant viruses are seed-transmitted in one or more host, and it is estimated that one-third of the plant viruses will eventually prove to be seed-transmitted in at least one host (Johansen et al. 1994; Khan and Dijkstra 2002; Albrechtsen 2006; Sastry 2013a).

Seeds are a very effective means of the entry of virus into the fields. Transmission through seeds is the most serious problem when a vector is present and it primarily depends on the initial virus inoculum load. For example, *Bean common mosaic virus* shows a very high percentage of transmission through seed and has a very efficient aphid vector. When even a small amount of infected seed is planted, aphids will spread the infection to healthy plants under field conditions. For example, cowpea seeds with 5 % of *Blackeye cowpea mosaic virus* have resulted in a yield loss of 34–53 %, while 0.5 % seed infection did not cause significant losses (Puttaraju et al. 2002).

The seed-borne plant virus diseases are distributed in 24 virus groups that include alfamo, bromo, carla, carmo, faba, furo, como, tobamo, tobra, tombus, sobemo, tricho, peclu, alfacrypto, betacrypto, ourmia, idaco, cucumo, hordei, ilar, poty and tymovirus groups (Sastry 2013a). It was also observed that high degree of seed transmission is noticed with viroid diseases like *Australian grapevine viroid*, *Avocado sunblotch viroid*, *Citrus exocortis viroid*, *Potato spindle tuber viroid*, and *Apple scar skin viroid* diseases. *Crypticviruses* which induce little or no disease symptoms are transmitted through seed at the highest percentage (Boccardo et al. 1983).

Throughout the world the majority of the germplasm resources of the cultivars of some crops like peas, peanuts, beans, cowpea and soybean, are generally contaminated with seed-borne viruses at some of the germplasm banks of international institutions (Hampton et al. 1982; Alconero and Hoch 1989). For breeding programmes, generally even the seed of wild species is exchanged. The studies carried out by Olukemi and Lava Kumar (2013) have indicated that wild relatives of *Vigna* spp. which are asymptomatic, carry viruses through seed to the tune of 70 %, warranting the diagnosis of seed-borne viruses. To combat this type of situation, IBPGR and other organizations have developed suitable quick diagnostic techniques and measures to increase healthy seed lots to maintain healthy germplasm.

(b) **Transmission through pollen**

Virus/viroid infected pollen plays an integral role in the spread of viruses belonging to cavemo, tombus, nepo, idaeo, enamo, alfa and beta cryptovirus groups infecting some of the woody and herbaceous plants. Pollen transmission is recorded in bromo, nepo, crypto, alfalfa, cucumo, and ilarviruses and the virus particles are externally or internally pollen-borne. Pollen transmission takes place both vertically and horizontally. In case of vertical transmission of viruses, the pollination and fertilization of healthy ovules by infected pollen results in the formation of infected seeds which on germination produce infected seedlings. For effective horizontal spread through infected pollen, the viruses invade and systematically infect the ovule bearing mother plant and large quantities of infected pollen are released to infect the contemporary healthy plants. The virus transmission through pollen is economically important in cross pollinated woody perennial plants than with annual crops wherein both vertical and horizontal transmission takes place.

Even some of the plant viroid diseases which are seed transmitted are also pollen transmitted as in case of *Potato spindle tuber viroid* (PSTVd) in potato and tomato (Fernow et al. 1970; Kryczynski et al. 1988); *Chrysanthemum stunt viroid*, *Cucumber pale fruit viroid* in tomato (Kryczynski et al. 1988) and *Peach latent mosaic viroid* in peach (Barba et al. 2007).

Pollen carrying insects like thrips, bumble bees and honey bees play a major role in transfer of virus infected pollen to the healthy plants. The possible role of honey bees for long distance spread of PNRSV in sweet cherry orchards was well established (Mink 1983). Bumble bees cause wounding of flowers during their visits and introduce the *Tomato apical stunt viroid* (Antignus et al. 2007). From India, Prasada Rao et al. (2006) have recorded that TSV is introduced into susceptible crops like sunflower and groundnut through feeding injuries. It is presumed that TSV transmission under field conditions occur in two ways viz., (a) Pollen carrying thrips from flowers of TSV infected sunflower and weeds plants when colonizing susceptible crops transmit the virus; (b) Wind-blown pollen from TSV infected plants when falling on the susceptible crops, thrips during their feeding, cause injury to both leaf and tissue and the deposited pollen enters into the host through wounds caused by thrips feeding. The topic of transmission of virus and viroid diseases has been reviewed by number of workers viz. (Shepherd 1972; Hardtl 1978; Mandahar 1981, 1985; Mink 1993 and Sastry 2013b).

(c) **Insect vectors and virus-vector relationships**

The third major group of biological vectors of plant viruses is insects of different types. Virus transmission through insects has assumed enormous significance in plant health because of the ease with which they transmit the viruses in the field. Of the 991 plant viruses reported by 9th ICTV, nearly 76 % are transmitted by insects and other vectors (Hogenhout et al. 2008; King et al. 2012). Among the insect vectors, aphids transmit the maximum number of plant viruses followed by whiteflies.

Table 1.4 Plant virus vectors of some homoptera, coleoptera, and thysanoptera families

Order, suborder family	Common name of insect group	Approx. no. of species described	No. of vector species	No. of viruses transmitted
Homoptera				
Auchernorrhyncha				
Cicadidae	Cicada	3200	0	0
Membracidae	Treehopper	4500	1	1
Cercopidae	Spittlebug	3600	0	0
Cicadellidae	Leafhopper	20000	49	31
Fulgoroidea	Planthopper	19000	28	24
Sternorhyncha				
Psyllidae	Psyllid	2000	0	0
Aleyroididae	Whitefly	1200	3	200
Aphididae	Aphid	4000	192	475
Pseudococcidae	Mealybug	6000	19	10
Coleoptera				
Chrysomelidae	Leaf beetle	37000	48	30
Coccinellidae	Ladybird beetle	4500	2	7
Cucurlionidae	Weevil	50000	10	4
Meloidae	Blister beetle	3000	1	1
Thysanoptera				
Thripidae	Thrips	~5000	6	13
Arachnida				
Eriophyidae	Mites	>900	4	12

Source Hull (2002), Matthews' Plant Virology. Nault (1997) improved (with permission of Elsevier Publishers)

Aphids, leafhoppers, beetles, mites, thrips and other insect vectors help in the spread of virus diseases for long distances because of their flight behavior and help in the spread of virus diseases for long distances depending on the availability of susceptible crops. Among the different insect vectors listed in Table 1.4, aphids transmit maximum number of plant viruses (475) followed by whitefly vector (200). It is also clear from the table that leaf beetles, leafhoppers, planthoppers, and thrips play a major role in transmitting number of plant virus diseases. Since a large number of populations of insects exist in the tropics, some of them cause more direct damage to the crop as a pest.

Table 1.5 Certain genera of plant viruses and insect species that serve as vectors

Virus genera	Mode	Persistence	Presence in vector	Insects involved
Alfamovirus	N	Few hours	External	Aphids
Badnavirus	SP	Days	External	Mealybugs and leafhoppers
Begomovirus	C	Weeks	Internal	Whiteflies
Carlavirus	N	Few hours	External	Aphids
Caulimovirus	N	Many hours	External	Aphids
Closterovirus	SP	Many hours	External	Aphids or mealybugs
Comovirus	SP	Days	Internal	Beetles
Cucumovirus	N	Few hours	External	Aphids
Curtovirus	C	Weeks	Internal	Leafhoppers
Enamovirus	C	Weeks	Internal	Aphids
Fabavirus	N	Few hours	External	Aphids
Luteovirus	C	Weeks	Internal	Aphids
Machlomovirus	SP	Many days	External	Leafhoppers
Mastrevirus	C	Weeks	Internal	Leafhoppers
Polerovirus	C	Weeks	Internal	Aphids
Potyvirus	N	Few hours	External	Aphids
Sequivirus	SP	Few hours	External	Aphids
Sobemovirus	SP	Days	Not determined	Beetles
Tymovirus	SP	Days	Not determined	Beetles
Waikavirus	SP	Few days	External	Leafhoppers

C circulative, *N* nonpersistent, *SP* semi-persistent. Internal the virus crosses gut and salivary gland barriers. External virus does not cross the gut barrier; but remains attached to the foregut epithelium

Source Raccah and Fereres (2009)

The relationship of plant viruses with their insect vectors have been divided into four classes (Nault 1997) viz., (1) Nonpersistently transmitted, stylet-borne viruses; (2) Semi-persistent transmitted, foregut-borne viruses; (3) Persistently transmitted circulative viruses; (4) Persistently transmitted, propagative viruses. The virus-vector relationship of each vector was discussed separately in this chapter. More details are furnished on vector transmission in Chapter-4 of Volume I (Sastry 2013b) and also from the review articles of Nault (1997), Andret-Link and Fuchs (2005), Blanc et al. (2011), Bragard et al. (2013).

1.8.3 Arthropod Vector Transmission

Among the arthropod vectors aphids, leafhoppers, planthoppers, thrips, whiteflies and beetles are very important and they transmit a large number of plant virus diseases. The names of different viruses they transmit and the virus-vector relationship is provided in the Tables 1.5, 1.6.

Table 1.6 Means of spread of certain virus and virus-like causal agents of certain diseases of crops in the tropics

Crop	Disease/agent	Contact/sap transmission	Graft/vegetative propagule/seed transmission	Vector type (Mode of transmission[a])
Apple	*Apple mosaic virus*	Sap ?	Stem piece, pollen	Unknown
Avocado	*Avocado sunblotch viroid*	Contact	No	Unknown
Banana/plantain	*Banana bunchy top virus*	No	Suckers	Aphids (C)
	Banana streak virus	No	Suckers	Mealybugs (SP)
	Banana bract mosaic virus	No	Suckers	Aphids (NP)
	Cucumber mosaic virus	Sap	Suckers	Aphids (NP)
Barley	*Barley stripe mosaic virus*	Sap contact	Seed	Unknown
Black pepper	*Pepper yellow mottle virus*	No	Stem piece	Mealybugs (SP)
Cabbage, cauliflower	*Turnip mosaic virus*	Sap	No	Aphids (NP)
	Cauliflower mosaic virus	Sap	No	Aphids (NP)
Cardamom	*Cardamom mosaic virus*	Sap	Rhizome	Aphids (NP)
Cassava	*African cassava mosaic virus*	Sap ?	Stem piece	Whiteflies (C)
	East African cassava mosaic virus	Sap ?	Stem piece	Whiteflies (C)
	Indian cassava mosaic virus	Sap	Stem piece	Whiteflies (C)
	Srilankan cassava mosaic virus	Sap	Stem piece	Whiteflies (C)
	South African cassava mosaic virus	Sap ?	Stem piece	Whiteflies (C)
	Cassava common mosaic virus	Contact, sap	Stem piece	Unknown
	Cassava brown streak virus	No	Stem piece	Whiteflies (C)
Cocoa	*Cocoa swollen shoot virus*	No	Stem piece	Mealybugs (C, P)
Common bean	*Bean common mosaic virus*	Sap	Seed	Aphids (NP)
	Bean common mosaic necrosis virus	Sap	Seed	Aphids (NP)
	Bean golden mosaic viruses	Sap ?	No	Whitefly (C)
	Southern bean mosaic virus	Sap	Seed	Beetles
	Cucumber mosaic virus	Sap	Seed	Aphids (NP)
Cotton	*Cotton leaf curl viruses*	No	No	Whiteflies (C)

(continued)

Table 1.6 (continued)

Crop	Disease/agent	Contact/sap transmission	Graft/vegetative propagule/seed transmission	Vector type (Mode of transmission[a])
Chickpea	*Bean leaf roll virus*	No	No	Aphids (C)
Chilli/pepper	*Tomato spotted wilt virus*	Sap	No	Thrips (C, P)
	Cucumber mosaic virus	Sap	No	Aphids (NP)
	Potyviruses	Sap	No	Aphids (NP)
	Chilli leaf curl virus	No	No	Whiteflies (C)
Citrus	*Citrus tristeza virus*	No	Budwood	Aphids (SP)
	Citrus yellow mosaic virus	No	Budwood	Mealybugs (SP)
	Citrus exocortis viroid	Contact	Budwood	Unknown
	Citrus stubborn viroid	No	Budwood	Unknown
	Citrus psorosis virus	Sap	Budwood	Unknown
Coconut	*Coconut cadang–cadang viroid*	Contact	No	Unknown
	Coconut foliar decay virus	No	No	Vector ?????
Cowpea	*Cowpea mosaic virus*	Sap	Seed	Beetles
	Cowpea severe mosaic virus	Sap	Seed	Beetles
	Cucumber mosaic virus	Sap	Seed	Aphids (NP)
	Cowpea mottle virus	Sap	Seed	Beetles
	Cowpea golden mosaic virus	No	No	Whitefly (C)
Cucurbits	*Cucumber mosaic virus*	Sap	Seed	Aphids (NP)
	Watermelon mosaic virus	Sap	No	Aphids (NP)
	Papaya ring spot virus	Sap	No	Aphids (NP)
	Zucchini yellow mosaic virus	Sap	No	Aphids (NP)
	Squash mosaic virus	Sap	Seed	Beetles
	Squash leaf curl virus	No	No	Whiteflies (C)
	Watermelon bud necrosis virus	Sap	No	Thrips (C, P)
	Cucumber green mottle mosaic virus	Sap, Contact	Seed	Unknown

(continued)

Table 1.6 (continued)

Crop	Disease/agent	Contact/sap transmission	Graft/vegetative propagule/seed transmission	Vector type (Mode of transmission[a])
Eggplant/brinjal	*Cucumber mosaic virus*	Sap	No	Aphids (NP)
	Eggplant mottled dwarf virus	No	No	Leafhoppers (C)
	Potato virus Y	Sap	No	Aphids (NP)
Faba bean	*Bean leaf roll virus*	No	No	Aphids (C)
	Bean yellow mosaic virus	Sap	Seed	Aphids (NP)
Grapevines	*Grapevine fanleaf virus*	Sap	Stem piece	Nematodes
	Grapevine leafroll viruses	No	Stem piece	Unknown
Green gram/black gram	*Mungbean yellow mosaic virus*	Sap?	No	Whiteflies (C)
Groundnut	*Groundnut bud necrosis virus*	Sap	No	Thrips (C, P)
	Tomato spotted wilt virus	Sap	No	Thrips (C, P)
	Tobacco streak virus	Sap	Pollen	Thrips
	African and Indian peanut clump viruses	Sap	Seed	Fungus
	Groundnut rosette viruses	Sap	No	Aphids (C)
	Peanut mottle virus	Sap	Seed	Aphids (NP)
	Peanut stripe virus	Sap	Seed	Aphids (NP)
Lentil	*Bean leafroll virus*	No	No	Aphids (C)
	Bean yellow mosaic virus	Sap	Seed	Aphids (NP)
Maize/sorghum	*Maize streak virus*	No	No	Leafhoppers (C)
	Maize dwarf mosaic virus	Sap	Seed	Aphids (NP)
	Maize rough dwarf virus	No	No	Planthopper (C)
	Maize stripe virus	No	No	Planthopper (C, P)
	Maize rayado fino virus	No	No	Leafhopper (C)
	Maize mosaic virus	No	No	Planthopper (C)

(continued)

Table 1.6 (continued)

Crop	Disease/agent	Contact/sap transmission	Graft/vegetative propagule/seed transmission	Vector type (Mode of transmission[a])
Okra/bhendi	*Okra leaf curl virus*	No	No	Whiteflies (C)
	Bhendi yellow vein mosaic virus	No	No	Whiteflies (C)
Onion/garlic	*Onion yellow dwarf virus*	Sap	Bulbs	Aphids (NP)
	Iris yellow spot virus	Sap	Bulbs	Thrips (C, P)
Papaya	*Papaya ringspot virus*	Sap	No	Aphids (NP)
	Papaya leaf curl virus	No	No	Whiteflies (C)
Pineapple	*Pineapple wilt virus*	No	????	Mealybugs
Pea	*Pea seed-borne mosaic virus*	Sap	Seed	Aphids (NP)
	Bean yellow mosaic virus	Sap	Seed	Aphids (NP)
Pigeonpea	*Pigeonpea sterility mosaic virus*	Sap	No	Mites (C)
Potato	*Potato leafroll virus*	No	Tuber	Aphids (C)
	Potato virus Y	Sap	Tuber	Aphids (NP)
	Potato virus X	Contact	Tuber	Unknown
	Potato virus S	Sap	Tuber	Aphids (NP)
	Potato spindle tuber viroid	Sap, Contact	Tuber	Unknown
Rape seed and mustard	*Turnip mosaic virus*	Sap	No	Aphids (NP)
Rice	*Rice tungro spherical and bacilliform viruses*	No	No	Leafhoppers (SP)
	Rice ragged stunt virus	No	No	
	Rice grassy stunt virus	No	No	
	Rice stripe virus	No	No	Planthopper (C)
	Rice hoja blanca virus	No	No	Planthopper (SP)
	Rice yellow mottle virus	Sap	Unknown	Beetles

(continued)

Table 1.6 (continued)

Crop	Disease/agent	Contact/sap transmission	Graft/vegetative propagule/seed transmission	Vector type (Mode of transmission[a])
Soybean	*Soybean mosaic virus*	Sap	Seed	Aphids (NP)
	Bean pod mottle virus	Sap	Seed	Beetles
	Tobacco ring spot virus	Sap	Seed	Nematode/Mites/Aphids
Sugar beet	*Beet curly top virus*	No	No	Leafhoppers (C)
Sugarcane	*Sugarcane mosaic virus*	Sap	Stem piece	Aphids (NP)
	Sugarcane streak mosaic virus	Sap	Stem piece	Unknown
	Sugarcane Fiji disease virus	No	Stem piece	Planthoppers (C, P)
Sunflower	*Tobacco streak virus*	Sap	Pollen	Thrips
Sweet potato	*Sweet potato feathery mottle virus*	Sap	Stem piece	Aphids (NP)
	Sweet potato mild mottle virus	Sap	Stem piece	Whitefly (C)
	Sweet potato sunken vein virus	No	Stem piece	Whitefly (SP)
Taro/coca yam	*Dasheen mosaic virus*	Sap	Corm	Aphids (NP)
	Taro bacilliform virus	No	Corm	Mealybugs (SP)
Tobacco	*Tobacco leaf curl virus*	No	No	Whiteflies (C)
	Tomato spotted wilt virus	Sap	No	Thrips (C, P)
	Tobacco mosaic virus	Sap, Contact	Seed	Unknown
Tomato	*Tomato mosaic virus*	Sap, Contact	Seed-borne	Unknown
	Tomato leaf curl viruses	No	No	Whiteflies (C)
	Tomato spotted wilt virus	Sap	No	Thrips (C, P)
	Groundnut bud necrosis virus	Sap	No	Thrips (C, P)
	Cucumber mosaic virus	Sap	No	Aphids (NP)
Wheat	*Barley yellow dwarf virus*	No	No	Aphids (C)
Yam	*Yam mosaic virus*	Sap	Tuber	Aphids (NP)
	Yam mild mosaic virus	Sap	Tuber	Aphids (NP)
	Diascorea bacilliform virus	No	Tuber	Mealybugs

[a] *C* circulative, *NP* nonpersistent, *SP* semi-persistent, *P* propagative

These insect vectors and to a much lesser extent eriophyid mites play a crucial role in the transmission of plant viruses, and there are many tropical examples. They include viruses transmitted by six different insect taxa, and there are several different mechanisms of transmission (Tables 1.5, 1.6). An important feature of active transmission by arthropods is that it compensates for the almost complete immobility of rooted plants and the limited opportunity for contact spread.

The mobility of insect vectors, their ability to colonize and exploit new habitats and the effectiveness with which they transmit viruses can account for the current prevalence of many of the most important viruses of crops in the tropics. Many *Cicadellids* transmit cereal viruses and have two contrasting types of spreads i.e., the long fliers with short bodies predominating the main flight period, and the short fliers with relatively long bodies. This type of structural polymorphism has obvious survival value for leafhoppers, facilitating colonization and subsequent exploitation provides an efficient means of virus spread. This situation will continue as new vector species or biotypes are introduced inadvertently to tropical countries, exacerbating existing problems and creating new ones. Moreover, current trends in agriculture facilitate the introduction, survival and reproduction of vectors and the opportunity for them to acquire and transmit viruses.

Because of the changing environmental factors, in future the arthropod-borne viruses will become even more important than they are at present. Furthermore, they will be difficult to control, in part because of increasing restrictions on the use of pesticides on environmental grounds and also to avoid hazards to the health of farm operators and consumers. There is abundant evidence from developed countries that effective integrated pest management (IPM) and Integrated crop management (ICM) procedures can be developed to control arthropod-borne viruses. However, they require a substantial and long-term investment in research, of a type that has seldom been achieved in developing countries. Even where solutions are identified which are suitable for farming systems in tropics, they may be expensive in labor, or capital or both, particularly for smallholder farmers.

Progress will be facilitated by developments in the detection and identification of both viruses and vectors. These are already leading to improved methods of disease monitoring and could provide effective early warning systems of a type not yet available. Additionally, more appropriate disease and vector management strategies can be developed for smallholder farmers if researchers gain a better understanding of farmers' perceptions of virus diseases and vectors, their attitude to risk and the resources that are available to them for disease management.

(a) **Aphid vectors**

Among the virus vectors, aphids constitute the most important group which transmits more viruses than any others. About 50 % of plant viruses are aphid transmitted and belong to carla-, clostero-, cucumo-, luteo-, enamo-, cavemo-, soymo-, babu-, nano-, polero-, alfamo-, caulimo-, faba-, poty- viruses and also certain plant rhabdoviruses. Both the acquisition and inoculation probes of 15–60 s each are for optimal transmission in a nonpersistent relationship. Some aphid vectors transmit the virus it acquired only after a lapse of few days to few weeks and the virus is considered to be circulative-persistent in the vector and there is a latent period. Examples for this category include *Pentalonia nigronervosa* for *Banana bunchy top virus, Aphis craccivora* for *Groundnut rosette virus,* and *Myzus persicae* for *Potato leafroll virus.* Even some aphids have persistent-propagative relationship with certain viruses and take long periods of acquisition and inoculation periods and the virus is not lost during moulting and the virus replicates in the aphid vectors. The virus examples here are BYNV, LNYV, SCV, etc. Both winged and wingless single aphids can transmit the virus. High percent of transmission is achieved by the preliminary fasting period with optimal aphid numbers; however, the inoculative capacity of the aphid decreases as the period between acquisition and inoculation of the virus increases. Aphids almost always probe in the anticlinal grooves of adjacent epidermal cells, which they locate via receptors mechano pegs present on the labial tip. Aphids always secrete saliva at the start of the probe as well as during stylet penetration forming a sheath. Stylets move fairly rapidly within this sheath but subsequently extend beyond the sheath for ingestion of food material from host cells. During the process of feeding, the virus present in the sap usually adheres to their mouth parts and is introduced subsequently into another plant when the viruliferous aphids feed; which is termed as mechanical contamination hypothesis. Forbes (1977) has extensively reviewed the feeding mechanism of aphids. In general, the nonpersistent plant viruses have low level of vector specificity because they are transmitted by several aphid species, such as AMV by 14 aphid species (Crill et al. 1970); CMV by 60 aphid species (Kennedy et al. 1962) and SMV by 32 aphid species (Irwin and Goodman 1981).

(b) **Leafhopper vectors**

Leafhoppers are long movers and belong to the vector family *Cicadellidae* and *Delphacidae* which transmit a number of virus and phytoplasma diseases of economically important crops (Weintraub and Beanland 2006). Around 20,000 leafhopper (*Cicadellidae*) species have been described, but estimates suggest 100,000 species may exist (Dietrich 2005). Of the 60 subfamilies, *Agallinae* and *Deltocephalinae* have vectors of plant viruses whereas planthoppers occur in families *Fulgoroidea* and *Delphacidae* and include vectors of cereal viruses. Leafhoppers transmit the viruses in semi persistently or persistently. The persistent leafhopper transmitted viruses are of circulative and propagative type. There are 12 circulative type viruses in two genera *Mastrevirus* and *Curtovirus,* and there are more than 70 viruses which are propagatively persistently transmitted. Generally, nymphs are more efficient in transmitting the virus. Viruses transmitted by leafhoppers mainly cause yellowing and leaf rolling symptoms in the infected host plants and only a few are mechanically sap transmitted. The vectors feed only on the phloem tissues as the viruses are concentrated in phloem tissues. Some of the leafhopper-transmitted viruses are *Rice tungrovirus* by *Nephotettix nigropictus*, *Maize streak virus* by *Cicadulina mbila*, *Beet curly top virus* by *Agalina albidula*, *Wheat streak virus* and other *African streak viruses* by *Javesella pellucida*, *Maize rayado fino virus* by *Dalbulus maidis* and *Potato yellow dwarf virus* by

Aceratagallia curvata. For more details, one can refer to Maramorosch and Harris (1979), Nault (1997).

Leafhopper vectors are so active that even a few infected plants (1 %) in the field are enough for secondary spread and cause 100 % disease incidence. Nymphs are more efficient vectors. Leafhoppers will have several generations in a year and thus acquire the virus for number of times from source plants and transmit the virus to new plants in the same field or elsewhere. Some of the leafhoppers pass the virus through the eggs to the progeny (transovarially transmission).

(c) **Planthopper vectors**

Planthoppers have received far less attention from vector researchers than the aphids and leafhoppers. Currently 13 genera and 23 vector species are recorded, and these are responsible for the transmission of 24 viruses. The transmissions are of circulative, and most, if not all, of the viruses also appear to be propagative. Planthoppers occur in only family *Delphacidae* and has vectors that feed on poaceae members like rice, wheat and maize.

Some of the examples of planthopper-transmitted viruses are viz., *Maize mosaic virus* (*Peregrinus maidis*), *Rice stripe virus* and *Rice black streaked dwarf virus* (*Laodelphaxstriatellus*), *Sugarcane fiji disease* fiji virus (*Perkinsiella saccharicida*), *Oat sterile dwarf fijivirus* (*Javasella pellucida*), *Ramu stunt disease* (*Eumetopina flavipes*); *Rice hoja blanca virus* (*Sogatodes orizicola*), *Rice ragged stunt virus* (*Nilaparvata lugens*), *Sorghum stripe virus* (*Peregrinus maidis*). In general, nymphs and female adults transmit these viruses in a more efficient manner (Narayana and Muniyappa 1996; Falk and Tsai 1998; Ammar and Nault 2002; Anderson et al. 2007).

(d) **Whitefly vectors**

In the tropics, whitefly-transmitted virus diseases infect a wide range of plants including both monocotyledonous and dicotyledonous species and *Bemisia tabaci* is the vector. This vector has been recorded from more than 1,000 crop and weed host species in tropics and sub-tropics, and occupies a great diversity of niches with variable ecological conditions. The family *Geminiviridae* is classified into four genera. Nearly 90 % of the plant viruses transmitted by *B. tabaci* belong to the genus *Begomovirus*. Whiteflies also transmit the viruses belonging to remaining three different genera, *Crinivirus*, *Carlavirus*, and *Ipomovirus* (Fauquet et al. 2000, 2005; Morales and Anderson 2001; Jones 2003; Varma et al. 2011).

Whiteflies are small, piercing and sucking insects belonging to the family *Aleyrodidae* in the order *Homoptera*. One can recognize the adults, nymphs and pupae of the whitefly, *Bemisia tabaci*, as they occur on the undersides of leaves. Adults of the whitefly are small (about 1 mm long). The males are slightly smaller than the females. The wings are brilliant white, as are those of the spiraling whitefly, *Aleurodicus disperus*. Only the first instar of *B. tabaci* is mobile to find feeding sites, it does not move far, moults and settles down to feed. Many generations of the whiteflies occur in a year. Both nymphal and adults acquire and transmit the virus. *Begomoviruses* are persistently and circulatively transmitted by whiteflies with 4–48 h latent period. Females were found to transmit the virus more frequently than males. A female *B. tabaci* can lay up to 160 eggs on the undersides of leaves in its life time of up to two months. The eggs hatch into nymphs (larvae) within a week. Newly hatched nymphs (or crawlers) are the only mobile nymph stage of the insect. Crawlers move to, and settle at, suitable feeding locations on lower leaf surfaces and become sessile throughout the remaining nymphal stages. The *B. tabaci* whitefly can produce 11–15 generations in a year.

Only whiteflies in the *Bemisia* and *Trialeurodes* genera are virus vectors. In the genus *Bemisia*, only *B. tabaci* has been shown to be a vector of over 200 viruses (Jones 2003; Morales 2007); *Bemisia* was considered as vector of *Cassava brown streak virus* and *Sweet potato chlorotic stunt virus* (Malathi et al. 2004; Gamarra et al. 2010); whereas in the *Trialeurodes* genus, *Trialeurodes vaporariorum, T. abutilonea* and *T. ricini* transmit viruses like *Potato yellow vein virus* in potato (Salazar et al. 2000) and *Tomato infectious chlorosis* and *Tomato torrado virus* in tomato (Wisler et al. 1998; Duffus et al. 1996). There are biotypes of whitefly (*B. tabaci*) like B, Q and A. Naming of the biotypes has been based on the esterase profile, dendrograms produced from genetic similarities based on randomly amplified polymorphic DNA (RAPD)-PCR and amplified fragment length polymorphism (AFLP) analysis (Guirao et al. 1997). According to Stanley et al. (2005) and also of Fauquet and Nawaz-ul-Rehman (2008), in the VIIIth report ICTV on virus taxonomy, more than 178 species of whitefly-transmitted viruses are grouped in the genus *Begomovirus* and family *Geminiviridae.* Whitefly instar nymphs and adults feed by inserting their proboscises into the leaf, penetrating into the phloem and withdrawing sap. It is during this feeding process that plant viruses are acquired. Adult whiteflies will disperse and transmit the virus to new plants while feeding. In general whiteflies are poor fliers and their long distance movements are likely assisted by wind and human involvement (Byrne and Bellows 1991). The whitefly-transmitted viruses do not normally multiply within the vectors i.e., they are not propagative and transovarian transmission through the eggs has not been shown. However Ghanim et al. (1998) have provided the evidence for transovarial transmission by its whitefly vector, *B. tabaci.* Hunter et al. (1998) have studied the location of geminiviruses in *B. tabaci* and they detected both ToMoV and CabLCV in the anterior region of the midgut and filter chamber of adult whiteflies.

Geminiviruses have no special insect transmission factor like helper component. All properties required for geminivirus transmission, including GroEL interaction rely on the viral capsid protein. The concern for the whitefly-transmitted virus family *Geminiviridae* is increased due to the emergence of new geminiviruses through recombination or pseudo-recombination among strains and/or species (Navas-Castillo et al. 2000).

Since geminiviruses are causing enormous yield losses throughout the globe on different economically important crops like cassava, sweet potato, tomato, capsicum, okra and others, a lot of epidemiological models were developed and are implemented in day to day agriculture (Chellaiah and Murugesan 1976; Sastry et al. 1978; Mazyad et al. 1979; Muniyappa 1980, 1983; Ioannau and Iordanou 1985; Saikia and Muniyappa 1989; Valand and Muniyappa 1992; Cohen and

Antignus 1994; Polston et al. 1996; Ramappa et al. 1998; Holt et al. 1999; Pun and Doraisamy 2000; Gupta and Keshwal 2003; Morales 2004; Morales and Jones 2004; Anderson and Morales 2005; Kalita et al. 2005).

(e) **Thrips vectors**

Thrips belong to *Thysanoptera* and are generally reproduce parthenogenitically. Only the first instar larvae acquire the virus and both larvae and adult transmit the virus. The main genera transmitted by thrips vectors are: *Tospovirus*, *Ilarvirus*, *Carmovirus* and *Sobemovirus*. *Frankliniella*, *Thrips* and *Scirtothrips* species are the major vectors, which are polyphagous and have piercing-sucking mouth parts. *Thrips tabaci*, the most common species and efficient virus vector feeds on 140 plant species in over 40 families. Similarly, *Frankliniella occidentalis* has 148 hosts (Ullman et al. 2002). In tropical and sub-tropical Asia, *Thrips palmi* is the predominant vector which transmits several different *Tospoviruses* viz., *Peanut (groundnut) bud necrosis virus*, *Capsicum chlorosis virus*, *Melon yellow spot virus*, *Watermelon bud necrosis virus*, *Watermelon silver mottle virus* and *Lily chlorotic spot virus*. *T. palmi* is widely distributed in several countries including India, Indonesia, Japan, the Philippines and Thailand (Jones 2005a; Cannon et al. 2007; Pappu et al. 2007), while *Frankliniella occidentalis*, another important thrips vector, is prevalent in Australia, Europe and Middle East, North and South America and transmits *Tospoviruses* like *Tomato spotted wilt virus*, *Tomato chlorotic spot virus*, *Groundnut ring spot virus*, *Impatiens necrotic spot virus* and *Chrysanthemum stem necrosis virus* (Jones 2005a; Pappu et al. 2009). Because of the wide host range of virus and the vector, the *Tospo*- and *Ilarviruses* can affect a number of economically important crops resulting in heavy yield losses.

Except in certain Asian countries, *Tomato spotted wilt virus* (TSWV) is an economically important virus in majority of the countries and is transmitted by eight thrips species viz., *Frankliniella occidentalis* (western flower thrips), *F. schultzei*, *F. fusca* (tobacco thrips), *Thrips tabaci* (onion thrips), *T. setosus*, *T. moultoni*; *F. tenuicornis*, and *Scirtothrips dorsalis*. The first four are considered to be the most important vectors because of their wide distribution and the overlapping host ranges of these thrips species and TSWV (Whitefield et al. 2005). TSWV infects over 800 different plant species from 82 families. Food crops like peanut, watermelon, capsicum, tomato, zucchini, celery, eggplant, cucumber, lettuce, pineapple, grape, many legumes as well as ornamental species (gladiolus, dahlia, lily, impatiens, chrysanthemums, and iris) are affected by TSWV and effective management measures are to be worked out.

In India *Peanut* or *Groundnut bud necrosis virus* (GBNV/PBNV) is wide spread in peanut, potato, tomato, taro, capsicum and a number of legumes and is transmitted by different thrips vectors like *T. palmi*, *F. schultzei* and *Scirtothrips dorsalis* (Mandal et al. 2012). Similarly in onion *Iris yellow spot virus* is transmitted by *T. tabaci* has worldwide distribution. In India, potato stem necrosis disease caused by a strain of PBNV-Po is seen in potato growing states. Jain et al. (1998) have reported *Watermelon bud necrosis tospovirus* on watermelon to be economically important in south India.

As discussed above, tospoviruses in India, Australia, Taiwan, Japan, Thailand and USA are transmitted by different types of thrips species and the details are provided in the table given below.

Thrips transmission of tospoviruses in South and Southeast Asia						
Thrips	**PBNV**	**CaCV**	**WBNV**	**WSMV**	**MYSV**	**TSWV**
T. palmi	Yes[1]	Yes[2]	Yes?	Yes[3]	Yes[4]	?
S. dorsalis	Yes[1]	?	?	?	?	?
C. claratris	?	Yes[5]	?	?	?	?
F. schultzei	Yes[1]	?	?	?	?	?
T. tabaci	?	?	?	?	?	?
F. occidentalis	?	?	?	?	?	Yes[6]
[1]India, [2]Australia, [3]Taiwan, [4]Japan?, [5]Thailand, [6]USA						

Courtesy RA Naidu; USAID-IPM CRSP project

Sunflower necrosis disease caused by *Tobacco streak virus* (TSV) belonging to *Ilarvirus*, is transmitted by *Thrips palmi* and is prevalent in India, Australia and other sunflower growing countries. In India, epiphytotics of TSV have occurred in certain parts of Andhra Pradesh (India) and the total sunflower crop was wiped out during 1999–2000 (Prasada Rao et al. 2000). The virus spreads through infected pollen while thrips feeding. Both the virus and thrips have a wide host range and are responsible for heavy crop losses in a number of crops like mungbean, cotton, peanut, sunnhemp, etc., Regarding virus-vector relationship, TSV is persistently and propagatively transmitted by thrips. The first instar larvae of thrips acquire the virus while feeding on virus-infected tissue. The virus crosses through the midgut barrier and enters the salivary glands. The virus must be acquired by the larval stage of thrips as the adult thrips cannot acquire the virus (Pappu et al. 2009; Riley et al. 2011). Thrips lay their eggs in plant tissues and the eggs will hatch within 2–3 days. The two feeding larval and two non-feeding pupal stages and winged adult phase constitute the life cycle of thrips, which is about 15–30 days depending on the temperature. The virus passes from larvae to adult thrips as it undergoes pupation and changes associated with maturity. This is known as transstadial passage (Whitefield et al. 2005). Thrips retain infectivity for their life time and the virus titer has been shown to increase as the virus replicates in the thrips. There is no evidence of passage of virus through the egg. Extensive work was carried out on the epidemiological aspects of tospoviruses, which has helped in developing IDM to manage TSWV incidence in different vegetable crops. For more details of thrips vectors and TSV, one can refer the reviews of Francki et al. (1981), German et al. (1992), Jones (2005a), Whitefield et al. (2005), Pappu et al. (2009), Peters (2008), Riley et al. (2011).

(f) **Beetle vectors**

Another active vector group of plant viruses are beetles, which belong to Coleoptera whose members transmit viruses belonging to genera *Bromovirus*, *Carmovirus*, *Comovirus*, *Machlomovirus*, *Sobemovirus*, and *Tymovirus* (Fulton et al. 1980, 1987; Meier et al. 2008). There is a high degree of specificity between the beetle vectors and the viruses they transmit. *Cowpea mottle* and *Southern bean mosaic viruses* are transmitted by the *Ootheca mutabilis* and *Medythia quarterna*, both belonging to *Chrysomelidae* (Allen et al. 1981). Several species of *Chrysomelidae* beetles are vectors of *Rice yellow mottle virus* (RYMV), viz., *Sessielia pusilla*, *Chaetocnema pulla*, *Trichispa serica* and *Dicladispa viridicyanea* (Bakker 1974; Abo et al. 2000). *Maize chlorotic mottle virus* in maize is also transmitted by six species of beetles of the family *Chrysomelidae* (Scheets 2008).

The beetle vectors are confined to families of *Chrysomelidae*, *Coccinellidae* and *Meloidae*. They have biting-chewing type mouth parts and transmit viruses mechanically by carrying them on the mouth parts. The beetles become viruliferous and transmit the virus in a single bite but a longer acquisition access period and inoculation access period make the probability of transmission higher. These vectors acquire the virus during acquisition feeding periods ranging from few minutes to 24 h and transmit them immediately. The length of time a beetle retains and transmits the virus primarily depends on the type of virus, its host and environmental conditions. Both Walters (1969) and Selman (1973) suggested that virus retention time by the vector is characteristic of a particular virus. The viruses transmitted for 1 or 2 days fall into one group while those of longer periods into the other group. The Mexican bean beetle retains different viruses and transmits only for short periods of time, whereas the bean leaf beetle retains the same viruses for extended periods. Beetles can retain viruses up to a period of 8 days and can be detected in hemolymph, but apparently do not multiply in their beetle vectors. Virus-vector relationship appears to be similar to semi-persistent type.

There are also certain examples where some beetle-borne viruses are circulative, as they were found to move into the insect hemolymph immediately after injection. Beetles can also be rendered viruliferous by injecting virus into the hemolymph (Gergerich 2001; Raccah et al. 2001). The beetles may transmit viruses even if they are not carried in the hemolymph. The retention of inoculativity of beetles differs for different beetle vectors; thus, *Epilachna varivestis* retains *Cowpea severe mosaic virus* (CPSMV) for 1 day, while *Cerotoma trifurcata* transmitted the same virus for several days. The virus does not propagate in the beetle as the virus titre declines with time.

Markham and Smith (1949) have also pointed out that beetles regurgitate during feeding which contain viruses over several days. The viruses are also detected in the fecal matter of beetles which have fed on infected plants. It has been established that the regurgitant fluid is a key factor in determining whether a virus will or will not have beetle vectors (Gergerich et al. 1983).

(g) **Mealybug vectors**

Mealybugs belong to Coccidea and Pseudococcidea. When compared to aphids and leafhoppers, the mealybug vectors are less mobile on the plant and hence relatively inefficient vectors. Viruses belonging to the genera *Ampelovirus*, *Badnavirus*, *Trichovirus* and *Vitivirus* have plant viruses which have mealybugs as the vectors. Mealybugs feed in the phloem and spread through wind currents and also carried by ants. The virus is acquired within a 20 min period and it requires 16 min to inoculate to the host plant and the virus persists in the vectors for less than 3 h. The mealybug vector can retain the virus for a maximum period of 3–4 days and nymphs are the more efficient vectors than adults. The virus-vector relationship is of the semi-persistent type. Only the nymphs of the first, second and third larval stages and the adult females are able to transmit the virus. Among the mealybug-transmitted viruses, *Cocoa swollen shoot virus* (CSSV) is economically important disease in African countries and is transmitted by fourteen species of mealybugs and the primary vectors being *Planococcoides njalensis* and *Planococcus citri*. CSSV is also transmitted by seed to the extent of 34–54 % (Quainoo et al. 2008), and virus was located in testa, cotyledons and embryo; the virus is also pollen-borne. Extensive research was conducted on the epidemiological aspects of CSSV (Thresh 1991; Fargette et al. 1993, 1994a, b; Holt et al. 1997; Legg et al. 1997; Calvert and Thresh 2002).

Several other mealybug-transmitted diseases which are important in the tropics are *Grapevine leafroll associated virus* is transmitted by *Planococcus ficus* and *Pseudococcus longispinus* (Tsai et al. 2008, 2010). Another virus which is transmitted by mealybug in sugarcane is *Sugarcane bacilliform virus* and the vector is *Saccharicoccus sacchari*. Even *Sugarcane mild mosaic virus* is transmitted by *Saccharicoccus sacchari* (Lockhart et al. 1992). The citrus mealybug, *Planococcus citri* is the vector of *Citrus mosaic badnavirus* (CMBV) (Garnsey et al. 1998). *Planococcus citri* and *Saccharicoccus sacchari* transmits *Banana streak virus* (Lockhart and Jones 1999).

Pineapple mealybug wilt associated virus-3 infecting pineapple in Cuba is transmitted by *Dysmicoccus brevipes* (Hernandez et al. 2010). Earlier *Pineapple mealybug associated virus-1*, which was transmitted by the same mealybug vector in a semipersistent manner and was first reported in Hawaii. In India, *Piper yellow mottle virus* of black pepper belonging to genus *Badnavirus* spreads by mealybug vectors like *Ferrsia virgata* and *Planococcus citri* (Bhat et al. 2003). Among the aerial vectors, mealybugs are less mobile. The crawling nymphs spread the virus better than the adults. The first-instar nymphal stages of mealybug vectors are small and well-adopted to dispersal by wind. Although adult male mealybugs have wings, they do not have functional mouth parts and so cannot transmit viruses.

(h) **Mites**

Mite vectors belong to Phylum Arthropoda and class *Arachnida*. Members of the mite families *Eriophyidae* and *Tetranychidae* are the plant virus vectors. Mites are very small, about 0.2–0.8 mm long; they can retain infectivity through moults and may remain infective for 6–9 days. Mites become infective as nymphs but not as adults when feeding on infected plants. They pierce the plant cell wall with their stylet and suck the cell sap. They complete their life cycle in 6–14 days. As early as 1963, Thresh had demonstrated that *Cecidophioposis ribis*, was the vector of black currant reversion. Some of the mite-transmitted viruses are: *Wheat streak mosaic virus* and *Triticum mosaic virus* transmitted by *Aceria tulipae; Triticum mosaic virus* by *Aceria tosichella* (McMechan 2012); *Peach mosaic virus* by *Eriophyes insidiosus; Cherry mottle leaf virus* by *E. inaequalis*; *Garlic virus C* and *Garlic Virus D* (*Alexiviruses*) by *Aceria tulipae;* and *Onion mosaic virus* by *Aceria tulipae* (Brakke 1971a, b). From Brazil, Rodrigues et al. (2000) have reported that *Brevipalpus phoenicis* to be the mite vector of *Citrus leprosis virus* in mandarins. *B. phoenicis* is the vector for *Coffee ringspot nucleorhabdo virus* (Chagas et al. 2003).

For the first time, Khetarpal (1989) reported the transmission of the *Pea seed-borne mosaic virus* in pea by the mite *Tetranychus urticae* under glass house conditions, but requires confirmation. *Wheat streak mosaic virus,* which is also a seed-borne virus in wheat, is transmitted by mite vector, *Aceria tosichella* (Scifers et al. 1997; Jones et al. 2005a). This mite vector also transmits *High plains virus* in maize. *Pigeon pea sterility mosaic virus* (PPSMV) in pigeon pea (*Cajanus cajan*) is transmitted by eriophyid mite *Aceria cajani*. Kulkarni et al. (2002) have reported the transmission efficiency of a single *A. cajani* to be up to 53 %, but was 100 % when <5 mites per plant were used. *A. cajani* acquired PPSMV after a minimum acquisition access period (AAP) of 15 min and inoculated virus after a minimum inoculation access period (IAP) of 90 min. No latent period was observed. In general, wind is the main means of mites dispersal in nature, because of their small size and are essentially carried like dust particles. Epidemiology of pigeon pea sterility was studied by Singh and Rathi (1997), Kumar et al. (2008a). More information on mite-transmitted viruses can be obtained from the review articles of Slykhuis (1960), Oldfield (1970), Agrios (2005), Nayudu (2008).

1.8.4 Non Arthropod Transmission

(a) **Nematode vectors**

The soil-inhabiting nematodes are responsible as vectors for transmission of tobra and nepo- virus groups, which are also seed-borne, and were reviewed exhaustively (Taylor 1980, 2002). Nematodes that transmit plant viruses occur in two families: Longidoridae, which comprise the three vector genera *Longidorus, Paralongidorus,* and *Xiphinema,* and *Trichodoridae*, comprising two genera, *Trichodorus* and *Paratrichodorus.* The longidorid nematodes vector transmits polyhedral-shaped or isometric viruses called nepoviruses, while the trichodorid nematodes vector transmits straight, tubular viruses called tobraviruses (Cadman 1963; Harrison et al. 1971; Taylor and Robertson 1975). There are 22 longidorids and 14 trichodorids that are known to vector plant viruses (Lamberti and Roca 1987).

Nematode vectors measure 2–12 mm long and are migratory ectoparasites feeding mainly on root tips. Regarding their habitat, *X. diversicaudatum* is generally associated with moist and shady sites in the vicinity of water or medium soils with high moisture levels (Fritzsche et al. 1972). *L. elongatus* is found in light medium soils and *L. attenuatus* in light sandy soils (Harrison 1977). *T. pachydermatus* usually occurs in light rather than heavy soils (Van Hoof 1962). Longidorids feed at or above the root tip. *Longidorus* and *Paratrichodorus* species feed on the apical meristems of the roots. On the other hand, Trichodorid nematode feeding is restricted to the epidermal cells of the root (Christie and Perry 1951). Viruses can be transmitted both nonpersistently and persistently by nematode vectors. They may acquire the virus as soon as cell contents are ingested from infected roots and there is no latent period within the nematode before transmission takes place. The virions attach to the stylet (feeding organ) or to the gut when they feed on the infected plant, and they can transmit viruses while feeding on healthy susceptible plants. The virus retention as a mono layer adhering to the cuticular lining of the oesophagus track was observed in *X. diversicaudatum* carrying *Arabis mosaic virus* and also in *X. index* carrying GFLV (Taylor and Robertson 1973). Both adult and larval stages can transmit the viruses with equal efficiency. *L. elongatus* retains the virus for 8–9 weeks and *Xiphinema sp.* for many months; there is no evidence that viruses multiply inside the nematode vector and virions are not transmitted to its progeny.

The spread of these viruses is very slow since the vector nematodes move over short distances of approximately 50 cm in a year. Transport of vector nematodes over longer distances in soil is probably less feasible since they are susceptible to desiccation, but movement of moist soil on agricultural implements, plant roots and the feet of animals and birds may play some role in vector distribution. Some of the important nematode-transmitted viruses with their vectors are: *Grapevine fanleaf virus* (GRLV)- *Xiphinema index; Arabis mosaic virus- X. diversicaudatum*; *Tobacco rattle virus- Trichodorus pachydermis; Raspberry ringspot virus- Longidorus elongates;* and *Pea early browning virus- Paratrichodorus* and *Trichodorus spp* (Hull 2002). Demangeat et al. (2010) have reported that even a single female *Xiphinema index* has the capacity to transmit GFLV. More details on nematode vectors can be had from Lamberti and Roca (1987), Brown et al. (1995), Taylor and Brown (1997), Macfarlane (2003), and Nayudu (2008).

(b) **Fungal vectors**

Certain fungi like *Polymyxa sp.*, *Olpidium sp.*, and *Synchytrium sp.* are implicated as vectors of plant viruses belonging to necro-, furo-, peclu- virus groups. *Plasmodiophoroids* (*Polymyxa*, *Sporosphaera* and *Spongospora* species) are only known to transmit the viruses internally in zoospores and these are ssRNA (+) viruses with divided genomes. Some have rod-shaped virions (genera: *Benyvirus*, *Furovirus*, *Pecluvirus*, *Pomovirus*) while others have filamentous particles (genus: *Bymovirus*, Family: *Potyviridae*). Polyhedral virions are seen in the genera *Tombusvirus, Carmovirus, Necrovirus, Dianthovirus* and *Satellitevirus.* A number of virus genera are transmitted both persistently and nonpersistently by these soil-borne zoosporic protozoa. These are not phytopathogenic themselves but parasitic. Transmission of the virus takes place when they become associated with the plant roots. All five species of fungal vectors are obligate parasites of plant roots and have similar developmental stages. They can survive from crop to crop as resting spores that produce zoospores and infect the host. These soil-borne viruses spread very slowly and continuous soil moisture is required for further spread of the virus. Depending on the conditions in the soil, resting spores germinate and release the primary motile zoospores that move to roots. Depending on the fungus vector involved, the zoospores may be uniflagillate or biflagellate.

In some of the South African countries, *Peanut clump virus* in peanut transmitted by *Polymyxa graminis* as reported by Thouvenel and Fauquet (1981); Dieryck et al. (2011) have studied the transmission and acquisition aspects. The epidemiological data on this virus has been developed by Delfosse (2000).

In India, *Indian peanut clump virus* (IPCV) belonging to *Pecluvirus* genus is present to a limited extent in sandy soils of Punjab, Rajasthan and Andhra Pradesh states and is transmitted both by *Polymyxa graminis* and also through seed (Reddy et al. 1983a, b; 2008; Delfosse et al. 1999). The life cycle of *Polymyxa graminis*, the vector of IPCV in its graminaceous hosts has been studied (Ratna et al. 1991) and also the epidemiological aspects Delfosse (2000) and Delfosse et al. (2002). Another species of *Polymyxa, P. betae* transmits the *Beet necrotic yellow vein virus* in sugar beet throughout the world (Abe and Tamada 1986). *Soil-borne wheat mosaic virus* (SBWMV) is transmitted by *P. graminis* and the virus is seed-transmitted in rye and wheat crops (Delfosse et al. 1999). Viruses like *Tobacco necrosis virus* is transmitted by *Olpidium brassicae* of the order *Chytridiales* and *Melon necrotic spot carmovirus* transmitted by *Olpidium bornovanus* which is seed-transmitted in melon both externally and internally (Campbell et al. 1996). *Mirafiori lettuce big vein virus* and *Tobacco stunt virus* are transmitted by *Olpidium virulentus* (Sasaya and Koganezawa 2006). *Olpidium* species transmit some small monopartite ssRNA (+) viruses that are carried on the outside of zoospores and resting spores, and the examples are members of the family *Tombusviridae* (some of the genera are *Aureusvirus*, *Carmovirus*, *Necrovirus* and *Tombusvirus*). Some viruses like *Mirafiori lettuce big-vein virus* (MLBVV) transmitted by *O. brassicae* and *Soil-borne wheat mosaic virus,* transmitted by *P. graminis,* and *Beet necrotic yellows virus* by *P. betae,* the virus was noticed within the zoospores when released from the vegetative sporangia in fungal transmission; a 'read through domain' (RTD) of the CP, CP-RT was implicated in virus transmission by the fungus. The RT-protein was 75 kD and was required for virus assembly and transmission (Schmitt et al. 1992). *Olpidium* species also transmit some ssRNA viruses with divided negative sense genomes that are members of the genera *Ophiovirus* and *Varicosavirus*. These genera are not currently assigned to any family and the viruses are internally borne. *Potato virus X*, transmitted by the fungus, *Synchytrium endobioticum* is seed-borne in potato to an extent of 0.6–2.3 %, but requires confirmation (Darozhkin and Chykava 1974). Another virus infecting potato viz., *Potato mop top virus*, for which the vector is *Spongospora subterranea* (de Bokx and vander Want (1987). In Germany, for grapevine viruses, *Sporosphaera viticola* is the most potential vector (Kirchmair et al. 2005). In South Africa, *Brome mosaic virus* is transmitted by rust spores, *Puccinia graminis tritici* (Von Wechmar 1980). Different aspects of fungus transmission of plant viruses have been reviewed by Teakle (1980), Campbell (1996) and Nayudu (2008).

(c) **Role of irrigated water in virus spread**

Since Koenig's (1986) review that certain plant viruses were shown to be present in the environmental water that contain considerable amount of plant viruses and survive and spread to other healthy plants. The use of circulating nutrient solutions in hydroponic systems hold the potential for rapid and effective spread of plant viruses. Additionally, the surface wash-out of locally scattered and infected decaying plant residues and of the related soil surface layer, including animal feces, bring the plant viruses into water. Virus contaminating-seeds and virus transmitting nematodes, zoospores or, more so resting sporangia of fungi, may also be washed away by running soil water or drainage water. For example, Yakovleva (1965) reported that *Cucumber green mottle mosaic virus* (CGMMV) spreads with the surface water used for watering, and also water flowing from waste heaps at dumping sites. From India, Vani and Varma (1988) have isolated CMMV from Jamuna river water. For more information, one can refer the review articles (Koenig 1986; Mehle and Ravnikar 2012).

1.9 Dispersal and Migration of Insect Vectors

Insect vectors have to move in search of food, a change of habitat, place for reproduction and to overcome adverse conditions. A few insect vectors crawl in the field, for example wingless aphids, but many virus vectors enter by winged-flights. The flying process also depends upon the physiology of the host, climatic and meteorological conditions. Generally the dispersal of vectors is relatively for short distances, sometimes up to a few kilometers. Certain insect vectors may migrate from one climatic region to another by their own flight activity and are also blown by wind. Insects can fly at all times of the day and night depending upon the species.

The barley yellow dwarf (BYD) virus transmitted by the greenbug (*Schizaphis graminum*), the grain aphid (*Macrosiphum avenae*) and many other viruliferous insects spread over long distances along the '*Puccinia* path' in U.S. Thresh (1983) has reviewed the long range dispersal of plant viruses by arthropod vectors. For aphid-borne viruses, he presented data on the migration of aphids carrying persistent and semi-persistent viruses like BYDV, PLRV, SCSV, GRV, and BYV.

Similarly, in the Philippines, it has been well documented that cyclones spread the rice tungro leafhopper vector (*Nephotettix* spp.) over long distances (Rosenberg 1981). Within the lowest 1,500 m of the atmosphere, vectors have been found to be viable and in a viruliferous state even after being in transit for 30 h. These vectors spread between islands in the Philippines covering a distance of 500–600 km. The spread of such viruliferous vectors from Taiwan to far off places like Liliw, in the Philippines is well documented. The leafhoppers have been found to spread during October - a transition period between south-west and north-east monsoons. The dispersal tracts of the vector demonstrate the influence of cyclones or typhoons in vector migration. Similarly the spread of brown planthoppers (*Nilaparvata lugens*) influences both epidemiology of the grassy stunt disease and appearance of new races of the planthopper population.

The brown planthopper is a migrant and moves northward over China each spring on spells of southerly winds. In July 1967, mass flight of rice planthoppers was witnessed from a weather ship on the Pacific Ocean far from Japan (Ashina and Tsuruka 1968). This species can overwinter only in the extreme south, where the insect is present throughout the year (Jiang et al. 1981).

The dispersal of vectors like nematode and fungi takes place through water and also by the transport of the infested soil to other areas during tractor operations. Through some of the agricultural practices like contaminated machinery, farm worker's foot wear and implements, certain viruses are spread from one area to another.

1.10 Factors of Vector Transmission

Insect vectors of plant viruses are found in 7 of the 32 orders of the class Insecta. Vectors of plant viruses are taxonomically very diverse and can be found among arthropods, nematodes, fungi, and plasmodiophorids (Froissart et al. 2002; Hull 2002). Arthropod vectors that transmit most plant viruses are aphids, whiteflies, leafhoppers, thrips, beetles, mealybugs, mirids, and mites (Andret-Link and Fuchs 2005; Spence 2001), with the most common being aphids were more than 200 vector species have been identified (Ng and Perry 2004). More than half of the nearly 550 vector transmitted virus species recorded so far are disseminated by aphids (55 %), 11 % by leafhoppers, 11 % by beetles, 9 % by whiteflies, 7 % by nematodes, 5 % by fungi and plasmodiophorids, and the remaining 2 % by thrips, mites, mirids, or mealybugs (Astier et al. 2001). More details on the vector transmission is furnished in Chapter-4 of Volume I.

Most plant viruses are transmitted by vectors from one host to another. Virus transmission by a vector is often characterized by some degree of specificity. Many studies have revealed that the per cent disease incidence is directly proportional to vector population (Sastry et al. 1978; Anjaneyulu et al. 1994; Kumar et al. 2008a). The multiplication of the vector and their flight behavior are influenced by number of environmental factors which has been discussed elsewhere in this chapter.

The majority of vectors are found in the two orders of insects with pierce-sucking mouthparts: Thysanoptera (6) and Hemiptera (300). Fewer vector species are found in 5 orders of chewing insects (number in parenthesis): Orthoptera (10), Dermaptera (1), Coleoptera (30), Lepidoptera (4) and Diptera (2). Apparently, the feeding organs of the Hemiptera are the principal reasons for their successful role as vectors. Description of the biology and flight behavior of insect vectors is beyond the scope of this chapter. Plant viruses demonstrate a high level of specificity for the group of insects that may transmit them. A virus that is transmitted by one type of vector will not be transmitted by another (Raccah and Fereres 2009).

The importance of vector movement is likely to be a key feature in plant virus epidemics (Power 1992). Virus-vector associations can be categorized according to different transmission classes, dependent in part on the time-scales involved in acquisition and transmission, and whether the virus moves through and even multiplies in the insect vector.

(a) **Principal reasons for plant viruses to depend on vectors for their survival**

An impermeable cuticle coats of the plant epidermis, preventing entry of virus particles (animal viruses enter readily through natural openings). Most vectors are insects (non-insect vectors include mites, nematodes and fungi). Several plant viruses may spread by contact, through pollen and seed or by vegetative reproduction. Many insects such as hemipterans are well adapted to their role as vectors by their capacity to pierce the epidermis and delicately deposit the virus in the cytoplasm without risking the integrity of the plant cell.

Plants are rooted and lack independent mobility. Therefore, many viruses depend on insects for transport among hosts (unlike animals that, by their own mobility, transport the virus to new niches).

Insect-borne plant viruses may cause severe or even crippling losses to many annual and perennial crops. On occasion, insects are responsible for transition from a non-spreading form to the epidemic form of diseases. Outbreaks of disease caused by insect vectors are demonstrated in two examples: In perennials, the almost total destruction of the citrus industry in the 1930s in Argentina and Brazil was attributed to the aphid *Toxoptera citricidus*. This aphid was recently found in Portugal and Spain threatening the Mediterranean citriculture. The second example is that in annuals, where outbreaks of *Tomato spotted wilt virus* (TSWV) in the last two decades is attributed to the spread of the thrips vector *Frankliniella occidentalis*.

(b) **Vector species**

A number of major insect vectors like aphids, whiteflies, leafhoppers are present as species or biotypes. In sub-Saharan Africa, rosette virus in groundnut is most destructive and is transmitted by *Aphis craccivora*. Storey and Ryland (1955) reported that not all cultures of *A. craccivora* transmit rosette disease. Nutman et al. (1964) found that colonies of *A. craccivora* collected from beans (*Vicia faba* and *Phaseolus multiflorus*) did not transmit rosette, and rosette disease was transmitted only in aphid colonies from and then maintained on groundnut. It has also been shown that a Nigerian race of *A. craccivora* transmitted rosette isolates from Nigeria, Kenya and Uganda, whereas races of *A. craccivora* from Kenya and Uganda transmitted East African isolates, but not those from West Africa (Watson and Okusanya 1967). Biochemical techniques for examining aphid genomes may prove to be more informative. For example, the whitefly vector *Bemisia* although transmitting a number of geminiviruses, there is evidence showing biotic variation among *B. tabaci.* Host preferences, host range, host adaptations and gene recombinations are some of the factors for biotic variation of the whitefly vector. It is manifested primarily through host specialization or host range and through differences in fecundity, dispersal behavior and virus transmission efficiency. The B type of *B. tabaci* is characterized by a wide host range and its ability to induce phytotoxic disorders and to transmit new and old world begomoviruses. Sseruwagi (2005) has studied B and Q biotypes of whitefly vector in Uganda. The biotype Q showed fewer numbers of probes than B biotype. In China Q biotype demonstrated superior acquisition and transmission capacity than B biotype. Q biotype whitefly acquired significantly more viral DNA than the B biotype (Pan et al. 2011). Cervera et al. (2000) have grouped *B. tabaci* biotypes into four clusters which comprised:

(i) Near East and Indian subcontinent biotypes;
(ii) B and Q biotypes plus a Nigerian population from cowpea;
(iii) New World A biotype; and
(iv) S biotype and a Nigerian population from cassava.

These results were consistent with a previous grouping of biotypes based on RAPD-PCR analysis. Even in aphid vectors biotypes were identified. For example, pea aphid biotypes are differing in transmitting *Bean yellow mosaic virus* (Sohi and Swenson 1964). Soybean aphid biotypes were also identified by Kim et al. (2008) based on molecular markers, antibiosis, antixenosis body size and fecundity.

Some plant virus species are transmitted by multiple species, especially the non-persistently transmitted *Potyviruses* (Madden et al. 1983). For some *Potyviruses*, 30 or more aphid species have been identified as vectors, with 10 or more occurring in the same field at the same time (Halbert et al. 1981). However, statistical analysis generally show that only a few species at a given location are significantly associated with infected plants (Madden et al. 1983; Raccah et al. 1988). To fully account for these vector species, the components involving *P* (population size of vector) for disease change in the differential equations would have to be replaced by a sum of components, one for each vector species with corresponding parameters for each species, e.g. for mobility. There would also have to be additional equations for the population dynamics of each species. This approach was taken by Marcus and Raccah (1986) for the situation in which virus disease dynamics were related to observed densities of aphid species. In general, however, the model would become intractable with more than two aphid species, especially for determining solutions for steady-state and thresholds. One approach for simplification is to represent aphid density with a weighted summation of the individual species densities, with weights determined based on statistical analysis (Madden et al. 1983) or laboratory assays for transmission (Halbert et al. 1981). The other approach is to utilize information on all vector species, and work only with numerical solutions of the SEIR models (Chan and Jeger 1994) or with computer simulations of virus epidemics (Ruesink and Irwin 1986; Irwin et al. 2000).

(c) **Vector behavior**

Plant viruses may show plant-mediated manipulation of vector behavior in ways that enhance virus transmission. This needs to be investigated in an evolutionary, comparative context, especially where more subtle behavioral changes are involved. Non-propagative viruses are unlikely to affect their vectors directly, but exert plant-mediated influences on vector biology, as has been postulated for whitefly-transmitted diseases. Holt and Colvin (2001) have modeled the vector-virus infected plant interaction in terms of nutrient availability, fecundity, crowding and migration and potential spread of new whitefly biotypes. Viruses cause characteristic symptoms in infected plants that are associated with numerous physiological/biochemical changes. Vector colonization behavior, development and activity will be altered by these plant changes. Holt and Colvin (2001) concluded that a vector biotype with higher fecundity would accelerate epidemic development more than an increase in vector fecundity only on virus-infected hosts. For aphids, studies have indicated that aphid reproductive performance (Castle and Berger 1993; Blua et al. 1994; Mowry 1994) and the development of winged forms (Gildow 1983; Blua and Perring 1992) are enhanced on virus-infected plants, but the effects of infection on aphid ability to locate plants and subsequent behavior has been a largely neglected area of research. Viruses corresponding to the different transmission classes exploit different stylet penetration activities and may therefore influence vector behavior and performance differently. Moreover, the enhanced performance of aphids on virus-infected plants may benefit aphid natural enemies, possibly having a detrimental effect on vector populations and virus spread. However, it was suggested that a luteovirus (circulative-persistent) has a beneficial, protective effect on its vector by directly disrupting the development of a braconid parasitoid within the common aphid vector/host (Christiansen-Weniger et al. 1998). The mechanisms and consequences of virus manipulation of vector and natural-enemy behavior within a crop species, taking a comparative perspective to test predictions for the different transmission classes need to be investigated. The general model can then be used to scale up from individual virus-vector-parasitoid interactions and their effects on model parameters to the consequences for virus fitness at the population level (Jeger et al. 1998).

Studies of the pattern and sequence of virus spread have shown considerable differences between the different groups of arthropod-borne viruses. These differences can be related to their mode of transmission and to the behavior of their vectors. However, in evaluating the available results it is important to appreciate that a disproportionate amount of research has been on aphid vectors and particularly on a few well known species such as *Myzus persicae*. Other aphid species are now receiving more attention and some that are inefficient vectors of non-persistently transmitted viruses have been shown to be of considerable importance because they fly in very large numbers. Intensive work is also being done on other types of vector, especially whiteflies and thrips that seem to have become increasingly important during last three decades in the tropics and sub-tropics. This increased effort has led to a better appreciation of the main features of spread by the different types of vectors and provided a rational basis for devising the most appropriate virus disease management measures.

(d) **Role of viral capsid protein in insect transmission**

Numerous studies have suggested the involvement of a virus-ligand interaction in transmission specificity. The coat protein (CP) and its derivatives through CP and minor CP, and nonstructural proteins, such as a helper component (HC) or a transmission factor, are major viral determinants of transmission specificity. A number of virion-binding vector proteins have been identified as potential receptors.

Protein subunits are important for the specificity of transmission of circulative viruses. Rochow (1969), Brault et al. (2001) showed that strains of *Barley yellow dwarf virus* (BYDV) that are transmitted by one aphid species become transmitted by another aphid species if co-infected with another strain of BYDV. This phenomenon was attributed by Rochow to heteroencapsidation, where the non-transmissible RNA is encapsidated with some protein subunits of the transmissible strain.

1.11 Virus Survival and Spread

1.11.1 Strategies of Virus Spread

Considering the great range of crop species and habitats, it is hardly surprising that there is equivalent diversity among viruses in host range, means of spread, and perennation. Cereal yellow dwarf and several other important viruses are "specialists" in an ecological sense, as they have a very restricted host range amongst a few closely related species, or they are transmitted by only one or a few vector species of the same taxonomic group. By contrast, *Cucumber mosaic virus* and *Alfalfa mosaic virus* are examples of "generalists" as they have a very wide host range, and are transmitted by many different aphid species. Generalist viruses have less opportunity than specialists to become closely adapted to particular hosts or vectors in ways that facilitate survival. The different means of spread of some important virus diseases of tropics is listed in the Table 1.6.

Similar considerations could account for the great differences between viruses in mode of spread by contact, through pollen and/or seed and by arthropod, nematode, or fungus vectors. No one virus is known to be transmitted by each of these routes and it is unlikely that the typically small viral genomes can carry sufficient information for this to be possible. Thus each virus utilizes only one or a combination of some of the possible modes of spread to achieve an effective distribution of inoculum over short distances within plantings and also over greater distances between localities. This is of great epidemiological importance in enabling viruses to reach and then exploit new habitats as they arise (Vander plank 1963). The actual methods employed greatly influence the pattern and sequence of virus spread, and hence the overall dynamics of disease progress. They are discussed in detail by Thresh 1974b, 1978, 1983.

(a) Cultural practices and virus spread

The cultural practices like wide spacing in planting, type of fertilizers used, date of planting, vector population and favourable environmental conditions are favourable for more epidemics to occur. Some times in certain crops like tomato, potato etc., even short-term rotations and monocropping will also stimulate virus epidemics. Among the most important of these is the practice of planting new crops alongside a similar crop that has been a general practice followed. The old mature crop may contain infected plants that act as reservoirs of virus infection.

The use of irrigation to extend the growing season can also lead to major virus epidemics. In areas around the Mediterranean, irrigated crops may be planted following the natural wet season in March and April. At this time, aphid vectors may migrate in large numbers from the senescing winter crops and natural vegetation, to the young, irrigated crops as in case of epidemics of *Cucumber mosaic, Maize dwarf mosaic, Potato virus* Y and *Watermelon mosaic viruses* in the irrigated vegetable crops (Quiot 1980; Thresh 1982).

Although increased irrigation and protected cropping in plastic tunnels to extend the growing season may greatly increase the cropping potential of many areas, these examples show that these practices may increase the severity of many existing virus problems and cause the initiation of diseases in new areas. Such problems necessitate the careful management of new cultural practices, and, in some cases, may be a major restraint in their introduction.

1.12 Disease Forecasting

Many environmental and other factors influence the complex interactions between virus, host and vector, in many crops because of advanced technologies and molecular diagnostics tests, the disease forecasting systems have been developed against major virus diseases. In crops like potato, beetroot, tomato, capsicum, lettuce, peanut, and some other economically important crops, effective forecasting systems are being regularly applied for effective management of major virus diseases. There are also great potential benefits to growers, processors, and advisory officers who can anticipate losses and apply appropriate chemical or other control measures. Through disease forecasting system, it is possible to operate an early warning system where infection is likely to occur, or to select growing seasons or areas for special crops where infection is unlikely.

Disease forecasting is necessary to predict the appearance of a disease in a certain set of conditions based on environmental factors which vary from season to season and place to place. Thus forecasting is possible only for those diseases where the weather disease relationship is fully understood. For any virus disease spread the vector population is very important and this will depend on the favourable environmental factors. For example the peak period of leafhopper population which is the vectors of tungro and other virus diseases of rice which vary from place to place as seen from the studies carried out by different researchers (Mukhopadhyay and Chowdhury 1973; Chancellor and Thresh 1997; Chowdhury and Mallick 2001).

Some important virus diseases where sufficient information on disease forecasting system is available are as follows:

Tomato spotted wilt virus is the type member of *Tospovirus* that is found worldwide causing catastrophic yield losses in tomato, potato, peanut, other members belonging to *solanaceae, leguminosae* and other families. This virus is transmitted by thrips vector and exhaustive epidemiological data is available. For example even GBNV, WBNV and WSMoV which are members of *Tospoviruses* and transmitted by *Thrips palmi* are found only in the subtropics and known to occur only in Southeast Asia. A number of epidemiological aspects are studied in different countries have given clues for framing suitable disease management measures (Jones 2004; Pappu et al. 2009). In India GBNV is quite prevalent on green gram and Sreekanth et al. (2002) have reported that the thrips populations were always higher in *Kharif* (August) than in *Rabi* (November); and lower in summer (April). Similar trends have been reported by earlier workers in different virus host combinations (Reddy et al. 1983a; Singh 1989; Sreenivasulu 1994). In 2000, in spite of considerable differences in the occurrence of thrips numbers in both *Kharif* and *Rabi seasons, T. palmi* population and GBNV incidence reminded almost similar, indicating a certain relationship between the *T. palmi* densities and GBNV incidence (Reddy et al. 1983a; Cho et al. 1987). The effects of host plant and temperature on the growth and reproduction of *T. palmi* was studied by Tsai et al. (1995). Su and Chen (1986) also recorded higher *T. palmi* population during June to September, while Tsai et al. (1995) found higher *T. palmi* population in winter and spring than in summer. On the other hand, Kenchaiah and Porte (1989) reported that incidence of thrips in groundnut in general remained the same throughout the survey period during *Kharif* and *Rabi* seasons in several districts of Karnataka (India). These findings suggest the need to examine the susceptible crops of tospoviruses regularly to monitor thrips population, so as to initiate timely management steps to contain the thrips as well as viruses.

Sunflower necrosis virus of the *Ilarvirus* genus and family *Bromoviridae* is quite prevalent worldwide and at more intensity in India and Australia. It is transmitted by pollen-associated with *Thrips tabaci*, *Frankliniella schulzei* and *Macrocephalothrips abdominalis. Tobacco streak virus* belonging to genus *Illarvirus* is the causal agent of sunflower necrosis disease. The epidemiology of this virus was studied by Nagaraju et al. (2003), Shivasharanayya and Nagaraju (2003), Upendhar et al. (2006), Lakhmod et al. (2007), Kumar et al. (2008a), Lokesh et al. (2008). In India and central Queensland of Australia, *Parthenium hysterophorus* is a symptomless carrier of TSV and is also seed-transmitted in this weed host (Sharman et al. 2009). This virus is also naturally recorded on lettuce, cotton, tobacco, chickpea, bottlegourd, groundnut, cowpea, cucumber, chilies, soybean, tomato, mungbean, groundnut (Sharman et al. 2008, Jain et al. 2008).

Disease forecasting methods have been standardized and are implemented at field level viz., *Potato virus* Y^0 in potato (Sigvald 1986); *Papaya ringspot virus* in papaya

(Mora-Aguilera et al. 1993); *Barley yellow dwarf virus* in barley (Plumb et al. 1986); Fiji virus disease in sugarcane (Egan and Ryan 1986; Ryan 1988) and other crops.

Analysis of vector numbers and activity in relation to weather variables have also led to the development of practical forecasting schemes for some viral diseases, e.g. beet yellows (Watson and Healy 1953; Watson et al. 1975; Werker et al. 1998) and *Barley yellow dwarf virus* (A'Brook 1983). Multivariate methods were used to associate disease incidence of nonpersistent viruses with the range of aphid vector species trapped (Madden et al. 1983). Madden et al. (1990b) have modeled increase of nonpersistent virus diseases as a function of cumulative numbers of vectors trapped. Similarly Singh et al. (2010) have developed weather based prediction model for forecasting cotton leaf curl disease in the American cotton.

Computer models were developed for many plant diseases in the 1970s. Virus epidemics were simulated based mostly on vector population dynamics and weather (Gutierrez et al. 1974a, b; Frazer 1977; Kiritani and Sasaba 1978; Irwin et al. 2000). Some explicit process-based computer models continue to be developed for virus epidemics (Ruesink and Irwin 1986; Irwin and Kampmeier 1989; Morgan 1990; Head and Morgan 1996; Holt and Chancellor 1996; Bertschinger 1997).

Jones et al. (2010) in their review on principles of predicting plant virus disease epidemics have discussed about Internet-based decision support systems (DSSs) that determine when and where control measures are needed. Finally, case studies are provided of eight pathosystems representing different scenarios in which modeling approaches have been used with varying degrees of effectiveness to forecast virus epidemics in parts of the world with temperate, Mediterranean, subtropical, and tropical climates.

1.13 The Cyclical Nature of Plant Disease

The plant disease epidemics can be classified into two basic types: (1) Monocyclic and (2) Polycyclic. The monetary analogy has been used to clarify the differences between "simple interest" (monocyclic) and "compound interest" (polycyclic) diseases (Vander plank 1963; Thresh and Fargette 2003) (Fig. 1.8). Diseases caused by monocyclic pathogens are analogous to investment with simple interest; on the other hand the diseases caused by polycyclic pathogens are analogous to investment with compound interest. With simple interest, capital grows at a constant rate (the interest bearing capital remains unchanged); with compound interest, invested capital grows at an increasing rate over time as the accrued interest is reinvested.

(a) **The Monocyclic diseases**

Monocyclic diseases spread mainly or exclusively from initial sources of inoculum (= starting capital) and the newly infected plants (= interest) do not

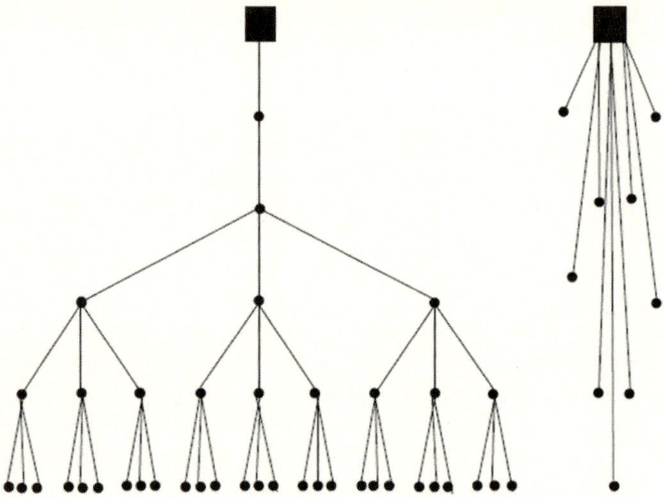

Fig. 1.8 Diagrammatic representation polycyclic ("compound interest") disease (*left*) and monocyclic ("simple interest") disease (*right*). *Source* Thresh and Fargette (2003)

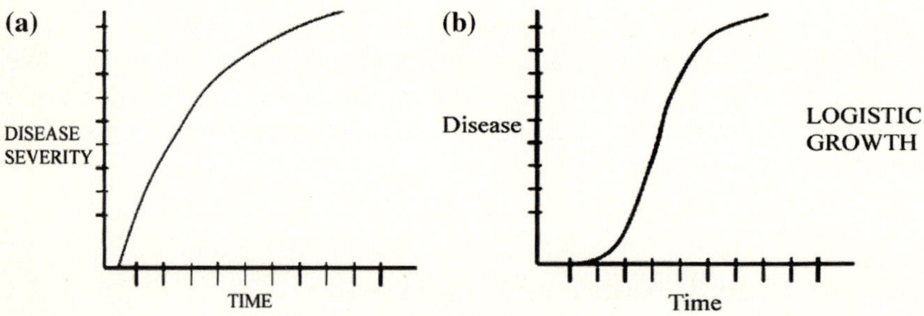

Fig. 1.9 **a** Disease progress *curve* for a typical monocyclic pathogen. **b** Disease progress *curve* for a typical polycyclic pathogen is an *S-shaped curve*

contribute to further spread (= monetary growth). In this case, infection occurs intermittently. The pathogens produce only one cycle of development (one infection cycle). In the absence of secondary spread, the infection is usually random. There is a limit to the amount of host tissue available and thus to amount of disease. Occurrence of infected individuals is one of the best indicator of monocyclic (occurring at random) and polycyclic diseases occur at non-random. The early stages of monocyclic epidemics can be described quite well by a linear model in the sense that disease increases steadily with the passage of time during the initial stages, but the newly infected host tissue does not immediately become infective (Fig. 1.9a). In contrast, newly affected plants in polycyclic epidemics contribute to the disease spread and there are repeated cycles of infection depending on the duration of the crop and the length of the incubation period.

A monocyclic pathogen completes the disease cycle per season. Soil-borne plant virus diseases are usually monocyclic due to physical constraints. Inoculum is not dispersed within the growing season. One of the monocyclic diseases that have been reported in sub-Saharan Africa, is "kromnek" disease of tobacco which spreads into crops by thrips vectors, dispersing from infected weed hosts and there is no subsequent spread within the crops. This led to the realization that the incidence of kromnek disease is decreased if thinning the plant stand is delayed until the main influx of infective vectors has occurred. A ringspot disease of sunflower in eastern and Southern Africa also behaves monocyclically as no inoculum is carried into crops by aphids from a Compositae weed host *Tridax procumbens* and there is no spread between sunflower plants (Theuri et al. 1987). Even soybean blight caused by *Tobacco streak virus* is monocyclic under Brazilian conditions (Almeida et al. 1994). Several diseases caused by *Tomato spotted wilt virus* in lettuce, pepper, tobacco, tomato and pineapple and other crops are of monocyclic diseases (Coutts et al. 2004a). Infection is carried into susceptible plantings by adult thrips that acquired the virus when they were nymphs. The monocyclic diseases are controlled by reducing the amount of inoculum or the efficacy of inoculum.

(b) **The polycyclic diseases**

Polycyclic epidemics result from the pathogen multiplication through successive generations in the course of the epidemic. Plant virus diseases are usually polycyclic and many plant viruses which have efficient flying insects as vector fall under this group of viruses transmitted by vectors like whiteflies, aphids, leafhoppers, and thrips. Exhaustive epidemiological work was carried out on cocoa swollen shoot disease as a well-known African example (Thresh 1958; Vander plank 1963). Some of the diseased or latently infected plants are removed as stands are thinned to provide what was referred to as "the mathematical solution to the problem of disease" (Vander plank and Anderssen 1944a, b).

Maize streak differs in that it seems to behave as a monocyclic disease in some situations and as a polycyclic disease in others (Rose 1974, 1978). The behavior of *Bean common mosaic virus* is also complex as it usually causes a polycyclic disease, but not in stands of bean cultivars that develop a severe hypersensitive necrosis when infected. These plants with necrosis become poor sources of inoculum from which little or no further spread can occur. Consequently, the incidence of infection in a hypersensitive cultivar increases monocyclically at a rate dependent on the extent of infection in non-hypersensitive cultivars nearby that act as sources of inoculum to the aphid vectors.

<u>Monocyclic Model</u>

1 2 3
↓ ↓ ↓

$$x = QRt$$

<u>Polycyclic Model</u>

$$x = \overset{\overset{1}{\downarrow}}{x_0}\, e^{\overset{\overset{2}{\downarrow}}{r}\overset{\overset{3}{\downarrow}}{t}}$$

Examining these models, we can see that in both models, there are three ways in which we can reduce disease (**x**) at any point in the epidemic:

1. Reduce the initial inoculum (**Q** in the monocyclic model and $\mathbf{x_0}$ in the polycyclic model). (Actually $\mathbf{x_0}$ is the initial incidence of disease, which is proportional to the initial inoculum.)
2. Reduce the rate of infection (**R** in the monocyclic model and **r** in the polycyclic model)
3. Reduce the duration of the epidemic (the time, **t**, at the end of the epidemic)

It is clear from the above model of a monocyclic epidemic that **Q**, **R**, and **t** have equal weight in their effect on **x**. A reduction in the initial inoculum or the rate of infection will result in a reduction in the level of disease by the same proportion at any time, **t**, throughout the epidemic. If **t** can be reduced, for example, by shortening the season, disease will be reduced proportionately.

These, then, can be used as three major strategies for managing plant disease epidemics, and we can organize our plant disease control tactics under one or more of these overall strategies. Furthermore, by means of the model we can assess the quantitative impact of each strategy, not only by itself, but in its interaction with others.

Tobacco streak virus (TSV) in sunflower is transmitted through pollen and wounds caused by thrips (*Frankliniella schulzei* and *Macephalothrips abdomanalis*); feeding helps for the entry of the virus carried by pollen externally, into the susceptible sunflower host (Sdoodee and Teakle 1993; Kumar et al. 2008a).

TSV is monocyclic virus in highly susceptible species (e.g., sunflower and groundnut) and polycyclic in asymptomatic weed species (e.g., *Parthenium*) Fig. 1.10. Wind-blown pollen of *Parthenium* contaminates the leaves and the thrips arriving independently may well contribute to TSV infection. Due to premature death of infected plants, virus is monocyclic symptomatic (Kumar et al. 2008a).

1.14 Disease Progress Curves

Disease progress curves show the epidemic dynamics over time (Thresh 1985; Agrios 2005). The mathematical tool can be used to obtain information about the appearance and amount of inoculum, changes in host susceptibility during growing period, weather events and the effectiveness of cultural and control measures.

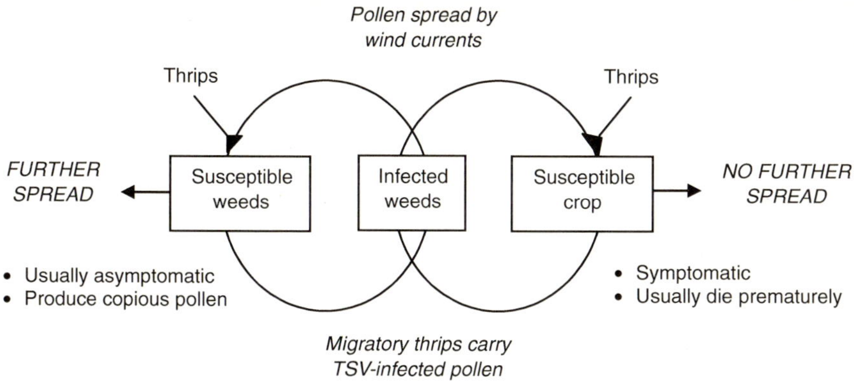

Fig. 1.10 Probable TSV cycle. Virus is polycyclic in asymptomatic weeds that host the virus as well as thrips, and act as a primary source of inoculums initiating several disease cycles a crop season. Due to pre-mature death of infected plants, virus is monocyclic (highly susceptible) in crops. *Source* Lavakumar P

1.14.1 Analyzing Disease Progress

Disease progress is the disease level that is assessed periodically. The time interval is chosen carefully by considering the host growth and the speed at which the disease is increasing. The disease progress curve is obtained by plotting the duration of the spread of the disease in days or weeks or months or years on the 'X' axis and the percent infection on the 'Y' axis. It is usually a sigmoid curve which indicates four phases of the disease cycle (Fig. 1.9b).

(1) An initial 'Lag' phase from the initial appearance of symptoms in a crop to the onset of a spread of the disease.
(2) The 'Log' or exponential phase during which the disease increases rapidly in time to a particular level.
(3) The period of decreasing spread or stationary phase when 'multiple' infections occur due to super infection of the already infected plants, and the availability of healthy plants or host tissue to infect is limited or nil. The curve reaches a plateau (a flat or horizontal shape). It also indicates the saturation phase of the disease in the absence of healthy plants for infection.
(4) The last phase when virus spread does not occur due to the absence of healthy plants, and the environmental conditions also preclude further infection.

Disease progress curves are often sigmoid in shape as the incidence of many diseases tends to increase slowly at the outset and also in the later stages of crop growth as crops mature, or when almost all plants become diseased and "saturation" occurs. Spread is usually most rapid at the intermediate stages of crop growth, as indicated by plotting successive increments of disease against time, which tend to increase to a maximum and then decline.

Disease progress curves merit detailed scrutiny to obtain the maximum possible information from the observations made. The various possible approaches and the mathematical procedures used are discussed in the general epidemiology texts of Vander plank (1963), Zadoks and Schein (1979), Campbell and Madden (1990) and Madden et al. (2007) and in reviews dealing specifically with plant virus diseases (Thresh 1974b, 1983; Nutter 1997). These publications discussed the derivation and utility of the various means of quantifying spread including:

- time of onset (first appearance of disease)
- duration of spread (time: *t*)
- rate of spread (*r*)

In the monetary analogy used by Vander plank (1963), the initial incidence of disease at the outset (*xo*) is regarded as the original capital sum invested and *r* is the rate of interest. The two variables are independent, but interrelated in their effects and both must be considered in evaluating the increase in capital (= incidence) that occurs over the entire investment period (= crop duration: *t*). Clearly, the greatest disease problems occur when there is a rapid spread over a prolonged period and from many initial foci. However, numerous initial foci, or a prolonged period of spread, can lead to high incidences of disease despite low rates of disease progress. All three variables (*xo*, *r* and *t*) can be manipulated to influence the final outcome and to evaluate control measures. These variables are also helpful to distinguish phytosanitation or other means of decreasing *xo* and "rate reducing" measures such as host plant resistance, or the use of pesticides to control vectors and so decrease *r* (Zadoks and Schein 1979; Thresh 1983).

1.15 Growth Models for Disease Progress Studies

Growth models provide a range of curves that are often similar to disease progress curves (Van Maanen and Xu 2003) and represent one of the most common mathematical tools to describe temporal disease epidemics (Xu 2006). The growth models commonly used are: Monomolecular, Exponential, Logistic and Gompertz (Zadoks and Schein 1979; Nutter 1997; Campbell and Madden 1990; Nutter and Parker 1997; Xu 2006) and exhaustive information is available. A brief description of each growth model is as follows:

(a) **Exponential:** This simplest model is also known as the Malthusian model that was used to predict increase in human population over time. The exponential model may be more appropriate when there is no limitation on disease increase or when newly diseased (infected) individuals lead to more diseased (infected) individuals and has been used to model changes in disease prevalence on a geographic scale. This can be applied to describe the very early stages of most of the polycyclic epidemics (Nutter 2007).

(b) **Monomolecular:** This growth model is appropriate for modeling epidemics where there is no secondary spread within a growing season, meaning that the plant disease has a single cycle during growing season (Nutter 2007). This model is also called negative exponential model (Campbell and Madden 1990). The rate of disease progress or *dy/dt* is the change in disease (*y*) with an infinitesimal change in time (*dt*). The quantity of change in disease (*dy/dt*) over time from a maximum at the beginning of epidemic and has a negative exponential probability density function.

(c) **Logistic:** This model was first proposed by Veshulst in 1838 to represent human population growth. A second type of logistic model was proposed by Vander plank (1963), being more appropriate for most polycyclic diseases, meaning that there is a secondary spread within a growing season (Nutter 2007). This growth model is the most widely used for temporal analyses and for describing epidemics of plant diseases (Segarra et al. 2001; Jeger 2004). When the rate of infection is constant, the plot of *dy/dt* versus t increases initially, reaches a maximum (at 50 % disease or infection point), and declines to zero. This curve that is symmetrical about the infection point can be linearized by using logarithmic transformation. This model incorporates features of both the exponential and monomolecular models and this model has been widely used in plant pathology over the past 45 years are so.

(d) **Gompertz:** This growth model is appropriate for polycyclic diseases as an alternative to logistic models. Gompertz model has an absolute rate curve that reaches a maximum more quickly and declines more gradually than the logistic models (Nutter 2007). The curve for *dy/dt* versus t for Gompertz, unlike logistic model, is not symmetrical around the infection point. The rate of infection approaches infection point more rapidly and declines more slowly than in the logistic model and the curve is skewed positively i.e. a larger portion of the area under curve is to the right of the infection point. Even the results obtained by Moreno et al. (2007b) in the temporal analysis of the spread of *Lettuce mosaic virus* (LMV) in lettuce, suggest that Gompertz model is the most appropriate to describe the disease progress curve (DPC) of LMV epidemics.

Figure 1.11 shows examples of disease progress curves represented by growth models, where it can be seen that Gompertz and logistic models have a characteristic sigmoid form and an inflection point meaning secondary inoculation or plant-to-plant spread within the crop in contrast to monomolecular model, which does not have inflection point. The exponential model presents a very small value at the beginning compared to other models and latter it increases exponentially.

In general, growth models that incorporate a few variables to describe temporal disease dynamics have a good performance; however, these kinds of models sometimes do not satisfy the acquiring process of key characteristics because they frequently ignore relevant variables that affect the epidemic development (Xu 2006),

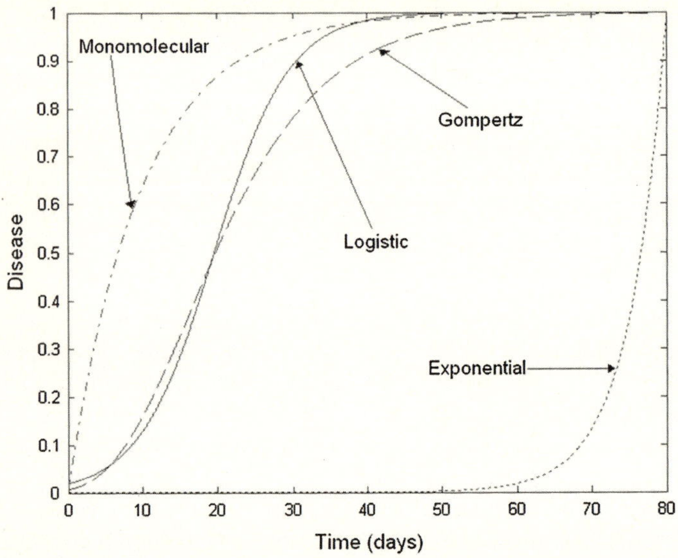

Fig. 1.11 Examples of disease progress *curves* represented by monomolecular, exponential, logistic and Gompertz models

like host growth, fluctuating environmental condition, length of latent and infectious period, etc. Nevertheless, advances in statistical and computing technologies have allowed incorporating several of these kinds of characteristics in order to obtain a more reliable model. It is important to mention that the researchers should be aware of some violations presented in these models by checking if some assumptions about the epidemic are not met and if there are some inevitable violations; they must try to find means to reduce such violations in order to diminish the bias and to correctly interpret results (Xu 2006). Vander plank (1960) used exponential, monomolecular and logistic models to describe the development of epidemics. Xu (1999) used a logistic model to forecast and model the apple powdery mildew provoked by *Podosphaera leucotricha*. The work presented by Mersha and Hau (2008) used logistic and Gompertz models to study the effects of bean rust on host dynamics of common bean in controlled greenhouse experiments with and without fungicide sprays.

1.16 Spatial Dynamics and Metapopulations

Very limited work has been done on the spatial dynamics of plant-virus-vector systems. A model of virus epidemics in combined field-nursery systems by Jeger et al. (2002) incorporates a spatial separation of the production field and the nursery. Madden et al. (2000) considered vector emigration and immigration from an in-principle infinite vector population with constant proportions of healthy, latent and viruliferous vectors in the immigrant population. Although this study gave insight into the effect of vector migration on disease dynamics, various important aspects were not considered in the model. The vectors migrating from one field may migrate to another field where a similar viral disease epidemic is proceeding. Therefore, the underlying vector pool is itself dynamic and the fractions of latent and viruliferous vectors depend on disease dynamics in other parts of a spatially extended system (Holt and Chancellor 1996, 1997). A system of fields, each with a viral disease, connected by migration of viruliferous vectors resembles a meta-population. Disease management in individual crop fields will influence the population dynamics and proportions of healthy, latent and viruliferous vectors and consequently disease dynamics in other fields. A better understanding of plant-virus-vector systems will be obtained with the development of spatially explicit models incorporating crop fields and vector populations with migration of vectors between fields.

(a) **Temporal patterns of spread**

Temporal patterns of spread or disease increase are concerned with changes in disease incidence with time. Within the life span of the crop, the amount of disease and the rapidity with which spread occurs are important in determining the losses caused. Moreover, such data are required in evaluating the effectiveness of host plant resistance and other approaches to control. Temporal patterns of spread are apparent from the disease progress curves produced by plotting cumulative incidence against time (Fig. 1.12).

The incidence of virus diseases is usually recorded as the percentage or proportion (P) of diseased plants within the stand being assessed on scales of 0 to 100 (%) or 0 to 1 (P) (Thresh 1974b).

(b) **Temporal and Spatial Distribution of Virus Spread**

Spatial distribution is one of the most characteristic ecological properties of species and is now broadly accepted by epidemiologists that the spatial component of plant disease is as important as the temporal component (Campbell and Noe 1985). The importance of describing the spatial distribution of virus-diseased plants has been recognized for over the past six decades (Bald 1937; Vander plank 1946). The spatial pattern of infected hosts has major consequences for disease development and therefore, analysis of spatial patterns of plant virus diseases has received much attention (Madden et al. 1995). The spatial pattern of individuals in a population of diseased plants may be described as random (independent dispersion of individuals), or as aggregated (clustered dispersion of individuals). Based on the type of data acquired, spatial patterns of diseased plants fall into three types of analysis: position of healthy or diseased plants within a row or a series of rows, quadrat or plot counts, and distance measurements. Since 1946, most investigators have quantified virus patterns by assessing the sequence of diseased and healthy plants in each row using a form of doublet or runs analysis (Madden and Campbell 1986; Vander plank 1946)

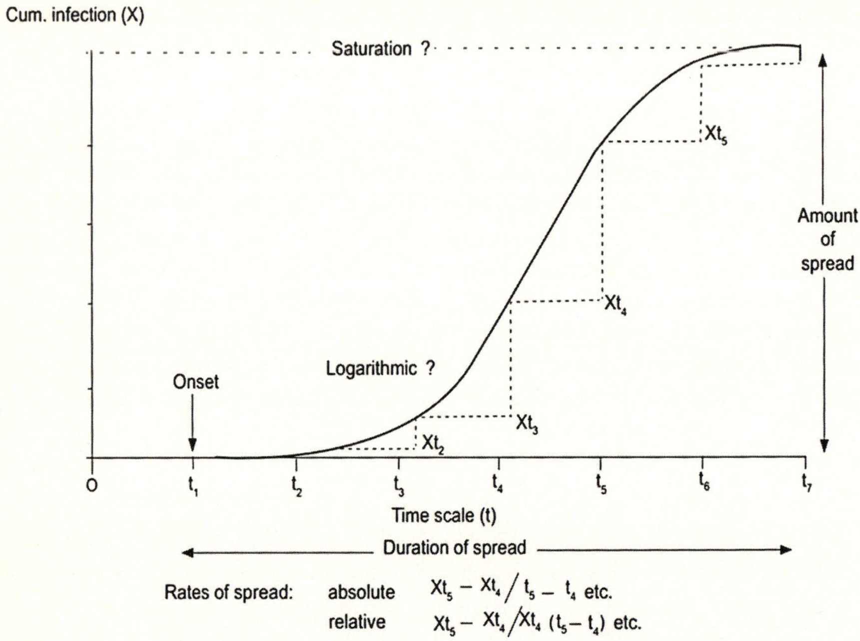

Fig. 1.12 Generalized curve of disease progress obtained by plotting cumulative incidence (x) against time (t) and the derivation of absolute and relative spread. *Source.* Thresh and Fargette (2003);

A full understanding of plant virus epidemiology requires studies at different scales of integration: from within-plant cell processes to vector population dynamics, behaviour and broader ecological interactions. Vectors respond to cues derived from plants (both healthy and virus-infected), from natural enemies and from other environmental influences, and these directly affect the temporal and spatial patterns of disease development. The key element in linking these scales is the transmission process and the determining factors involved (Jeger et al. 2011).

The distribution of viruses within a plot is described in several ways: presentation on the actual layout of infected and non-infected plants, progress curves in which the number of infections or the respective rate of infection by plotting it against the time or distance; and mathematical expressions which incorporate and quantity parameters or major variables related to infection that are involved in the spread of disease in time and space. Accordingly, the terms used are "temporal" to describe the changes over time and "spatial" to describe the changes over space in infected and disease-free plants.

Spatial and temporal analysis of viral diseases can be used to test hypotheses on the dynamics of virus spread, including those concerned with the importance of primary inoculum and the mechanisms involved in the spread of the pathogen.

Plant disease management practices can be improved by putting epidemiological information in the same format as other farm information using a geographic information system (GIS). A GIS is a computer system capable of assembling, storing, manipulating, and displaying data referenced by geographic coordinates (Star and Estes 1990). GIS can now be installed on any model desktop computer (e.g., a personal computer with at least 32 MB of RAM is adequate for most applications) and does not require an in-depth understanding of the statistical and mathematical basis of the technology and new tools, including global positioning systems (GPS) and geostatistics have been applied in plant virus epidemiology (Rossi et al. 1992; Nelson et al. 1999). By following these systems certain plant virus epidemiological data from large areas has been analyzed (Lecoustre et al. 1989; Nelson et al. 1994).

A new method, spatial analysis by distance indices (SADIE), was developed to quantify spatial patterns (Perry 1995, 1998; Perry and Nixon 2002). The concept of SADIE is to quantify the spatial pattern in a sampled population by measuring the distance (D) individuals must move to give an extreme arrangement in which the individuals in the samples are spaced as uniformly or regularly as possible. The distance to be moved thus relates to the effort required to be expended by the individuals to attain this theoretical arrangement. An assessment of the non-regularity of the observed spatial pattern can be made. In contrast to geostatistical methods (Fleisher 2002), SADIE offers spatial information of the data independently of their numeric properties allowing other aspects of spatial analysis to be investigated such as the spatial association between samples of two different populations (Perry et al. 1996, 1999). For this reason, the SADIE methodology has been used in many studies together with mathematical models for analysis of the disease progress curves (DPCs) to understand the driving forces involved in the spread of plant viruses in several crops viz., *Lettuce mosaic virus* (Moreno et al. 2007b; Legarrea et al. 2009) and Lettuce big-vein disease in lettuce (Coutts et al. 2004b; Latham et al. 2004); *Cucumber mosaic virus* in chickpea (Jones et al. 2008); *Cucumber mosaic virus* and *Cucumber aphid-borne yellows virus* in cucumber (Dader et al. 2012); *Cucumber mosaic virus* and *Watermelon mosaic virus*-2 in melon (Alonso Prados et al. 2003); *Tobacco etch virus* in tobacco (Eckel and Lampert 1993); *Maize dwarf mosaic virus* in maize (Madden et al. 1987); Rice tungro virus in rice (Kondaiah et al. 1976); *Tomato yellow mottle virus* in tomato (Jovel et al. 2000); *Bean dwarf mosaic geminivirus* in french beans (Wang et al. 1996); *Tomato spotted wilt virus* in pepper and lettuce (Coutts et al. 2004a); Aphid-borne viruses like Bean yellow mosaic and Cucumber mosaic viruses in lupin (Cheng et al. 2002; Jones 2005b); *Iris yellow spot virus* in onion (Boateng and Schwartz 2013); *Papaya ringspot virus* in papaya (Kumar et al. 2008b); Tristeza virus in *Citrus* spp. (Marcus et al. 1984); Sharka (*Plum pox virus*) in apricot and peach (Gottwald et al. 1995; Dallot et al. 2003).

The SADIE methodology along with mathematical models have proved to be one of the most powerful tools and are being attempted in other economically important virus-host combinations.

(c) **Dispersal and spatial dynamics**
Gregory (1968) was instrumental in laying many of the foundations of modern plant epidemiology. He examined in great detail how different forms of disease gradients in space arose and how these developed over time.

(d) **Plant virus epidemics**

A number of plant viruses like *Cocoa swollen shoot virus*, *African cassava mosaic virus*, *Tomato mosaic virus*, *Tomato yellow leaf curl virus*, *Citrus tristeza virus*, *Rice tungro virus*, *Rice yellow mottle virus*, *Peanut bud necrosis virus*, *Potato leafroll virus*, *Sunflower necrosis virus*, *Watermelon bud necrosis virus*, *Beet curly top virus* and *Lettuce mosaic virus* are responsible for heavy yield losses and this is evident from the data presented in the Chapter-3 of Volume I (Sastry 2013b). A majority of these viruses have caused epidemics and the virus diseases have faced the favourable conditions for causing epidemics. Epidemics have been defined as any increase of disease in a population. The dynamics of changes occur in plant disease over time and space. Plant disease epidemics sometimes called as ephiphytotics occur annually on most crops in many parts of the world. The epidemics may cause minor to moderate losses and occasionally, the epidemics may also get out of control and become extremely wide spread.

Disease development depends on three factors: the available host plant population should be susceptible; the pathogen should be virulent to cause disease; and environmental factors are favorable for disease spread. Several environmental factors contribute to the disease development by altering host metabolism and or virulence of the pathogen.

Epidemic development is influenced by: Environmental factors, Rate of pathogen reproduction, Mode of pathogen dispersal, Efficiency of pathogen survival, Level of aggressiveness of the pathogen, and Level of host plant resistance.

For developing a sound plant virus and viroid disease management strategy requires enough knowledge of the biology of the pathogen and host to select the appropriate epidemiological model. It also requires at least "ball-park" estimates of the model parameters and the magnitude of the impact of each specific tactic on the initial inoculum or the apparent infection rate. Failure to adopt such a quantitative approach can lead to some embarrassing or even very costly errors.

1.17 Disease Gradients and Progress Curves

In plant viruses, infection gradients are associated with spread from initial foci of infection which may be weeds or wild plant species or volunteers of previous crops. The incidence of virus diseases tends to decrease with increasing distance from the source of inoculum (Rose 1973; Thresh and Lister 1960; Fargette et al. 1993). Gradients are visualized by plotting disease incidence against distance using appropriate scales. The gradients tend to be curvilinear and concave as disease incidence decreases most rapidly near the source and then declines to a low 'horizon of infection', which is the distance beyond which spread is so infrequent or any further infection becomes insignificant.

For any particular disease, gradients tend to be shallower and over longer distances when conditions favour spread, compared with situations or seasons when condition are less favourable. For viruses with wind-borne vectors, the disease gradients tend to be steeper and more circumscribed up-wind than down-wind from the source. Consequently it is seldom possible to give a simple unqualified advice on the most appropriate isolation distance for adoption to avoid serious risk of infection.

(a) **Spatial patterns of spread**

Virus spread can occur from infected sources within the crops and from sources that are nearby or remote. After a period of time, a disease that started with a single infected plant, which may be through virus infected seed or through vector, will appear as a 'focus' and act as initial source of inoculum. It will spread gradually through vectors or mechanically or by any other means depending on the virus. Virus diseases in plants start at a low level, a small numbers of plants affected and these numbers of infected plants are increased with time.

Valuable information can be gained from observations on the distribution of diseased plants within crop stands. For example, it may facilitate identification of the initial source(s) of inoculum from which spread occurs. It may also be possible to infer the means of spread and the most effective control strategy to adopt. This has been apparent from observations made over many years on a wide range of virus diseases of temperate and tropical crops.

Thresh and Fargette (2003) have explained clearly that weeds are one of the sources of virus spread and the infected weeds or wild plants within the crop and alongside of the crop will lead to high virus incidence if the vector population exists. On the other hand, if the infected weeds are present remote from the crop then the chances of virus spread into the newly planted crops will be very less. The rate of spread will depend on the level of mobility of vectors. The distinctions between local and distant, and between crop and non-crop hosts can be somewhat arbitrary, as spread may occur from multiple and diverse sources within and/or outside the crop(s). Nevertheless, the distinctions are helpful in elucidating the role of the different sources, the magnitude of the threat they pose, and the likely effectiveness of control measures.

In most fields, it is common to observe the diseased plants in patches. Patches of disease often occur within crops as a consequence of local spread around initial foci of infection due to the debris of previous crops, or to the use of infected seed or vegetative propagules, or to the occurrence of infected weed or wild hosts within the stand. Patches can also occur as a consequence of soil-borne viruses like *Peanut clump virus, Soil-borne wheat mosaic virus, Tobacco ring spot virus* etc.,

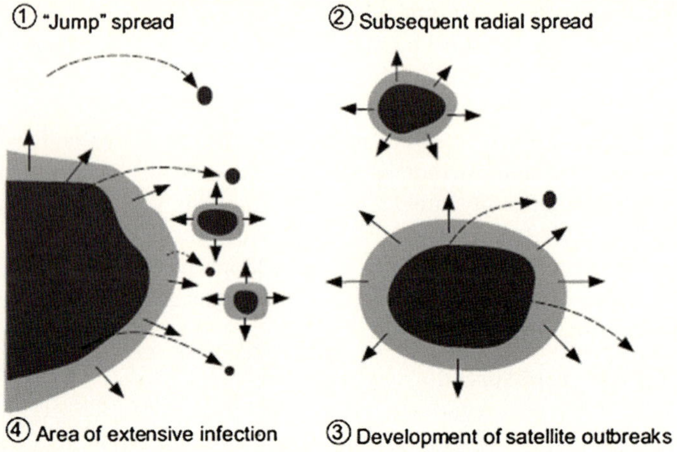

Fig. 1.13 Successive stages in the spread of cocoa swollen shoot disease by wind-borne mealybug ("jump-spread") and mealybugs walking between the canopy branches of adjacent trees (radial spread). *Source* Thresh and Fargette (2003)

which are transmitted by soil-borne vectors like fungi and nematodes. In case of *Groundnut rosette virus,* the disease occurs in patches through the incoming aphid vectors (*Aphis craccivora*) (Evans 1954; Naidu et al. 1998). With soil-borne vectors either fungal or nematodes, virus spread will be slow and will be in patches as seen in grape orchards infected with *Grape fanleaf virus* and groundnut fields with *Indian peanut clump virus*.

The temporal sequence of spread of *Cacao swollen shoot virus* (CSSV), is associated with virus transmitted by several species of mealybug (Thresh 1958). CSSV is transmitted by at least 13 species of mealybugs of the family *Pseudococcidae* and *Coccoidae* (Roivainen 1976). The nymphs of both sexes and adult females of mealybugs spread the disease radiating between the adjacent trees by crawling through the canopy from infected to healthy trees or by being carried by attendant ants. Adult male mealybugs have only rudimentary non-functional mouth parts and do not feed. An occasional jump spread may occur when infective mealybugs are blown by the wind and infect trees at some distance away from the original site of infection. New outbreaks are initiated by "jump spread" of viruliferous, wind-borne, first-instar mealybug nymphs (Cornwell 1960). Subsequent "radial spread" is by mealybugs which walk only short distances between the canopy branches of adjacent trees to form obvious patches of disease within plantations (Cornwell 1958; Thresh 1988). The patches continue to expand and give rise to additional 'satellite' out breaks that ultimately coalesce and form continuous 'areas of mass infection' (Fig. 1.13).

Such patterns of spread have been reported with many other pathogens and Fig. 1.11 has been generalized so as to apply to a wide range of virus diseases (Vander plank 1963). However, there can be big differences between diseases in their rate of progress and in the relative importance of "jump" and "radial" spread of *Cocao swollen shoot disease* by mealy bug vector (Thresh and Fargette 2003).

(b) **Disease gradients**

A feature of many pathosystems is that there are clear trends in the incidence of disease with increasing distance along one or more transects across the planting being assessed. Such changes are referred to as "disease gradients" and they are of two main types (Gregory 1968). The first type is "Environmental gradients" which are associated with changes in soil fertility or microclimate and also with features such as windbreaks, buildings, or other physical obstacles that influence the vulnerability of host plants or the distribution and deposition of vectors and inoculum. For example, *Cassava mosaic disease*, which tend to be the greatest at the margins of plantings, especially those facing the direction of the prevailing wind. This is associated with the tendency of incoming whitefly vectors to alight preferentially and become most numerous on plants in the outer rows, as observed in field studies in Côte d'Ivoire where there were pronounced edge effects on the distribution of disease (Fargette et al. 1985; Colvin et al. 1998).

The second type of disease gradient is "infection gradients" that are associated with spread from initial foci of infection and intensities are high. Generally disease incidence tends to decrease with increasing distance from the source of inoculum. Infection gradients are obtained by plotting disease incidence on the vertical axis against distance on the horizontal axis using appropriate scales. The gradients tend to be curvilinear and concave as disease incidence decreases most rapidly near the source and then declines to zero, or reaches a low "horizon of infection" (Fig. 1.14). Beyond certain distance when spread is so infrequent that any further infection becomes insignificant, or so unimportant that it can be disregarded (Vander plank 1949). Such gradients have been reported with many virus diseases of tropical crops (Thresh and Lister 1960; Rose 1973; Thresh 1976; Atiri and Varma 1991; Fargette et al. 1993).

For example, for any particular virus disease, gradients tend to be shallower and over longer distances when conditions favor spread, compared with situations or seasons when conditions are less favorable. Moreover, for viruses with wind-borne vectors, the disease gradients tend to be steeper and more circumscribed up-wind than downwind from the source. As mentioned earlier, it is seldom possible to give a simple, unqualified recommendation on the most appropriate isolation distance to adopt. The "safe" distance depends not only on site and season but also on the size and potency of the source of inoculum and the degree of risk and the extent of the losses that are considered acceptable. Clearly, the smaller the source and the greater the isolation, then the less the risk, but it may be necessary to compromise because the extreme isolation may be difficult to achieve and may also be very inconvenient. This explains why substantial isolation may be justified and appropriate for producing "elite", specially selected, virus-free propagules for further propagation or distribution to farmers, but not for routine crop production.

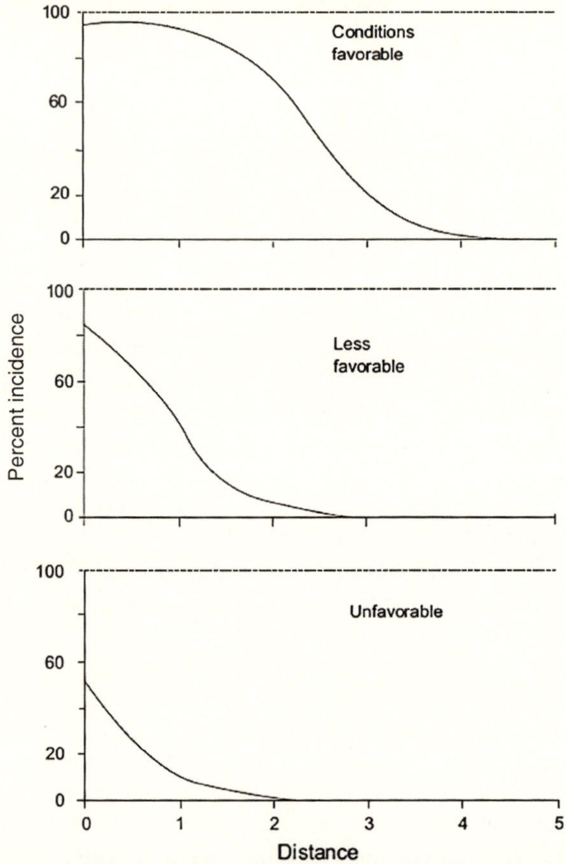

Fig. 1.14 Generalized gradients of disease obtained by plotting incidence (%) on the vertical scale against distance on the horizontal scale (arbitrary units) for spread under favorable (*top*), less favorable (*middle*), and unfavorable conditions (*bottom*). *Source* Thresh and Fargette (2003)

The mode of spread and type of vector are other important considerations. Shallow gradients that spread over great distances are due solely to the most mobile vectors and their occurrence may be restricted to certain clearly defined migration periods. Spread at other times and by less active individuals or development stages of the vector species result in a relatively steep gradient over mainly short distances around the source. Thus, the initial patterns of infection by migrants moving considerable distances into or between plantings may be quite different from those due to subsequent local movement by the incoming vectors, or their less mobile progeny.

Moreno et al. (2007a) have reported that the viruliferous winged aphid species like *Myzus persicae*, flying into the lettuce crop are responsible for the secondary spread of LMV. Relationship between the winged transient aphid populations of various species and virus spread has also been established for several potyviruses viz., LMV in lettuce (Nebreda et al. 2004; Moreno et al. 2007b). PVY in potato (Katis et al. 1998); CMV in pepper (Raccah et al. 1985); *Soybean mosaic virus* in soybean (Halbert et al. 1981) and *Zucchini yellow mosaic virus* in zucchini (Katis et al. 2006).

Where different forms of the vector coexist and each contributes to spread, the observed gradients of infection represent the superimposition of two or more separate distributions. However, there is seldom any discontinuity or inflexion in the disease gradients observed to make it possible to distinguish distinct zones or means of spread. The zones tend to merge imperceptibly and there is considerable uncertainty as to the relative importance in local spread of wind-borne and crawling mealybugs and of the short- and long-flying forms of certain leafhopper vectors, including the *Cicadulina* spp. that transmit *Maize streak virus* (Rose 1972). Conflicting claims have also been made on the performance of wingless (apterous) and winged (alate) aphids as vectors and their relative importance seems to differ between viruses and between regions (Broadbent 1959; Hodgson 1991). Nevertheless, there is general agreement that in some crops, apterous aphids or the relatively immobile forms of vector, often reach plants that have been infected already by the more active members of the vector population and so make little direct contribution to the local spread of disease.

1.18 Mathematical Modeling Techniques

The most common mathematical tools used in plant disease epidemiology are: disease progress curves, linked differential equations, area under disease progress curve and computer simulations. Nevertheless, there are other tools to evaluate the disease progress like statistical tools, visual evaluations and pictorial assessment.

Understanding how disease levels increase or decrease over time is one of the most basic elements of plant disease epidemiology and ecology. Statistical models are often applied in order to summarize and describe this complexity, so that disease processes can be more readily understood (Arneson 2006; Madden et al. 2007). For example, comparisons between patterns of disease progress for different diseases, cultivars, management strategies, or environmental settings can help in determining how plant diseases may best be managed.

Understanding how populations of different types of plant pathogens increase in plant pathosystems is critical to developing appropriate management methods. Bacteria and viruses are capable of building up enormous populations in a very short time, only days or even hours under the right conditions. Fungi also can reproduce quickly, within days or weeks, which is important for fungal pathogens that cause polycyclic diseases, i.e., fungi that reproduce multiple times per season. Nematodes, meanwhile, often take a year or more to complete their reproductive cycle. These different characteristics will greatly influence disease progress over time.

Mathematical tools have been employed to create models which give a description of epidemic dynamics. The commonly used mathematical tools are: Disease progress curves, Linked Differential Equation (LDE), Area Under disease

Progress Curve (AUDPC) and computer simulation. Nevertheless, there are other tools that have been employed in epidemiology of plant disease like: statistical tools, visual evaluations and pictorial assessment.

LDE is one of the most commonly used mathematical tools in modeling epidemics because it allows new terms to be easily included as needed in order to answer the question about the details of the pathosystem (Madden 2006). In general, it is too difficult to obtain analytical solutions for this mathematical tool, but numerical solutions can be obtained. An advantage of this mathematical tool is the intrinsic relationship with other tools here mentioned like disease progress curves and computer simulations. The important components of disease dynamics can be modeled by using growth models that latter could be linked and simulated using computer programs (Mersha and Hau 2008). A disadvantage of LDE is that the formulated equations using this tool are extremely troublesome for mathematical analysis, which makes it difficult to explore different biological properties of host and pathogens or to know the results of different control strategies, on long term disease development (Madden 2006).

The area under disease progress curve (AUDPC) is extensively used in analyzing disease progress curves to make comparisons between treatments, to know the relationship between yield losses and disease, to make a genetic analysis, to evaluate the effectiveness of chemical, biological and cultural controls over disease progress and also to know the resistance of certain species to plant pathogens (Jeger 2004). The disadvantage of AUDPC is that it depends on a good fitted disease progress curve to obtain correct results and this approach also can yield misleading results when it is summarized over a particular period instead of over a complete period (Xu 2006).

The variables used by mathematical tools are chosen by relevance, pathosystem features, and epidemiologist's necessities according to the model objective. The epidemiologist can decide on the mathematical tool which solely or jointly with other different mathematical tools to provide a better description of the reality in order to obtain a more accurate evaluation of the plant disease epidemics. The possible combination of the tools here presented can be made by the conjunction of several disciplines like image processing which could serve to obtain a better objective estimation of variables that can feed to the model allowing for obtaining a better fitness. Acceptable accuracy and precision of visual disease assessments have often been achieved using disease scales, particularly because of the time they allegedly save, and the ease with which they can be learned. But recent work suggests there can be some disadvantages to their use. New technologies offer opportunity to assess disease with greater objectivity (reliability, precision, and accuracy). One of these, visible light photography and digital image analysis has been increasingly used over the last 30 years, as software has become more sophisticated and user-friendly. Indeed, some studies have produced very accurate estimates of disease using image analysis. In contrast, hyper spectral imagery is relatively recent and has not been widely applied in plant pathology. Nevertheless, it offers interesting and potentially discerning opportunities to assess disease. As plant disease assessment becomes better understood, it is against the back drop of concepts of reliability, precision and accuracy in plant pathology and measurement science (Bock et al. 2010).

Mathematical models can be classified into three partially overlapping types; analytical, statistical and simulation. The analytical models are of interest in theoretical considerations of ecological concepts but quickly become too complex if applied to field populations. In many cases, knowledge and insight of the ecological processes are lacking, or at best are of a rudimentary nature. Because of their abstract character, analytical models are usually difficult to validate in the real world. The use of properly developed models, which incorporate experimentally collected parameters, to predict the likely effects of certain control measures, can improve decision making (Jeger and Chan 1995).

Analytical models are used to analyze epidemics on a theoretical basis without taking into account the effects of external variables (Jeger 1986). The well-known analytical models are those proposed by Vander plank (1963) to explain disease growth under non-restrictive environment, apparent rate of growth of infections, and basic rate of growth. The equations developed for the various fitness and survival of pathogens or for the impact of multilines also come under this category since the analytical models are tools for better inference.

Statistical models, usually regression type, are purely empirical not relying on causal relationships and hence, describe rather than explain processes. They are rarely consistent in the long-term and when they do fail, the model itself cannot be used for explanation. They are, therefore, of limited use in plant protection. They do provide useful information on the extent of possible yield losses that might occur when the disease intensity and severity are known. Empirical models also thus help to decide on the economical utility of any plant protection intervention.

(a) **Mathematical model for monocyclic diseases**

Modeling is a simplified representation of reality, or a representation or analogy of something that exists. Models do not replicate reality but are much simpler representation of it. Qualitative models such as disease cycle represent essential qualities or attributes of the original, whereas quantitative models use mathematical notations and symbols to represent an abstraction of the original. Quantitative models are essentially mathematical model which are also simplifications of reality.

Dynamic processes, including the change in the quantity of disease intensity in a plant population over time, are defined by their rate of change with time. A mathematical model for a monocyclic disease can be written as follows:

$$r_m = \frac{1}{t_2 - t_1}\left(\ln\frac{1}{1-x_2} - \ln\frac{1}{1-x_1}\right)$$

r_m infection rate (monocyclic)
t time
x_2 proportion disease at t_2
x_1 proportion disease at t_1

While assessing the disease on a set of plant population periodically in fields or under artificial conditions, estimated values on disease are converted into natural logarithmic values to linearize the disease progress.

(b) **Mathematical model for the polycyclic diseases**

In case of polycyclic disease, the virus multiplies for a period of a few hours in the invaded cells and moves into adjacent cells in a few hours after infection. Its reproductive cycle recurs at regular intervals of time. The recurring sequential infections eventually lead to infection of the whole plant. Thus in the life time of a plant or crop, a virus will show many cycles of multiplication (reproduction), infection and disease; and this will become a source of inoculum for further spread by vectors. The infection cycles are repeated in the crop, until saturation is reached and no more host plants or tissues are available for further infection. This is how repeated infections take place in a polycyclic disease. The vectors with their short life cycles completes many generations in the life time of a crop and thus spread the virus and induce disease in a crop in totality and even across fields and regions and countries by their mobility and through wind dispersal.

- If **r** is very high, the apparent effect of reducing $\mathbf{x_0}$ is to delay the epidemic.
- If **r** is very high, $\mathbf{x_0}$ must be reduced to very low levels to have a significant effect on the epidemic.
- Reducing **r** has a relatively greater effect on the epidemic than reducing $\mathbf{x_0}$.
- Reducing $\mathbf{x_0}$ makes good strategic sense only if **r** is low or if **r** is also being reduced.

It is easier to understand and remember these concepts if we actually select different values for $\mathbf{x_0}$ and **r**, plug them into the model, and graph the outcome. This can be done easily with a calculator that has an exponential function, or with the accompanying simulation.

(c) **Exponential Disease Increase—Compound-interest Disease**

Polycyclic Pathogen

$$X = X_0 e^{rt}$$

X amount of disease at time t
X_0 initial disease/primary inoculum (at time t = 0)
e 2.73… (base of natural log)
r rate of disease increase
t time period

In reality, polycyclic diseases increase exponentially only during the very early stages of an epidemic (<5 % disease) (Fig. 1.15).

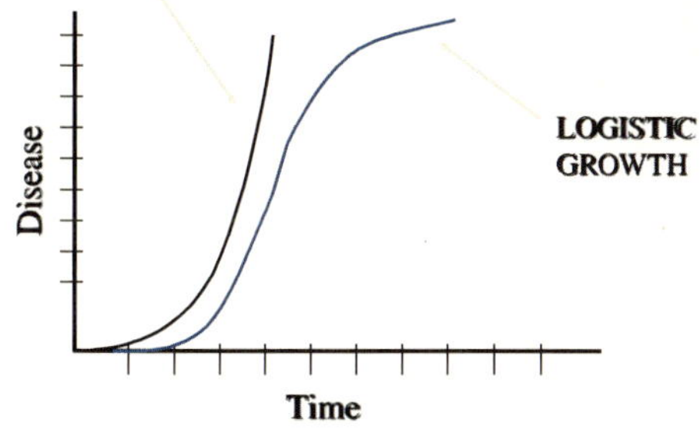

Fig. 1.15 Exponential growth

Reasons why disease does not increase exponentially for long:

- As the amount of healthy tissue becomes limiting, the rate of disease increase slows down.
- Infection occurs intermittently, not continuously.
- Newly infected tissue is not immediately infectious.
- Disease tends to occur in foci, not evenly throughout the field.

1.19 Virus-Vector Population Model

A model for the epidemiology of plant-virus-vector system was combined with a model for the within-plant dynamics of the virus. Mathematical models have been used to study the effects of plant, plant virus and vector characteristics on the development of disease epidemics (Chan and Jeger 1994; Holt and Chancellor 1996; Jeger et al. 1998; Madden et al. 2000; Zhang et al. 2000a, b; Jeger et al. 2002; Escriu et al. 2003). One of the significant shortcomings of these models is that most of the virologists ignore the evolutionary dynamics of the plant virus, even though evolution is likely to have a major impact on disease control. van den Bosch et al. (2006) used the theory of evolutionary stable strategies (ESS), as introduced by Maynard Smith (1982), such that if a resident population engages in an evolutionary stable strategy it cannot be invaded by a population with a different strategy.

Although van den Bosch et al. (2006) model is of a generic nature, and have used the specific system against whitefly-transmitted geminiviruses (begomoviruses) infecting tomato as the key example. Both experimental and modeling approaches have been used to study the epidemiology of tomato diseases (Holt et al. 1999; Moriones and Navas-Castillo 2000). Holt et al. (1999) studying tomato begomovirus diseases in India, reported that disease spread will occur for very low vector populations, and hence varietal resistance will be an important component of disease management.

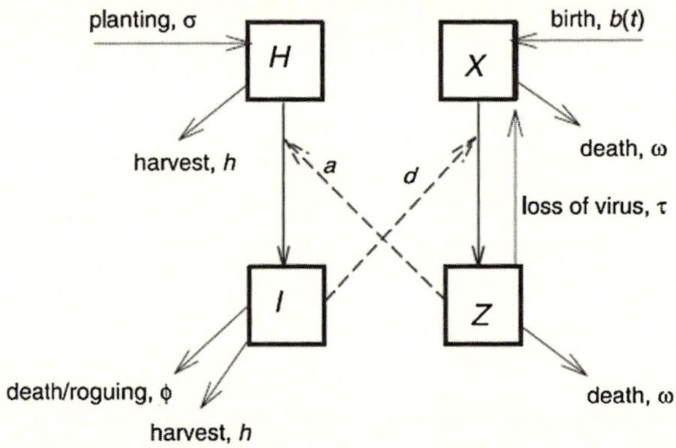

Fig. 1.16 Graphical representation of the virus-vector model. Boxes are the state variables: *H*, density of healthy plants; *I*, density of infectious plants; *X*, density of non-viruliferous vectors; *Z*, density of viruliferous vectors. *Source*. van den Bosch et al. (2006)

(a) **The model and the evolutionary stable strategy**

Virus-vector population model

The flow diagram Fig. 1.16 has been used to describe the model, which is a simplified version of the model by Jeger et al. (1998) and Madden et al. (2000).

The plant population

The model describes the dynamics of the density (plants per units area) of healthy, $H(t)$, and infected, $I(t)$, plants through time. Healthy plants are planted at a constant rate, σ, reflecting a continuous cropping system. The productive lifetime of a plant is denoted by $1/h$, implying that the number of healthy and infected plants removed per unit area per time unit equals $hH(t)$ and $hI(t)$, respectively. Infected plants have an additional removal rate, ϕ, which can either describe a death rate as a result of the virus infection or a roguing sanitation rate as a means of disease management. A healthy plant can become infected when a viruliferous vector inoculates the plant while feeding. To calculate the rate with which healthy plants become inoculated, van den Bosch et al. (2006) have applied the commonly used mass action assumption, meaning that the rate is proportional to the density of healthy plants and viruliferous vectors, with a proportionality constant α. The parameter α will be called the inoculation rate parameter. It is assumed that an infected plant is infectious and a vector can acquire the virus during feeding. This implies that van den Bosch et al. (2006) have omitted a latent period in this model.

The vector population

The model describes the density of healthy vectors, $X(t)$, and viruliferous vectors, $Z(t)$, through time. van den Bosch et al. (2006) assumed that total vector density, P, is constant and that thus $X(t) + Z(t) = P$. This assumption implies that the population birth rate, $b(t)$, equals the population death rate that is calculated from $\omega X(t) + \omega Z(t)$, where ω is the probability per time unit for a vector to die. Healthy vectors can acquire the virus when feeding on an infectious plant.

Using similar assumptions as for the plant inoculation rate, the rate with which healthy vectors become viruliferous is calculated from $\delta H(t)Z(t)$, where δ will be called the acquisition rate parameter. After acquisition of the virus, the vectors stay on average $1/\tau$ time units viruliferous and then return to the healthy state.

The basic reproductive number

A key quantity in this analysis will be the basic reproductive number, R_0, which for this model is given by:

$$R_0 = \left(\frac{\infty \hat{H}}{\omega + \tau}\right) \cdot \left(\frac{\delta P}{h + \phi}\right)$$

where $\hat{H} = \sigma/h$ is the density of plants in the absence of the disease. This expression has a clear biological interpretation. Consider a disease-free crop. In this situation all plants are healthy, implying that $\hat{H} = \sigma/h$ and also $X = P$. Now, consider the situation where one viruliferous vector whitefly enters into the crop. This viruliferous vector will infect αH plants per time unit. A viruliferous vector remains alive and viruliferous for, on average, $1/(\omega + \tau)$ time units (i.e. it either dies or loses the virus). Therefore, the total number of plants one viruliferous vector infects before it reaches the end of its life time or viruliferous period equals the first part on the left-hand side of equation 1. Next van den Bosch et al. (2006) considered the infected plants. Each infected plant will be visited by healthy vectors that then can become viruliferous themselves. The number of vectors that become viruliferous per time unit per infectious plant equals δP. An infectious plant remains infectious for, on average, $1/(h + \phi)$ time units (i.e. either the plant is harvested or is rogued/removed). Therefore, the total number of vectors per infectious plant that acquire the virus before the plant is removed as a result of harvesting or roguing equals the second part on the left-hand side of equation 1. The product of the two terms thus has the interpretation of the number of viruliferous vectors resulting from one viruliferous vector, with infected plants as an intermediate step, when the viruliferous vector is surrounded by an entirely healthy plant population and an entirely non-viruliferous vector population. It is clear that if this number of new 'cases' per case is larger than unity an epidemic will develop, whereas if it is smaller than unity no epidemic will develop. The basic reproductive number, R_0, thus is the threshold for epidemic development in a virus-vector system.

1.20 Systems Analysis and Simulation Models

A considerable number of simulation models have been developed for plant disease epidemiology since the late 1960s and early 1970s. It was thought that these models could serve to organize available knowledge about a plant pathosystem, and provide a better understanding of that system. It is controversial whether or not simulation models have answered the initial expectations (Campbell and Madden 1990). The information available on simulation models indicates that plant disease epidemics can be treated in a quantitative manner and that collections of equations can be utilized to describe disease development to understand and compare epidemics.

Insect-borne virus disease epidemiology is concerned with the complex interactions among viruses, vectors, plants and the environment. Furthermore, several factors governing these interactions are very difficult to measure and almost impossible to monitor on a regular basis. Approaches using simulation models are therefore a reasonable way of evaluating how insect-borne virus disease spread is affected by particular factor, including those which may be difficult or impossible to measure experimentally. Simulation models are also a useful way of evaluating the effects of different control measures aimed to reduce the spread of virus infection.

The first computer based simulators of plant diseases are the EPIDEM (Waggoner and Horsfall 1969) and EPIMAY (Waggoner et al. 1972). Here, starting from the infection chain various phases of the epidemic growth are estimated taking into account the externally influencing variables such as the weather conditions. Models to understand the spatial spread of diseases were subsequently developed such as the EPIMUL (Kampmeijer and Zadoks 1977). Later, several simulators have been designed for various diseases and situations.

Madden et al. (1990b) have developed some simple analytical models for viruses transmitted in a non-persistent manner. Simulation models have a more complex structure and are especially useful in mimicking quantitatively the course of epidemics, and the response of a specific pathosystem to external variables such as environmental effects, control practices, etc. Jeger (1986) has thoroughly compared analytical and simulation approaches in plant disease epidemiology.

Simulation models are undoubtedly necessary tools for understanding the dynamics of insect-borne virus diseases at a field level. Although great efforts have been made to develop simulation models for fungus diseases, considerably less research has been done on modeling virus disease epidemiology (Campbell and Madden 1990). Probably the disciplinary isolation of scientists studying various aspects of plant virus epidemiology is the main reason why modeling has been used so seldom.

The models of Berger and Ferriss (1989) and Miyai (1991) are partially stochastic, whereas the other models are wholly deterministic. In deterministic simulation models, computations are based on the expected values of the parameters, and the model need be run only once with any particular set of parameter values. Simulation models that incorporate stochastic elements require a sequence of random numbers, and the model need be run as many times as possible by changing the sequence of random numbers.

All the models except those of Berger and Ferriss (1989) and Miyai (1991) ignore the spatial aspects of epidemic spread but concentrate on temporal aspects of the disease in a single population. This approach simplifies considerably the resultant model, at the expense of understanding the importance of spatial factors in disease transmission. However, the models of Ruesink and Irwin (1986) and Marcus and Raccah (1986) do take into consideration the spatial distribution of virus infection (random or clumped) caused by vectors moving within the field.

Table 1.7 Simulation models developed for insect-borne plant virus diseases

Host	Virus	Vector	Virus-vector relationship	References
Barley	*Barley yellow dwarf virus*	Aphid	Persistent	Morgan (1990)
Beet	*Beet mosaic virus*	Aphid	Nonpersistent	Dusi (1999)
Citrus	*Citrus tristeza virus*	Aphid	Semi-persistent	Fishman et al. (1983)
Potato	*Potato virus Y*	Aphid	Nonpersistent	Sigvald (1992), Nemecek (1993), Bertschinger et al. (1995)
	Potato leafroll virus	Aphid	Persistent	Bertschinger et al. (1995)
Rice	*Rice dwarf virus*	Leafhopper	Persistent	Nakasuji and Kiritani (1972), Nakasuji et al. (1985), Miyai et al. (1986), Miyai (1993)
	Rice stripe virus	Planthopper	Persistent	Muramatsu (1979), Kisimoto and Yamada (1986), Miyai and Hokyo (1992)
	Rice tungro virus	Leafhopper	Semi-persistent	Holt and Chancellor (1996)
Soybean	*Soybean mosaic virus*	Aphid	Nonpersistent	Ruesink and Irwin (1986)
	Soybean dwarf virus	Aphid	Persistent	Miyai (1991)
Subterranean clover	*Subterranean clover stunt virus*	Aphid	Persistent	Gutierrez et al. (1974a, b)

The simulation models used by Nemecek (1993) and Sigvald (1992), to predict Potato virus Y spread in potatoes, included some behavioral characteristics and a more detailed description of the aphid species composition. These models could be used to simulate the final disease incidence in crops, which were initially infected with different numbers of virus sources. The studies in *Soybean mosaic virus* presented by Ruesink and Irwin (1986) also included some behavioral aspects of the vector and could be used to predict yield and level of seed transmission. The complementary information added by Dusi (1999) was an experimental demonstration that spread of *Beet mosaic virus* (BtMV) is related to the major migrating aphid flight. The calibration studies using a simulation model confirmed that this spread could be described by one absolute rate parameter and can be concluded that management strategies to control virus spread have to be focused on a delay of virus introductions in the field, or alternatively, to restrain aphid dispersal early in the season.

Simulation models so far developed for insect-borne virus epidemiology are listed in Table 1.7. The vector species concerned are leafhoppers, planthoppers and aphids and they model in various ways the effects of population dynamics of these vectors on disease spread.

Simulation models are dynamic and can either be deterministic (based on mean values) or stochastic (based on probability distributions). The majority of pest simulation models are deterministic, as the use of stochastic models is very expensive in computing time and are used against specific virus epidemiological studies.

Generally, there are two reasons to use stochastic models: (1) when a reliable estimate of the expected value is desired and when there exists a curvilinear relationship between a stochastic characteristic and one of the state variables, and (2) when interest lies in the variance of model output. Unfortunately, many biological properties show curvilinear relationship and the high variation in these characters often necessitates the use of stochastic models.

From the above cited examples it is very clear that the computers remain indispensible tools for quickly determining the solutions to model equations over the time and space dimensions, for graphing results, and for statistically quitting models to data and comparing the quit of models for multiple epidemics.

All information about simulation models and their application in plant virus epidemiology and management has been reviewed by Frazer (1977), Carter (1986), Jeger (1986), Ferriss and Berger (1993), Miyai (1993) and Bertschinger (1997).

1.21 Conclusions

Global food supply depends on cereals like wheat, maize, rice, barley, sorghum, oats, millet and rye to provide much of the world's food and drink. Even fruits and vegetables play a major role in nutrition of the human beings. It is well established that many different viruses cause diseases and yield loss on these crops and emphasis is made to manage these virus diseases for achieving higher yields and in turn self-sufficiency. To achieve the self-sufficiency in food production, the information on epidemiology of plant virus diseases plays a major role. It deals with how and why a virus spreads by vectors in an ecosystem and is fundamental to the choice and improvement of virus management methods in agriculture.

The problem of managing plant virus disease can be approached from two different time-scales. From an epidemiological and ecological perspective, an understanding of the population dynamics of host, vector and virus can contribute to decisions on how best to deploy control options involving host resistance, cropping practices, phytosanitation and chemical control of vectors; or if these options are not available, whether and in what form they should be developed. From an evolutionary perspective, considered over a longer time-scale, an understanding of the genetic variation present in host, vector and virus populations provides information on sources of variation, selective forces in operation and the population structures that result. For developing sustainable methods for managing future epidemics of plant virus disease the knowledge of ecology and epidemiology of virus diseases plays integral role.

References

A'Brook J (1983) Forecasting the incidence of aphids using weather data. EPPO Bull 13:229–333

Abad JA, Moyer JW (1992) Detection and distribution of sweet potato feathery mottle virus in sweet potato by in vitro-transcribed RNA probes (riboprobes), membrane immunobinding assay, and direct blotting. Phytopathology 82:300–305

Abarshi MM, Mohammed IU, Jeremiah SC, Legg JP, Kumar PL, Hillocks RJ, Maruthi MN (2012) Multiplex RT-PCR assays for the simultaneous detection of both RNA and DNA viruses infecting cassava and the common occurrence of mixed infections by two cassava brown streak viruses in East Africa. J Virol Methods 179(1):176–184

Abe H, Tamada T (1986) Association of beet necrotic yellow vein virus with isolates of polymyxa betae keskin. Ann Phytopathol Soc Jpn 52:235–247

Abo ME, Alegbejo MD, Sy A, Misari S (2000) An overview of the mode of transmission and host plants and methods of detection of rice yellow mottle virus in Africa. J Gen Virol 84:733–743

Aboul-Ata AE, Awad MAE, Abdel-Aziz S, Peters D, Megahed H, Sabik A (2000) Epidemiology of tomato yellow leaf curl begomovirus in the Fayium area, Egypt EPPO Bulletin 30:297–300

Adams DB, Kuhn CW (1977) Seed transmission of *Peanut mottle virus* in peanuts. Phytopathology 67:1126–1129

Adkins S, Webb SE, Baker CA, Kousik CS (2008) Squash vein yellowing virus detection using nested polymerase chain reaction demonstrates that the weed *Momordica charantia* is a reservoir host. Plant Dis 92:1119–1123

Adsuar J (1964) *Solanum nigrum* L. a wild host of *Pepper mosaic virus*. J Agric Univ P.R. 48:352–353

Agindotan B, Perry KL (2008) Macroarray detection of eleven potato-infecting viruses and potato spindle tuber viroid. Plant Dis 92:730–740

Agindotan B, Perry KL (2007) Macroarray detection of plant RNA viruses using randomlyprimed and amplified complementary DNAs from infected plants. Phytopathology 97:119–127

Agrios GN (2005) Plant pathology, 5th edn. Elsevier Academic Press, London, p 922

Ahangaran A, Mohammadi GM, Habibi MK, Khezri S, Shahraeen N (2009) Use of rapid serological and nucleic acid based methods for detecting Soybean mosaic virus. J Agric Sci Technol 11:91–97

Ahlawat YS (2010) Diagnosis of plant viruses and allied pathogens. Studium Press (India) Pvt. Ltd, New Delhi, p 224

Alabi OJ, Lava Kumar P, Naidu RA (2008) Multiplex PCR method for the detection of *African cassava mosaic virus* and *East African cassava mosaic Cameron virus* in cassava. J Virol Methods 154:111–120

Alam Z, Islam A (1959) Biology of the rice leafhopper, *Nephotettix bipunctatus* Fab. in East Pakistan. Pak J Sci Res 2:20–28

Albrechtsen SE (2006) Testing methods for seed transmitted viruses: principles and protocols. CAB1 publishing, Oxfordshire, pp 1–268

Alconero R, Hoch JG (1989) Incidence of pea seed borne mosaic virus pathotypes in the US national Pisum germplasm collection. Ann Appl Biol 114:311–315

Alconero R, Santiago A (1973) *Phaseolus lathyroides* as a reservoir of *Cowpea mosaic virus* in Puerto Rico. Phytopathology 63:120–123

Alegbejo M, Ogunlana M, Banwo O (2008) Survey for incidence of *Okra mosaic virus* in northern Nigeria and evidence for its transmission by beetles. Span J Agric Res 6(3):408–411

Alibi OJ, Ogbe FO, Bandyopadhyay R, Lava Kumar P, Dixon AGO, Hughes JD, Naidu RA (2008) Alternate hosts of *African cassava mosaic virus* and *East African cassava mosaic Cameron virus* in Nigeria. Arch Virol 153:1743–1747

Alicai T, Adipala E, Gibson RW (1999) Seasonal changes in whitefly numbers and their influence on incidence of sweet potato chlorotic stunt virus and sweet potato virus disease in sweet potato in Uganda. Int J Pest Manag 45:51–55

Allen RN (1978) Epidemiological factors influence in the success of roguing for the control of bunchy top disease of banana in New South Wales. Aus J Agric Res 29:534–544

Allen DJ, Anno-Nyako FO, Ochieng RS, Mar-kandu Ratinam (1981) Beetle transmission of cowpea mottle and southern bean mosaic viruses in West Africa. Trop Agric (Trinidad) 58:171–175

Almeida AMR, Yuki W, Val WMC, Harada A, Pola JN, Turkiewsky L (1994) Epidemiological studies on soybean mosaic virus in Brazil. Fitopatologia Brasileira 19:401–407

Alonso Prados JL, Luis Arteaga M, Alvarez JM, Moriones E, Batlle A, Lavina A, Garcia-Arenal F, Fraile A (2003) Epidemics of aphid-transmitted viruses in melon crops in Spain. Eur J Plant Pathol 109:129–138

Ammar ED, Nault LR (2002) Virus transmission of leafhoppers, plant hoppers and tree hoppers (Auchenorrhyncha, Homoptera). Adv Bot Res 36:141–167

Ampofo ST (1997) The current Swollen shoot virus disease situation in Ghana. In: Proceedings of 1st international Cocoa pests and diseases seminar, Accra, Ghana, 1995, pp 175–178

Anderson CW (1959) Vigna and Crotalaria viruses in Florida, Comparative tramission tests with aphids and beetles. Phytopathology 49:117

Anderson PK (2005) Ecological epidemiology: Review, synthesis and application of models for insect transmitted viruses. Paper presented at 9th International Epidemiology Symposium held at Lima, Peru, 3–7, April, 2005

Anderson PK, Morales FJ (2005) Whitefly and whitefly-borne viruses in the tropics: building a knowledge base for global action. pp 158–159

Anderson PK, Cunningham AA, Patel NG, Morales FJ, Epstein PR, Daszak P (2004) Emerging infectious diseases of plants: pathogen pollution, climate change and agrotechnology drivers. Trends Ecol Evol 19:535–544

Anderson KL, Sallam M, Congdon BC (2007) Long distance dispersal by Eumetopina flavipes (Hemiptera: Delphacidae), vector of Ramu stunt: is culture contributing? Proc Aust Soc Sugar Cane Tech 29:226–234

Andret-Link P, Fuchs M (2005) Transmission specificity of plant viruses by vectors. J Plant Pathol 87:153–165

Anjaneyulu A (1975) Rice stubbles and self sown rice seedlings, the reservoir hosts of tungro during off seasons. Sci Cult 41:298–299

Anjaneyulu A, Bhaktavatsalam G, Mohanty SK (1994) Prevention and control of rice tungro disease through new synthetic pyrethroids. In: Rishi N, Ahuja KL, Singh BP (eds) Virology in the tropics. Malhotra Publishing House, New Delhi, pp 660–671

Antignus Y, Lachman O, Pearlsman M (2007) The spread of tomato apical stunt viroid (TAS Vd) in green house tomato crops is associated with seed transmission and bumble bee activity. Plant Dis 91:47–50

Arguello-Astorga G, Ascencio-Ibanez JT, Dallas MB, Orozco BM, Hanley-Bowdoin L (2007) High-frequency reversion of geminivirus replication protein mutants during infection. J Virol 81:11005–11015

Arneson PA (2006) Plant disease epidemiology: temporal aspects. Plant Health Instructor. Online. doi:10.1094/PHI-A-2001-0524-01

Arnold WN, Bald JG (1960) *Turnip mosaic virus* from two weed hosts. Phytopathology 50:578–581

Asanzi CM, Bosque-Perez NA, Buddenhagen IW, Gordon DT, Nault LR (1994) Interactions among Maize streak virus disease, leafhopper vector populations and maize cultivars in forest and savanna zones of Nigeria. Plant Pathol 43:145–157

Ashina S, Tsuruka Y (1968) Records of insects which visited a weather ship located at the ocean weather station 'Tango' on the Pacific II. Kontyu 36:190–202

Astier S, Albouy J, Maury Y, Lecoq H (2001) Principes de Virologie Vegetale. INRA, Paris

Atiri GI (1984) The occurrence of Okra mosaic virus in Nigerian weeds. Ann Appl Biol 104:261–265

Atiri GI, Varma A (1991) Some climatic considerations in the epidemiology and control of two okra viruses in southern Nigeria. In: Proceedings of CTA/IFS seminar on the influence of climate on the production of tropical crops, Burkina Faso, pp 328–335

Atiri GI, Njukeng AP, Ekpo EJA (2000) Climate in relation to plant virus epidemiology and sustainable disease management in West Africa. J Sustain Agric 16:17–30

Attafuah A, Tinsley TW (1958) Virus diseases of *Adansonia digitata* (Bombacaceae) and their relations to cacao in Ghana. Ann Appl Biol 46:20–22

Ausher R, Ben-Zeev IS, Black R (1996) The role of plant clinics in plant disease diagnosis and education in developing countries. Ann Rev Phytopathol 34:51–66

Awan AR, Haq IU, Babar ME, Nasir IA (2010) Molecular detection of potato leaf roll polero virus through reverse transcription polymerase chain reaction in dormant potato tubers. Pak J Bot 42:3299–3306

Azzam O, Chancellor TCB (2002) The biology, epidemiology and management of Rice tungro disease in Asia. Plant Dis 86:88–100

Badami RS, Kassanis B (1959) Some properties of three viruses isolated from a diseased plant of *Solanum jasminoides* Paxt. from India. Ann Appl Biol 47:90–97

Bagherian SAA, AmidMotlagh MH, Izadapanah K (2009) A new sensitive method for detection of viroids. Iran J Virol 3:7–11

Bakker W (1970) Rice yellow mottle, a mechanically transmissible virus disease of rice in Kenya. Neth J Plant Path 76:53–63

Bakker W (1974) Characterization and ecological aspects of *Rice yellow mottle virus* in Kenya. Agric Res Rep, p 829

Bald JG (1937) Investigations on spotted wilt of tomatoes. III. Infection in field plots. Commonw Aust, Council Sci Ind Res Bull 106:32

Ball EM (1974) Serological tests for the identification of plant viruses. Ann Phytopathol Soc Monogr 31:1–7

Banttari EE, Khurana SMP (1998) Serological procedures in plant virology. In: Khurana SMP (ed) Pathological problems of economic crop plants and their management. Scientific Publishers, Jodhpur, pp 603–624

Baranwal VK, Ahlawat YS (2008) Techniques in diagnosis of plant viruses. Studium Press LLC, Houston, pp 141–151

Baranwal VK, Majumder S, Ahlawat YS, Singh RP (2003) Sodium sulphite yields improved DNA of higher stability for PCR detection of Citrus yellow mosaic virus from citrus leaves. J Virol Methods 112:153–156

Barba M, Hadidi A (2007) DNA microarrays: technology, applications, and potential applications for the detection of plant viruses and virus-like pathogens. In: Rao GP, Valverdi RA, Dovas CI (eds) Techniques in diagnosis of plant viruses. Studium Press LLC, Houston, pp 227–247

Barba M, Ragozzino E, Faggioli F (2007) Pollen transmission of peach latent mosaic viroid. J Plant Pathol 89(2):287–289

Barbara DJ, Morton A, Spence NJ, Miller A (1995) Rapid differentiation of closely related isolates of two plant viruses by polymerase chain reaction and restriction fragment length polymorphism analysis. J Virol Methods 55:121–131

Barbarossa L, Savino V (2006) Sensitive and specific digoxigenin-labeled RNA probes for routine detection of Citrus tristeza virus by dot-blot hybridisation. J Phytopathol 154:329–335

Bar-Joseph M, Garnsey SM (1981) Enzyme-linked immunosorbent assay (ELISA): principles and applications for diagnosis of plant viruses. In: Maramoroch K, Harri KF (eds) Plant diseases and vectors, ecology and epidemiology. Academic Press, New York, pp 35–59

Bar-Joseph M, Roistacher CN, Garnsey SM, Gumpf DJ (1981) A review on tristeza, an ongoing threat to citriculture. Proc Int Soc Citric 1:419–423

Bar-Joseph M, Lee RF, Pappu HR (1995) The closteroviruses. In: Singh RP, Singh US, Kohmoto K (eds) Pathogenesis and host-specificity in plant diseases. Pergamon Press, Oxford, pp 65–85

Bass HW, Nagar S, Hanley-Bowdoin L, Robertson D (2000) Chromosome condensation induced by geminivirus infection of mature plant cells. J Cell Sci 113:1149–1160

Beebee T, Rowe G (2004) An introduction to molecular ecology. University Press, Oxford, p 346

Benda GTA (1970) *Sugarcane mosaic virus* from *Arundinaria gigantean*, a bamboo. Plant Dis Rep 54:815–816

Benscher D, Pappu SS, Niblett CL, de Veron Agudelo F, Morales F, Hodson E, Alvarez E, Acosta O, Lee RF (1996) A strain of soybean mosaic virus infecting *Passiflora* spp. in Columbia. Plant Dis 80:255–262

Bercks R, Koenig R, Querfurth G (1972) Plant virus serology. In: Kado CI, Agrawal HO (eds) Principles and techniques in plant virology. Van Nostrand Reinhold, New York, pp 446–490

Berger PH, Ferriss RS (1989) Mechanisms of arthropod transmission of plant viruses: implications for the spread of disease. In: Jeger MJ (ed) Spatial components of plant disease epidemics. Prentice Hall, Englewood Cliffs, pp 40–84

Bernard P, Drogat J, Maure JF, Dheur S, Vaur S, Genier S, Javerzat JP (2006) A screen for cohesion mutants uncovers Ssl3, the fission yeast counterpart of the cohesin loading factor Scc4. Curr Biol 16:875–881

Bertschinger L (1997) Modelling plant virus disease epidemics: development and use of simulation models. International Potato Center (CIP), Lima, Peru

Bertschinger L, Keller ER, Gessler C (1995) Development of EPIVIT, a simulation model for contact- and aphid-transmitted potato viruses. Phytopathology 85:801–814

Bhargava KS (1975) Sugar cane mosaic—retrospect and prospects. Indian Phytopathol 28:1–11

Bhargava KS, Joshi RD, Rishi N (1969) Elephant grass (*Pennisetum purpureum*) a natural host of *Sugarcane mosaic virus*. Sugarcane Path Newsl 3:15

Bhargava B, Bhargava KS, Joshi RD (1975) Perpetuation of *Watermelon mosaic virus* in Eastern Utter Pradesh India. Plant Dis Rep 59:635–636

Bhat AI, Devasahayam S, Sarma YR, Pant RP (2003) Association of a badnavirus in black pepper (*Piper nigrum* L.) transmitted by mealy bug (*Ferrsia virgata*) in India. Curr Sci 84:1547–1550

Blanc S, Uzest M, Drucker M (2011) New research horizons in vector-transmission of plant viruses. Curr Opin Microbiol 14:483–491

Blua MJ, Perring TM (1992) Alatae production and population increase of aphid vectors on virus-infected host plants. Oecologia 92:65–70

Blua MJ, Perring TM, Madore MA (1994) Plant virus-induced changes in aphid population development and temporal fluctuations in plant nutrients. J Chem Ecol 20:691–707

Boateng CO, Schwartz HF (2013) Temporal and localized distribution of *Iris yellow spot virus* within tissues of infected onion plants. Southwestern Entomologist 38(2):183–200

Boccardo G, Lisa V, Milne RG (1983) Cryptic virus in plants. In: Company RW, Bishop DHL (eds) Double-stranded RNA viruses. Elsevier, Amsterdam, pp 425–430

Bock KR (1983) Epidemiology of cassava mosaic disease in Kenya. In: Plumb RT, Thresh JM (eds) Plant virus epidemiology. Blackwell Scientific Publications, Oxford, pp 337–347

Bock CH, Poole GH, Parker PE, Gottwald TR (2010) Plant disease severity estimated visually, by digital photography and image analysis, and by hyperspectral imaging. Crit Rev Plant Sci 29:59–107

Bonants P, De Weerdt M, Van Beckhoven J, Hilhorst R, Chan A, Boender P, Zijlstra C, Schoen C (2002) Multiplex detection of plant pathogens by microarrays: an innovative tool for plant health management. In: Proc Agric Biomarkers Array Technol, Mgt Comm Mt, Wadenswil, p 20

Bonants PJM, Schoen CD, Szemes M, Speksnijder A, Klerks MM, van den Boogert PHJF, Waalwijk C, van der Wolf JM, Zijlstra C (2005) From single to multiplex detection of plant pathogens: pUMA, a new concept of multiplex detection using microarrays. Phytopathol Polonica 35:29–47

Boonham N, Smith P, Walsh K, Tame J, Morris J, Spence N, Benison J, Barker I (2002) The detection of tomato spotted wilt virus (TSWV) in individual thrips vectors using real time fluorescent RT-PCR (TaqMan). J Virol Methods 101:37–48

Boonham N, Walsh K, Smith P, Madagan K, Graham I, Barker I (2003) Detection of potato viruses using micro-array technology: towards a generic method for plant viral disease diagnosis. J Virol Methods 108:181–187

Boonham N, Perez LG, Mendez MS, Peralta EL, Blockley A, Walsh K, Barker I, Mumford RA (2004) Development of a real-time RT-PCR assay for the detection of potato spindle tuber viroid. J Virol Methods 116:139–146

Boonham N, Tomlinson J, Mumford R (2007) Microarrays for rapid identification of plant viruses. Ann Rev Phytopathol 45:307–328

Borah BK, Anthony Johnson AM, Saigopal DVR, Dagupta I (2008) A comparison of four DNA extraction methods for the detection of Citrus yellow mosaic badna virus from two species of citrus using PCR and Dot-Blot hybridization. J Virol Methods 151:321–324

Borah BK, Sai Gopal DVR, Dasgupta I (2009) PCR–RFLP shows high genetic diversity of citrus yellow mosaic virus in southern India. Indian J Virol 20:59–63

Borkhardt B, Vongsasitorn D, Albrechtsen SE (1994) Chemiluminescent detection of potato spindle tuber viroid in true potato seed using a digoxigenin labeled DNA probe. Potato Res 37:249–255

Bos L (1981) Wild plants in the ecology of virus diseases. In Maramorosch K, Harris KF (eds) Plant diseases and vectors: ecology and epidemiology. Academic Press, New York, pp 1–33

Bos L (1992) New plant virus problems in developing countries: a corollary of agricultural moderization. Adv Virus Res 41:349–407

Bosque-Perez NA, Buddenhagen IW (1999) Biology of cicadulina leafhoppers and epidemiology of Maize streak virus disease in West Africa. South Afr J Plant Soil 16:50–55

Bottenberg H, Litsinger JA, Barrion AT, Kenmore PE (1990) Impact of cropping intensity and asynchrony on the epidemiology of rice tungro virus in Malaysia. J Plant Prot Tropics 7:103–116

Bragard C, Caciagli P, Lemaire O, Lopez-Moya JJ, MacFarlane S, Peters D, Susi P, Torrance L (2013) Status and prospects of plant virus control through interference with vector transmission. Annu Rev Phytopathol 51:177–201

Brakke MK (1971a) Wheat streak mosaic virus. Descriptions of plant viruses No 48. Kew, Surrey, UK, CMI/AAB

Brakke MK (1971b) Soil-borne wheat mosaic virus. Descriptions of plant viruses No 77. Kew, Surrey, UK, CMI/AAB

Brault V, Ziegler-Graff V, Richards KE (2001) Viral determinants involved in luteovirus-aphid interactions. In: Harris KF, Smith OP, Duffus JE (eds) Virus-insect-plant interactions. Academic Press, New York, pp 207–232

Broadbent L (1959) Insect vector behaviour and the spread of plant viruses in the field. In: Hotton CS, Fischer GW, Fulton RW, Hant H, McCallan SEA (eds) Pages 539–547 in Plant pathology: problems and progress 1908–1959. University of Wisconsin Press, Madison

Brown JK (2002) The molecular epidemiology of Begomoviruses. pp 279–316. In: Khan JA, Dijkstra J (eds) Plant viruses as molecular pathogens. The Haworth Press, Inc, New York, p 537

Brown PO, Botstein D (1999) Exploring the new world of the genome with DNA microarrays. Nat Genet 21:33–37

Brown DJF, Robertson WM, Trudgill DL (1995) Transmission of viruses by plant nematodes. Ann Rev Phytopathol 33:223–249

Browning JA (1981) The agro-ecosystem—natural ecosystem dichotomy and its impact on phytopathological concepts. In: Thresh JM (ed) Pests, pathogens and vegetation. Pitman, London, pp 159–172

Bruckart WL, Lorbeer JW (1976) Cucumber mosaic virus in weed hosts near commercial fields of lettuce and celery. Phytopathology 66:253–259

Bruehl GW (1961) Barley yellow dwarf. Am Phytopathol Soc Monog 1:1–52

Buddenhagen IW (1983) Crop improvement in relation to virus diseases and their epidemiology. In: Plumb RT, Thresh JM (eds) Plant virus epidemiology. Blackwell, Oxford, pp 25–37

Budzanivska I, Usko L, Gospodaryk A, Melnyk M, Polischuk V (2011) Epidemiology of sharka disease in Ukraine. Acta Hortic (ISHS) 899:57–63

Burges AJ, Harington R, Plumb RT (1999) Barley and cereal yellow dwarf virus epidemiology and control strategies. In: Smith HG, Barker H (eds) The Luteoviridae. CABI, Wallingford, pp 249–279

Byamukama E, Gibson RW, Aritua V, Adipala E (2004) Within-crop spread of sweet potato virus disease and the population dynamics of its whitefly and aphid vectors. Crop Prot 23:109–116

Byrne DN, Bellows TS Jr (1991) Whitefly biology. Ann Rev Entomol 36:431–457

Bystricka D, Lenz O, Mraz I, Dedic P, Síp M (2003) DNA microarray: parallel detection of potato viruses. Acta Virol 47:41–44

Bystricka D, Lenz O, Mraz I, Piherova L, Kmoch S, Sip M (2005) Oligonucleotide-based microarray: a new improvement in microarray detection of plant viruses. J Virol Methods 128:176–182

Cadman CH (1963) Biology of soil-borne viruses. Ann Rev Phytopathol 1:143–172

Cadman CH, Chambers J (1960) Factors affecting the spread of aphid-borne viruses in potato in Eastern Scotland. III. Effects of planting date, rogueing and age of crop on the spread of potato leaf roll and Y viruses. Ann Appl Biol 48:729–738

Calvert L, Thresh JM (2002) The viruses and virus diseases of cassava. In: Hillocks RJ, Thresh JM, Bellotti A (eds) Cassava: biology, production and utilization. CAB International, Wallingford, pp 237–260

Campbell RN (1996) Fungal transmission of plant viruses. Ann Rev Phytopathol 34:87–108

Campbell CL, Madden LV (1990) Introduction to plant disease epidemiology. Wiley, New York, p 532

Campbell CL, Noe JP (1985) The spatial analysis of soil-borne pathogens and root diseases. Annu Rev Phytopathol 23:129–148

Campbell RN, Wipf–Scheibel C, Lecoq H (1996) Vector—associated seed transmission of melon necrotic spot virus in melon. Phytopathology 86:1294–1298

Cannon RJC, Matthews L, Collins DW (2007) A review of the pest status and control options for *Thrips palmi*. Crop Prot 26:1089–1098

Carette JE, Guhl K, Wellink J, Van Kammen A (2002) Coalescence of the sites of cowpea mosaic virus RNA replication into a cytopathic structure. J Virol 76:6235–6243

Carter N (1986) Simulation modelling. In: McLean GD, Garrett RG, Ruesink WG (eds) Plant virus epidemics: monitoring, modelling and predicting outbreaks. Academic Press, Sydney, pp 193–215

Castle SJ, Berger PH (1993) Rates of growth and increase of *Myzus persicae* on virus-infected potatoes according to type of virus-vector relationship. Entomol Exp Appl 69:51–60

Cervera MT, Cabezas JA, Simon B, Martinez Zapater JM, Beitia F, Cenis JL (2000) Genetic relationships among biotypes of *Bemisia tabaci* (*Hemiptera: Aleyrodidae*) based on AFLP analysis. Bull Entomol Res 90:391–396

Chagas CM, Vicente M, Alba APC, July JR (1977) *Solanum atropurpureum* Schrank, a natural reservoir of potato virus Y (PVY). Phytopath Z 90:147–151

Chagas CM, Kitajima EW, Rodrigues JCV (2003) Coffee ringspot virus vectored by *Brevipalpus phoenicis* (Acari: Tenuipalpidae) in coffee. Exp Appl Acarol 30:203–213
Chamberlin JR, Todd JW, Beshear RJ, Culbreath AK, Demski JW (1992) Overwintering hosts and wing forms of thrips (*Frankliniella* spp.) in Georgia: implications for management of spotted wilt disease. Environ Entomol 21:121–128
Chan M-S, Jeger MJ (1994) An analytical model of plant virus disease dynamics with roguing and replanting. J Appl Ecol 31:413–427
Chancellor TCB, Thresh JM (eds) (1997) Epidemiology and management of rice tungro disease. Natural Resources Institute, Chatham
Chancellor TCB, Holt J, Villarreal S, Tiongco ER, Venn J (2006) Spread of plant virus diseases to new plantings: a case study of rice tungro disease. Adv Virus Res 66:1–30
Chandra Mohan DS, Narayana YD, Yelshetty SD (2006) Epidemiology of sunflower necrosis disease in Northern Karnataka. In: XVI annual convention of IVS and international symposium on management of vector-borne viruses, held at ICRISAT, Patancheru, Hyderabad (India). Abstract OP 9/22
Chatzivassiliou EK, Peters D, Katis NI (2007) The role of weeds in the spread of tomato spotted wilt virus by *Thrips tabaci* (Thysanoptera: Thripidae) in tobacco crops. J Phytopathol 155:699–705
Chellaiah S, Murugesan S (1976) Seasonal incidence of *Bhendi yellow vein mosaic*. Annamalai Univ Agric Res Annual 6:167–168
Cheng Y, Jones RAC, Thackray DJ (2002) Deploying strain specific hypersensitive resistance to diminish temporal virus spread. Annals Appl Biol 140:69–79
Chiykowski LN, Chapman RK (1965) Migration of the six spotted leafhopper in central North America. Res Bull Wis Agric Exp Stn 261:21–45
Cho JJ, Mitchell WC, Mau RF, Land Sakimura K (1987) Epidemiology of *Tomato spotted wilt virus* disease on crisphead lettuce in Hawaii. Plant Dis 71:505–508
Chowdhury AK, Mallick SC (2001) Spatial distribution of *Rice tungro virus* on some rice varieties in West Bengal. J Agric Sci Soc Northeast India 14:15–19
Christiansen-Weniger P, Powell G, Hardie J (1998) Plant virus and parasitoid interactions in a shared insect vector/host. Entomol Exp Appl 8:205–213
Christie JR, Perry VG (1951) A root disease of plants caused by a nematode of the genus *Trichodorus*. Science 113:491–493
Chu PWS, Francki RL (1979) The chemical subunit of tobacco ringspot virus protein. Virology 93:398–412
Clark MF, Adams AN (1977) Characteristics of the microplate method of enzyme-linked immunosorbent assay for the detection of plant viruses. J Gen Virol 34:475–483
Clark CA, Derrick KS, Pace CS, Watson B (1986) Survey of wild *Ipomoea* species as potential reservoirs of *Sweetpotato feathery mottle virus* in Louisiana. Plant Dis 70:931–932
Close RC, Tomlinson AI (1975) Dispersal of the grain aphid *Macrosiphum miscanthi* from Australia to New Zealand. N. Z. Entomol 6:62–65
Cock MJW (1986) *Bemisia tabaci*, A literature survey on the cotton whitefly with an annotated bibliography. FAO of the U.N., C.A.B. Int. Inst. Biol. Cont. 121 pages
Cockbain AJ (1969) Aphid studies: overwintering of pea aphids. Rep Rothamsted Exp Stn for 1968, 202
Cohen S, Antignus Y (1994) *Tomato yellow leaf curl virus*—a whitefly-borne geminivirus of tomatoes. Adv Dis Vector Res 10:259–288
Cohen S, Kern J, Harpez I, Ben-Joseph R (1988) Epidemiological studies of the tomato yellow leaf curl virus (TYLCV) in the Jordan Valley, Israel. Phytoparasitica 16:259–270
Colvin J, Fishpool LDC, Fargette D, Sherington J, Fauquet C (1998) *Bemisia tabaci* (Hemiptera: Aleyrodidae) trap catches in a cassava field in Cote d'Ivoire in relation to environmental factors and the distribution of *African cassava mosaic disease*. Bull Entomol Res 88:369–378
Conover RA, Fulton RW (1953) Occurrence of potato Y virus on tomatoes in Florida. Plant Dis Rep 37:460–462

Conti M, Milne RG, Luisonx E, Boccardo G (1980) A closterovirus from a stem-pitting-diseased grapevine. Phytopathology 70:394–398

Cooper I, Jones RAC (2006) Wild plants and viruses: under-investigated ecosystems. Adv Virus Res 67:1–47

Cornwell PB (1958) Movement of vectors of virus disease of cacao in Ghana I. Canopy movement in and between trees. Bull Entomol Res 49:613–630

Cornwell PB (1960) Movements of the vectors of virus diseases of cocoa in Ghana II. Wind movements and aerial dispersal. Bull Entomol Res 51:175–201

Coutts BA, Thomas-Carroll ML, Jones RAC (2004a) Patterns of spread of *Tomato spotted wilt virus* in field crops of lettuce and pepper: spatial dynamics and validation of control measures. Ann Appl Biol 145:231–245

Coutts BA, Thomas-Carroll ML, Jones RAC (2004b) Analysing spatial patterns of spread of *Lettuce necrotic yellows virus* and Lettuce big-vein disease in lettuce field plantings. Ann Appl Biol 145:339–343

Crill P, Hagedorn FJ, Hanson EW (1970) Alfalfa mosaic the disease and its virus incitant. University of Wisconsin Res Bull No 280:39

Cropley R (1968) The identification of plum pox (Sharka) virus in England. Plant Pathol 17:66–70

Crosslin JM, Hamlin LL (2011) Standardized RT-PCR conditions for detection and identification of eleven viruses of potato and potato spindle tuber viroid. Am J Potato Res 88:333–338

Culbreath AK, Todd JW, Brown SL (2003) Epidemiology and management of tomato spotted wilt in peanut. Ann Rev Phytopathol 41:53–75

Czosnek H (2007) Tomato yellow leaf curl virus disease. Springer, New York

Dader B, Moreno A, Vinuela E, Fereres A (2012) Spatio-temporal dynamics of viruses are differentially affected by Parasitoids depending on the mode of transmission. Viruses 4:3069–3089

Dale WT (1962) Diseases and pests of Cocoa, A. Virus diseases. In: Wills JB (ed) Agriculture and land use in Ghana. Oxford University Press, London, pp 286–316

Dallot S, Gottwald T, Labonne G, Quiot J-B (2003) Spatial pattern analysis of sharka disease (*Plum pox virus* strain M) in peach orchards in southern France. Phytopathology 93:1543–1551

Darozhkin MA, Chykava SIG (1974) Da pytannya peredachy X-virusa nasennem bul Oby. (Transmission of virus X by potato seeds). Vestsi Akad Navuk BSSR Biyal Navuk 5:80–85

de Assis Filho FM, Sherwood JL (2000) Evaluation of seed transmission of Turnip yellow mosaic virus and Tobacco mosaic virus in *Arabidopsis thaliana*. Phytopathology 90:1233–1238

de Bokx JA, Vander want JPH (1987) Viruses of potato and sweet potato production, 2nd edn. Wageningen, The Netherlands

Dean GJ, Luuring BB (1970) Distribution of aphids in cereal crops. Ann Appl Biol 66:485–496

Debrot EA, Lastra R, de Vzcategin RC (1977) *Solanum seaforthianum*, a new weed host of *Egg plant mosaic virus* in Venezuela. Plant Dis Rep 61:628–631

Delfosse P (2000) Epidemiology and management of the *Indian peanut clump virus*. Ph.D, Thesis, Faculty des Sciences Agronomiques, Universite Catholique De Louvain. Louvain-la-Neuve, Belgique, pp 199

Delfosse P, Reddy AS, Legreve A, Devi PS, Devi KT, Maraite H, Reddy DVR (1999) Indian peanut clump virus (IPCV) infection on wheat and barley: symptoms, yield loss and transmission through seed. Plant Pathol 48:278–282

Delfosse P, Reddy AS, Tirumala Devi K, Legreve A, Risopoulas J, Doucet D, Devi PS, Maraite H, Reddy DVR (2002) Dynamics of *Polymyxa graminis* and *Indian peanut clump virus* (IPCV) infection on various monocotyledonous crops and groundnut during the rainy season. Plant Pathol 51:546–550

Demangeat G, Komar V, Van-Ghelder C, Voisin R, Lemaire O, Esmen Jaud D, Fuchs M (2010) Transmission competency of single female *Xiphinema index* lines for Grapevine fanleaf virus. Phytopathology 100:384–389

Demski JW, Reddy DVR, Sowell G Jr (1984) Stripe disease of groundnuts. FAO Plant Prot Bull 32:114–115

Derrick RS, Brlansky RH (1976) Assay for viruses and mycoplasmas using serologically specific electron microscopy. Phytopathology 66:815–820

Dias HF (1977) Incidence and geographic distribution of *Tomato ring spot virus* in De Chaunac vineyards in the Niagara Peninsula. Plant Dis Rep 61:24–28

Dieryck B, Weyns J, Doucet D, Bragard C, Legreve A (2011) Acquisition and transmission of *Peanut clump virus* by *Polymyxa graminis* on cereal species. Phytopathology 101:1149–1158

Dietrich CH (2005) Keys to the families of *Cicadomorpha* and subfamilies of cicadellidae (Hemiptera: Auchenorrhyncha). Fla Entomol 88:502–517

Dietzgen RG, Sander E (1982) Monoclonal antibodies against a plant virus. Arch Virol 74:197–204

Dietzgen RG, Xu Z, Teycheney P-Y (1994) Digoxigenin-labeled cRNA probes for the detection of two potyviruses infecting peanut (Arachis hypogaea). Plant Dis 78:708–711

Dietzgen RG, Callaghan B, Higgins CM, Birch RG, Chen K, Xu Z (2001) Differentiation of peanut seed-borne poty viruses and cucumoviruses by RT-PCR. Plant Dis 85:989–992

Domingo E, Holland JJ (1997) RNA virus mutations and fitness for survival. Ann Rev Microbiol 51:151–178

Doncaster JP, Gregory PH (1948) The spread of virus diseases in the potato crop. Agric Res Council Rep Series No. 7, London, H.M.S.O

Dovas CI, Spinthiropoulou H, Stavrakakis N, Katis NI (2003) Sanitary status of wine grape varieties (*Vitis vinifera* L.) in northern Greece originating from clonal selection for preservation. In: Extended abstracts 14th meeting ICVG, Locorotondo, Italy, 13–17 Sept 2003. Department of Plant Protection and Applied Microbiology, University Bari (Italy), pp 169–170

Duffus JE (1964) Host relationships of *Beet western yellows virus* strains. Phytopathology 54:736–738

Duffus JE (1971) Role of weeds in the incidence of virus diseases. Ann Rev Phytopathol 9:319–340

Duffus JE (1973) The yellowing virus diseases of beet. Adv Virus Res 18:347–386

Duffus JE, Liu HY, Wisler GC (1996) Tomato infectious chlorosis virus—a new clostero-like virus transmitted by Trialeurodes vapovariosum. Eur J Plant Pathol 102:591–596

Dusi A (1999) Beet mosaic virus: epidemiology and damage. University of Wageningen, The Netherlands, p 137, ISBN: 9058080757

Dzahini-Obiatey H, Domfeh Owusu, Amoah FM (2010) Over seventy years of a viral disease of cocoa in Ghana: from researchers perspective. Afr J Agric Res 5:476–485

Eckel RVW, Lampert EP (1993) Spatial and temporal analysis of *Tobacco etch virus* distribution and its relationship to aphid (Homoptera: Aphididae) vectors in flue-cured tobacco. J Econ Entomol 86:1534–1545

Egan BT, Ryan CC (1986) Predicting disease incidence and yield losses in sugarcane in a Fiji disease epidemic. In: McLean RG, Garrett RG, Ruesink WD (eds) Plant virus epidemics. Academic Press, Sydney, pp 443–457

Ercolini D, Villani F, Aponte M, Mauriello G (2006) Fluorescence in situ hybridization detection of *Lactobacillus plantarum* group on olives to be used in natural fermentations. Int J Food Microbiol 112:291–296

Escriu F, Fraile A, Garcia-Arenal F (2003) The evolution of virulence in a plant virus. Evolution 57:755–765

Evans AC (1954) Groundnut rosette disease in Tanganyika I. Field studies. Ann Appl Biol 41:189–206

Fajinmi AA, Odebode CA, Fajinmi OB (2011) Incidence and distribution of aphid vectors of *Pepper veinal mottle virus* on bell pepper. Int J Entomol 2:11–22

Falk BW, Tsai JH (1998) Biology and molecular biology of viruses in the genus Tenuivirus. Ann Plant Pathol 36:139–163

Fang S, Yu J, Feng J, Han C, Li D, Liu Y (2001) Identification of rice black-streaked dwarf fijivirus in maize with rough dwarf disease in China. Arch Virol 146:167–170

Fargette D, Fauquet C, Thouvenel C Jr (1985) Field studies on the spread of *African cassava mosaic*. Ann Appl Biol 106:285–294

Fargette D, Muniyappa V, Fauquet C, Guessan PN, Thouvenal C (1993) Comparative epidemiology of three tropical white-fly transmitted *Geminiviruses*. Biochimie 75:547–554

Fargette D, Fauquet C, Thresh JM (1994a) Analysis and modelling of the temporal spread of *African cassava mosaic virus* and implications for disease control. Afr Crop Sci J 2:449–458

Fargette D, Jeger M, Fauquet C, Fishpool L (1994b) Analysis and modelling of the temporal spread of *African cassava mosaic virus*. Phytopathology 84:91–98

Fargette D, Konate G, Fauquet C, Muller E, Peterschmitt M, Thresh JM (2006) Molecular ecology and emergence of tropical plant viruses. Ann Rev Phytopathol 44:235–260

Fauquet CM, Fargette D (1990) *African cassava mosaic virus*: etilogy, epidemiology, and control. Plant Dis 74:404–411

Fauquet CM, Nawaz-ul-Rehman MS (2008) Emerging gemini viruses In: Desk encyclopedia of plant and fungal virology. Academic Press, New York, pp 404–411

Fauquet CM, Fargette D, Thouvenel JC (1988) Some aspects of the epidemiology of African cassava mosaic geminivirus in Ivory Coast. Trop Pest Manag 34:92–96

Fauquet CM, Maxwell DP, Gronenborn B, Stanley J (2000) Revised proposal for naming geminiviruses. Arch Virol 145:1743–1761

Fauquet CM, Bisaro DM, Briddon RW, Brown J, Harrison BD, Rybicki EP, Stenger DC, Stanley J (2003) Revision of taxonomic criteria for species demarcation in the family *Geminiviridae*, and an updated list of *begomovirus* species. Arch Virol 148:405–421

Fauquet CM, Mayo MA, Maniloff J, Desselberger U, Ball IA (2005) Virus taxonomy. In: Eighth report of the International committee on taxonomy of viruses, Elsevier, San Diego

Feldman JM, Gracia O (1970) Two new natural hosts of alfalfa mosaic virus. Plant Dis Rep 54:722–723

Fereres A, Thresh JM, Irwin ME (2000) Plant virus epidemiology: challenges for the 21st century. Virus Res (Special issue): 1–267

Fernow KH, Peterson LC, Plaisted RL (1970) Spindle tuber virus in seeds and pollen of infected potato plants. Am Potato J 47:75–80

Ferriss RS, Berger PH (1993) A stochastic simulation models of epidemics of arthropod-vectored plant viruses. Phytopathology 83:1269–1278

Fessehaie ASH, Boer D, Levesque CA (2003) An oligonucleotide array for the identification and differentiation of bacteria pathogenic on potato. Phytopathology 93:262–269

Fishman S, Marcus R, Talpaz H, Bar-Joseph M, Oren Y, Salomon R, Zohar M (1983) Epidemiological and economic models for spread and control of citrus tristeza virus disease. Phytoparasitica 11:39–49

Fleisher SJ (2002) GIS and GPS systems in pest control. In: Pimentel D (ed) Encyclopedia of pest management, Marcel Dekker Inc, New York, pp. 338–340

Floresca ET, Nishimoto RK, Ishii M (1978) Susceptibility of Emilia species and types to *Tomato spotted wilt virus*. Plant Dis Rep 62:40–43

Fondong VN, Thresh JM, Zok S (2002) Spatial and temporal spread of *Cassava mosaic virus* disease in cassava grown alone and when intercropped with maize and/or cowpea. J Phytopathol 150:365–374

Forbes AR (1977) The mouth parts and feeding mechanism of aphids. In: Harris KF, Maramorosch K (eds) Aphids as virus vectors. Academic Press, London, pp 83–104

Francki RIB, Kitajima EW, Peters D (1981) Tomato spotted wilt virus. In: Kurstak E (ed) Handbook of plant virus infections. Elsevier/North-Holland, London, pp 455–490

Francl LJ (2001) The disease triangle: a plant pathological paradigm revisited. The Plant Health Instructor. doi:10.1094/PHI-T-2001-0517-01

Franke-Whittle IH, Klammer SH, Insam H (2005) Design and application of an oligonucleotide microarray for the investigation of compost microbial communities. J Microbiol Methods 62:37–56

Frazer BD (1977) Plant virus epidemiology and computer simulation of aphid populations. In: Harris KF, Maramorosch K (eds) Aphids as virus vectors. Academic Press, New York, pp 413–431

Freeland J (2005) Molecular ecology. New York J Wiley, p 400

Frison EA, Putter CAJ (1989) FAO/IBPGR technical guidelines for the safe movement of germplasm, food and agriculture organisation of the United Nations/International Board for Plant Genetic Resources, Rome

Fritzsche R, Karl E, Lehmann W, Proesler G (1972) Tierische Vektoren Pflanzen Pathogener viren. Veb Gustav. Fischer Verlag, Jena, p 521

Froissart R, Michalakis Y, Blanc S (2002) Helper component-transcomplementation in the vector transmission of plant viruses. Phytopathology 92:576–579

Fukuta S, Iida T, Mizukami Y, Ishida A, Ueda J, Kanbe M, Ishimoto Y (2003a) Detection of Japanese yam mosaic virus by RT-LAMP. Arch Virol 148:1713–1720

Fukuta S, Kato S, Yoshida K, Mizukami Y, Ishida A, Ueda J, Kanbe M, Ishimoto Y (2003b) Detection of tomato yellow leaf curl virus by loop-mediated isothermal amplification reaction. J Virol Methods 112(1–2):35–40

Fulton JP, Scott HA, Gamez R (1980) Beetles. In: Harris KF, Maramorosch K (eds) Vectors of plant pathogens. Academic Press, New York, pp 115–132

Fulton JP, Gergerich RC, Scott HA (1987) Beetle transmission of plant viruses. Ann Rev Phytopathol 25:111–123

Gallitelli D (2004) Nucleic acid-based assay for the diagnosis of viral pathogens. Phytopathol Mediterr 43:221–227

Gamarra HA, Fuentes S, Morales FJ, Glover R, Malumphy C, Barker I (2010) Bemisia afer sensu lato, a vector of sweet potato chlorotic stunt virus. Plant Dis 94:510–514

Garcia-Arenal F, Fraile A, Malpica JM (2003) Variation and evolution of plant virus populations. Int Microbiol 6:225–232

Garnsey SM, Permar TA, Cambra M, Henderson CT (1993) Direct tissue blot immunoassay (DTBIA) for detection of citrus tristeza virus (CTV). In: Moreno P, Da Graca J, Timmer L (eds) Proceedings of the 12th conference on international organization of citrus virologists. IOCV, Riverside, pp 39–50

Garnsey SM, Behe CG, Lockhart BEL (1998) Transmission of citrus yellow mosaic badnavirus by the citrus mealybug. Phytopathology 9. Supplement Abstract S43

Gergerich RC (2001) Mechanism of virus transmission by leaf feeding beetles. In: Harris KF, Smith OP, Duffus JE (eds) Virus-insect-plant interactions. Academic Press, New York, pp 133–142

Gergerich RC, Scott HA, Fulton JP (1983) Regurgitant as a determinant of specificity in the transmission of plant viruses by beetles. Phytopathology 73:936–938

German TL, Ullman DE, Moyer JW (1992) Tospo viruses: diagnosis, molecular biology, phylogeny and vector relationships. Ann Rev Phytopathol 30:315–348

Ghanim M, Morin S, Zeidan M, Czosnek H (1998) Evidence for transovarial transmission of *Tomato yellow leaf curl virus* by its vector, the whitefly *Bemisia tabaci*. Virology 240:295–303

Ghanim M, Brumin M, Popovski S (2009) A simple, rapid and inexpensive method for localization of *Tomato yellow leaf curl virus* and *Potato leafroll virus* in plant and insect vectors. J Virol Methods 159:311–314

Ghosh D, Das AK (2012) Duplex PCR-based assay for simultaneous detection of tristeza and greening diseases in citrus. Indian Phytopathol 65:328–333

Ghosh DK, Aglave B, Baranwal VK (2008) Simultaneous detection of one RNA and one DNA virus from naturally infected citrus plants using duplex PCR technique. Curr Sci 94:1314–1318

Gibson RW, Mpembe I, Alicai T, Carey EE, Mwanga ROM, Seal SE, Vetten HJ (1998) Symptoms, aetiology and serological analysis of sweet potato virus disease in Uganda. Plant Pathol 47:95–102

Gilbertson RL, Rojas MR, Russell D, Maxwell DP (1991) The use of the asymmetric polymerase chain reaction and DNA sequencing to determine genetic variability among isolates of bean golden mosaic geminiviruses in the Dominican Republic. J Gen Virol 72:2843–2848

Gilbertson RL, Hidayat SH, Paplomatas EJ, Rojas MR, Hou Y-M, Maxwell DP (1993) Pseudorecombination between infectious cloned DNA components of tomato mottle and bean dwarf mosaic geminiviruses. J Gen Virol 74:23–31

Gildow FE (1983) Influence of *Barley yellow dwarf virus*-infected oats and barley on morphology of aphid vectors. Phytopathology 73:1196–1199

Gillaspie AG Jr, Pio-Ribeiro G, Andrade GP, Pappu HR (2001) RT-PCR detection of seed-borne cowpea aphid-borne mosaic virus in peanut. Plant Dis 85:1181–1182

Gil-Salas FM, Peters J, Boonham N, Cuadrado IM, Janssen D (2011) Yellowing disease in zucchini squash produced by mixed infections of *Cucurbit yellow stunting disorder virus* and *Cucumber vein yellowing virus*. Phytopathology 101:1365–1372

Gitaitis RD, Dowler CC, Chalfant RB (1998) Epidemiology of tomato spotted wilt in pepper and tomato in Southern Georgia. Plant Dis 82:752–756

Givord L (1978) Alternative hosts of *Okra mosaic virus* near plantings of okra in southern Ivory Coast. Plant Dis Rep 62:412–416

Goldbach R, de Haan P, Berduin BJM (1992) Gene sequences in viruses, viroids and prokaryotes, and the development of diagnostic tools. In: Moss JP (ed) Biotechnology and crop improvement in Asia. ICRISAT, Hyderabad, pp 265–279

Goldberg KB, Brakke MK (1987) Concentration of *maize chlorotic mosaic virus* increased in mixed infections with *maize dwarf mosaic virus* strain B. Phytopathology 77:162–167

Gooding GV Jr, Young JB, Lapp NA (1975) Physalis virginiana, a perennial weed host for *Potato virus Y* in North Carolina. Proc Am Phytopathol Soc 2:134 (Abstract)

Goodman RM, Oard JH (1980) Seed transmission and yield losses in tropical soybeans infected by soybean mosaic virus. Plant Dis 64:913–914

Gottwald TR, Avinent L, Llacer G, Hermoso de Mendoza A, Cambra M (1995) Analysis of the spatial spread of sharka (plum pox virus) in apricot and peach orchards in eastern Spain. Plant Dis 79:266–278

Gottwald TR, Villalobos W, Rivera C (2002) Comparative epidemiology of citrus tristeza virus in plantings of various citrus species in costa rica, and long distance spread by the brown citrus aphid. In: 15th IOCV conference, Citrus Tristeza Virus, pp 102–115

Gracia O, Feldman JM (1972) Studies of weed plants as sources of virus 1. *Datura ferox* as a source of *Beet curly top virus*. Phytopathol Z 73:69–74

Gregory PH (1968) Interpreting plant disease dispersal gradients. Ann Rev Phytopathol 6:189–212

Groves, RL, Walgenbach, JF, Moyer, JF, Kennedy, GG (2002) The role of weed hosts and tobacco thrips, *Frankliniella fusca* in the epidemiology of *Tomato spotted wilt virus*. Plant Disease 86:573–582

Guirao P, Beitia F, Cenis JL (1997) Biotype determination in Spanish populations of *Bemisia tabaci* (Hemiptera: Aleyrodidae). Bull Entomol Res 87:587–593

Gupta KN, Keshwal RL (2003) Epidemiological studies on yellow mosaic disease of soybean. Ann Pl Proc Sci 11:324–328

Gupta KN, Buddi P, Arya M, Pant RP, Gopal K, Baranwal VK (2007) Simultaneous detection of Citrus mosaic virus and Indian citrus ringspot virus by duplex polymerase chain reaction. In: Abstracts of the papers presented at the international conference of Indian Virological Society held at IARI, New Delhi, India, 11–14 Dec 2007

Gutierrez AP, Havenstein DE, Nix HA, Moore PA (1974a) The ecology of *Aphis craccivora* Kosch and *Subterranean clover stunt virus* in south-east Australia. II. A model of cowpea aphid populations in temperate pastures. J Appl Ecol 11:1–20

Gutierrez AP, Nix HA, Havenstein DE, Moore PA (1974b) The ecology of *Aphis craccivora* Koch and *Subterranean clover stunt virus* in south-east Austgralia. III. A regional perspective of the phenology and migration of the cowpea aphid. J Appl Ecol 11:21–35

Hadidi A, Barba M (2008) DNA microarrays: 21st century pathogen detection. Acta Hortic 781:331–339

Hadidi A, Yang X (1990) Detection of pome fruit viroids by enzymatic cDNA amplification. J Virol Methods 30:261–270

Hadidi A, Hansen AJ, Parish CL, Yang X (1991) Scar skin and dapple apple viroids are seed borne and persistant in infected apple trees. Res Virol 142:289–296

Hadidi A, Terai Y, Powell CA, Scott SW, Desvignes JC, Ibrahim LM, Levy L (1992) Enzymatic cDNA amplification of hop stunt viroid variants from naturally infected fruit crops. Acta Hortic 309:339–344

Hadidi A, Davis I, Minafra A, Boscia D, Postman JD (1995) A virus affecting apples and pears is related to grapevine virus as revealed by IC-RT-PCR and molecular hybridization. Acta Hortic 386:37–44

Hadidi A, Giunchedi L, Shamloul AM, Poggi-Pollini C, Amer MA (1997) Occurrence of peach latent mosaic viroid in stone fruits and its transmission with contaminated blades. Plant Dis 81:154–158

Hadidi A, Czosnek H, Barba M (2004) DNA microarrays and their potential applications for the detection of plant viruses, viroids and phytoplasmas. J Plant Pathol 86:97–104

Haider MS, Tahir M, Latif S, Briddon RW (2006) First report of *Tomato leaf curl New Delhi virus* infecting *Eclipta prostrata* in Pakistan. Plant Pathol 55:285

Halbert SE, Irwin ME (1981) Effect of soybean canopy closure on landing rates of aphids with implications for restricting spread of *Soybean mosaic virus*. Ann Appl Biol 98:15–19

Halbert SE, Irwin ME, Goodman RM (1981) A late aphid Homoptera: aphididae species and their relative importance as field vectors of *Soybean mosaic virus*. Ann Appl Biol 97:1–9

Halk EL, De Boer SH (1985) Monoclonal antibodies in plant-disease research. Ann Rev Phytopathol 23:321–350

Hampton RO, Braverman SW (1979) Occurrence of pea seed borne mosaic virus in North American pea breed ing lines, and new virus-immune germplasm in the plant introduction collection of *Pisum sativum*. Plant Dis Rep 79:631–633

Hampton RO, Waterworth H, Goodman RM, Lee R (1982) Importance of seed-borne viruses in crop germplasm. Plant Dis 66:977–978

Hampton R, Ball E, De Boer S (1990) Serological methods for detection and identification of viral and bacterial plant pathogens. APS Press, St. Paul 389 pp

Hardtl H (1978) Pollen as carrier of virus. Gartanban Wissenschaft 43:34–47

Harpaz I (1972) Maize rough dwarf. Israel University Press, Jerusalem

Harper K, Creamer R (1995) Hybridization detection of insect-transmitted plant viruses with digoxigenin-labeled probes. Plant Dis 79:563–567

Harper G, Ganesh D, Thottappilly G, Hull R (1999) Detection of episomal banana streak badna virus by IC-PCR. Virol Methods 79:1–8

Harrison BD (1977) Ecology and control of viruses with soil-inhabiting vectors. Annu Rev Phytopathol 15:331–360

Harrison BD (1981) Plant virus ecology: ingredients, interactions and environmental influences. Ann Appl Biol 99:195–209

Harrison BD (2002) Virus variation in relation to resistance—breaking in plants. Euphytica 124:181–192

Harrison BD, Robinson DJ (1999) Natural genomic and antigenic variation in whitefly transmitted geminiviruses (*Begomoviruses*). Ann Rev Phytopathol 37:369–398

Harrison BD, Finch JT, Gibbs AJ, Hollings M, Shepherd RJ, Valenta V, Wetter C (1971) Sixteen groups of plant viruses. Virology 45:356–363

Hartleb H, Bauer E (1977) The significance of field weeds for the overwintering of the *Beet mild yellowing virus* in the German Democratic Republic from 1972 to 1976. Arch Phytopathologic Pflanzen Schutz 13:153–162

Hassan M, Rysanek P, Alioto D, Malfitano M (2008) Identification and molecular characterization of peach latent mosaic viroid isolate infecting peach in Egypt. Acta Hortic 781:551–556

Head J, Morgan D (1996) Computer models in plant health campaign management: their use in *Bemisia tabaci* eradiction. In: Proceedings of British crop protection conference, pests and diseases, 18–21 November 1996 Brighton, U.K. pp 1047–1052

Heathcote GD (1972) Influence of cultural factors on incidence of aphids and yellows in beet. I.I.R.B. 6:6–14

Hema M, Kirthi N, Sreenivasulu P, Savithri HS (2003) Development of recombinant coat protein antibody based IC-RT-PCR for detection and discremination of sugarcane streak mosaic virus isolates from Southern India. Arch Virol 148:1185–1193

Hernandez L, Ramos PL, Rodriguez M, Pena I, Perez JM (2010) First report of pineapple mealybug wilt associated virus-3 infecting pineapple in Cuba. New Dis Rep 22:18

Herranz MC, Sanchez-Navarro JA, Aparicio F, Pallas V (2005) Simultaneous detection of six stone fruit viruses by non-isotopic molecular hybridization using a unique riboprobe or 'polyprobe'. J Virol Methods 124:49–55

Heslop-Harrison JS, Schwarzacher T (2002) In situ hybridization to plant chromosomes and DNA fibres. In: Gilmartin PM, Bowler C (eds) Molecular plant biology: vol 1. A practical approach. Oxford University Press, Oxford, pp 159–178

Hewitt WB, Chiarappa L (1977) Plant health and quarantine in international transfer of genetic resources. CRC Press, Cleveland

Hewitt WB, Goheen AC, Raski DJ, Gooding GV (1962) Studies on virus diseases in the grapevine in California. Vitis 3:57–83

Hibino H (1996) Biology and epidemiology of rice viruses. Ann Rev Phytopathol 34:249–274

Hodgson C (1991) Dispersal of apterous aphids (Homoptera: Aphididae) from their host plant and its significance. Bull Entomolog Res 81:417–427

Hogenhout SA, Ammar ED, Whitefield AE, Redinbaugh MG (2008) Insect vector transmissions with persistently transmitted viruses. Ann Rev Phytopathol 46:327–359

Hollings M, Stone OM (1975) Serological and immunoelectrophoretic relationships among viruses in the tombusvirus group. Ann Appl Biol 80:37–48

Holmes FO (1939) Handbook of phytopathogenic viruses. Burgess, Minneapolis

Holt J, Chancellor TCB (1996) Simulation modeling of the spread of *Rice tungro virus* disease: the potential for management by rogueing. J Appl Ecol 33:927–936

Holt J, Chancellor TCB (1997) A model of plant virus disease epidemics in asynchronously-planted cropping systems. Plant Pathol 46:490–501

Holt JK, Colvin J (2001) Observation and theory of whitefly-borne virus disease epidemics. In: Jeger MJ, Spence NJ (eds) Biotic interactions in plant–pathogen interactions. CAB International, Wallingford, pp 331–343

Holt J, Jeger MJ, Thresh JM, Otim-Nape GW (1997) An epidemiological model incorporating vector population dynamics applied to *African cassava mosaic virus* disease. J Appl Ecol 34:793–806

Holt J, Colvin J, Muniyappa V (1999) Identifying control strategies for *Tomato leaf curl virus* disease using an epidemiological model. J Appl Ecol 36:625–633

Hooftman R, Arts MJ, Shamloul AM, Van Zaayen A, Hadidi A (2001) Detection of chrysanthemum stunt viroid by reverse transcription-polymerase chain reaction and by tissueblot hybridization. Acta Hortic 432:88–94

Hosseini S, Mosahebi GH, Habibi MK, Okhovvat SM (2007) Characterization of the Zucchini yellow mosaic virus from squash in Tehran Province. J Agric Sci Technol 9:137–143

Howell WE, Mink GI (1971) The relationship between volunteer sugarbeets and occurrence of beet mosaic and beet western yellow viruses in Washington beet fields. Plant Dis Rep 55:676–678

Howell WE, Mink GI (1977) The role of weed hosts, volunteer carrots, and overlapping growing seasons in the epidemiology of *Carrot thin leaf* and *Carrot motley dwarf viruses* in central Washington. Plant Dis Rep 61:217–222

Hu JS, Li HP, Barry K, Wang M (1995) Comparison of dot blot, ELISA and RT-PCR assays for the detection of two cucumber mosaic virus isolates infecting banana in Hawaii. Plant Dis 79:902–906

Hull R (1969) The spray warning scheme for the control of virus yellows. Br Sugar Beet Rev 37:169–172

Hull R (1993) Nucleic acid hybridization procedures. In: Matthews REF (ed) Diagnosis of plant virus diseases. CRC Press, Boca Raton, pp 253–272

Hull R (2002) Matthews' plant virology, 4th edn. Academic Press, San Diego

Hunter WB, Hiebert E, Webb SE, Tsai JH, Polston JE (1998) Location of geminiviruses in the whitefly *Bemisia tabaci* (Homoptera: Aleyrodidae). Plant Dis 82:1147–1151

Hurtt SS, Podleckis EV (1995) Apple scar skin viroid is not seed transmitted or transmitted at a low rate in oriental pear. Acta Hortic 386:544–547

Ioannau N, Iordanou N (1985) Epidemiology of *Tomato yellow leafcurl virus* in relation to the population density of its whitefly vector (*Bemisia tabaci* Genn.). Tech Bull 71. Agricultural Research Institute, Nicosia, p 7

Irwin ME, Goodman RM (1981) Ecology and control of *Soybean mosaic virus* In: Maramorosch K and Harris KF (eds) Plant diseases and vectors: ecology epidemiology. Academic Press, New York. pp 181–220

Irwin ME, Kampmeier GE (1989) Vector behavior, environmental stimuli, and the dynamics of plant virus epidemics. In: Jeger MJ (ed) Spatial components of plant disease epidemics. Prentice-Hall, Inglewood Cliffs, pp 14–39

Irwin ME, Thresh JM (1990) Epidemiology of barley yellow dwarf virus: a study in ecological complexity. Ann Rev Phytopathol 28:393–424

Irwin ME, Ruesink WG, Isard SA, Kampmeier GE (2000) Mitigating epidemics caused by non-persistently transmitted aphid-borne viruses: the role of the pliant environment. Virus Res 71:185–211

Ito T, Ieki H, Ozaki K (2002) Simultaneous detection of six citrus viroids and apple stem grooving virus from citrus plants by multiplex reverse transcription polymerase chain reaction. J Virol Methods 106:235–239

Ivars P, Alonso M, Borja M, Hernandez C (2004) Development of a non-radioactive dot-blot hybridisation assay for the detection of Pelargonium flower break virus and Pelargonium line pattern virus. Eur J Plant Pathol 110:275–283

Jain RK, Pappu SS, Pappu HR, Culbreath AK, Todd JW (1998) Molecular diagnosis of tomato spotted wilt tospo virus infection of peanut and other field and greenhouse crops. Plant Dis 82:900–904

Jain RK, Vemana K, Bag S (2008) *Tobacco streak virus* an emerging virus in vegetable crops. In: Rao GP, Kumar PL, Holuguin-Pena RJ (eds) Characterization, diagnosis and management of plant viruses Vol. 3: vegetable and pulse crops. Studium Press LLC, Texas, pp 203–212

James D, Jelkmann W, Upton C (1999) Specific detection of Cherry mottle leaf virus using digoxigenin-labeled cDNA probes and RT-PCR. Plant Dis 83:235–239

James D, Varga A, Pallas V, Candresse T (2006) Strategies for simultaneous detection of multiple plant viruses. Can J Plant Pathol 28:16–29

Jeger MJ (1986) The potential of an analytic compared with simulation approaches to modelling in plant disease epidemiology. In: Leonard KJ, Fry WE (eds) Plant disease epidemiology, vol 1. Macmillan, New York, p 372

Jeger MJ (2004) Analysis of disease progress as a basis for evaluating disease management practices. Ann Rev Phytopatol 42:61–82

Jeger MJ (2009) Epidemiology of plant disease. doi:0.1002/9780470015902.a0021268

Jeger MJ, Chan MS (1995) Theoretical aspects of epidemics: uses of analytical models to make strategic management decisions. Can J Plant Pathol 17:109–114

Jeger MJ, van den Bosch F, Madden LV, Holt J (1998) A model for analyzing plant-virus transmission characteristics and the epidemic development. J Math Appl Med Biol 15:1–18

Jeger MJ, Dutmer MY, Van Den Bosch F (2002) Modelling plant virus epidemics in a combined field-nursery system. IMA J Math Appl Med Biol 19:1–20

Jeger MJ, Holt J, Van Den Bosch F, Madden LV (2004) Epidemiology of insect-transmitted plant viruses: modelling disease dynamics and control interventions. Physiol Entomol 29:291–304

Jeger MJ, Chen Z, Powell G, Hodge S, van den Bosch F (2011) Interactions in a host plant-virus-vector-parasitoid system: modelling the consequences for virus transmission and disease dynamics. Virus Res 159(2):183–193

Jenkins GM, Rambaut A, Pybus OG, Holmes EC, Rambaut A, Pybus OG, Holmes EC, Pybus OG, Holmes EC, Holmes EC (2002) Rates of molecular evolution in RNA viruses: a quantitative phylogenetic analysis. J Mol Evol 54:156–165

Jeremiah SC, Maruthi MN, Ijumba JN, Legg JP (2013) Field epidemiology of cassava brown streak viruses in Tanzania. In: 12th International symposium on plant virus epidemiology. 28th Jan–1st Feb 2013, Arusha Tanzania. (Abstract OP53, p 78)

Jiang GW, Tan HQ, Shen WZ, Cheng X-N, Chen RC (1981) The relationship between long distance northward migration of brown planthopper (*Nilaparvata lugens*) and synoptic weather conditions. Acta Entomol Scinica 24:251–261

Johansen E, Edwards MC, Hampton RO (1994) Seed transmission of viruses: current perspectives. Ann Rev Phytopathol 32:363–386

Johansen IE, Keller KE, Dougherty WG, Hampton RO (1996) Biological and molecular properties of a pathotype P-1 and a pathotype P-4 isolate of *Pea seed-borne mosaic virus.* J Gen Virol 77:1329–1333

John VT, Rao RDVJP (1971) Factors favouring the incidence of tungro and methods of control. Oryza 8:365–369

Jones RAC (1981) The ecology of viruses infecting wild and cultivated potatoes in the Andean region of South America. In: Thresh JM (ed) Pests, pathogens and vegetation. Pitman, London, pp 89–107

Jones DR (2003) Plant viruses transmitted by whiteflies. Eur J Plant Pathol 109:195–219

Jones RAC (2004) Using epidemiological information to develop effective integrated virus disease management strategies. Virus Res 100:5–30

Jones DR (2005a) Plant viruses transmitted by thrips. Eur J Plant Pathol 113:119–157

Jones RAC (2005b) Patterns of spread of two non-persistently aphid-borne viruses in lupin stands under four different infection scenarios. Ann Appl Biol 146:337–350

Jones RAC (2009) Plant virus emergence and evolution: origins, new encounter scenarios, factors driving emergence, effects of changing world conditions, and prospects for control. Virus Res 141:113–130

Jones RAC (2013) Influence of climate change on plant virus disease infections and epidemics. In: 12th international symposium on plant virus epidemiology. 28th Jan–1st Feb 2013, Arusha, Tanzania. (Abstract OP5, p 30)

Jones RAC, Barbetti MJ (2012) Influence of climate change on plant disease infections and epidemics caused by viruses and bacteria. CAB Rev 7(22):1–32 (On-line Publication)

Jones RAC, Torrance L (1985) Developments and applications in virus testing. Assoc Appl Biol (AAB), Wellesbourne, pp 1–300

Jones RAC, Coutts BA, Mackie AE, Dwyer GI (2005) Seed transmission of wheat streak mosaic virus shown unequivocally in wheat. Plant Dis 89:1048–1050

Jones RAC, Coutts BA, Latham LJ, McKirdy SJ (2008) Cucumber mosaic virus infection of chickpea stands:Temporal and spatial patterns of spread and yield-limiting potential. Plant Pathol 57:842–853

Jones RAC, Salam MU, Maling TJ, Arthur JD, Thackray DJ (2010) Principles of predicting plant virus disease epidemics. Ann Rev Phytopathol 48:179–203

Joshi RD, Dubey LN (1975) Some weed reservoirs of *Cucumber mosaic virus* in Gorakpur and adjacent areas. Indian Phytopathol 28:568

Joshi RD, Suteri BD, Dubey LN (1977) Desmodium trifolium mottle—A virus disease. Indian J Mycol Pl Pathol 7:189

Jovel J, Kleinn C, Hilje L, Ramfrez P (2000) Distribution of spatio-temporal spread of yellow mottle virus (ToYMoV) in tomato plots, in Turrialba, Costa Rica. Manejo Integrado de Plagas 57:35–44

Kaczmarek U (1976) Preliminary results of investigation of weeds as sources of infection of potato fields by viruses. Zesz Probl Postepow Nauk Roln 174:109–115

Kahn RP (1989) Plant protection and quarantine: selected pests and pathogens of quarantine significance, vol II. CRC Press Inc, Florida 265 p

Kahn RP, Mathur SB (1999) Containment facilities and safeguards for exotic plant pathogens and pests. The American Phytopathological Society, St. Paul

Kalita MK, Dhawan P, Rishi N (2005) Influence of weather on whitefly population and incidence of *Okra yellow vein mosaic* and *Okra leaf curl begomo-virus* diseases. J Mycol Pl Pathol 35:114–116

Kampmeijer P, Zadoks JC (1977) EPIMUL a simulator of foci and epidemics in mixtures, multilines and mosaics of resistant and susceptible plants. PUDOC, Wageningen. The Netherlands, pp 50

Karyeija RF, Gibson RW, Valkonen JPT (1998) Resistance to *Sweetpotato virus disease* (SPVD) in wild East African Ipomoea. Ann Appl Biol 133:39–44

Katis N, Tsitsipis JA, Avgelis A, Gargalianou J, Papanayotou A, Milla S (1998) Aphid populations and Potato virus Y potyvirus (PVY) spread in potato fields. In: Aphids in Natural and Managed Ecosystems pp 585–593

Katis NI, Tsitsipis JA, Lykouressis DP, Papapanayotou A, Margaritopoulos JT, Kokinis GM, Perdikis DC, Manoussopoulos IN (2006) Transmission of *Zucchini yellow mosaic virus* by colonizing and non-colonizing aphids in Greece and new aphid species vectors of the virus. J Phytopathol 154:293–302

Kato-Maeda M, Gao Q, Small PM (2001) Microarray analysis of pathogens and their interaction with hosts. Technorev Cell Microb 3:713–719

Katono K, Alicai T, Bagume Y, Edema R, Bua A, Omongo CA (2013) Spread of cassava brown streak disease in Uganda as influenced by host-resistance and prevailing disease pressure. In: 12th international symposium on plant virus epidemiology, 28th Jan–1st Feb 2013, Arusha, Tanzania. (Abstract, pp 74, p 175)

Kemper A (1962) Zum anfttreten *Von Salat mosaik virus* an salt (Lactuca sativa L.) und an Krenz krant (*Senecio vulgaris* L.). Z pfl kranckh pfl schutz 69:653–663

Kenchaiah RPM, Porte BS (1989) Some observations on population fluctuation of insect pests of groundnut in Karnataka. Plant Prot Bull 41:7–10

Kennedy JS, Day MF, Eastop VP (1962) A conspectus of aphids as vectors of plant viruses. Commonwealth Institute of Entomology, London

Khan JA, Dijkstra J (2002) Seed transmission of viruses: biological and molecular insights. In: Khan JA, Dijkstra J (eds) Plant viruses and molecular pathogens. Haworth Press, New York, pp 105–126

Khan MA, Mukhopadhyay S (1985) Effect of different pesticide combinations on the incidence of yellow vein mosaic virus disease of okra (Abelmoschus esculentus) and its whitefly vector *Bemisia tabaci* Genn. Indian J Virol 1:147–151

Khan S, Timmermann C, Hoque MI, Sarker RH, Muhlbach HP (2009) Detection of potato spindle tuber viroid (PSTVd) in minute amounts of potato (Solanum tuberosum L.) leaf tissue by hybridization techniques and, together with potato viruses, by multiplex RT-PCR. J Plant Dis Prot 116:97–105

Khatab Eman AH, Zein Salwa N, Ahmed Amal A (2012) Purification, serology and prevalence of Broad bean true mosaic comovirus (BBTMV). Int J Virol 8:224–233

Khetarpal RK (1989) Contribution A L'etude des relations pea seed-borne mosaic virus-Pois Universite De Paris Sud. Centre D'orsay, Paris

Khurana SMP (1970) *Trianthema monogyna* and *Boerhaavia diffusa* unrecorded hosts of brinjal mosaic disease. Plant Dis Rep 54:437–438

Khurana SMP, Garg ID (1993) New techniques for detection of viruses and viroids. In: Chadha KL, Grewal JS (eds) Advances in horticulture: potato, vol 7. Malhotra Publishing House, New Delhi, pp 529–566

Kim K-S, Hill CB, Hartman GL, Mian MAR, Diers BW (2008) Discovery of soybean aphid biotypes. Crop Sci 48:923–928

King AMQ, Adams MJ, Carstens EB, Lefkowitz EJ (2012) Virus taxonomy: nineth report of the international committee on taxonomy of viruses. Elsevier Academic Press, San Diego

Kiranmai G, Sreenivasulu P, Nayudu MV (1998) Epidemiology of cucumber mosaic cucumovirus isolates naturally infecting three solanaceous vegetable crops around Tirupathi. Indian Phytopathol 51:315–318

Kirchmair M, Neuhauser S, Scholz C, Huber L (2005) *Sorosphaera viticola* nom. prov., a newly discovered plasmodiophorid, a potential vector for grapevine viruses? In: Rush C (ed) Proceedings of IWGPVFV 6th symposium on American Society of Sugar Beet Technologists, Denver, pp 108–111

Kiritani K, Sasaba T (1978) An experimental validation of the systems model for the prediction of *Rice dwarf virus* infection. Appl Entomol Zool 13:209–214

Kisimoto R, Yamada Y (1986) A planthopper—rice virus epidemiology model: rice stripe and small brown planthopper, Laodelphax striatellus Fallén. In: McLean GD, Garrett RG, Ruesink WG (eds) Plant virus epidemics: monitoring, modelling and predicting outbreaks. Academic Press, Sydney, pp 327–344

Klerks MM, Leone G, Lindner JL, Schoen CD, van den Heuvel JFJM (2001) Rapid and sensitive detection of apple stem pitting virus in apple trees through RNA amplification and probing with fluorescent molecular beacons. Phytopathology 91:1085–1091

Koenig R (1986) Plant viruses in rivers and lakes. Adv Virus Res 31:321–333

Kokkinos CD, Clark CA (2006a) Interactions among sweet potato chlorotic stunt virus and different potyviruses and potyvirus strains infecting sweet potato in the United States. Plant Dis 90:1347–1352

Kokkinos CD, Clark CA (2006b) Real-time PCR assays for detection and quantification of sweet potato viruses. Plant Dis 90:783–788

Kokkinos CD, Clark CA, McGregor CE, LaBonte DR (2006) The effect of sweet potato virus disease and its viral components on gene expression levels in sweet potato. J Am Soc Hortic Sci 131:657–666

Komm DA, Agrios GN (1978) Incidence and epidemiology of viruses affecting cucurbit crops in Massachusetts. Plant Dis Rep 62:746–750

Konate G, Traore O (1992) Reservoir hosts of *Maize streak virus* (MSV) in the sudan-sahel zone: Identification and spatio-temporal distribution. Phytoprotection 73:111–117

Kondaiah A, John VT (1977) Recovery of rice Tungro virus from dried infected rice leaves. Phytopathol Z 90:311–314

Kondaiah A, Rao AV, Srinivasan TE (1976) Factors favoring spread of rice "tungro" disease under field conditions. Plant Dis Rep 60(9):803–806

Kontzog HG, Kleinhempel H, Kegler H (1988) Detection of plant pathogenic viruses in waters. Arch Phytopathol Pflanzenschutz 24:171–172

Kranz J (1990) Epidemics their mathematical analysis and modeling. An introduction. In: Epidemics of plant diseases: mathematical analysis and modeling. 2nd edn. Springer, Berlin

Kreuze JF, Karyeija RF, Gibson RW, Valkonen JPT (2000) Comparisons of coat protein gene sequences show that East African isolates of sweet potato feathery mottle virus form a genetically distinct group. Arch Virol 145:567–574

Krishnaveni D, Laha GS, Shankar Chitra, Katti G, Murali Dharan K (2011) Virus-vector composition analysis in *Rice tungro virus* epidemicsin Telangana region of Andhra Pradesh. India J Mycol Plant Pathol 41:441–444

Kryczynski S, Paduch-Cichal E, Skrzeclzkowski LJ (1988) Transmission of three viroids through seed and pollen of tomato plants. J Phytopathol 121:51–57

Kuan C-P, Wu M-T, Lu Y-L, Huang H-C (2010) Rapid detection of squash leaf curl virus by loop-mediated isothermal amplification. J Virol Methods 169:61–65

Kuhn CW, Dawson WO (1973) Multiplication and pathogenesis of *Cowpea chlorotic mottle virus* and *Southern bean mosaic virus* in single and double infections in cowpea. Phytopathology 63:1380–1385

Kulkarni NK, Kumar PL, Muniyappa V, Jones AT, Reddy DVR (2002) Transmission of Pigeon pea sterility mosaic virus by the Eriophyid mite, *Aceria cajani* (Acari: Arthropoda). Plant Dis 86:1297–1302

Kumar NKK, Kalleshwaraswamy CM, Ranganath HR, Kashinath BL (2008b) Role of aphis gossypii glover (aphididae: homoptera) on temporal and spatial spread of *Papaya ringspot virus*-W on pumpkin (*cucurbita moschata* L.). Pest Management in Horticultural Ecosystems 14:44–50

Kumar PL, Prasada Rao RDVJ, Reddy AS, Madhavi KJ, Anitha K, Waliyar F (2008a) Emergence and spread of *Tobacco streak virus* menace in India and control strategies. Indian J Plant Prot 36:1–8

Kumar PL, Sharma K (2010) DNA barcodes for pathogens of African food crops. R4D Rev 4:51–53

Kviscala BA (1975) Some natural weed hosts of *Alfalfa mosaic virus*. Zentralblatt fur Bakteriologie Parasitenkunde infecktionskran—Kheitan and Hygiene 130:704–708

Laird EF Jr, Dickson RC (1972) *Turnip mosaic virus*—vector relationships in field grown static, Limonium perizii. Plant Dis Rep 56:722–725

Lakhmod LK, Suryawanshi AP, Wadje AG, Magar SA, Totawad RJ, Ghube NB (2007) Epidemiology of *Sunflower necrosis disease*. J Pl Dis Sci 2:115–116

Lamberti F, Roca F (1987) Present status of nematodes as vectors of plant viruses. In: Veech JA, Dickson DW (eds) Vistas on nematology. 25th Anniv. Publ J Nematol, pp 321–328

Lange L (1986) The practical application of new developments in test procedures for the detection of viruses in seed. In: Jones RAC, Torrance L (eds) Developments and applications in virus testing. Lavenham Press, Sadbury, pp 269–281

Lareu M, Phillips C, Torres M, Brion M, Carracedo A (2003) Typing Y-chromosome single nucleotide polymorphisms with DNA microarray technology. Int Congr Ser 1239:21–25

Latham LJ, Jones RAC, McKirdy SJ (2004) Lettuce big-vein disease: sources, patterns of spread and losses. Australian J Exper Agri 55:125–130

Leam Khang S, Riangwong L, Chatchawankanphanich O (2005) Detection of tomato yellow leaf curl Thailand virus by PCR without DNA extraction. Mol Biotechnol 31:233–238

Lecoustre R, Fargette D, Fauquet C, de Reffye P (1989) Analysis and mapping of the spatial spread of *African cassava mosaic virus* using geostatistics and the kriging technique. Phytopathology 79:913–920

Lee GP, Min BE, Kim CS, Choi SH, Harn CH, Kim SU, Ryu KH (2003) Plant virus cDNA chip hybridization for detection and differentiation of four cucurbit-infecting Tobamoviruses. J Virol Methods 110:19–24

Lee SH, Lee DG, Woo HS, Lee KW, Kim DH (2006) Production of transgenic orchardgrass via Agrobacterium-mediated transformation of seed-derived callus tissues. Plant Sci 171:408–414

Legarrea S, Fraile A, Garcia-Arenal, F, Vinuela E, Fereres A (2009) Spatio-temporal dynamics of aphid-transmitted viruses in lettuce crops under UV-absorbing nets. IOBC/wprs Bulletin 49:253–258

Legg JP, James B, Cudjoe A, Saizonous S, Gvaguidi B, Ogbe F, Ntonifor N, Ogwal S, Thresh J, Hughes J (1997) A regional collaborative approach to the study of ACMD epidemiology in sub-saharan Africa. In: Adipala E, Tenywa JS, Ogenga-latijo NW (eds) Proceedings of African crop science conference, Pretoria, 13–17 Jan 1997. pp 1021–1033

Lesemann DE, Paul HL (1980) Conditions for the use of protein-A in combination with the Derrick method of immuno-electron microscopy. Acta Hortic 110:119–128

Lesemann DE, Bozarth RF, Koenig R (1980) The trapping of tymovirus particles on electron microscope grids by adsorption and serological binding. J Gen Virol 48:257–264

Levy L, Hadidi A, Garnsey SM (1992) Reverse transcription polymerase chain reaction assays for the rapid detection of citrus viroids using multiplex primer sets. Proc Int Soc Citric 2:800–803

Lim GS (1969) The bionomics and control of *Nephotettix impicticeps* Ishihara and transmission studies on its associated viruses in West Malaysia. Malays Min Agr Coop Bull 121:1–62

Lima JAA, Nelson MR (1977) Etiology and epidemiology of cowpea in ceara. Brazil Plant Dis Rep 61:864–867

Ling KC (1972) Rice virus diseases. International Rice Research Institute, Los Bajios, 142 pp

Ling KC, Palomar MK (1966) Studies on the rice plants infected with the tungro virus at different arias. Philipp Agriculturist 50:165–177

Liu Y, Cady NC, Batt CA (2007) A plastic microchip for nucleic acid purification. Biomed Microdevices 9:769–776

Liu F, Feng L, Chen X, Han Y, Li W, Xu W, Cai B, Lin M (2012) Simultaneous detection of four banana viruses by multiplex PCR. J Phytopathol 160:622–627

Lockhart BEL, Jones DR (1999) Banana streak, in diseases of banana. In: Jones (ed) Abaca and enset. CABI Publishing, Wallingford, pp 263–274

Lockhart BEL, Autrey LJC, Comstock JC (1992) Partial purification and serology of sugarcane mild mosaic virus, a mealybug transmitted clostero like virus. Phytopathology 82:691–695

Lockhart DJ, Dong H, Byrne MC, Follettie MT, Gallo MV, Chee MS, Mittmann M, Wang C, Kobayashi M, Horton H, Brown EL (1996) Expression monitoring by hybridisation to highdensity oligonucleotide arrays. Nat Biotechnol 14:1675–1680

Loconsole G, Fatone MT, Savino V (2009) Specific digoxigenin-labelled riboprobes for detection of citrus psorosis virus and citrus variegation virus by molecular hybridization. J Plant Pathol 91(2):311–319

Lokesh BK, Nagaraju Jagadish KS, Shadakshari YG (2008) Effects of different dates of sowing and weather parameter on *Sunflower necrosis virus disease* and its thrips vector. Int J Agric Sci 4:576–579

Lopez MM, Bertolini E, Marco-Noales E, Llop P, Cambra M (2006) Update on molecular tools for detection of plant pathogenic bacteria and viruses. In: Fleming CC, Moore JE (eds) Molecular diagnostics: current. Horizon Bioscience, Wymondham, pp 1–46

Louie R (1968) Epidemiology of *Onion yellow dwarf*. Ph.D. thesis Cornell University, Ithaca, pp 112

Louws FJ, Rademaker JLW, Brujin FJ (1999) The three Ds of PCR-based genomic analysis of phytobacteria: diversity, detection, and disease diagnosis. Ann Rev Phytopathol 37:81–125

Loy A, Lehner A, Lee N, Adamczyk J, Meier H, Ernst J, Schleifer KH, Wagner M (2002) Oligonucleotide microarray for 16S rRNA genebased detection of all recognized lineages of sulfate-reducing prokaryotes in the environment. Appl Environ Microbiol 68:5064–5081

Loy A, Schulz C, Lucker S, Schopfer-Wendels A, Stoecker K, Baranyi C, Lehner A, Wagner M (2005) 16S rRNA gene-based oligonucleotide microarray for environmental monitoring of the betaproteobacterial order ''Rhodocyclales''. Appl Environ Microbiol 71:1373–1386

Lwoff A, Horne R, Tournier P (1962) A system of viruses. Cold Spring Harb Symp Quant Biol 27:51–55

Macfarlane SA (2003) Molecular determinants of the transmission of plant viruses by nematodes. Mol Plant Pathol 4:211–215

Madden LV (2006) Botanical epidemiology: some key advances and its continuing role in disease management. Eur J Plant Pathol 115:3–23

Madden LV, Campbell CL (1986) Descriptions of virus disease epidemics in time and space. In: Plant virus epidemics; monitoring, modelling and predicting outbreaks. In: McLean GD, Garrett RG, Ruesink WG (eds) Academic Press, Sydney, pp 550

Madden LV, Knoke JK, Louie R (1983) Classification and prediction of maize dwarf mosaic intensity. In: Gordon DT, Knoke JK, Nault LR, Ritter RM (eds) Proceedings of international maize virus disease colloquium and workshop, 2–6 Aug 1982. Ohio Agricultural Research and Development Centre, Wooster, pp 238–242

Madden LV, Louie R, Knoke JK (1987) Temporal and spatial analysis of maize dwarf mosaic epidemics. Phytopathology 77:148–156

Madden LV, Knoke JK, Louie R (1990a) Spread of *Maize chlorotic dwarf virus* in maize fields by its leafhopper vector, *Graminella nigrifrons*. Phytopathology 80:291–298

Madden LV, Nault LR, Murral DJ, Apelt MR (1995) Spatial pattern analysis of incidence of aster yellows disease in lettuce. Research on Population Ecology 37:279–289

Madden LV, Jeger MJ, van den Bosch F (2000) A theoretical assessment of the effect of vector virus transmission mechanism on plant virus disease epidemics. Phytopathology 90:576–594

Madden LV, Hughes G, van den Bosch F (2007) The study of plant disease epidemics. APS Press, St. Paul MN pp 421

Madden LV, Raccah B, Pirone TP (1990b) Modeling plant disease increase as a function of vector numbers: Nonpersistent viruses. Res Popul Ecol 32:47–65

Magenya OEV, Mueke J, Omwega C (2008) Significance and transmission of *Maize streak virus* disease in Africa and options for management: a review. Afr J Biotechnol 71:4897–4910

Magyarosy AC, Duffus JE (1977) The occurrence of highly virulent strains of the *Beet curley top virus* in California. Plant Dis Rep 61:248–251

Makesh Kumar T, Sankar Anoop, Nair RR, Edison S (2005) Detection of Cassava mosaic virus in India: using polymerase chain reaction and nucleic acid hybridization technique. J Root Crops 31:1–6

Makkouk KM, Comeau A (1994) Evaluation of various methods for the detection of barley yellow dwarf virus by the tissue blot immunoassay and its use for virus detection in cereals inoculated at different growth stages. Eur J Plant Pathol 100:71–80

Makkouk KM, Kumari SG (1996) Detection of ten viruses by the tissue-blot immunoassay (TBIA). Arab J Plant Prot 14:3–9

Makkouk K, Kumari S (2006) Molecular diagnosis of plant viruses. Arab J Plant Prot 24:135–138

Makkouk KM, Kumari SG (2009) Epidemiology and integrated management of persistently transmitted aphid borne virus of legume and cereal crops in West Asia and North America. Virus Res 141:209–218

Malathi VG, Usharani KS, Sivalingam PN, Rouhibakhsh A, Padma Latha KV, Periasamy M (2004) Diversity and complexity of begomoviruses. Ann Rev Plant Pathol 3:225–270

Mandahar CL (1981) Virus transmission through seed and pollen. In: Harris HF, Maramorsch K (eds) Plant diseases and vectors. Academic Press, New York, pp 241–292

Mandahar CL (1985) Vertical and horizontal spread of plant viruses through seed and pollen: an epidemiological view. In: Gupta BM, Singh BT, Nerma HN, Srivastava KM (eds) Perspectives in plant virology, vol I. Print House (India), Lucknow, pp 23–44

Mandal B, Jain RK, Krishnareddy M, Krishna Kumar NK, Ravi KS, Pappu HR (2012) Emerging problems of Tospoviruses (Bunyaviridae) and their management in the Indian subcontinent. Plant Dis 96:468–479

Mansoor S, Amin I, Iram S, Hussain M, Zafar Y, Malik KA, Briddon RW (2003) Breakdown of resistance in cotton leaf curl disease in Pakistan. Plant Pathol 52:784

Maramorosch K, Harris KF (eds) (1979) Leafhopper vectors and plant disease agents. Academic Press, New York, p 654

Maramorosch K, Harris KF (1981) Plant diseases and vectors: ecology and epidemiology. Academic Press, London, p 368

Marcus R, Raccah B (1986) A model for spread of non-persistent virus diseases. J Appl Stat 13:167–175

Marcus R, Fishman S, Talpaz H, Salomon R, Bar-Joseph M (1984) On the spatial distribution of citrus tristeza virus disease. Phytoparasitica 12:45–52

Mariappan V, Narayanaswamy P (1977a) Characterisation of viruses affecting weeds-1. Mosaic diseases. Madras Agric J 64:106–112

Mariappan V, Narayanaswamy P (1977b) Characterization of viruses affecting weeds II. leaf curl disease. Madras Agric J 64:740–744

Markham R, Smith KM (1949) Studies on the virus of turnip yellow mosaic. Parasitology 39:330–342

Mas P, Sanchez-Navarro JA, Sanchez-Pina MV, Pallas V (1993) Chemiluminescent and colorigenic detection of cherry leaf roll virus with digoxigenin-labeled RNA probes. J Virol Methods 45:93–102

Mason G, Caciagli P, Accotto AC, Noris E (2007) Real time PCR for the quantification of tomato yellow leaf curl Sardinia virus in tomato plants and in Bemisia tabaci. J Virol Methods 147:282–289

Maule AJ, Hull R, Donson J (1983) The application of spot hybridization to the detection of DNA and RNA viruses in plant tissues. J Virol Methods 6:214–225

Maynard Smith J (1982) Evolution and the theory of games. Cambridge University Press, Cambridge

Mazyad HM, Omar F, Al-Taher K, Salha M (1979) Observations on the epidemiology of *Tomato yellow leaf curl disease* on tomato plants. Plant Dis Rep 63:695–698

McLean GD, Garrett RG, Ruesink WG (eds) (1986) Plant virus epidemics. Academic Press, Sidney

McMechan AJ (2012) Transmission of Triticum mosaic virus and its impact on the biology of the wheat curl mite *Aceria tosichella* Keifer (Eriophyidae), and an evaluation of management tactics for the wheat curl mite and the wheat-mite-virus complex. Dissertation and student research in entomology. Paper 16. http://www.digitalcommons.unl.edu/entomologydiss/16

McNamara DG, Flegg JJM (1981) The distribution of virus-vector nematodes in Great Britain in relation to past and present natural vegetation. In: Thresh JM (ed) Pests, pathogens and vegetation. Pitman, London, pp 225–235

Mehle N, Ravnikar M (2012) Plant viruses in aquaseous environment-survival water mediated transmission and detection. Water Research 46:4902–4917

Meier M, Olspert A, Sarmiento C, Truve E (2008) Sobemovirus. In: Mahy BWJ, Van Regenmortel MHV (eds) Desk encyclopedia of plant and fungal virology. Academic Press, Washington, pp 312–320

Mersha Z, Hau B (2008) Effect of bean rust (*Uromyces appendiculatus*) epidemics of host dynamics of common beans (*Phaseolus bulgaris*). Plant Pathol 57:674–686

Mesfin T, Bosque-Perez NA, Buddenhagen IW, Thottappilly G, Olojede SO (1992) Studies of *Maize streak virus* isolates from grass and cereal hosts in Nigeria. Plant Dis 76:789–795

Milne RG (1980) Some observations and experiments on immunosorbent electron microscopy of plant viruses. Acta Hortic 110:129–135

Milne RG, Lesemann OE (1978) An immuno-electron microscopic investigation of oat sterile dwarf and related viruses. Virology 90:299–304

Milne RG, Luisoni E (1977) Rapid immune electron microscopy of virus preparations. In: Maramorosch K, Kaprowski H (eds) Methods in virology, vol VI. Academic, New York, pp 265–281

Mink GI (1983) The possible role of honeybees in long distance spread of prunus necrotic ringspot virus from California into Washington sweet cherry orchards. In: Plumb RT, Thresh JM (eds) Plant virus epidemiology. Black Well Scientific Publications, Oxford, pp 85–91

Mink GI (1993) Pollen and seed-transmitted viruses and viroids. Annu Rev Phytopathol 31:375–402

Miyai S (1991) Modelling the effects of aphid vector (Aulacorthum solani) movement on the spread of soybean dwarf virus. In: Proceedings of international seminar on migration and dispersal of agricultural insects. National Institute of Agro-Environmental Sciences, Tsukuba, Japan, pp 145–154

Miyai S (1993) Simulation approach to the control of insect-borne virus diseases. Tech Bull Food Fertil Technol Center 137:1–7

Miyai S, Hokyo N (1992) Modelling approach to simulate the progress of insect-borne rice virus diseases in paddy fields. In: Shiyomi M, Yano E, Koizumi H, Andow DA, Hokyo N (eds) Ecological processes in agro-ecosystems. National Institute of Agro-Environmental Sciences, Tsukuba, Japan, pp 139–154

Miyai S, Kiritani K, Nakasuji F (1986) Models of epidemics of rice dwarf. In: McLean GD, Garrett RG, Ruesink WG (eds) Plant virus epidemics: monitoring, modelling and predicting outbreaks. Academic Press, Sydney, pp 459–480

Moghal SM, Francki RIB (1976) Towards a system of identification and classification of potyviruses. Virology 73:350–362

Moore BJ, Scott HA, Walters HJ (1969) *Desmodium paniculatum*, a perennial host of *Bean pod mottle virus* in nature. Plant Dis Rep 53:154–155

Mora-Aguilera G, Nieto-Angel D, Teliz D, Campbell CL (1993) Development of a prediction model for papaya ringspot in Veracruz, Mexico. Plant Dis 77:1205–1211

Morales FJ (2004) Common bean. In: Loebenstein G, Thottappilly G (eds) Virus and virus-like diseases of major crops in developing countries. Kluwer Academic Publishers, New York, pp 425–446

Morales FJ (2006) History and current distribution of the *Begomoviruses* in Latin America. Adv Virus Res 67:128–163

Morales FJ (2007) Tropical whitefly IPM project. Adv Virus Res 69:249–306

Morales FJ, Anderson PK (2001) The emergence and dissemination of whitefly-transmitted geminiviruses in Latin America. Arch Virol 146(3):415–441

Morales FJ, Jennings PR (2010) Rice hoja blanca: a complex plant virus vector pathosystem perspectives in agriculture. Vet Sci, Nutr Nat Resour 5:1–16

Morales FJ, Jones PG (2004) The ecology and epidemiology of whitefly-transmitted viruses in Latin America. Virus Res 100:57–65

Morales FJ, Araya CM, Hernandez JC, Arroyave JA, Cuervo M, Velasco AC, Castano M (1999) Etiologia del "amachamiento" del frijol comun en Costa Rica. Manejo Integrado de Plagas (CATIE, Costa Rica) 52:42–48

Moreno A, Bertolini E, Olmos A, Cambra M, Fereres A (2007a) Estimation of vector propensity for *Lettuce mosaic virus* based on viral detection in single aphids. Span J Agric Res 5:376–384

Moreno A, Nebreda M, Diaz BM, Garcia M, Salas F, Fereres A (2007b) Temporal and spatial spread of lettuce mosaic virus in lettuce crops in central Spain: factors involved in Lettuce mosaic virus epidemics. Ann Appl Biol 150: 351–360

Morgan D (1990) Simulation model of *Barley yellow dwarf virus* epidemiology. World perspectives on Barley yellow dwarf. In: Burnett PA (ed) Proceedings of international workshop, 6–11 July 1987, Udine, Italy, pp 300–304

Moriones E, Garcia-Andres S (2008) Diagnosis of begomo viruses. In: Rao GP, Valverde RA, Dovas CI (eds) Techniques in diagnosis of plant viruses. Studium Press LLC, Houston, pp 165–187

Moriones E, Navas-Castillo J (2000) Tomato yellow leaf curl virus, an emerging virus complex causing epidemics worldwide. Virus Res 71:123–134

Mowry TM (1994) Russian wheat aphid (Homoptera, Aphididae) survival and fecundity on *Barley yellow dwarf virus*-infected wheat resistant and susceptible to the aphid. Environ Entomol 23:326–330

Mukasa SB, Rubaihayo PR, Valkonen JPT (2003) Incidence of viruses and virus like diseases of sweet potato in Uganda. Plant Dis 87:329–335

Mukasa SB, Rubaihayo PR, Valkonen JPT (2006) Interactions between a crinivirus, an ipomovirus and a potyvirus in co-infected sweet potato plants. Plant Pathol 55:458–467

Mukhopadhyay S (1980) Ecology of Nephotettix spp. And its relation with *Rice tungro virus*. Final report "Ecology of Indian Council of Agricultural Research, New Delhi"

Mukhopadhyay S (1986) Ecosystem analysis of rice green leafhoppers and *Rice tungro virus* epidemiology in West Bengal, India. Paper presented at *Rice Tungro Virus* technical meeting. 24–27 Sept, Maros, Indonesia, p 33

Mukhopadhyay S, Chowdhury AK (1973) Some epidemiological aspects of tungro virus disease of rice in West Bengal. Int Rice Commn Newsl 22:44–57

Mulholland V (2009) Immunocapture-PCR for plant virus detection. Plant Pathol Methods Mol Biol 508:183–192

Mumford RA, Barker I, Wood KR (1996) An improved method for the detection of tospoviruses using the polymerase chain reaction. J Virol Methods 57:109–115

Mumford RA, Walsh K, Boonham N (2000) A comparison of molecular methods for the routine detection of viroids. EPPO Bull 30:431–435

Muniyappa V (1980) Viral diseases transmitted by whiteflies. In: Harris KF, Maramorosch K (eds) Vectors of plant pathogens. Academic Press, New York, pp 39–85

Muniyappa V (1983) Epidemiology of yellow mosaic disease of horsegram (*Macrotyla uniflorum*) in southern India. In: Plumb RT, Thresh JM (eds) Plant virus epidemiology. Blackwell, London, pp 331–335

Muralidharan K, Krishnaveni D, Neeraja CN, Prasad GSV (2003) Reflections on the complexity of rice tungro virus disease. Ann Rev Plant Pathol 2:379–436

Muramatsu Y (1979) A system dynamic model for epidemic of rice stripe virus. Bull Shizuoka Agric Exp Station 24:1–13 (In Japanese)

Murant AF (1981) Nepoviruses. In: Kurstak E (ed) Hand book of plant virus infections and comparative diagnosis. Elsevier/North Holland, Amsterdam, pp 197–238

Murant AF, Lister RM (1967) Seed-transmission in the ecology of nematode borne viruses. Ann Appl Biol 59:63–76

Murphy JF, Bowen KL (2006) Synergistic disease in pepper caused by the mixed infection of Cucumber mosaic virus and pepper mottle virus. Phytopathology 96:240–247

Nagamine K, Hase T, Notomi T (2002) Accelerated reaction by loop-mediated isothermal amplification using loop primers. Mol Cell Probes 16:223–229

Nagaraju, Jagadeesh BN, Jayaramaiah H, Channakrishnaiah KM (2003) Effect of seasonal variations on th appearance of sunflower necrosis virus disease in Bangalore. Plant Dis Res 18:83–84

Naidu RA (2013) Epidemiology and management of intractable virus diseases in subsistence agriculture: the case of a thrips—transmitted tospovirus in South Asia. In: 12th international symposium on plant virus epidemiology, 28th Jan–1st Feb 2013, Arusha, Tanzania. (Abstract OP27, p 52)

Naidu RA, Hughes JDA (2003) Methods for the detection of plant viral diseases in plant virology in Sub-Saharan Africa. In: Hughes JDA, Odu B (eds) Proc plant virology. I.I.T.A, Ibadan, pp 233–260

Naidu RA, Bottenberg H, Subrahmanyam P, Kimmins FM, Robinson DJ, Thresh JM (1998) Epidemiology of groundnut rosette virus disease: current status and future research needs. Ann Appl Biol 132:525–548

Naidu RA, Poojari S, Alabi OJ, Karthikeyan G, Damayanti T, Kodetham G, Lavakumar P (2013) Use of FTA® Classic Card technology for building epidemiologic intelligence of plant viruses. In: 12th international symposium on plant virus epidemiology, 28th Jan–1st Feb 2013, Arusha, Tanzania. (Abstract, pp 100, p 201)

Nakasuji F, Kiritani K (1972) Descriptive models for the system of the natural spread of infection of rice dwarf virus (RDV) by the green rice leafhopper, Nephotettix cincticeps Uhler (Hemiptera: Deltocephalidae). Res Popul Ecol 14:18–35

Nakasuji F, Miyai S, Kawamoto H, Kiritani K (1985) Mathematical epidemiology of rice dwarf virus transmitted by green rice leafhoppers: a differential equation model. J Appl Ecol 22:839–847

Narayana YD, Muniyappa V (1996) Virus-vector relationships of a planthopper (*Peregrinus maidis*)—borne sorghum stripe tenuivirus. Intl J Pest Manag 42:165–170

Narayanasamy P (2011) Microbial pathogens-detection and disease diagnosis: viral and viroid pathogens, vol 3. Springer, Dordrecht, pp 321. ISBN: 978-90-481-9753-8

Nault LR (1997) Arthropod transmission of plant viruses: a new synthesis. Ann Entomol Soc Am 90:521–541

Navas-Castillo J, Sanchez-Campos, Noris E, Louro D, Accotto GP, Moriones E (2000) Natural recombination between *Tomato yellow leaf curl virus*-Is and *Tomato leaf curl virus*. J Gen Virol 81:2797–2801

Navot N, Zeidan M, Pichersky E, Zamir D, Czosnek H (1992) Use of polymerase chain reaction to amplify tomato yellow leaf curl virus DNA from infected plants and viruliferous whiteflies. Phytopathology 82:1199–1202

Nayudu MV (2008) Plant viruses. Tata McGraw-Hill Publishing Company Limited, New Delhi. pp 1249

NCS-TCP (2008) National certification system for tissue culture raised plants: standard operating procedures for recognized tissue culture production facility. Department of Biotechnology, Government of India

Ndunguru J, Tairo F, Gwandu C, Sseruwagi P, Kayeke J (2013) A tree shrub in the family *Sapindaceae* is a natural host of a *Cassava mosaic begomovirus* in Tanzania. In: Presented at 12th international symposium plant virus epidemiology, held at Arusha, Tanzania. (Abstract)

Nebreda M, Moreno A, Perez N, Palacios I, Seco-Fernandez V, Fereres A (2004) Activity of aphids associated with lettuce and broccoli in Spain and their efficiency as vectors of *Lettuce mosaic virus*. Virus Research 100:83–88

Nelson MR, Wheeler RE (1972) A new virus disease of pepper in Arizona. Plant Dis Rep 56:731–735

Nelson MR, Felix-Gastelum R, Orum TV, Stowell LJ, Myers DE (1994) Geographic information systems and geostatistics in the design and validation of regional plant virus management programs. Phytopathology 84:898–905

Nelson MR, Orum TV, Jaime-Garcia R, Nadeem A (1999) Applications of geographic information systems and geostatistics in plant disease epidemiology and management. Plant Disease 83:4

Nemecek T (1993) The role of aphid behavior in the epidemiology of potato virus Y: a simulation study. Ph.D. thesis, Swiss Federal Institute of Technology Ziirick, Obersiggental, Switzerland, 232 pp

Ng JCK, Perry KL (2004) Transmission of plant viruses by aphid vectors. Mol Plant Pathol 5:505–511

Nicolaieff A, Van Regenmortel MHV (1980) Detection of rota virus by serological trapping on antibody-coated electron microscopic grids. J Clin Microbiol 12:101–104

Nie X, Singh RP (2001) A novel usage of random primers for multiplex RT–PCR detection of virus and viroids in aphids, leaves and tubers. J Virol Methods 91:37–49

Nome CF, Nome SF, Guzman F, Conci LR, Laguna IG (2007) Localization of *Sweet potato chlorotic stunt virus* (SPCSV) in synergic infection with potyviruses in sweet potato. Biocell: official journal of the Sociedades Latinoamericanas de Microscopia Electronica 31:23–31

Notomi T, Okayama H, Masubuchi H, Yonekawa T, Watanabe K, Amino N, Hase T (2000) Loop-mediated isothermal amplification of DNA. Nucleic Acids Res 28:E63

Nutman FJ, Roberts FM, Williamson JG (1964) Studies on varietal resistance in the groundnut (*Arachis hypogaea* L.) to rosette disease. Rhod J Agric Res 2:63–77

Nutter FW (1997) Quantifying the temporal dynamics of plant virus epidemics: a review. Crop Prot 16:603–618

Nutter FW Jr (2007) The role of plant disease epidemiology in developing successful integrated disease management programs. In: Ciancio A, Mukerji KG (eds) General concepts in integrated pest and disease management. Springer, The Netherland, pp 45–79

Nutter FW, Parker SK (1997) Fitting disease progress curves EPIMODEL. In: Francl LJ, Neher DA (eds) Exercises in plant disease epidemiology. APS Press, St. Paul, pp 24–28

Ogbe FO, Dixon AGO, Jd'A Hughes, Alabi OJ, Okechukwu R (2006) Status of *Cassava begomoviruses* and their new natural hosts in Nigeria. Plant Dis 90:548–553

Oldfield GF (1970) Mite transmission of plant viruses. Ann Rev Entomol 15:343–380

Olmos A, Bertolini E, Cambra M (2007) Isothermal amplification coupled with rapid flow through hybridization for sensitive diagnosis of Plum pox virus. J Virol Methods 139:111–115

Okazaki S, Okuda M, Komi K, Yoshimatsu H, Iwanami T (2007) Overwintering viruliferous *Frankliniella occidentalis* (thysanoptera: thripidae) as an infection source of *Tomato spotted wilt virus* in green pepper fields. Plant Dis 91:842–846

Olukemi OO, Lava Kumar P (2013) Assessment seed transmitted viruses in wild *Vigna*. In: International symposium on plant virus epidemiology. 28th Jan–1st Feb 2013, Arusha, Tanzania. (Abstract pp 048, p 149)

Osmat AJ (2000) Methods for evaluating resistance to rice tungro disease. 38. International Rice Research Institute, Makati City, Philippines

Otim-Nape GW, Bua A, Baguma Y, Thresh JM (1997) Epidemic of severe cassava mosaic disease in Uganda and efforts to control it. Afr J Root Tuber Crops 2:42–43

Otim-Nape GW, Bua A, Thresh JM, Baguma Y, Ogwal S, Semakula GN, Acola G, Byabakama B, Colvin J, Cooter RJ, Martin A (2000) The current pandemic of *Cassava mosaic virus* disease in East Africa and its control. NARO, NRI, DFID, Chatham, UK. Natural Resources Institute, 100 pp

Owens RA, Diener TO (1981) Sensitive and rapid diagnosis of potato spindle tuber viroid by nucleic acid spot hybridization. Science 213:67–672

Padidam M, Sawyer S, Fauquet CM (1999) Possible emergence of new geminiviruses by frequent recombination. Virology 265:218–225

Pallas V, Mas P, Sanchez-Navarro JA (1998a) Detection of plant RNA viruses by nonisotopic dot-blot hybridization. Methods Mol Biol 81:461–468

Pallas V, Sanchez-Navarro JA, Mas P, Canizares MC, Aparicio F, Marcos JF (1998b) Molecular diagnostic techniques and their potential role in stone fruit certification schemes. Options Mediterr 19:191–208

Pallas V, Gomez G, Duran-Vila N (2003) Viroids in Europe. In: Hadidi A, Flores R, Randles JW, Semanick JS (eds) Viroids. CS/RO, Collingwood, pp 268–274

Pan H, Chu D, Ge D, Wang S, Wu Q, Xie W, Jiao X, Liu B, Yang X, Yang N, Su Q, Xu B, Zhang Y (2011) Further spread of and domination by *Bemisia tabaci* (Hemiptera: Aleyrodidae) Biotype Q on field crops in China. J Econ Entomol 104:978–985

Pappu HR, Sevgan S (2013) Ever increasing diversity of tospoviruses: implications for Africa. In: 12th international symposium on plant virus epidemiology, 28th Jan–1st Feb 2013. Arusha, Tanzania (Abstract, OP-72), p 97

Pappu HR, Niblet CL, Lee RF (1995) Application of recombinant DNA technology for plant protection: molecular approaches to engineering virus resistance in crop plants. World J Microbiol Biotechnol 1:426–437

Pappu HR, Sampangi R, Krishna Mohan S, Schwartz HF, Rondon SI (2007) Thrips-transmitted Iris yellow spot tospovirus epidemics in the US: understanding the epidemiological factors behind the outbreaks in onion seed and bulb crops. In: 10th international plant virus epidemiol symposium: controlling epidemics of emerging and established plant virus diseases-the way forward. 15–19 Oct 2007, ICRISAT, Hyderabad, India

Pappu HR, Jones RAC, Jain RK (2009) Global status of tospovirus epidemics in diverse cropping systems: success achieved and challenges ahead. Virus Res 141:219–236

Pasquini G, Barba M, Hadidi A, Faggioli F, Negri R, Sobol I, Tiberini A, Caglayan K, Mazyad H, Anfoka G, Ghanim M, Zeidan M, Czosnek H (2008) Oligonucleotide microarray-based detection and genotyping of Plum pox virus. J Virol Methods 147:118–126

Patil BL, Fauquet CM (2011) Ecology of plant viruses with special reference to geminiviruses. In: Hurst C (ed) Studies in viral ecology, vol I. Wiley & Sons, Hoboken, pp Chapter 11, pp 273–306

Paul HL, Querfurth G, Huth W (1980) Serological studies on the relationships of some isometric viruses of Gramineae. J Gen Virol 47:67–77

Peplies J, Glockner FO, Amann R (2003) Optimization strategies for DNA microarray-based detection of bacteria with 16S rRNA-targeting oligonucleotide probes. Appl Environ Microbiol 69:1397–1407

Perry JN (1995) Spatial analysis by distance indices. J Animal Ecology 64:303–314

Perry JN (1998) Measures of spatial pattern for counts. Ecology 79:1008–1017

Perry JN, Nixon PA (2002) New method to measure spatial association for ecological data. Eco science 9:133–141

Perry JN, Bell ED, Smith RH, Woiwod IP (1996) SADIE: software to measure and model spatial pattern. Modelling in applied biology: spatial aspects. Aspec Appl Biol, 46:95–102

Perry JN, Winder L, Holland JM, Alston RD (1999) Red-blue plots for detecting clusters in count data. Ecology Letters, 2:106–113

Pesic Z, Hiruki C (1986) Different rates of Alfalfa mosaic virus incidence in seed coat and embryo of alfalfa seed. Can J Plant Pathol 8:39–42

Peters D (2008) Thrips as unique vectors of tospovirues. Entomol Berichten 68(5):182–186

Plumb RT, Thresh JM (1983) Plant virus epidemiology: the spread and control of insect borne viruses. Blackwell Scientific Publications, Oxford, p 377

Plumb RT, Lennon EA, Gutteridge RA (1986) Forecasting barley yellow dwarf virus by monitoring vector populations and infectivity. In: McLean RG, Garrett RG, Ruesink WD (eds) Plant virus epidemics. Academic Press, Sydney, pp 387–398

Polak Z (1967) Impatiens parviflora—a natural host of cabbage black ring spot and cucumber mosaic viruses. Boil Plant, Praba 9:354–359

Polston JC, Bois D, Serra CA, Concepcion S (1994) First report of a tomato yellow leaf curl-like geminivirus in the Western Hemisphere. Plant Dis 78:831

Polston JE, Chellemi DO, Schuster DJ, Mc Govern RJ, Stansly PA (1996) Spatial and temporal dynamics of tomato mottle geminivirus and *Bemisia tabaci* (Genn.) in Florida tomato fields. Plant Dis 80:1022–1028

Pontis RE, Feldman JM (1963) A common weed, *Physalis viscose*, new host for *Potato virus Y*. Plant Dis Rep 47:22

Porto MCM, Aiedu A, Dixon A, Hahn SK (1994) An Agroecologically-oriented introduction of cassava germplasm from Latin America into Africa. In: Ofori F, Hahn SK (eds) Tropical root crops in a developing economy. Proceedings of 9th symposium international society tropical root crops, pp 118–129

Posnette AF (1947) Virus diseases of cacao in West Africa. I. Cacao viruses. 1A, 1B, 1C, and 1D. Ann Appl Biol 34:388–402

Posnette AF (1981) The role of wild hosts in cocoa swollen shoot disease. In: Thresh JM (ed) Pests, pathogens and vegetation. Pitman, London, pp 71–78

Posnette AF, Robertson NF, Todd JM (1950) Virus diseases of Cocao in West Africa V alternative host plants. Ann Appl Biol 37:229–240

Poudel B, Wintermantel WM, Cortez AA, Thien Ho, Khadgi A, Tzanetakis IE (2013) Epidemiology of blackberry yellow vein associated virus. Plant Disease 97:1352–1357

Power AG (1992) Host plant dispersion, leafhopper movement and disease transmission. Ecol Entomol 17:63–68

Prakash S, Singh SJ (2010) Molecular based techniques for diagnosis of viruses, viroids and phytoplasmas. In: Gangawane LV, Khilare VC (eds) Molecular biology of plant pathogens

Prakash DP, Ramakrishnappa K, Viswanath M, Sujatha NT, Nagaraj SK, Gangadhara Swamy S (2010) Virus diagnosis in suckers and tissue culture of banana. Acta Hortic 865:241–246

Pramesh D, Mandal B, Phaneendra C, Muniyappa V (2013) Host range and genetic diversity of *Croton yellow vein mosaic virus*, a weed-infecting monopartite begomovirus causing leaf curl disease in tomato. Arch Virol 158(3):531–542

Prasada Rao RDVJ, John VT (1974) Alternate hosts of *Rice tungro virus* and its vector. Plant Dis Rep 58:856–860

Prasada Rao RDVJ, Yaraguntaiah RC (1976) Natural occurrence of *Potato virus-Y* on *Datura metel*. Curr Sci 45:467

Prasada Rao RDVJ, Reddy AS, Chander Rao S, Varaprasd KS, Thirumala Devi K, Nagaraju V, Muniyappa V, Reddy DVR (2000) *Tobacco streak ilarvirus* as causal agent of sunflower necrosis disease in Indian. J Oil Seeds Res 17:401–402

Prasada Rao RDVJ, Reddy AS, Reddy SV, Tirumala Devi K, Chander Rao S, Manoj Kumar V, Subramanyam K, Yellamanda Reddy T, Nigam SN, Reddy DVR (2003a) The host range of *Tobacco streak virus* in India and transmission by thrips. Ann Appl Biol 142:365–368

Prasada Rao RDVJ, Reddy DVR, Nigam SN, Reddy AS, Waliyar F, Yellamanda Reddy T, Subramanian K, Johnsudheer M, Naik KSS, Bandopadhyay A, Desai S, Ghewande MP, Basu MS, Somasekhar (2003b) Peanut stem necrosis: a new disease of groundnut in India. Inf Bull No. 67 ICRISAT, Patancheru-502324, Andhra Pradesh, India, 16 pp

Prasada Rao RDVJ, Sarath Babu K, Anitha SK, Chakrabarty SK, Varaprasad KS, Khetarpal RK (2006) Pollen associated tobacco streak virus transmission to groundnut by various thrips species. Abstract OP 9/09. In: XVI annual conference IVS and international symposium on management of vector-borne viruses, held at ICRISAT, Patancheru, Hyderabad

Puchta H, Sanger HL (1989) Sequence analysis of minute amounts of viroid RNA using the polymerase chain reaction (PCR). Arch Virol 106:335–340

Pun KB, Doraiswamy S (2000) Influence of weather factors on the incidence of yellow vein mosaic virus disease. J Agric Sci Soc Northeast India 13:91–96

Punja ZK, De Boer S (2007) Biotechnology and plant disease management. CABI Publications, Wallingford, 624 pp

Purcifull DE, Batchelor DL (1977) Immunodiffusion tests with sodium dodecyl sulfate (SDS)-treated plant viruses and plant viral inclusions. Florida Agric Exp Sta Tech Bull 788, 39 pp

Purcifull DE, Zitter TA, Hiebert E (1975) Morphology, host range and serological relationships of *Pepper mottle virus*. Phytopathology 65:559–562

Puttaraju HR, Prakash HS, Shetty HS (2002) Contribution of seed-borne blackeye cowpea mosaic potyvirus to disease dynamics and loss of yield. Trop Sci 42:147–152

Quainoo AK, Wetten AC, Allainguillaume J (2008) Transmission of Cocoa swollen shoot virus by seeds. J Virol Methods 150:45–49

Quiot JB (1980) Ecology of *Cucumber mosaic virus* in the Rhone Valley of France. Acta Hortic 88:9–21

Raccah B, Fereres A (2009) Plant virus transmission by insects, Encyclopedia of life sciences (ELS). Wiley, Chichester

Raccah B, Gal-On A, Eastop VF (1985) The role of flying aphid vectors in the transmission of *Cucumber mosaic virus* and *Potato virus Y* to peppers in Israel. Annal Appl Biol 106:451–460

Raccah B, Pirone TP, Madden LV (1988) Correlation between the incidence of aphid species and the incidence of two nonpersistent viruses in tobacco. Agric Ecosyst Environ 21:281–292

Raccah B, Blanc S, Huet H (2001) Molecular basis of vector transmission: potyvirus. In: Harris K, Duffus JE, Smith OP (eds) Virus-insect-plant interactions. Academic Press, San Diego, pp 181–206

Ram RD, Chatterjee SN (1977) Mosaic virus of Panicum crusgalli L. Sci Cult 43:337–386

Ramappa HK, Muniyappa V, Colvin J (1998) The contribution of tomato and alternative host plants to *Tomato leaf curl virus* inoculum pressure in different areas of South India. Ann Appl Biol 133:187–198

Randles JW, Carver M (1971) Epidemiology of lettuce necrotic yellows virus in South Australia. II. distribution of virus in South Australia. II. Distribution of virus, host plants and vectors. Aust J Agric Res 22:231–237

Ranganath HR, Krishna Kumar NK, Krishnareddy M, Venugopalan R, Kalleshwaraswamy CN (2006) Population dynamics of *Thrips palmi* (carny) and *Frankliniella schultzei* (tribom) in relation to epidemiology of *Watermelon bud necrosis virus*. Abstract OP 9/15. In: XVI annual convention of IVS and international symposium on management of vector-borne viruses, held at ICRISAT, Patancheru, Hyderabad

Rao GP, Singh Maneesha (2008) Techniques in diagnosis of plant viruses. In: Rao GP, Varverde RA, Dovas CI (eds) Techniques in diagnosis of plant viruses. Studium Press LLC, Houstan, pp 1–47

Rao GP, Kumar PL, Holguin-Pena RJ (2008) Characterization, diagnosis and management of plant viruses, vol III: vegetable and pulse crops. Studium Press LLS, Houstan 408 p

Ratna AS, Rao AS, Reddy AS, Nolt BL, Reddy DVR, Vijaylakshmi M, McDonald D (1991) Studies on transmission of *Peanut clump virus* disease by *Polymyxa graminis*. Ann Appl Biol 118:71–78

Reckhaus PM, Andriamasintscheno (1997) Rice yellow mottle virus in Madagascar and its epidemiology in the north west of the Island. J Plant Dis Prot 104:289–295

Reddy DVR (1991) Groundnut viruses and virus diseases: distribution, identification and control. Rev Plant Pathol 70:665–678

Reddy DVR, Amin PW, Mc Donald D, Ghanekar AM (1983a) Epidemiology and control of groundnut bud necrosis and other diseases of legume crops in India caused by *Tomato spotted wilt virus*. In: Plumb RT, Thresh JM (eds) Plant virus epidemiology. Blackwell Scientific Publications, Oxford, pp 93–102

Reddy DVR, Rajeshwari R, Iizuka N, Lesemann DE, Nolt BL, Goto T (1983b) The occurrence of *Indian peanut clump*, a soil-borne virus disease of groundnuts (*Arachis hypogaea*) in India. Ann Appl Biol 102:305–310

Reddy DVR, Bragard C, Sreenivasulu P, Delfosse P (2008) Pecluvirus. In: Brain WJ, Mahy, Van Regenmortel MHV (eds) Desk encyclopedia of plant and fungal virology. Academic Press, New York, pp 257–263

Revers F, Legall O, Candresse T, Leromancer M, Dunez J (1996) Frequent occurrence of recombinant potyvirus isolates. J Gen Virol 77:1953–1965

Rezaian MA, Krake LR, Golino DA (1992) Common identity of grapevine viroids from USA and Australia revealed by PCR analysis. Intervirology 34:38–43

Richmond CS, Glasner JD, Mau R, Jin H, Blattner FR (1999) Genome-wide expression profiling in Escherichia coli K-12. Nucleic Acids Res 27:3821–3835

Riley DG, Joseph SV, Srinivasan R, Diffie S (2011) Thrips vectors of tospoviruses. J Integr Pest Manag 1:2011. doi:10.1603/IPM10020

Robert SL, Kerst AJ (2001) Nucleic acid sequence-based amplification assays for rapid detection of West Nile and St. Louis encephalitis viruses. J Clin Microbiol 39:4506–4513

Roberts IM, Tamada T, Harrison BD (1980) Relationship of potato leafroll virus to luteoviruses: evidence from electron microscope serological tests. J Gen Virol 47:209–213

Robertson RS Jr, Klostermeyer EC (1962) The role of alternate plant hosts in the aphid transmission of bean mosaics in Central Washington. J Econ Entomol 55:460–462

Rochow WF (1969) Biological properties of four isolates of *Barley yellow dwarf virus*. Phytopathology 59:1587–1589

Rodrigues JCV, Machado MA, Kitajima EW, Muller GW (2000) Transmission of citrus leprosis virus to mandarins by Brevipalpus phoenicis (Acari: Tenuipalpidae). Proc Conf Int Org Citrus Virol 14:174–178

Rodriguez PPE, Hanada K, Laguna IG, Zerbini FM, Ducasse DA (2011) Molecular characterization and relative incidence of bean- and soybean- infecting begomoviruses in North Western Argentina. Ann Appl Biol 158:69–78

Roivainen O (1976) Transmission of cocoa viruses by mealy bugs (Homoptera: Pseudococcidae). J Sci Agric Soc Finl 48:433–453

Rojas MR, Gilbertson RL, Russell DR, Maxwell DP (1993) Use of degenerate primers in the polymerase chain reaction to detect whitefly-transmitted geminiviruses. Plant Dis 77:340–347

Rose DJW (1972) Times and sizes of dispersal flights by Cicadulina species (Homoptera Cicadellidae), vectors of maize streak disease. J Anim Ecol 41:495–506

Rose DJW (1973) Field studies in rhodesia on Cycadullina spp. (Hem. Cicadellidae) vectors of maize streak disease. Bull Entomol Res 62:477

Rose DJW (1974) The epidemiology of maize streak disease in relation to population densities of Cicadulina spp. Ann Appl Biol 76:199–207

Rose DJW (1978) Epidemiology of *Maize streak disease virus* spread by several species of leafhoppers. Cicadulina Ann Rev Entomol 23:259–282

Rosenberg LJ (1981) Potential and assisted migration by plant hoppers of the Philippines. Int Rice Res Newsl 6:16–17

Ross JP (1963) Transmission of bean pod mottle virus in soybeans by beetles. Plant Dis Rep 47:1049–1050

Rossi RE, Mulla DJ, Journel AG, Franz EH (1992) Geostatistical tools for modeling and interpreting ecological spatial dependence. Ecol Monogr 62:277–314

Ruesink WG, Irwin ME (1986) Soybean mosaic virus epidemiology: a model and some implications. In: McLean GD, Garrett RG, Ruesink WG (eds) Plant virus epidemics. Monitoring, modelling and predicting outbreaks. Academic Press, Australia, pp 295–313

Ruiz-Garcia AB, Olmos A, Arahal DR, Antunez O, Llop P, Perez-Ortín JE, Lopez MM, Cambra M (2004) Biochip electronico para la deteccion y caracterizacion simultanea de los principales virus y bacterias patogenos de la patata. XII Congreso de la Sociedad Espanola de Fitopatologia. Lloret de Mar, p 12

Ryan CC (1988) Epidemiology and control of Fiji disease virus of sugarcane. Adv Dis Vect Res 5:163–176

Rybicki EP (2008) Vaccine production in plants. In: Mahy BWJ, Van Regenmortel MHV (eds) Desk encyclopedia of plant and fungal virology. Elsevier and Academic Press, New York, pp 67–71

Sacristan S, Fraile A, Garcia-Arenal F (2004) Population dynamics of *Cucumber mosaic virus* in melon crops and in weeds in central Spain. Phytopathology 94:992–998

Saikia AK, Muniyappa V (1989) Epidemiology and control of *Tomato leaf curl virus* in Southern India. Trop Agric (Trinidad) 66:350–354

Saki RK, Scharf S, Faloona F, Mullis KB, Horn GT, Erlich HA, Arnheim N (1985) PCR—technique. Science 230:1350–1354

Salazar LF, Muller G, Querci M, Zapata JL, Owens RA (2000) Potato yellow vein virus: its host range, distribution in South America and identification as a crinivirus transmitted by Trialeurodes vaporariorum. Ann Appl Biol 137:7–19

Salem NM, Ehlers JD, Roberts PA, Ng JCK (2010) Biological and molecular diagnosis of seed borne viruses in cowpea germplasm of geographically diverse sub-saharan origins. Plant Pathol 59:773–784

Sanchez-Campos S, Navas-Castillo J, Comero R, Soria C, Diaz JA, Moriones E (1999) Displacement of *Tomato yellow leafcurl virus* (TYLCV)-Sr by TYLCV-Is in tomato epidemics in Spain. Phytopathology 89:1038–1043

Sander E, Dietzgen RG (1984) Monoclonal antibodies against plant viruses. Adv Virus Res 29:131–168

Sanz AI, Fraile A, Gallego JM, Malpica JM, Garia-Arenal F (1999) Genetic variability of natural populations of cotton leaf curl geminivirus, single-stranded DNA virus. J Mol Evolut 49:672–681

Sarra S, Oevering P, Guindo S, Peters D (2004) Wind-mediated spread of Rice Yellow Mottle Virus (RYMV) in irrigated rice crops. Plant Pathol 53(2):148–153

Sarovar B, Saigopal DVR (2010) Development of a probe-based blotting technique for the detection of tobacco streak virus. Acta Virol 54:221–224

Sarovar B, Sivaprasad Y, Saigopal DVR (2010) Detection of Tobacco streak virus by immunocapture reverse transcriptase-polymerase chain reaction and molecular variability analysis of a part of RNA 3 of Sunflower, Gherkin and Pumpkin from Andhra Pradesh, India. Sci Asia 36:194–198

Sasaya T, Koganezawa H (2006) Molecular analysis and virus transmission tests place Olpidium virulentus, a vector of Mirafiori lettuce big-vein virus and tobacco stunt virus, as a distinct species rather than a strain of Olpidium brassicae. J Gen Plant Pathol 72:20–25

Sastry KS (1984) Role of weed hosts in the perpetuation of virus. In: Misra A, Polosa H (eds) Virus ecology. South Asian Publishers, New Delhi, India, p 239, pp 59–90

Sastry KS (2013a) Seed-transmitted plant virus diseases. Springer, Berlin, p 327

Sastry KS (2013b) Plant virus and viroid diseases in the tropics. Volume-1: Introduction of plant viruses and sub-viral agents, classification, assessment of loss, transmission and diagnosis. Springer (In Press)

Sastry KS, Sastry KSM, Singh SJ (1974) Influence of different insecticides on *Tomato leaf curl virus* incidence in the field. Pesticides 8:41–42

Sastry KSM, Singh SJ, Sastry KS (1978) Studies on epidemiology of *Tomato leaf curl virus*. Indian J Hortic 35:269–277

Scheets K (2008) Machlomovirus (Tombusviridae). In: Mahy BWJ, Van Regenmortel MHV (eds) Encyclopedia of virology, 5 vols, 3rd edn. Elsevier, Oxford, pp 259–263

Schmelzer K, Schmelzer A (1974) Euphorbia peplus L., a possible reservoir of cabbage black ring virus. Archiv Phytopathologie Pflanzenschutz 10:217–220

Schmitt C, Balmori E, Guilley H, Richards K, Jonard G (1992) *In vitro* mutagenesis of biologically active transcripts of *Beet necrotic yellow vein virus* RNA 2: evidence that a domain of the 75 kD read through protein is important for efficient virus assembly. PNAS, USA, 89:5715–5719

Schnell RJ, Kuhn DN, Ronning CM, Harkins D (1997) Application of RT-PCR for indexing avocado sunblotch viroid. Plant Dis 81:1023–1026

Schoen C, De Weerdt M, Hillhorst R, Chan A, Boender P, Zijlstra C, Bonants P (2002) Use of novel 3D microarray flow through system for plant pathogen multiplex detection. In: Proceedings agricultural biomarkers for array technology, management committee meeting, Wadenswil, 2002, p 11

Schoen C, De Weerdt M, Hillhorst R, Boender P, Szemes M, Bonants P (2003) Multiple detection of plant (quarantine) pathogens by micro-arrays: innovative tool for plant health management. In: Proceedings of 19th international symposium on virus and virus-like diseases of temperate fruit crops, Valencia, 2003, p 108

Scifers DL, Harvey TL, Martin TJ, Jensen SG (1997) Identification of the wheat curl mite as the vector of the high plains virus in corn and wheat. Plant Dis 81:1161–1166

Scuderi G, Golmohammadi M, Lopez MM, Cirvilleri G, Llop P (2007) Optimization of methods for evaluation of viability of Xanthomonas axonopodis pv. citri in fresh fruits. In: Abstracts of the XIII international congress on molecular plant-microbe interactions, Sorrento

Sdoodee R, Teakle DS (1993) Studies on the mechanism of transmission of pollen-associated Tobacco streak ilarvirus by *Thrips tabaci*. Plant Pathol 42:88–92

Seal SE, van den Bosch F, Jeger MJ (2006) Factors influencing begomovirus evolution and their increasing global significance: implications for sustainable control. Crit Rev Plant Sci 25:23–46

Segarra J, Jeger MJ, Van den Bosch F (2001) Epidemic dynamics and patterns of plant diseases. Phytopathology 91:1001–1010

Selman BJ (1973) Beetles-phytophagus coleoptera. In: Gibbs AJ (ed) Virus and invertabrates. North-Holland Publishing Co, London, pp 157–177

Selvarajan R, Balasubramaniam V, Dhayakar S, Uma S, Ahlawat YS, Sathiamoorthy S (2007) Molecular diagnosis of *Banana bunchy top nano virus* (Indian isolate) coat protein gene by polymerase chain reaction. In: Singh HP, Uma S (eds) Banana-technological advancements. AIPUB, Trichus, pp 431–434

Selvarajan R, Sheeba MM, Balasubramanian V (2011) Simultaneous detection of episomal *Banana streak mysore virus* and *Banana bunchy top virus* using multiplex RT-PCR. Current Sci 100:31–34

Shahid Ali, Kadian MS, Akhtar M, Ortiz O (2013) Epidemiology of aphid vectors of potato viruses in North-Eastern hills of India. In: 12th international symposium on plant virus epidemiology, 28th Jan–1st Feb 2013, Arusha, Tanzania. (Abstract, pp 062, p 163)

Shamloul AM, Minafra A, Hadidi A, Giunchedi L, Waterworth HE, Allam EK (1995) Peach latent mosaic viroid: nucleotide sequence of an Italian isolate, sensitive detection using RTPCR and geographic distribution. Acta Hortic 386:522–530

Shamloul AM, Faggioli F, Keith JM, Hadidi A (2002) A novel multiplex RT-PCR probe capture hybridization (RT-PCR-ELISA) for simultaneous detection of six viroids in four genera: apscaviroid, hostuviroid, pelamoviroid and pospiviroid. J Virol Methods 105:115–121

Sharma Manju, Thakur PD, Gupta Dharmesh, Thakur AK (2013) Identification of viruses and screening of summer squash (Cucurbita pepo) germplasm against viral disease under controlled conditions. Indian J Agr Sci 83(4):426–430

Sharman M, Thomas JE, Persley DM (2008) First report of *Tobacco streak virus* in sunflower (*Helianthus annuus*), cotton (*Gossypium hirsutum*), chickpea (*Cicer arietinum*) and mung bean (*Vigna radiata*) in Australia. Australas Plant Dis Notes 3:27–29

Sharman M, Persley DM, Thomas JE (2009) Distribution in Australia and seed transmission of *Tobacco streak virus* in *Parthenium hysterophorus*. Plant Dis 93:708–712

Shepard JF, Secor G, Purcifull DE (1974) Immunological cross-reactivity between the dissociated capsid proteins of PVY group plant viruses. Virology 58:464–475

Shepherd RJ (1972) Transmission of viruses through seed and pollen. In: Kado CI, Agrawal HO (eds) Principles and techniques in plant virology. Van Nostrand-Reinhold Co., New York, pp 267–292

Shivasharanayya, Nagaraju (2003) Relationship among weather parameters, thrips population and incidence of sunflower necrosis virus disease. Plant Dis Res 18:4–47

Shukla DD, Gough KH (1979) The use of protein A, from *Staphylococcus aureus*, in immune electron microscopy for detecting plant virus particles. J Gen Virol 45:533–536

Sigvald R (1986) Forecasting the incidence of potato virus Y^0. In: McLean GD, Garrett RG, Ruesink WD (eds) Plant virus epidemics. Academic Press, Sydney, pp 419–441

Sigvald R (1992) Progress in aphid forecasting systems. Neth J Plant Pathol 98(suppl 2):55–62

Simmons HE, Holmes EC, Gildow FC, Bothe-Goralczyk ME, Stephensonx AG (2011) Experimental verification of seed transmission of Zucchini yellow mosaic virus. Plant Dis 95:751–754

Simons JN (1956) The *Pepper vein banding mosaic virus* in the Everglades area of South Florida. Phytopathology 46:53–57

Singh TVK (1989) Certain aspects of insect pest management in groundnut. *Arachis hypogaea* Linn. Ph.D Thesis Indian Agricultural Research Institute, New Delhi, India, 345 pp

Singh CA, Bhargava KS (1975) Crow foot grass (*Dactyloctenium aegyptiacum* Willd) another natural host of *Sugarcane mosaic virus* from Gorakpur, India. Sugarcane Path Newsl 14:12

Singh RP, Nie X (2002) Nucleic acid hybridization for plant virus and viroid detection. In: Khan JA, Dijkstra J (eds) Plant viruses as molecular pathogens. Food Products Press/Haworth Press, New York, pp 443–469

Singh AK, Rathi YPS (1997) Epidemiology of *Pigeonpea sterility mosaic virus*. Indian J Virol 13:143–145

Singh SP, Sharma JR (1989) Genetic improvement of pyrethrum. Theor Appl Genet 78:841–846

Singh M, Singh RP (1996) Factors affecting detection of PVY in dormant tubers by reverse transcription polymerase chain reaction and nucleic acid spot hybridization. J Virol Methods 60:47–57

Singh RP, Boucher A, Seabrook JEA (1988) Detection of the mild strains of potato spindle tuber viroid from single true potato seed by return electrophoresis. Phytopathology 78:663–667

Singh Z, Jones RHC, Jones MGK (1995) Identification of cucumber mosaic virus subgroups-1 isolates from banana plants affected by infection chlorosis using RT-PCR. Plant Dis 79:713–716

Singh RP, Dilworth AD, Ao X, Singh M, Baranwal VK (2009) Citrus exocortis viroid transmission through commercially-distributed seeds of impatiens and Verbena plants. Eur J Plant Pathol 124:691–694

Singh D, Singh P, Gill JS, Brar JS (2010) Weather based prediction model for forecasting cotton leaf curl disease in American cotton. Indian Phytopath 63:87–90

Sipahioglu HM, Ocak M, Usta M (2007) Comparison of three conventional methods for the detection of plant virus/viroid RNAs from heat dried-high phenolic host leaves. Asian J Plant Sci 6:102–107

Sivalingam PN, Varma Anupam (2007) Non-tomato natural hosts of tomato infecting begomoviruses in north-western India. Indian J Virol 18:20–27

Skaljac M, Ghanim M (2010) Tomato yellow leaf curl disease and plant-virus vector interactions. Israel J Plant Sci 58:103–111

Slykhuis JT (1960) Current status of mite transmitted plant viruses. Proc Entomol Soc Ont 90:22–30

Smith EA, Di Tommaso A, Fuchs M, Shelton AM, Nault BA (2011) Weed hosts for onion thrips (thysanoptera: thripidae) and their potential role in the epidemiology of *Iris yellow spot virus* in an onion ecosystem. Environ Entomol 40:194–203

Smith MC, Holt J, Kenyon L, Foot C (1998) Quantitative epidemiology of banana bunchy top virus disease and its control. Plant Pathol 47:177–187

Smith MC, Page WW, Holt J, Kyetere D (2000) Spatial dynamics of *Maize streak virus* disease epidemic developement in maize fields. Int J Pest Manag 46:55–66

Sohi SS, Swenson KG (1964) Pea aphid biotypes differing in *Bean yellow mosaic virus* transmission. Entomol Exp Appl 7:9–14

Somani AK, Kushwah VS, Khurana SMP, Garg ID (2007) Integrated management of stem necrosis disease of potato in India. Potato J 34:231–234

Sosnowski RG, Tu E, Butler WF, O'Connell JP, Heller MJ (1997) Rapid determination of single base mismatch mutations in DNA hybrids by direct electric field control. Proc Natl Acad Sci USA 94:1119–1123

Spence NJ (2001) Virus-vector interactions in plant virus disease transmission and epidemiology. In: Jeger MJ, Spence NJ (eds) Biotic interactions in plant-pathogen associations. Oxford University Press, New York, pp 15–26

Spence N, Barbara D (2000) Viroids and other sub-viral pathogens of plants: the smallest living fossils. Microbiol Today 20:168–170

Sreekanth M, Sreeramulu M, Prasada Rao RDVJ, Sarath Babu B, Ramesh Babu T (2002) Occurrence and distribution of thrips population and *Peanut bud necrosis virus* (PBNV) incidence on greengram (*Vigna radiata* L. Wilczek) in Andhrapradesh. In: Sarath Babu B,

Varaprasad KS, Anitha K, Prasadarao RDVJ, Chakrabarty SK, Chandurkar PS (eds) Resources management in plant protection, vol 2. Published by Pl. Prot. Assoc. Hyderabad, India, pp 116–120

Sreenivasulu A (1994) Effect of certain management practices on the occurrence of thrips and leaf curl virus on blackgram (*Vigna mungo* L. hepper). M.Sc. Thesis Acharya N.G. Ranga Agricultural University, Rajendranagar, Hyderabad, Andhra Pradesh, India, 90 pp

Sreenivasulu P, Ramesh B, Subba Reddy ChV, Srinivas KP, Padmavathi M (2012) Recombinant DNA technology based antibodies for detection and diagnosis of plant viruses. In: Rao GP, Baranwal VK, Mandal B, Rishi N (eds) Recent trends in plant virology. Studium Press, Texas, pp 17–46

Sserubombwe WS, Thresh JM, Otim-Nape GW, Osiru DOS (2001) Progress of cassava mosaic virus disease and whitefly vector populations in single and mixed stands of four cassava varieties grown under epidemic conditions in Uganda. Ann Appl Biol 138:161–170

Sseruwagi P (2005) Report on the IXth international plant virus epidemiology symposium—applying epidemiological research to improve virus disease management, Lima, Peru, 3–7 April 2005 (Abstract)

Stanley J, Bisar DM, Briddon RW, Brown JK, Fauquet CM, Harrison BD, Rybicki EP, Stenger DC (2005) Geminiviridae. In: Fauquet CM, Mayo MA, Maniloff J, Desselberger U, Ball LA (eds) Virus taxonomy. VIIIth report of ICTV. Academic Press, New York, pp 301–326

Star J, Estes J (1990) Geographic information systems: an introduction. Prentice Hall, Englewood Cliffs, NJ

Staub U, Polivka H, Gross HJ (1995) Two rapid microscale procedures for isolation of total RNA from leaves rich in polyphenols and polysaccharides: application for sensitive detection of grapevine viroids. J Virol Methods 52:209–218

Steven WF (1937) A new disease of cocoa in the gold coast. Trop Agric Trin 14:84

Stevenson WR, Hagedorn DJ (1973) Further studies on seed transmission of *Pea seed-borne mosaic virus* in *Pisum sativum*. Plant Dis Rep 57:248–252

Storey HH, Bottomley AM (1928) The rosette disease of peanuts (Arachis hypogaea). Ann Appl Biol 15:26–45

Storey HH, McClean APD (1930) The transmission of streak disease between maize, sugarcane and wild grasses. Ann Appl Biol 17:691–719

Storey HH, Ryland AK (1955) Transmission of *Ground rosette virus*. Ann Appl Biol 45:319–326

Strickland AH (1950) The dispersal of Pseudococcidae (Hemiptera: Homoptera) by air currents in the Gold Coast. Proc Royal Entomol Soc, London 25:1–9

Stubbs LL, Grogan RG (1963) Necrotic yellows: a newly recognized virus disease of lettuce. Aust J Agric Res 14:439–459

Su P, Chen LS (1986) Thrips associated with peppers and their control. Bull Hualien dais 2:73–85

Subba Reddy ChV, Sreenivasulu P, Sekhar G (2011) Duplex-immunocapture RT-PCR for detection and discrimination of two distinct potyviruses naturally infecting sugarcane (Saccharum spp. hybrid). Indian J Exp Biol 49:68–73

Suteri BD, Joshi GC, Bala S (1979) Some ornamentals and weeds as reservoirs of *Potato virus Y* and *Cucumber mosaic virus* in kumaon. Indian Phytopathol 32:640

Swanson MM, Harrison BD (1994) Properties, relationships and distribution of cassava mosaic geminiviruses. Trop Sci 34:15–25

Swarnalatha P, Mamatha M, Manasa M, Singh RP, Krishnareddy M (2013) Molecular identification of *Ageratum enation virus* associated with leaf curl disease of tomato (*Solanum lycopersicum*) in India. Australasian Plant Disease Notes, DOI: 10.1007/s13314-013-0097-7

Szemes M, Schoen CD (2003) Design of molecular beacons for AmpliDet RNA assay—characterization of binding stability and probe specificity. Anal Biochem 315:189–201

Takaichi M, Yamamoto M, Nagakubo T, Oeda K (1998) Four garlic viruses identified by reverse transcription polymerase chain reaction and their regional distribution in Northern Japan. Plant Dis 82:694–698

Takaichi M, Nagakubo T, Oeda K (2001) Mixed virus infections of garlic determined by a multivalent polyclonal antiserum and virus effects on disease symptoms. Plant Dis 85:71–75

Tarafder P, Mukhopadhyay S (1979) Potential of stubbles in spreading tungro in West Bengal, India. Int Rice Res Newsl 4:18
Tarafder P, Mukhopadhyay S (1980) Further studies on the potential of rice stubble for spreading tungro in West Bengal, India. IRRN 5:12
Tarr SAJ (1951) Leaf curl disease of cotton. Commonwealth Mycol Inst, Kew, 55 pp
Taylor CE (1980) Nematodes. In: Harris KF, Maramorosch K (eds) Vectors of plant pathogens. Academic Press, New York, pp 375–416
Taylor MJ (2002) Wolbachia endosymbiotic bacteria of filarial nematodes. A new insight into disease pathogenesis and control. Arch Med Res 33:422–424
Taylor CE, Brown DJF (1997) Nematode vectors of plant viruses. CAB International, Wallingford, p 286
Taylor CE, Robertson WM (1973) Sites of virus retention. In: Rep. Scott. Hort. Res. Inst. pp 77–78
Taylor CE, Robertson WM (1975) Acquisition, retention and transmission of viruses by nematodes. In: Lamberti F, Taylor CW, and Seinhorst JW (eds) Nematode vectors of plant viruses. NATO ASI., Ser. A. Life Sci. Vol. 2. Plenum Press, New York and London, pp 253–275
Teakle DS (1980) Fungi. In: Harris KF, Maramorsch K (eds) Vectors of plant pathogens. Academic Press, London, pp 151–192
Teakle DS, Grylls NE (1973) Four strains of *Sugarcane mosaic virus* infecting cereals and other grasses in Australia. Aust J Agric Res 24:465–477
Thackray DJ, Smith LJ, Cheng Y, Perry JN, Jones RAC (2002) Effect of strain-specific hypersensitive resistance on spatial patterns of virus spread. Ann Appl Biol 141:45–59
Thackray DJ, Diggle AJ, Berlandier FA, Jones RAC (2004) Forecasting aphid outbreaks and epidemics of cucumber mosaic virus in lupin crops in Mediterranean type environment. Virus Res 100:67–82
Theuri JM, Bock KR, Woods RD (1987) Distribution, host range and some properties of a virus disease of sunflower. Trop Pest Manag 33:202–206
Thouvenel JC, Fauquet C (1980) *Polymyxa graminis* on new Sorghum species in Africa. Plant Dis 64:957–958
Thouvenel JC, Fauquet C (1981) Further properties of *Peanut clump virus* and studies on its natural transmission. Ann Appl Biol 97:99–107
Thresh JM (1958) The spread of virus diseases in Cacao. West Afr Cacao Res Inst Tech Bull 5:1–36
Thresh JM (1963) A vein pattern of black currant leaves associated with reversion disease. Rept East Malling Res Stat pp 97–98
Thresh JM (1974a) Vector relationships and the development of epidemics: the epidemiology of plant viruses. Phytopathology 64:1050–1056
Thresh JM (1974b) Temporal patterns of virus spread. Ann Rev Phytopathol 12:111–128
Thresh JM (1976) Gradients of plant virus diseases. Ann Appl Biol 82:381–406
Thresh JM (1978) The epidemiology of plant virus diseases. In: Scott PR, Bainbridge A (eds) Plant disease epidemiology. Blackwell Scientific Publications, Oxford, pp 79–91
Thresh JM (1980) The origins and epidemiology of some important plant virus diseases. Ann Appl Biol 5:1–65
Thresh JM (1981) The role of weeds and wild plants in the epidemiology of plant virus disease. In: Thresh JM (ed) Pests pathogens and vegetation. Pitman, London. pp 53–70
Thresh JM (1982) Cropping practices and virus spread. Ann Rev Phytopathol 20:193–218
Thresh JM (1983) Progress curves of plant virus diseases. Adv Appl Biol 8:1–85
Thresh JM (1985) Progress curves of plant virus diseases. In: Coaker (ed) Applied Biology. Adv Appl Biol 8:1–85
Thresh JM (1988) Eradication as a virus disease control measure. In: Clifford BC, Lester E (eds) Control of plant diseases: costs and benefits, pp 155–194. Blackwell Scientific Publications, Oxford

Thresh JM (1989) Insect—borne viruses of rice and the green revolution. Trop Pest Manag 35:264–272

Thresh JM (1991) The ecology of tropical plant viruses. Plant Pathol 40:324–339

Thresh JM (2003) Control of plant virus diseases in sub-saharan Africa: the possibility and feasibility of an integrated approach. Afr Crop Sci J 11:199–223

Thresh JM (2006a) Control of tropical plant virus diseases. Adv Virus Res 67:245–295

Thresh JM (2006b) Crop viruses and virus diseases: a global perspective. In: Cooper JI, Kuehne T, Polischuk VP (eds) Virus diseases and crop biosecurity. IOS Press, Amsterdam, pp 9–32

Thresh JM, Lister RM (1960) Copping experiments on the spread and control of cacao swollen shoot disease in Nigeria. Ann Appl Biol 48:65–74

Thresh JM, Fargette D (2003) The epidemiology of African plant viruses: basic principles and concepts. In: Plant virology in Sub-Saharan Africa. Ibadan, Nigeria: International Institute for Tropical Agriculture, pp 61–111

Thresh JM, Fargette D, Otim-Nape GW (1994) The viruses and virus diseases of cassava in Africa. African Crop Sci J 2:459–478

Thresh JM, Owusu GK, Boamah A, Lockwood G (1988) Ghanain cocoa varieties and swollen shoot virus disease. Crop Prot 7:219–231

Thresh JM, Fargette D, Jeger MJ (2003) Epidemiology of tropical plant viruses. In: Loebenstein G, Thottappilly G (eds) Viruses and virus-like diseases of major crops in developing countries. Kluwer Academic Publications, Dordrecht, pp 55–77

Thresh JM, Kuehne T, Jones RAC (2004) Plant virus epidemiology: First steps into new millennium virus research., 100:1–3

Thurston HD (1977) International crop development centers: a pathologist's perspective. Ann Rev Plant Pathol 15:223–247

Tomlinson JA, Walkey DGA, Hughes DE, Watson DH (1965) Multiple transverse breakage of the filamentous particles of turnip mosaic virus by ultrasonic vibration. Nature, London 207:495–497

Tomlinson JA, Carter AL, Dale WT, Simpson CJ (1970) Weed plants as sources of *Cucumber mosaic virus*. Ann Appl Biol 66:11–16

Torrance L, Jones RAC (1981) Recent developments in serological methods suited for use in routine testing of plant viruses. Plant Pathol 30:1–24

Tsai JH, Yue B, Webb SE, Funderburk JE, Hsu HT (1995) Effects of host plant and temperature on growth and reproduction of *Thrips palmi* (Thysanoptera: Thripidae). Environ Entomol 24:1598–1603

Tsai CW, Chau J, Fernandez D, Bosco D, Daane KM, Almeida RPP (2008) Transmission of grapevine leafroll associated virus-3 by the vine mealybug (Planococcus ficus). Phytopathology 98:1093–1098

Tsai CW, Rowhani A, Golino DA, Deane KM, Almeida RPP (2010) Mealybug transmission of grapevine leaf roll viruses: an analysis of virus vector specificity. Phytopathology 100:830–834

Tugume AK, Mukasa SB, Valkonen JPT (2008) Natural wild hosts of *Sweet potato feathery mottle virus* show spatial differences in virus incidence and virus-like diseases in Uganda. Phytopathology 98:640–652

Ullman DE, Meideros R, Campbell LR, Whitefield AE, Sherwood JL, German TL (2002) Thrips as vectors of tospovirus. In: Plumb RT (ed) Plant virus vector interactions. Adv Bot Res 36:113–140

Untiveros M, Fuentes S, Salazar LF (2007) Synergistic interaction of *Sweet potato chlorotic stunt virus* (*Crinivirus*) with carla-, cucumo-, ipomo-, and potyviruses infecting sweet potato. Plant Dis 91:669–676

Upendhar S, Singh TVK, Prasada Rao RDVJ (2006) Relationship between thrips population, *Sunflower necrosis disease* (SND) incidence and weather parameters. J Oilseeds Res 23:267–269

Uyemoto JK (1983) Biology and control of maize chlorotic mottle virus. Plant Dis 67:7–10

Valand GV, Muniyappa V (1992) Epidemiology of *Tobacco leaf curl virus* in India. Ann Appl Biol 120:257–267

Valenta V (1960) *Echinocystis lobata*—a reservoir plant of *Cucumber mosaic virus* in Slovakia. Biologia, Bratislava 15:217–220

van Beckhoven JRCM, Stead DE, van der Wolf JM (2002) Detection of Clavibacter michiganensis subsp. sepedonicus by AmpliDet RNA, a new technology based on real time monitoring of NASBA amplicons with a molecular beacon. J Appl Microbiol 93:840–849

van den Bosch F, Akudibilah G, Seal S, Jeger M (2006) Host resistance and the evolutionary response of plant viruses. J Appl Ecol 43:506–516

van der Walt E, Martin DP, Varsani A, Polston JE, Rybicki EP (2008) Experimental observations of rapid Maize streak virus evolution reveal a strand-specific nucleotide substitution bias. Virol J 5:104

Van der Wolf JM, Van Beckhoven JRCM, De Haan EG, Van den Bovenkamp GW, Leone GOM (2004) Specific detection of *Ralstonia solanacearum* 16S rRNA sequences by AmpliDet RNA. Eur J Plant Pathol 110:25–33

van Doorn R, Szemes M, Bonants P, Kowalchuk GA, Salles JF, Ortenberg E, Schoen CD (2007) Quantitative multiplex detection of plant pathogens using a novel ligation probe-based system coupled with universal, high-throughput realtime PCR on Open Arrays. BMC Genomics 14:276

Van Hoof HA (1962) Trichodorns pachydermus and I. teres vectors of early browning virus of peas. Tijdschr Plant Ziekt 68:391–396

Van Maanen A, Xu XM (2003) Modelling plant disease epidemics. Eur J Plant Pathol 109:669–682

Van Os B, Stancanelli G, Mela L, Lisa V (1993) Role of wild plants and weeds in the epidemiology of tospovirus in Liguria. Informatore Fitopatologico 43:40–44 (in Italian)

Van Regenmortel MHV (1967) Serological studies on naturally occurring strains and chemically induced mutants of tobacco mosaic virus. Virology 31:467–480

Vander plank JE (1946) A method for estimating the number of random groups of adjacent diseased plants in a homogenous field. Trans R Soc S Afr 3:269–278

Vander plank JE (1949) The relation between the size of the fields and the spread of plant—disease into them. Part III. Examples and discussion. Empire J Expt Agric 17:141–147

Vander plank JE (1960) Analysis of epidemics. In: Horsfall JG, Cowling EB (eds) Plant pathology. Academic Press, New York, pp 230–287

Vander plank JE (1963) Plant disease: epidemics and control. Acad Press, New York, 349 pp

Vander plank JE, Anderssen EE (1944a) Kromneck disease of tobacco. Farming S. Africa 19:391–394

Vander plank JE, Anderssen EE (1944b) "Kromnek disease of tobacco; a mathematical solution to a problem of diseases," pp. 6, Science Bulletin No. 240, Department of Agriculture and Forestry, South Africa

Vander plank JE, Anderssen EE (1945) Kromnek disease of tobacco; a mathematical solution to a problem of disease. Science Bulletin, Department of Agriculture, South Africa No. 240

Varga A, James D (2006) Use of reverse transcripton loop-mediated isothermal amplification for the detection of Plum pox virus. J Virol Methods 138:184–190

Vani S, Varma A (1988) Properties of *Cucumber green mottle mosaic virus* isolated from water of river Jamuna. Indian Phytopathol 46:118–122

Varma A, Niazi FR, Dasgupta I, Singh J, Cheema SS, Sokhi SS (1999) Alarming epidemic of rice tungro disease in North–West India. Indian Phytopathol 52:71–74

Varma A, Mandal B, Singh MK (2011) Global emergence and spread of whitefly Bemisia tabaci transmitted geminiviruses, In: Thompson WMO (ed) The whitefly, *Bemisia tabaci* (homoptera: aleyrodidae) interaction with geminivirus-infected host plants. Springer Netherlands, pp 205–292

Vemana K, Jain RK (2010) New experimental hosts of *Tobacco streak virus* and absence of true seed transmission in leguminous hosts. Indian J Virol 21:117–127

Verhoeven JThJ, Jansen CCC, Botermans M, Roenhorst JW (2010) Epidemiological evidence that vegetatively propagated, solanaceous plant species act as sources of potato spindle tuber viroid inoculum for tomato. Plant Pathol 59:3–12

Vicente M, Chagas CM, July JR (1979) Three wild solanaceae of a virus. Tres solanaceas da vegetacao espontanea como hospedeiras naturais de virus. Fitopatologia Bresileira 4:73–76

Vilasov Yu, Abdukarimova A (1965) Distribution of virus X in weeds. Kartofel' ovoshchi, 1965 (8):20

Vinayarani G, Madhusudhan KN, Deepak SA, Niranjana SR, Prakash HS (2011) Detection of mixed infection of tobamoviruses in tomato and bell pepper by using RT-PCR and Duplex RT-PCR. Int J Plant Pathol. ISSN 1996-0719/DOI: 10. 3923/IJPP

Viswanathan R, Balamuralikrishnan M, Karuppaiah R (2008) Duplex-reverse transcription-polymerase chain reaction (D-RT-PCR)-a technique for the simultaneous detection of viruses causing sugarcane mosaic. Sugar Tech 10:81–86

Viswanathan R, Karuppaiah R, Balamuralikrishnan M (2010) Detection of three major RNA viruses infecting sugarcane by multiplex reverse transcription-polymerase chain reaction (Multiplex-RT-PCR). Austr Plant Pathol 39:79–84

Von Wechmar MB (1980) Transmission of Brome mosaic virus by Puccinia graminis tritici. Phytopathology 5:239–250

Von Wechmar MB, Van Regenmortel MHV (1968) Serological studies on brome grass mosaic virus and its protein fragments. Virology 34:36–45

Vuittenez A (1970) Fanleaf of grapevines. In: Frazier NW et al (eds) Virus diseases of small fruits and grapevines. University of California Press, Berkeley, pp 217–228

Waggoner PE, Aylor DE (2000) Epidemiology, a science of patterns. Ann Rev Phytopathol 38:1–24

Waggoner PE, Horsfall JG (1969) Epidem—A simulator of plant disease written for a computer. In: Bull. Conn. Agric. Expt. Stn. No. 698. Connecticut, USA, pp 80

Waggoner PE, Horsfall JG, Likens RJ (1972) Epimay—A simulator of southern corn leaf blight. In: Bull. Conn. Agric. Expt. Stn. No. 729. Connecticut, USA, pp 84

Wah Y, Symons RH (1997) A high sensitivity RT-PCR assay for the diagnosis of grapevine viroids in field and tissue culture samples. J Virol Methods 63:57–69

Wallis RL (1967) Green peach aphids and the spread of beet western yellows in the Northwest. J Econ Entomol 60:313–315

Walters HJ (1969) Beetle transmission of plant viruses. Adv Virus Res 15:339–363

Wang HL, Gilbertson RL, Lucas WJ (1996) Spatial and temporal distribution of bean dwarf mosaic geminivirus in phaseolus vulgaris and nicotiana benthamiana. Phytopathology 86:1204–1214

Watson MA, Healy MJR (1953) The spread of beet yellows and beet mosaic viruses in the sugar-beet root crop. II. The effects of aphid numbers on disease incidence. Ann Appl Biol 40:38–59

Watson MA, Okusanya BAM (1967) Studies on the tranmission of *Groundnut rosette virus* by *Aphis craccivora* Koch. Ann Appl Biol 60:199–208

Watson MA, Heathcote GD, Lauckner FB, Sowray PA (1975) The use of weather data and counts of aphids in the field to predict the incidence of yellowing viruses of sugarbeet crops in England in relation to the use of insecticides. Ann Appl Biol 81:181–198

Wei Y, Lee JM, Smulski DR, LaRossa RA (2001) Global impact of sdiA amplification revealed by comprehensive gene expression profiling of *Escherichia coli*. J Bacteriol 183:2265–2272

Weintraub PG, Beanland L (2006) Insect vectors of phytoplasmas. Ann Rev Entomol 51:91–111

Wells ML, Culbreath AK, Todd JW, Csinos AS, Mandal B (2003) Dynamics of spring tobacco thrips (Thysanoptera: Thripidae) populations: implications for *Tomato spotted wilt virus* management. Environ Entomol 31:1282–1290

Werker AR, Dewar AM, Harrington R (1998) Modeling the incidence of virus yellows in sugar beet in the UK in relation to numbers of Myzus persicae. J Appl Ecol 35:811–818

Wetzel T, Tavert G, Teycheney PY, Ravelonandro M, Candresse T, Dunez J (1990) Dotblot hybridization detection of plum pox virus using 32P-labeled RNA probes representing nonstructural viral protein genes. J Virol Methods 30:161–172

Wetzel T, Candresse T, Macquaire G, Revelonandro M, Dunez J (1992) A highly sensitive immunocapture polymerase chain reaction method for plum pox potyvirus detection. J Virol Methods 39:27–37

Whitefield AE, Ullman DE, German TL (2005) Tospovirus—thrips interactions. Ann Rev Phytopathol 43:459–489

Wisler GC, Duffus JE, Lui HY, Li RH (1998) Ecology and epidemiology of whitefly transmitted closteroviruses. Plant Dis 82:270–280

Wren JD, Roossinck MJ, Nelson RS, Scheets K, Palmer MW (2006) Plant virus biodiversity and ecology. PLoS Biol 4:e80

WTO (1995) Agreement on the application of sanitary and phytosanitary measures. In: Results of the Uruguay round of multilateral trade negotiations: the legal texts. World Trade Organization, Geneva

Xu XM (1999) Modelling and forecasting epidemics of apple powdery mildew (*Podosphaera leucotricha*). Plant Pathol 48:462–471

Xu X (2006) Modelling and interpreting disease progress in time. In: Cooke BM, Gareth Jones D, Kaye B (eds) The epidemiology of plant disease. Springer, Dordrecht, pp 215–238

Yadav BP, Mishra MD (1990) Spread infection rate of RTV in susceptible and tolerant rice cultivars. Indian Phytopathol 43:431–434

Yakovleva N (1965) Borba s zelenoi mazaikoi Ogurtsov. (Control of green mosaic of cucumber). Z Rast Vredit Bolez 10:50–51

Yamamoto H, Sano T (2006) An epidemiological survey of Chrysanthemum chlorotic mottle viroid in akita prefecture as a model region in Japan. J Gen Plant Pathol 72(6):387–390

Yang X, Hadidi A, Garnsey SM (1992) Enzymatic cDNA amplification of citrus exocortis and cachexia viroids from infected citrus hosts. Phytopathology 82:279–285

Zadoks JC (2001) Plant disease epidemiology in the twentieth century. A picture by means of selected controversies. Plant Dis 85:808–816

Zadoks JC, Schein RD (1979) Epidemiology and plant disease management. Oxford University Press, London

Zechmann B, Zelling G (2009) Rapid TEM diagnosis of plant virus diseases. J Virol Meth 162:163–169

Zein HS, Nakazawa M, Ueda M, Miyatake K (2007) Development of serological procedures for rapid and reliable detection of Cucumber mosaic virus with dot-immunobinding assay. World J Agric Sci 3:430–439

Zhang X-S, Holt J, Colvin J (2000a) A general model of plant-virus disease infection incorporating vector aggregation. Plant Pathol 49:435–444

Zhang XS, Holt J, Colvin J (2000b) Mathematical models of host plant infection by helper-dependent virus complexes: why are helper viruses always avirulent? Phytopathology 90:85–93

Zhou X, Liu Y, Calvert L, Munoz C, Otim-Nape GW, Robinson D, Harrison BD (1997) Evidence that DNA-A of geminivirus associated with severe cassava mosaic disease in Uganda has arisen by inter specific recombination. J Gen Virol 78:2101–2111

Zhou GH, Wen JJ, Cai DJ, Li P, Xu DL, Zhang SG (2008) Southern rice black-streaked dwarf virus: a new proposed Fijivirus species in the family Reoviridae. Chin Sci Bull 53:3677–3685

Zimmer DP, Soupene E, Lee HL, Wendisch VF, Khodursky AB, Peterm BJ, Bender RA, Kustu S (2000) Nitrogen regulatory protein C-controlled genes of *Escherichia coli*: scavenging as a defense against nitrogen limitation. Proc Natl Acad Sci USA 97:14674–14679

Zitter TA (1977) Epidemics of aphid-borne viruses. In: Harris KF, Maramorosch K (eds) Aphids as virus vectors. Academic Press, New York, pp 385–412

Chapter 2
Management of Virus and Viroid Diseases of Crops in the Tropics

2.1 Introduction

Management of plant virus and viroid diseases is a matter of vital importance and concern to the farmer, horticulturist, forester and gardener. It is well established that the virus and viroid diseases in different crops cause enormous losses in terms of quantity and quality of products (see Chapter-3 of Volume-I) (Sastry 2013b). Because of the serious losses they cause to agriculture and horticulture, virus/viroid diseases have acquired great importance in the realm of plant pathology and call for effective management measures against them. These diseases are not amenable to control by direct methods. Unlike fungicides and bactericides, no commercial viricides have yet been developed. However a number of indirect control measures for different crop plant-virus/viroid pathosystems suitable to varied agro ecosystems have been developed with an aim to avoid the pathogen or to minimize their impact on crop yields (Thresh 2003, 2006). As these diseases routinely cause systemic infections and have very effective methods of transmission, preventing their spread has become a complicated problem. Some diseases like *Tobacco mosaic virus* (TMV), *Potato virus X* (PVX), *Cucumber green mottle virus* (CGMMV), *Citrus exocortis viroid* (CEVd) and *Potato spindle tuber viroid* (PSTVd) are highly contagious and affect large number of hosts belonging to several families.

2.2 Need for Diagnosis of Plant Virus and Viroid Diseases

An effective and applicable virus management strategy requires an accurate diagnosis and understanding the life and disease cycle of etiological agents. Recent developments in molecular techniques have revolutionized the field of diagnostics in agriculture (see Chapter-5 of Volume-I) (Sastry 2013b). High detectability, specificity, sensitivity and accuracy supported with simplicity, amenable to automation and low cost are the main characteristics of an ideal diagnostic strategy. However, the virus disease diagnostic methods started with serodiagnosis which

K. S. Sastry and T. A. Zitter, *Plant Virus and Viroid Diseases in the Tropics*,
DOI: 10.1007/978-94-007-7820-7_2,

includes precipitin tests in liquid mode, gel diffusion tests, immuno electrophoresis, agglutination tests, and radio immunology (Van Regenmortel 1982), which were used as early as 1960s before the advent of nucleic acid-based tests, which in recent years have become absolute and hence no advanced laboratory is following them. However, if the reader is interested to have more information on the earlier used techniques, one can go through the review articles and text books covering the basic principles and protocols of different techniques (Van Regenmortel 1982; Ahlawat 2010; Albrechtsen 2006; Sastry 2013b). The other techniques like biological, technological and economical perspectives of various techniques have been reviewed in many publications (Schaad 1979; Clark 1981; Hawkes et al. 1982; Miller and Martin 1988; Singh and Sharma 1989; Chu et al. 1989; Hampton et al. 1990; Duncan and Torrance 1992; Henson and French 1993; Khurana and Garg 1993; Torrance 1995; Banttari and Khurana 1998; Louws et al. 1999; Schaad et al. 2003; Craig et al. 2004; Khurana 2004; Akram et al. 2004; Jain et al. 2004a, b; Rao and Maneesha Singh 2008). The objective of the present chapter is not to cover all the details but to highlight the current status of various diagnostic techniques and their applications in plant virology.

In Chapter-3 of Volume-I, the information presented indicates that enormous yield losses both in quality and quantity of the produce varies depending on the stage of infection, virus strain and host plant cultivar involved besides overarching environmental factors. Almost all crop plants are raised for agricultural purposes or for research through sowing the true seed or vegetative propagule. It is the responsibility of the governmental agencies or seed companies or the farming community to use the pathogen-free seed for sowing purposes since large numbers of virus and virus-like diseases are transmitted through true seed/vegetative propagules which serve as the primary source of infection to cause epidemics in favorable agro-ecosystems.

2.3 Approaches for Virus and Viroid Diseases Management

The term pandemic can be used to label epidemics that occur over large areas with high levels of disease. The intensity of the virus and viroid diseases vary with the virus/viroid strain involved, their mode of transmission, host variety involved and environmental factors. Plant diseases usually matter when many plants or plant parts (such as flowers, fruits or leaves) in a field, forest or region are infected. "Intensity" is a general term that characterizes the magnitude of the disease or the infection. The disease intensity varies from time to time and from location to location during epidemics. An epidemics is the progress of disease in time and space. Therefore, the typical goal of the virus/viroid disease management is to maintain the intensity of the disease to a low level, but not necessarily zero.

There are essentially two approaches to manage virus and viroid diseases, similar to other plant diseases. The first approach is to decrease the sources of infection (reservoirs) and secondly to minimize the rate of spread by following some of the ways and means discussed in this chapter.

The particular choice of management strategies to follow depends on the nature of the particular virus and viroid, but the decision to use the control measures depends on the assessment of the economic risks involved in taking no action. This decision is often a difficult one to make and is particularly so, where the nature and magnitude of disease problems vary from season to season and where routine use of expensive control measures are uneconomical. In these circumstances one needs to develop methods which can be implemented on such occasions when monetary benefits are likely to be achieved. Alternatively, new control measures may be devised with low costs, so that their routine use can be justified in economic terms, where by the value of the crop saved should exceed the cost of control measure used. The recommended control measure should be sufficiently effective to reduce the disease to an acceptable level and should be simple, environmentally safe, and inexpensive to apply.

Many ingenious methods have been tried to manage virus/viroid diseases by breaking the disease cycle at inoculum source, and interfering with transmission or infection. The entire range of methods aimed at the control of plant virus and viroid diseases is generally divisible into the following distinct categories according to their nature and the subject is discussed in detail in order to understand the various management measures:

- Use of virus- or viroid-free propagules of crop plants
- Phytosanitation in managing the virus and viroid diseases
- Variation of the crop cultural practices
- Cross protection
- Control of vectors
- Interfering with vector transmission
- Breeding for disease resistance and development of transgenic plants
- Quarantines
- Integrated approaches

2.4 Use of Virus/Viroid-Free Vegetative Propagules and True Seed

When the annual (vegetable, oil seeds, cereals, pulses etc.,) or perennial (fruit) crops are established by using disease-free planting material with good genetic background, then the agriculturist can expect to achieve good yields from the crops. Even after planting the disease-free material, if one does not follow the observations on disease spread and subsequence precautionary measures, then also the expected yield will not be achieved. Hence one has to take continuous steps for restricting the spread of virus and virus-like diseases by cultural, chemical or biological means which will help in reducing the vector populations, which are often responsible for spread of the virus diseases. Hence virus-free seed and subsequent vector management are the most effective input in achieving higher yields.

The improvement of seed has been a key component of increasing agricultural productivity. Seed quality for production has several components: (1) physiological-germination rate and initial vigor; (2) genetic-heritable traits and purity; and (3) sanitary-presence of pests and diseases (Almekinders and Louwaars 1999). Improved yields can come from more uniform and vigorous crop stands, clean planting material free of pests and diseases and cultivars with greater regional yield potential or resistance to key pests and diseases.

Approximately 90 % of all food crops in the world are propagated through seeds (Schwinn 1994). The major concern of farmers since crop cultivation began has been the procurement of good quality seed for planting and it is a basic input requirement for increasing crop yields. There are many successful examples of increasing productivity through improving the quality of the seeds by conventional breeding to molecular breeding programmes (genetic engineering). Plant genetic engineering has played a major role in increasing agricultural production.

Plant pathogens like fungi, bacteria, virus, and viroids are carried "in", "on" or "within" the seeds and as a result, these pathogens are responsible for causing heavy crop losses. Nearly 231 virus and viroid diseases are seed-borne in different plants (Sastry 2013a).

Seed health testing is an essential management tool for the management of seed-borne and seed-transmitted pathogens and production of healthy seed continues to be an important activity through phytosanitary certification and quarantine programmes in domestic and international seed trade. Hence the role of seed testing in determining the seed quality was first initiated in Germany as early as 1869. Since then researchers were carried out in seed pathology and now there is greater international co-operation in seed testing and certification activities.

Virus transmission through vegetative plant material and true seed provides an ideal starting point for the establishment of a disease in field crops. Firstly, it enables infection to occur at the earliest possible time in the development of the young seedling, a factor that frequently governs the severity of virus infection in an individual plant. Secondly, vegetative planting material infection results in individual infected seedlings being scattered widely throughout a crop field, with each infected seedling providing a virus reservoir for subsequent secondary spread by vectors or contact.

The seed industry has a twofold responsibility in the area of seed health: To deliver healthy seed to farmers and seed producers to respect Intellectual Property Rights (IPR). Seed health issues are increasingly important in International Seed Trade (IST). The economic significance of seed-borne pests and diseases including virus diseases with the changing global scenario in the light of the General Agreement on Trade and Traffic (GATT) is being worked out to the maximum. Sanitary and Phyto Sanitary (SPS) issues in WTO are pressurizing the developing countries to give special attention of seed health testing. Both International Seed Testing Association (ISTA) and Seed Health Initiative-vegetables (ISHI-veg) are involved in standardizing the techniques for the detection of seed-borne pathogens.

It is the mission of these organizations to develop/secure the healthy seeds and delivery to customers on a worldwide basis by the private sector. Under the umbrella of ISHI, seed companies exchange information on seed-borne diseases and jointly developing methods for their control.

Seed certification is a scientific and systematically designed, legally sanctioned process for quality control of seeds through verification of seed sources, field inspection of seed crop, supervision of post-harvest stages including processing, packing, seed sampling, analysis, grant of certificate, certification tags, labels and sealing.

2.4.1 Approved Seed Certification Standards

Seeds are the foundation to crop production, and seed health is related to food production in various ways. Seed being the foundation of successful agriculture, the demand for quality seeds of improved cultivars/hybrids is growing fast and adoption of new technologies around the world by the farmers is happening at an amazing pace. Therefore, production and supply of high-quality seed of improved cultivars and hybrids to the growers are a high priority in agricultural growth and development.

Governments are strongly encouraged to implement a predictable, reliable, user-friendly and affordable regulatory environment to ensure that farmers have access to high-quality seed at a fair price. In particular, FAO member countries are participating in the internationally harmonized systems of the Organization for Economic Cooperation and Development (OECD), the International Union for the Protection of New Varieties of Plants (UPOV), the International Treaty on Plant and Genetic Resources for Food and Agriculture (ITPGRFA) and the International Seed Testing Association (ISTA). Participation in these systems will facilitate the availability of germplasm, new plant cultivars and high-quality seed for the benefit of their farmers, without which their ability to respond to the challenges ahead will be substantially impaired. It is inevitable that governments need to develop and maintain an enabling environment to encourage plant breeding and the production and distribution of high-quality seed.

2.4.2 Stages of Seed Multiplication

Genetically pure, disease-free seeds are the prerequisite for a healthy, vigorous and high yielding crop. There are three important stages of seed multiplication.

Nucleus seed also known as original parental material, is the product of scientific breeding by crop breeders who evolved the cultivar/hybrids.

Breeder seed is the first and is the product of scientific breeding by crop breeders who evolved the particular variety. Breeder seed (sexually/asexually propagating material) is directly controlled by the original plant breeder. The production is also personally supervised by the breeder. It should possess the maximum genetic purity (99.5–100 %). Breeder seed is made available in small quantities for further multiplication. The agencies involved in the production of breeder seeds are agricultural universities and agricultural research institutes in every country.

Foundation seed is the progeny of breeder seed, and it is supervised and certified by the Seed Certification Agency. It should possess the minimum genetic purity of 99 %. Expert handling is needed for producing foundation seed, and they are produced on seed farms where specialists and facilities are available. Agencies involved in the production of foundation seeds are the State Department of Agriculture, Horticulture, State Seed Corporation and the State Farms Corporations of their respective countries.

Certified seed is the progeny of foundation seed and certified by the certification agency. The agencies involved in the production of certified seeds are the State Seed Corporation through their Registered Certified Seed Growers and the Private Seed Companies. Its genetic purity is 98 % for cultivars/composites/synthetics, and for cotton and castor hybrids, it is 85 %.

The seed should meet minimum standards of germination and purity prescribed and labeled as per Seed Act 1966. Use of quality seed and planting material which are free from seed-transmitted diseases including viruses contributes to the enhancement in crop production by better realization of its own productivity potential (to the extent of 20–25 % depending on the crop) and also by contributing to enhance the utilization efficiency of other inputs to the extent of another 20–25 %. Therefore, availability of good quality disease-free seeds of superior cultivars/hybrids needs to be ensured for boosting the agricultural production. The central seed certification board, department of agricultural and cooperation, ministry of agricultural, and Governments of every country, have prescribed disease certification standards in foundation, and certified seed crops and are being implemented by seed companies and other agencies concerned with seed production.

2.5 Certification Schemes of Crops Having Virus/Viroid Transmission Through True Seed

Certain crops like tomato, chilies, brinjal, tobacco, lettuce, rice etc., are raised in nursery beds in micro/macro tunnels when the nursery is raised in a large scale by the seed companies/international agricultural institutions (Fig. 2.1) and the seedlings are transplanted at the required stage. In majority of the Asian countries tungro virus infection occurs in rice nursery through viruliferous leafhoppers coming from outside. Even in African countries, rice seed beds are one of the sources of inoculum for *Rice yellow mottle virus* (Traore et al. 2006). The nursery

Fig. 2.1 Macro tunnels: Raising nurseries. *Courtesy* Francisco J. Morales (CIAT)

should be carefully observed for the visual viral symptoms and such plants should be removal immediately whenever noticed. Suspected plants are to be tested whether the crop is symptomless carrier (like *Tomato yellow leaf curl virus* in certain *Capsicum* species), the nursery should be tested serologically by ELISA or some other easy diagnostic techniques.

Crops like legumes (pea, French beans, soybean, cowpea, and peanut), cereals (wheat, barley, sorghum, maize, etc.,) and cucurbits are raised by directly sowing the true seed in the field either with or without ridges and furrows. Some of these crops carry certain virus and viroid diseases through seed and certification schemes to produce virus free-seeds are required. Agarwal and Sinclair (1996), Maude (1996), Maury et al. (1998a), Albrechtsen (2006) and Sastry (2013a) have reviewed the aspects of seed transmission related to seed certification. Some of the crops where certification schemes are developed are discussed herein.

Among all the methods of obtaining healthy seed material, seed certification is the most dependable and prudent measure which helps to maintain reasonable health standards without affecting seed germination. Programmes against seed-borne viruses must start at the basic level of germplasm collection available to the plant breeder and continue through the subsequent development of varieties and the increase of seed through breeder seed (pre-basic seed), foundation seed (basic seed), registered seed (certified seed, 1st generation) and certified seed (2nd generation).

Regular inspection of all propagation sources by qualified inspectors is necessary if certification is to be meaningful. When virus-free material has been obtained, its dissemination and utilization may present several difficulties for the research center and for the nursery man. The problems are simplified if, as in certain countries like USA, Europe, South America and Asia, all nurserymen belong to an association which already collaborates with an official inspection service in the operation of a certification scheme. It is then practicable to supply nursery men according to their individual requirements with scion wood cut from selected mother trees which can be virus-tested periodically. The future of these certification schemes mainly depends on the continuous collaborative programmes between propagators and research establishments. Some details of positive and negative aspects in certification schemes for crops like lettuce, barley, beans, pea and cowpea are briefly illustrated below:

(a) **Lettuce**

A field applicable seed certification programme of lettuce (*Lactuca sativa*) seed production developed against *Lettuce mosaic virus* (LMV) was worked out by number of workers. Grogan et al. (1952) developed a system in which lettuce seed was produced from virus-free plants grown under glasshouse conditions. In isolated areas, only the healthy plants were grown to maturity after roguing out the infected plants. The tolerance levels of seed transmission varied from place to place depending on the ecology and epidemiology of the virus disease.

Researchers in California (Zink et al. 1956; Wisler and Duffus 2000) and Florida (Purcifull and Zitter 1971) have established that if seed transmission exceeded 0.1 %, control was not satisfactory because of large winter and spring aphid populations. In England, when lettuce seed with 0.1 % infection were sown only 0.5 % of the plants were infected (Tomlinson 1962). It was observed that the initial level of 1.6 % LMV infection in lettuce seed reached to 29 %, whereas zero starting infection reached only 3 % (Zink et al. 1956). The number of infected lettuce seedlings emerging per acre at the 0.1 % rate would be 200/acre or 20,000 infected seedlings/100 acres (Greathead 1966). This is based upon an estimated 400,000 seeds/lb, planting 1 lb/acre, and with 50 % emergence. Programmes in both the Salinas Valley and Imperial Valley of California have, however, stressed the use of seed which is double-tested-both by the seed companies and by a committee of lettuce growers to show zero mosaic/30,000 seeds tested (Kimble et al. 1975; Grogan 1980, 1983). This has reduced losses by an estimated 95–100 % resulting in considerable increase in yield and quality. In Florida, since 1974, the use of double indexed seed has eliminated LMV, a major limiting factor in lettuce production (Zitter 1977). However during 1995, in Florida, due to non-compliance with the lettuce mosaic certification rule 5B-38, an outbreak of LMV in commercial lettuce production was noticed (Raid and Nagata 1996).

(b) **Barley**

In barley (*Hordeum vulgare*), *Barley stripe mosaic virus* (BSMV) which has no biological vector, seed transmission is the major factor facilitating disease spread. The simplest way of ameliorating the losses caused by this virus is by planting uninfected seed. A seed certification programme to obtain barley seed free from BSMV was attempted at Kansas (Hampton et al. 1957), Montana and North Dakota in the U.S. (Carroll 1983). It began with the growing of samples of seed in glass houses during late autumn and winter to select those which are most free for further propagation. The control programme involved assaying seedlings for presence of BSMV by serology and roguing infected plants from certified seed plots which lead to a significant reduction in the incidence of the virus. In 1972, a zero tolerance for seed-borne BSMV was placed on all certified seed sold in Montana and as a result, losses due to BSMV declined dramatically.

The certification scheme against BSMV in Montana involved inspection of every barley field being considered for certification according to the field standards published by the Montana Seed Growers Association and Montana State University. When a field intended for the production of registered or certified barley seed was inspected for quality and purity, it was also carefully checked for plants exhibiting symptoms of BSMV. One or two inspections of each field were made by the Secretary-Manager of the Montana Seed Growers' Association or the Agronomist with the Montana Cooperative Extension Service, Bozeman-Montana or a member of his staff. If infected plants were found in such a field, that field was no longer eligible for certification. The barley harvested from an uncertifiable field was usually sold for feed purposes. The second procedure of the certification programme is the inspection of seed from each potentially certifiable barley field by employing serological tests. At Montana, because of the implementation of certification schemes, the percentage of virus infected seed lots declined appreciably from 1967 onwards. At North Dakota, the certification scheme against BSMV resembled the Montana scheme, which involved both a procedure for inspection of seed fields and seed lot examination. Only foundation seed determined by test to be BSMV-free was used to produce certified seed which was then made available to growers.

It was reported that by 1971, for all practical purposes BSMV was eliminated in North Dakota (Timian 1971). Similarly, at Montana, the two major barley cultivars viz., Compana and Uniton, suffered losses of approximately 4 bushels/acre from 1953–1967 but by 1967–1970 these losses declined to 2 bushels/acre. Since 1970, losses due to BSMV were reported to be about half a bushel per acre after using the certified seed (Carroll 1980; Jackson and Lane 1981).

(c) **Pea**

Among the viruses infecting peas (*Pisum sativum*), *Pea seed-borne mosaic virus* (PSbMV) has seed transmission up to 100 % and even 23 % of pea accessions are infected in USDA collection (Hampton and Braverman 1979b). Because of high seed transmission and potent aphid vectors, schemes to produce virus-free seed were developed. In Canada, the mother plants in dry (field) pea breeding programmes are propagated under aphid-free conditions and screened by visual inspection and a serological assay based on SDS gel immuno-diffusion tests (Hamilton and Nichols 1978). Subsequent seed increase plots are grown in isolation in areas of Manitoba and Saskatchewan where populations of aphid vectors are minimal. Before release to the commercial seed trade, the standing crop of a new cultivar is inspected visually and samples of harvested seed (200 seed/50 lb) are subjected to a growing on out test. Seedlings are examined and five randomly selected seedlings are assayed for PSbMV by SDS gel immuno-diffusion. Virus-free lots are then released to the Canadian Seed Growers Association or to SeCan, a federal government agency for propagation in a seed certification programme. However, no further testing for PSbMV is done. In USA, in the primary seed-producing areas (Washington and Idaho), there is a zero tolerance for PSbMV-induced symptoms observed during field inspections (10 days post-emergence and 10 days prior to harvest). Visual inspection is being supplemented by direct assay of seed (200 seeds/seed lot irrespective of size of seed lot) and there is increasing use of ELISA and other advanced diagnostic techniques for the testing of breeder seed, seed from breeding lines and seed for commercial growers. Official certification (phytosanitary certificate) of testing for PSbMV by ELISA is required for shipment of pea or lentil seed from Washington and Idaho to Australia, New Zealand or South Africa.

(d) **Beans**

Bean common mosaic virus (BCMV) in French beans (*Phaseolus vulgaris*) is seed transmitted up to 83 % and economic yield losses are encountered, whenever virus infected seed is used. Hence in the U.S. and Canada the certification programme of field and garden bean against BCMV is based only on field inspection of the crop. These crops are not allowed to contain BCMV-diseased plants when grown for foundation seed, 0.5 % mosaic diseased plants for registered seed and 1 % for certified seed (Anon 1971). In Brazil, bean lines are screened visually for major virus diseases, primarily BCMV and apparently virus-free lines are increased under insect-free conditions in greenhouses. In the next stage this seed is multiplied under field conditions in arid regions of Brazil for two generations, with periodic visual inspection for virus symptoms before being sold to certified seed growers. Yield increases of about 30–100 % have been reported following the use of such certified seed (Hamilton 1983).

(e) **Soybean**

In soybean (*Glycine max*), *Soybean mosaic virus* (SMV), *Cucumber mosaic virus* (CMV) (also syn. for *Soybean stunt virus*), *Tobacco ring spot virus* (TRSV), and *Tobacco streak virus* (TSV) are seed transmitted at higher percentage (up to 100 %) and are wide spread in occurrence. Despite the fact that SMV is present and can cause serious damage wherever soybean is cultivated, there appears to be no bonafide seed certification programmes against seed-borne viruses in this crop. Two major soybean improvement programmes (INTSOY at the University of Illinois, Urbana and the Asian Vegetable Research and Development Center, Shanhua, Taiwan) are screening soybean lines for resistance to some strains but neither of these agencies employed formal certification schemes. INTSOY has eliminated SMV from a number of tropical soybean lines by roguing standing crops and monitoring remaining plants by ELISA. Seed from such lines has been exported to breeding programmes in other international institutes but the continued maintenance of virus-free nursery becomes the responsibility of the recipient breeding programmes (Hamilton 1983).

(f) **Cowpea**

In cowpea (*Vigna unguiculata*), out of the 12 seed-borne viruses, *Cowpea aphid borne mosaic virus* (CABMV), *Cucumber mosaic virus* (CMV) and *Cowpea mottle virus* (CPMoV) are highly seed-transmitted and are worldwide in distribution. In Nigeria the International Institute of Tropical Agriculture (IITA) and the Institute of Agricultural Research and Training (IAR&T) at Ibadan, Nigeria are the main centers for cowpea improvement and it is their policy to test cowpea lines before being sent to other institutions for breeding purposes or for increase in other parts of Nigeria. Samples of 1000 seeds are planted and the resulting seedlings are examined visually for the presence of seed-transmitted viruses; differential hosts and serological tests are used to facilitate further identification. Seed lots showing 2 % or more seed transmission are not distributed; those with less than 2 % seed transmission are so indicated and the recipient is advised. Preliminary results indicate that the ELISA may be applicable in detecting CPMoV in seedlings (Hamilton 1983).

(g) **Peanut**

In peanut (*Arachis hypogaea*), *Peanut stripe virus* (PStV), now considered a strain of *Bean common mosaic virus* (BCMV) and *Peanut mottle virus* (PeMoV) are very important and seed transmission recorded for both the viruses was to the extent of 8.5–30 %. In Georgia (U.S.), a seed certification programme against PeMoV was successfully implemented (Kuhn and Demski 1975). Even against PStV, a well worked out certification programme was framed in the U.S. Inspection of the standing crop is the method used for detecting PStV infected plants in peanut seed fields in Virginia, a major producer of peanut seed for commercial peanut growers in Georgia, which demands zero tolerance for PStV. Infected plants are rogued from seed fields. The effectiveness of the detection system has been monitored by serological assays (ELISA) of suspicious plants; correlation between diagnosis and confirmation has been essentially 100 % (Demski and Lovell 1985). Even in Florida, Zettler et al. (1993) developed a scheme for production of peanut seed free of PStV and PeMoV by using virus-indexed greenhouse-grown seed.

2.6 Virus and Viroid Disease Transmission Through Vegetative Propagules

A majority of the crops can also be vegetatively propagated which involves the reproduction from vegetative parts of the plants. By in large, most fruit and some ornamental/spices crops are propagated by one or another vegetative propagation methods. From different specialized vegetative structures like runners, suckers, tubers, bulbs, rhizomes and also by grafting and budding/micropropagation processes the plants are propagated. Vegetatively propagated crops like potato, sugarcane, cassava, sweet potato, beet, taro, cocoyam, yam, onion, strawberry, banana, citrus, apple, strawberry and certain ornamental plants are generally multiplied continuously for many years and hence ample opportunities exist for the introduction for one or more virus/viroid pathogens. The vegetative propagules from such infected plants generally lead to reduced yields of subsequent plants.

2.6.1 Production of Virus-Free Plant Material

(a) **Meristem-tip culture**

New accessions released by the plant breeder may be virus-free initially but later will become infected with one or more viruses if grown in an area where virus sources and vectors are prevalent. The need for virus-free planting material has become essential. To supply the healthy planting materials, certification schemes have been developed. Special stocks that form the basis of certification schemes are built up by propagating from single virus-free plants. Such virus-free plants were identified by indexing on indicator hosts, serology, and by use of the appropriate molecular test. In many instances, a new clone with good agronomical and high yielding characters have been rejected due to virus infection of the stock, and presently such cases can be rescued by the method of meristem-tip culture. By removing a small (0.2–0.5 mm) piece of tissue from the meristematic area and culturing on a nutrient medium, it can result in a pathogen-free plantlet for regeneration. The explants sometimes are used from the apical dome, although in most cases one to several leaf primordia of the sub apical region are also included. The number of leaf primordia included depends upon the plant species, the stability of the virus being eradicated and the experience of the operator with respect to root regeneration. Generally apical meristems without having leaf primordia, have the highest probability of producing plant lets that are virus-free. However apical meristems lacking leaf primordia have the lowest probability of surviving in the culture medium.

Another aspect that warrants consideration is the effect of size of the explants and the rate of contamination of culture units. The risk of infection by a virus also increases with explant size. The elimination of the virus also depends on the concentration of the virus in plant tissue and the physiological condition of the mother plant from which a meristem has been isolated (Verma et al. 2004). Even the choice of the culture medium selected is a key step in meristem tip culture success. Besides the shoot tips, even root tissue is now considered to be suitable organ for meristem tip culture, since root meristem has a high potential of cell division. Hosokawa et al. (2004a, b) have produced virus-free plants of chrysanthemum, petunia, cabbage and carnation when root tissue was used for meristem tissue culture.

The combination of thermo- and chemo therapy in meristematic-tip culture were tried throughout the world and have eliminated the virus pathogen and produced virus-free plant material of many species of vegetables, fruits and ornamentals viz., potato (Mori 1971; Pennazio and Redolfi 1973; Faccioli 2001; Wang et al. 2006a; Al-Taleb et al. 2011); sweet potato (Jeeva et al. 2004; Gichuki et al. 2005; Ozturk et al. 2012); cassava (Kartha et al. 1974; Nair et al. 1979; Hegde et al. 2005; Maruthi et al. 2013); sugarcane (Leu 1978; Cha-um et al. 2006; Subba Reddy et al. 2011); peanut (Prasada Rao et al. 1995); impatiens (Milosevic et al. 2011); garlic (Walkey et al. 1987); banana (Berg and Bustamante 1974); chrysanthemum (Verma et al. 2004; Kumar et al. 2009); alstromeria (Chiari and Bridgen 2002); Impatiens hawkerii (Milosevic et al. 2011); strawberry (Cieslinska 2003; Biswas et al. 2007); pear (Cieslinska 2002); Patchauli (Singh et al. 2009).

Through tissue culture even viroid-free plants were developed through meristem tissue culture; examples include *Potato spindle tuber viroid* from potato (Stace-Smith and Mellor 1970), *Chrysanthemum stunt viroid* in chrysanthemum (Hosokawa et al. 2004b).

More information on meristem-tip culture can be obtained from review articles and text book chapters (Kassanis and Posnette 1961; Hollings 1965; Nyland and Goheen 1969; Mori 1971; Murashige 1974; Quak 1977; Walkey 1978; Wang and Hu 1980; Kartha 1981; Bhojwani and Razdan 1983; Grout 1990; Spiegel et al. 1993; Knapp et al. 1995; Jayasinghe and Salazar 1997; Faccioli and Marani 1998; Mink et al. 1998; Zilka et al. 2002; O'Herlihy et al. 2003; Laimer et al. 2006; Zhang et al. 2006; Panattoni et al. 2013).

(b) **Thermotherapy**

The pre-treatment of donor plant to enhance virus eradication by thermotherapy and chemotherapy has proven to be quite useful in achieving the maximum number of virus-free plants. Most viruses either do not replicate or replicate poorly in plants incubated at above 30 °C. Holding virus-infected plants continuously near the maximum temperature will often result into virus-free plants. Developments over the last 20 years in research aimed at investigating the metabolic processes involved in defense mechanisms of plants have suggested an interpretation of the heat treatment effects according to new metabolic "pathways" triggered by the natural antiviral response produced by the infected plant, with particular reference to Virus-Induced Gene Silencing (VIGS) induced by the presence of viral RNA in infected plants (Ruiz et al. 1998). RNA silencing was described as such an effective defense as to constitute an immunity mechanism at the genomic level and there is close relationship between temperature and RNA silencing which seems to act as a means to increase the degradation of virus RNA (Wang et al. 2008). Since Kunkel's first report of the elimination of peach yellows by dry heat and hot water, there have been several examples of eradication of viruses from whole plants. Thermotherapy is used to eliminate virus and viroid diseases from the infected budwood of fruit crops; some examples include, *Citrus tristeza closterovirus* from citrus bud-wood (Cheema et al. 1999; Arif et al. 2005), and *Plum pox virus* from plum, apricot and peach cultivars (Spiegel et al. 1995; Polak and Hauptmanova 2009). Some of the examples where thermotherapy along with meristem tip culture has helped to obtain virus-free plants include garlic with *Onion yellow dwarf virus* (Shiboleth et al. 2001); chrysanthemum with *Chrysanthemum virus B* (Hakkart and Quak 1964); sugarcane with *Sugarcane mosaic virus* (Waterworth and Kahn 1978; Balamuralikrishnan et al. 2002), and in pear with *Apple stem grooving virus* and *Apple chlorotic leaf spot virus* (Wang et al. 2006a).

(c) **Electrotherapy**

Electrotherapy technique has been applied for the elimination of viruses from the vegetative propagules and also from any part of the infected plant. Electrotherapy is a simple method of virus eradication without the need to use any special or expensive equipment. With this technique an electric current is applied to plant tissues in order to disrupt or degrade viral nucleoprotein and thus eliminating its virulence.

Blanchard (1974) obtained virus-free tissues using 1–4 A of direct current pulsed to 6500 V h^{-1}, after 2–3 days. These studies established a basis for electric current mediated elimination of plant RNA viruses. The use of electric current for the production of virus-free plants has been reported for a number of crop plants since the 1990s. In early studies of this kind, voltage of 500 V for 5–10 min was applied to viral infected *Cactanucia* tree stakes showing mosaic symptoms from which up to 90 % virus-free almond plants were obtained (Quacquerelli et al. 1980). This technique was first comprehensively employed for the elimination of *Potato virus X* (PVX) from potato plants in which an electric current of 15 mA for 5 min led to 60–100 % elimination of PVX in various cultivars (Lozoya-Saldana et al. 1996). Electrotherapy has also been used successfully for elimination of *Potato virus Y* (PVY), *Potato virus A* (PVA), *Potato virus S* (PVS) and *Potato leaf roll virus* (PLRV) (Pazhouhande and Mozafari 2001). Similar results were also reported for the elimination of *Onion yellow dwarf virus* and *leek yellow stripe virus* from garlic (Hernandez et al. 1997) and *Banana streak virus* from banana (Hernandez et al. 2002). Even similar technique was also applied to cocoyam (*Xanthosoma sagittifolia*) and grapevine (*Vitis vinifera*) for the elimination of *Dasheen mosaic virus* and *Grapevine leafroll virus*, respectively (Igarza-Castro et al. 2001; Guta et al. 2010).

Hormozi-Nejad et al. (2010) applied the electrotherapy technique to eliminate BCMV by using an electronic current of 15 mA for 10 min. from infected nodal cuttings of common bean and proved to be a most effective treatment. They have also reported that nodal cuttings treated by electricity and then culturing in a medium supplemented with Ribavirin (20 mg/l) resulted in greater elimination of the viruses. It has been postulated that viral nucleoproteins may be denatured by increasing temperature inside the tissue when exposed to electric current.

The electrotherapy technique can also be more effective in combination with chemotherapy in eliminating PVY and PLRV (Dhital et al. 2008; Mahmoud et al. 2009). In addition, the DNA virus known as *Tomato yellow leaf curl virus* (TYLCV) was successfully eliminated from tomato plants using electrotherapy technique (Fallah et al. 2007).

(d) **Chemotherapy**

Chemotherapy is the application of chemicals to an infected or diseased plant that stops (i.e., eradicates) the infection. Although many attempts have been made to utilize chemotherapy, few have been successful against plant virus and viroid diseases.

The requirements of a useful antiviral chemical include abilities to inhibit multiplication, spread, or symptom induction of the virus; be selective enough not to harm the host; have broad-spectrum activity against a number of virus diseases; move systemically in the host; and not have harmful effects on the environment. Numerous compounds have been identified that inhibit the replication of plant viruses. For example, the replication of TMV was inhibited by bitriazolyl compounds (Xia et al. 2006), tylophorine B (Xia et al. 2006), phanantherene-based tylophorine derivatives (Wang et al. 2010a), derivatives of thiadiazole acetamide (Zhao et al. 2006), cyanoaweylate derivates (Chen et al. 2008) and racemic phenanthroindolizidine alkaloids or pure alkaloids (Wang et al. 2010b). However none so far identified is selective enough to be useful against virus diseases of crops. Most inhibitory compounds are analogues of normal nucleoside or nucleic acid base precursors of viral RNAs. One has to know what modifications affect activity against virus multiplication and, in general, what types of alteration are likely to be more successful. There are some early reports that compounds like *ribavirin* and *vidrabine* were most promising against plant viruses (Schneider 1954). Even analogues like 2-azadenine, 8-azaguanine and 5-fluorouracil have also showed plant virus inhibition (Gupta 1985).

Some of the compounds were used *in vitro* studies for virus elimination from meristems, for example, *Ribavirin* (virazole) has inhibited virus multiplication of CMV, PVY, PMV, LSV and TBV in tobacco, apple and lily (Long and Cassells 1986; Hansen and Lane 1985); *Bamboo mosaic virus* in bamboo (Chen and Lu 2000); *Prunus necrotic ringspot virus* in prunus species (Hansen 1984); *Bean yellow mosaic virus* and *Cucumber mosaic virus* in gladiolus (Singh et al. 2007); PVS, PVM, in potato (Griffiths et al. 1990). Even studies of Veelar et al. (1992) indicated the inhibition of *Ornithogalum mosaic virus* in ornithogalum. In tobacco, 2-thiouracil was proved to be effective against PVY (Kassanis and Tinsley 1958). Panattoni et al. (2007) have reported the effectiveness of tiazofurin and mycophenolic acid against *Grapevine leaf roll associated virus 3* in grape explants. Studies with the above cited chemicals as foliar spray and also incorporating in the tissue culture medium have reduced the virus multiplication.

(e) **Cryotherapy**

The technique of cryotherapy of shoot tips has also been used to eliminate virus and virus-like pathogens from the vegetatively propagated plants like banana and temperate fruits (Wang et al. 2006a, b, 2009). In cryotherapy of infected shoot tips were exposed briefly to liquid nitrogen (−196 °C). The cryotherapy results in a high frequency of virus-free regenerates. Thermotherapy followed by cryotherapy of shoot tips can be used to enhance virus eradication.

In cryotherapy, plant pathogens such as viruses, phytoplasmas and bacteria are eradicated from shoot tips by exposing them briefly to liquid nitrogen. Uneven distribution of viruses and obligate vasculature-limited microbes in shoot tips allows elimination of the infected cells by injuring them with the cryo-treatment and regeneration of healthy shoots from the surviving pathogen-free meristematic cells. Cryotherapy of shoot tips is easy to implement. It allows treatment of large numbers of samples and results in a high frequency of pathogen-free regenerants. Difficulties related to excision and regeneration of small meristems is largely circumvented. To date, several pathogens in banana (*Musa* spp.), *Citrus* spp., grapevine (*Vitis vinifera*), *Prunus* spp., raspberry (*Rubus idaeus*), potato (*Solanum tuberosum*) and sweet potato (*Ipomoea batatas*) have been eradicated using cryotherapy. These pathogens include nine viruses (*Banana streak virus, Cucumber mosaic virus, Citrus tristeza virus,* Grapevine virus A, *Plum pox virus, Potato leaf roll virus, Potato virus Y, Raspberry bushy dwarf virus, Sweet potato feathery mottle virus* and *Sweet potato chlorotic stunt virus*), *Sweet potato little leaf phytoplasma* and *Huanglongbing* bacterium causing 'citrus greening'. Cryo-preservation protocols have been developed for a wide variety of plant species, including agricultural and horticultural crops and ornamental plants, and can be used as such or adjusted for the purpose of cryotherapy (Wang et al. 2009).

2.6.2 Production of Virus-Free Stock Material

For achieving the global food security, FAO has always recognizes the potential contribution for achieving higher yields from virus-free planting materials. To raise the healthy crops successfully for higher yields, quality developed planting material (QDPM) is a prerequisite. The expert consultation group under the guidance of FAO has developed programmes to guide the production of clean, disease-free planting materials of vegetatively reproduced crops (FAO 2010; Fajardo et al. 2010). These consultation groups have prepared QDPM protocols for the easy monitoring of production practices and the distribution process of quality planting material. Any QDPM scheme for plant multiplication must comply with prevailing national seed regulations.

The production of virus-free stock material depends on the following activities. (1) recognition of the virus diseases by studying their characteristics, (2) development of reliable indexing methods to retrieve any existing healthy stock, (3) establishment of clones of virus-free foundation stocks by employing different techniques of chemo, thermo therapy, and tissue culture, (4) propagation of the pathogen-free plants under maximum hygiene and distribution of the planting material. If the nucleus seed/vegetative propagule of the healthy stock can be obtained through the methods described earlier, then stocks can be multiplied and kept healthy for an indefinite period. As early as 1961, Kassanis and Posnette have stated that about half of the viruses affecting vegetatively propagated plants could be eliminated by heat treatment.

As plant propagators carefully select the best looking and excellent performing plants for propagating new plants, they may knowingly or unknowingly avoid many of the virus diseases. This visual selection process helps in diagnosis of the diseased plants based on external visual symptoms.

2.6.3 National Certification System for Tissue Culture Plants NCS-TCP in India

In India, the Department of Biotechnology (DBT), Govt. of India, under the "Seeds Act, 1966" of the Ministry of Agriculture has established the National Certification System for Tissue Culture Plants (NCS-TCP) for ensuring production and distribution of quality tissue culture planting materials and to regulate its genetic fidelity as prescribed (NCS-TCP 2008).

The Project Monitoring Unit (PMU), Biotech Consortium India Limited (BCIL) is facilitating coordination among the various components and monitoring the activities of Accredited Test Laboratories and Recognized Tissue Culture Production Facilities. In India more than 70 commercial tissue culture production units have been recognized by the Department of Biotechnology (DBT), Govt. of India under the National Certification System for Tissue Culture Raised Plants (NCS-TCP).

The work carried out at the NCS-TCP accredited centers located at Bangalore, Coimbatore, Calicut, Trivandrum for developing virus-free plants was conducted for the following crops: Banana, Citrus, Apple, Strawberry, Black Pepper, Cardamom large, Cardamom small, Vanilla, Potato, Sugarcane, Alocassia, Alstromeria, Anthurium, Asparagus, Caladium, Calla Lily, Carnation, Chrysanthemum, Colacasia, Cordyline, Geranium, Gerbera, Ginger, Lily (Asiatic/Oriental lily), Limonium, Orchids, Philodendron, Spathyphyllum and Syngonium. The virus, viroid and virus-like diseases present in these crops were determined by using protocols for ELISA, DIBA, RT-PCR and NASH.

In India even though several plant tissue culture laboratories are producing a large number of tissue culture raised plants of commercial crops, primarily due to limited facilities to perform tests to establish their quality and verification of their virus-free nature, another agency was required to overcome this limitation. The Department of biotechnology (GOI) has established a multi-institutional collaborative programme with the goal of ensuring production of virus-free plants of assured quality raised through tissue culture. The main center is at IARI, New Delhi (India) with five satellite centers at NCL, Pune; Teri, New Delhi; IHBT, Palampur; IIHR, Bangalore and Spic, Chennai.

The objectives of this project are: (1) to develop protocols for virus diagnosis and quality testing for clonal fidelity, and (2) to ensure that tissue culture-raised plants are virus-free and of an assured quality. Tissue culture-raised plants are tested for clonal fidelity using standard molecular marker techniques for DNA fingerprinting such as RAPD, RFLP, AFLP, and SSR.

2.7 Success Stories of Production of Virus-Free Plant Propagules

For any country to have a successful disease-free seed programme, two requirements must be met: (i) healthy plants of the proper type basic plants as selected from a field population of high-class seed, and (ii) the progeny of basic plants multiplied as separate clones under very low infection pressure of the seed-borne pathogens to provide pre-basic seed for further multiplication, to provide basic and certified seeds.

Crop plants have a greater potential for improved yield and quality when they are free from harmful diseases. Stocks of the vegetatively propagated crops like potato, sugarcane, cassava, sweet potato, beet, taro, cocoyam, yam, onion, strawberry, blueberry, mulberry, banana and certain ornamental plants are multiplied continuously for many years have ample chances for infection with one or more viruses/viroids and show a decline in growth and yield. Plant growers, especially nursery men and horticulturists have greatly contributed to the spread of these diseases inadvertently by using infected planting material (scions or root stock). Because of the catastrophic losses they cause, intensive attempts were made to produce pathogen-free planting material. The production and distribution of virus-free propagating materials has proved highly successful in controlling virus diseases in many crops and promise to be of wider application in others. Cropley (1979) has provided the details of production and practical benefits of virus-free propagating material of fruit crops. The term "virus-free" denotes material found to be free from known virus and virus-like agents. The term is also misused nowadays to label plants that have been treated to free them from some of these viruses. Yet alternatives such as "virus tested" or "free from the following viruses…." is appropriate. Prof. Hollings defined virus-free plants as "free from the known and specified viruses for which tests have been done". The majority of the plants are raised through true seeds. More than two hundred and thirty one (231) viruses and viroids are seed transmitted (Sastry 2013a). Similarly, whenever a crop is multiplied through vegetative propagules like tuber, rhizome, stem piece, runner, bulb, sucker, or corm, when these propagules are once infected with virus and viroids, then these infectious pathogens are vertically carried to the next generation through the vegetative planting materials. Exhaustive information is available on how true seed and vegetative propagative plant materials can transmit viruses, how these pathogens are diagnosed, and what procedures are required to produce disease-free materials (Sastry 2013a, b). Additional information on seed-borne viruses can be obtained from review and text book chapters (Neergaard

1977a, b; Agarwal and Sinclair 1996; Maude 1996; Albrechtsen 2006; Sastry 2013a).

Some of the successful stories of meristem tip culture are found in crops like sugarcane (Balamuralikrishnan et al. 2002; Mishra et al. 2010), potato (Faccioli 2001; Awan et al. 2007), sweet potato (Mervat and Ashoub 2009), cassava (Wasswa et al. 2010).

Throughout the world, meristem tip culture technique is being used for production of virus and virus-like diseases-free planting materials primarily for vegetative propagated plants. In majority of the countries the responsibility of production and distribution of the virus-free plant material lies on the state and central governments. The researches carried out on number of crops to develop virus-free plant materials and have proved that plant tissue culture with or without chemo and thermotherapy, is one of the most popular and successful techniques and is proved by the examples cited in Table 2.1.

For a number of crops, chemo- and thermo therapy along with meristem tip culture has helped in the production of virus-free plants and this has been reviewed by number of researchers (Kassanis and Posnette 1961; Nyland and Goheen 1969; Mori 1971; Quak 1977; Kartha 1981; Mink et al. 1998; Faccioli and Marani 1998).

2.7.1 Certification Schemes

According to the European and Mediterranean plant protection organization (EPPO), a certification scheme is defined as a system for the production of vegetatively propagated plants for planting, intended for further propagation or for sale, obtained from nuclear stock after several propagation stages under conditions ensuring that stated health standards are met. The filiation of the material is recorded throughout the scheme.

Certification is also defined as a procedure whereby candidate mother plants, to be used as source of material for propagation, must undergo required treatments to secure trueness-to-type and freedom from all number of pathogens affecting the crop, as specified by regulations officially issued, or endorsed, by competent governmental agencies. There are two major kinds of certification viz., (1) voluntary and (2) compulsory. In case of voluntary certification, there will be increased demand of planting material from growers', with known sanitary status in order to establish their plantations, especially if such plants are long lasting woody plants. Voluntary certification is regimented by regulations issued by a "certifying authority" i.e., usually a branch of the country's Ministry of Agriculture or the equivalent, but the implementation cannot be forcefully imposed. On the other hand, compulsory certification is enforced whenever the crops existence is threatened by pests, diseases and other factors and it becomes essential to prevent the dissemination of destructive pests and diseases at this juncture. The dispersal through propagative material may establish an initial infection foci and further spread by an efficient vector or some other effective means could cause heavy damage to the crop. Hence, the respective

Table 2.1 Production of virus-free plants through tissue culture

Crop	Virus	Reference
Apple	Apple mosaic virus	Hanson and Lane (1985), Gabova (1989), Yamana and Munakata (1991), Bhardwaj et al. (1998)
Apricot	Plum pox virus	Koubouris et al. (2007)
Banana	Cucumber mosaic virus	Helliot et al. (2004)
	Banana streak virus	Kenganal et al. (2008)
Begonia	Prunus necrotic ringspot virus	Verma et al. (2005)
Black currant	Raspberry ringspot virus	Kolbanova et al. (2004)
Blueberry	Blueberry scorch carlavirus	Postman (1997)
Cassava	Cassava brown streak virus	Kaiser and Teemba (1979), Adejare and Courts (1981), Ng et al. (1992), Wasswa et al. (2010)
	Indian cassava mosaic virus	Vasanthi et al. (2001)
Chrysanthemum	*Chrysanthemum virus B*	Hakkart and Quak (1964)
	Cucumber mosaic	Kumar et al. (2009)
	Tomato aspermy virus	Kumar et al. (2009)
Citrus spp.	Indian citrus ringspot virus	Sharma et al. (2008)
Cymbidium	Odontoglossum ringspot virus	Freitas and Rezende (1998)
Dahlia	Dahlia mosaic virus	Sediva et al. (2006)
Garlic	Garlic virus disease	Bruna (1997), Ghosh et al. (1997), Robert et al. (1998), Ayabe and Sumi (2001)
	Garlic mite-borne virus	Bertaccini et al. (2004)
	Onion yellow dwarf virus	Shiboleth et al. (2001)
Grapevine	Grapevine viruses	Milkus et al. (2000)
Horseradish	Turnip mosaic virus	Uchanski et al. (2002)
Lily	Lily symptomless virus	Nesi et al. (2009)
Peanut	Peanut mottle virus	Chen and Sherwood (1991)
Pear	Apple stem grooving virus	Wang et al. (2006a)
	Apple chlorotic leaf spot virus	Cieslinska (2002), Wang et al. (2006a)
	Pear viruses	Tan et al. (2010)
Plum	Prunus necrotic ring spot virus	Janeckova (1993), Dziedzic (2008)
	PPV, PDV	Janeckova (1993)
Potato	Potato virus Y	Nascimento et al. (2003)
	Potato viruses	Faccioli and Colombarini (1991), Faccioli (2001), Zaman et al. (2001), Nagib et al. (2003), Fang et al. (2005), Awan et al. (2007), Al-Taleb et al. (2011)
Shallot	Onion yellow dwarf virus	Fletcher et al. (1998)
	Shallot latent virus	Fletcher et al. (1998)
Strawberry	Strawberry mottle virus	Cieslinska (2003)
	Strawberry viruses	Biswas et al. (2007)
Sugarcane	*Sugarcane mosaic virus*	Waterworth and Kahn (1978), Ramgareeb et al. (2010)
	Sugarcane yellow leaf virus	Chatenet et al. (2001), Fitch et al. (2001), Balamuralikrishnan et al. (2002), Parmessur et al. (2002), Mishra et al. (2010)

(continued)

Table 2.1 (continued)

Crop	Virus	Reference
Sweet potato	Sweet potato feathery mottle virus	Love et al. (1989), Nagata (1991), Pozzer et al. (1995), Chen et al. (2004), Jeeva et al. (2004), Gichuki et al. (2005), Mervat and Ashoub (2009)
	Sweet potato leafcurl virus	Green et al. (1992), Torres et al. (1996)
Taro	Taro viruses	Li et al. (2002)
Yam	Yam viruses	Mantell et al. (1979)

state and central governments have made certification as an integral part of sanitary improvement programmes and is often the only way to curb certain devastating diseases and pests of primarily woody crops. It is a long standing endeavor that must continue indefinitely for attaining the desired health level of the crop continuously for longer periods. Efforts by the research projects supported by government and regional authorities are involved in majority of the countries. For example, the bylaws generated by the directives promulgated in European Union member states that the national certification schemes have been in operation for many years and created the impact of using compulsory certification (Martelli 1992). Continuing from 1998, Martelli and Walter (1998) have provided different virus certification schemes of grapevines in detail.

Certification safeguards both the nursery man who sells vegetatively propagated plant material and the grower who buys the nursery man's products. Hence, certification schemes have been established to comply with official standards/regulations set by the national and international authorities to guarantee the quality of propagative material to trueness-to-type (genetic purity) and absence from specified pathogens in the micropropagated plants (Golino and Savino 2005). Usually, it is a domestic program consisting of multiplication and distribution of plant material that involves a combination of activities. The process basically involves assessing the risks (pathogens and pests), selection of clean-planting material, virus testing, micropropagation and tests for genetic fidelity. A certificate is only provided to the plants that are produced as per the directive of the scheme.

Certification schemes depend on the objective of the program and vary considerably depending on crops and countries. Initially, certification schemes originated as a method to control virus diseases in seed potatoes and tree crops (Anon 1992). In the absence of effective control strategies and host resistance, cultivation of virus-free planting material has become the choice for growers and vigorously promoted by the public and private agencies associated with welfare of agriculture sector. Certification scheme usually involves several components to ensure success of the program (Hollings 1965; Meijneke 1982; NCS-TCP 2008). Some of the key components are: (i) standards that clearly define the purpose of certification and protocols/guidelines for producing certified planting material; (ii) guidelines for selection of mother-stocks to be used in the micropropagation program; (iii) availability of reliable indexing methods for the detection of regulated and unregulated quarantine pests and pathogens; (iv) protocols and procedures for production of virus-free planting material; (v) guidelines for maintaining the health of mother-stocks and propagated material; (vi) guidelines for labeling planting material for traceability; (vii) guidelines for national and international distribution of planting material (viii) guidelines for infrastructure and technical competence; and (ix) guidelines for supervision and monitoring compliance by the producers.

The systems of rules, regulations and check inspections designed for the production of virus-free material, constitute crop certification schemes. Selection is a basic step in programme for clean stock production. It determines that a specific clone possesses the desirable growth and yield characteristics of the cultivar and that the cultivar is correctly identified. Selection is the responsibility of horticulturists cooperating in the clean stock programme. It will be effective when all the virus and viroid diseases are recognized. Indexing tests suggests that many cultivars are wholly infected with one or more viruses and a successful program is achieved when elimination of pathogens by methods like thermotherapy, chemotherapy, cryotherapy, electrotherapy, and meristem tip culture. These methods have enabled plants to be freed from more than two hundred viruses. Disease recognition, indexing and therapy are the responsibility of plant virologists. The recent reviews catalogue far more successes than failures.

2.7.2 Schemes for the Production of Certified Propagative Material

Even though temperate region is the primary zone for cultivation of major fruit crops like apple, grapevine, rubus, strawberry, citrus species, plum, peach apricot, sweet cherry etc., and majority of these fruit crops are grown in certain parts of tropics but to a very limited extent. These fruit crops are affected with number of virus and viroid diseases which are responsible for severe yield losses and resulting in early death of the plants. Production of healthy planting material through certification schemes is the prime solution for success of cultivation of these fruit crops. Since the experiments carried out in temperate region have resulted in the production of healthy planting material by using techniques like thermotherapy, chemotherapy and tissue culture and they are included in the text to grew more clarity of the topic. In this chapter the certification schemes of deciduous fruit trees, EMLA scheme for virus-free fruit trees, strawberry certification programme and others were also discussed as the similar type of virus and viroid diseases are occurring in some parts of tropics in certain fruit crops.

As suggested by EPPO, the Italian legislation foresees a certification by filiations; the steps for multiplication are strictly linked allowing, in each stage of the scheme, the identification of the original source of the material under propagation. The following levels are considered for production of certified planting material.

(a) **Nuclear stock (prebasic plants)**

This is the first step of the certification scheme. The prebasic plants must show, varietal characteristics corresponding to those typical of that variety and must have been found free from the major virus and viroid diseases. They are maintained in a screen house, in containers with sterile growth medium, isolated from the soil. Each plant must be checked for trueness to type during its' vegetative stage. Each single plant has to be tested for virus and virus-like organisms, according to the specific annexes. Clonal origin must be identified by a number that must be maintained throughout the propagative production and commercialization process. Plants satisfying the above mentioned conditions are used for production of prebasic plant material.

(b) **Propagation stock (basic plants)**

Prebasic plant material is used for the production of basic plants. Basic plants are maintained in a screen house and must satisfy the conditions described for maintenance of prebasic plants. Material from these plants can be used as basic material.

(c) **Certified mother plants**

Basic plant material is used for the production of certified mother plants. *In vitro* micropropagation can be used to produce certified mother plants. Every year, at least 10 % of these plants must be tested for virus and virus-like diseases according to specific annexes. Mother plants for production of certified plant material must be separated by at least 100 m from any other non-certified plant material and maintained under continuous surveillance to control pests and diseases and the soil must be free of virus transmitting nematodes. These plants are maintained in open air under conditions ensuring freedom from reinfection and trueness to the variety and are used for the production of certified plant material.

(d) **Certified plant material**

Certified plant material is used for production of certified plants. These are kept in plots free of virus transmitting nematodes and separated at least 5 m from other non-certified material plots. Certified plants are visually checked for harmful organisms and diseases that impair quality, or any signs or symptoms thereof. The nurseries are officially inspected by the responsible authority at least once a year. The responsible authority issues the nursery men the appropriate number of labels for individual labeling of certified plants. Rowhani et al. (2005) have reviewed testing and certification schemes of *Vitis* and *Prunus* species.

Some of the outstanding examples of the crops where indexing and certification schemes which have helped the rehabilitation of the crop industry are citrus bud wood certification, stone fruit virus diseases certification, strawberry certification, clear stock programme of grapes, avocado certification schemes, etc. Even for the tuber crops like potato, sweet potato and beet root, the certification schemes are well worked out and are being implemented wherever these crops are grown. The quality of these vegetative planting materials depends on its trueness to cultivar type, its vigor and health. Each of these characters is amenable to a certain extent to control by certification, which may be regarded as an administrative means of quality control of planting materials. Based on the information provided in this chapter, one will get the broader idea that the success of certification schemes depends on the collaboration and coordination among different public and private institutions whose goal is to improve the quality of propagative plant material and has to be considered as a useful tool. Some of the widely adopted certification schemes are as follows:

2.7.3 Certification Schemes of Economically Important Crops

Since virus and viroid diseases cause heavy yield and economic losses, use of virus-free planting material raised through different methods discussed above are being used and are achieving satisfactory yields of the crops by managing the virus and viroid diseases. In almost all countries, research carried out for production of virus-free material have resulted in the development of different certification schemes which are developed with the involvement of National and International organizations. Some of the well-established certification schemes which are followed in different countries are presented herein:

2.7.4 Virus Certification of Deciduous Fruit Trees

In general, virus diseases in perennial fruit crop plants, such as apples, peaches, plums, raspberry etc. are potentially damaging. Viruses can remain latent, spreading through an orchard and inflicting damage, often without the growers' knowledge. Latent virus infection can produce small to moderate losses in fruit production. Often growers can maintain the productivity of diseased orchards at a profitable level by removing infected plant parts and replacing dead trees to limit the spread of the virus. Sometimes, however, losses are severe, and an acute viral infection can require tree removal.

Majority of the fruit crops like apple, peach, plum, strawberry, and raspberry are very extensively grown in temperate countries and to a very limited area in tropics. Since deciduous fruit crops occupy major areas in U.S., Canada, Europe, Mediterranean Region, the certification schemes were developed. Producing virus-free plants is important because viruses can stunt tree growth, reduce fruit yields, produce deformed, off color foliage, and increase susceptibility to other diseases. Virus-free trees grow faster, flower earlier and may be resistant to disease and environmental stresses.

In 1954, directors of U.S. Agricultural Experimental Stations established IR-2 supported by the regional research funds of the Hatch Act (US) and was officially activated in July 1955. One of the major objectives of IR-2 project is to obtain apparently virus-free valuable cultivars and clones of deciduous fruit trees, verifying their freedom from viruses, maintaining them in isolated repositories, and distributing small amounts of propagating materials to research or regulatory scientists for research or release to industry (Fridlund 1968). Through this project the generated virus-free plant materials have been distributed to 40 states, 5 Canadian provinces, and 40 other nations, the latter via the USDA in response to requests from USDA-AID, FAO of the UN, and other agencies. The details of IR-2 project have been reviewed by Fridlund (1980).

In 1992, the programme was renamed as National Research Support Project 5 (NRSP-5). The only facility in the United States that tests for viruses in fruit trees is the NRSP-5 which is located at the Washington State University Irrigated Research and Extension Center in Prosser, Washington and implements the nation's virus protection program for all deciduous fruit trees. NRSP-5 is responsible for providing sources of deciduous fruit tree propagation materials free of virus and virus-like diseases. During 1997–2002 NRSP-5 also developed, evaluated, and implemented new technologies for virus detection and the elimination of viruses and virus-like agents from commercially important cultivars (NRSP-5 1997). Prior to the establishment of NRSP-5, viruses abounded in every fruit-growing region in the US. It was responsible for providing sources of deciduous fruit tree propagation materials free of virus and virus-like diseases. The success of this project has resulted in a dramatic reduction in the incidence of viral diseases. Cembali et al. (2003) have reported the benefit estimates to 3 economic sectors-nurseries, producers, and consumers. The total benefits for all the 3 sectors were approximately $227.4 million a year, or more than 420 times the cost of the programme.

(a) **Meristem-tip culture in deciduous fruit trees**

For large scale production of virus-free planting materials of deciduous fruit crops is being done by meristem tip culture. For production of planting materials, virus-free mother plants, (root stocks and scion) are identified and tested regularly, utilizing ELISA, PCR and other molecular techniques, wherever is applicable and appropriate. Many samples from scion trees and layering beds or seed trees can be tested each year due to the sensitivity and accuracy of these tests and the shorter turnaround time to obtain the results.

Like the other crops discussed in this book, even the production of virus and virus-like disease-free plants of deciduous crops were produced by meristem tip culture from virus-free mother plants. This technique is widely used against crops like apple (Hansen and Lane 1985; Gabova 1989; Yamaga and Munakata 1991); stone fruits (Hadidi 1995). The general protocol of tissue culture used in the production of virus-free deciduous plants is presented in Fig. 2.2.

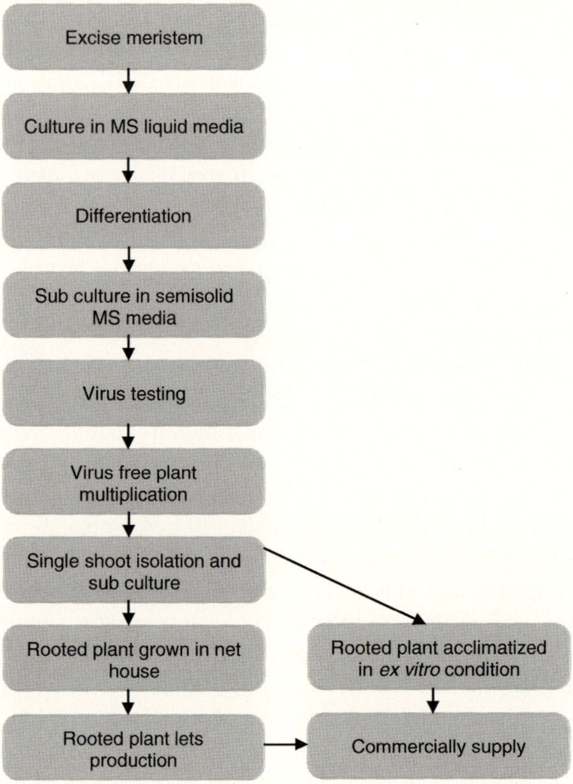

Fig. 2.2 General protocol for commercial production of virus-free deciduous fruit crops

In India also, virus-free certification programmes are designed to deal with general virus and virus-like problems. Virus indexed mother stock blocks of trees are established and these plants provide the nurseries with the basic materials used in producing the certified plants are sold to orchardists. Virus indexed, healthy material used as nuclear, foundation or registered mother block must be isolated from non-indexed plants. This is done to prevent possible disease spread into the virus tested material. Rouging of virus infected trees, when found, and the elimination of other plants that may severe as virus reservoirs are important in maintaining disease free mother blocks. In all planting stock certification programmes the basic procedures must be followed and adhered to by programme particulars for such schemes to succeed (Bhardwaj and Thakur 1993; Handa and Thakur 1999).

(b) **Micrografting**

Micrografting is an *in vitro* grafting technique and this procedure involves the placement of a meristem tip or shoot tip explant onto a decapitated rootstock that has been grown aseptically from seed or micropropagated (Hartmann et al. 2002;

Zilka et al. 2002). The *in vitro* micrografting of apple and pear was successfully achieved. A significant difference was noticed in apple micrografting success while using *in vitro* micropropagated shoots as rootstocks (60 %) when compared with the use of *in vitro* germinated seedlings (48 %). The same micrografting success (60 %) was achieved for pear micrografting with both kinds of rootstocks.

Some tree fruit crops are either difficult to establish with *in vitro* culture from excised meristems, or difficult to root. These conditions have been overcome in *Citrus* (Nauer et al. 1983), *Malus* (Huang and Millikan 1980), and *Prunus* (Deogratias et al. 1986) by 'micrografting' shoot tips or apical meristems onto *in vitro* grown seedling rootstocks, which are later transplanted to soil after the graft have become established. Pears, which grow easily from meristems, are difficult to root. Under such cases the *in vitro* plants could be removed from tissue culture when they had elongated to 1 cm and cleft-grafted onto potted seedlings. Selecting rootstock seedlings with distinct leaf morphology and/ or color can help differentiate micro grafts from rootstock sprouts. The advantage of this technique is the small micrografted trees are a convenient way to exchange germplasm between countries (Navarro et al. 1975).

(c) **EMLA Scheme for virus-free fruit trees**

At East Malling and Long Ashton research stations (UK), EMLA-certification schemes were developed for apples, pears, raspberry, cherries etc. In 1960s the two research stations set up the 'EMLA Scheme' so that virus-free material (free from all known viruses) could be made available to the fruit industry. There are three levels of certification 'A' and Special Stock (SS) for fruit free root stocks and the third the 'Elite'. S.S. Certification also applies to mother trees (Scion wood source) and nursery trees of the most used commercial cultivars. The 'A' certificate is mainly to ensure identity 'trueness-to-variety', but visible symptoms of diseases, including viruses, can fail the stock. No isolation or source requirements have to be met. The SS Scheme requires that mother trees and root stock stool-beds, layer beds or hedges shall be: (1) without symptoms of virus/phytoplasma diseases and with vegetative characters typical for the cultivar, (2) From an approved source obtained through the Nuclear Stock Association, (3) Grown in isolation; 100 m for plum and cherry Mother Trees and hedges (vulnerable to infection via pollen), 50 m for plum and cherry layer beds and stool beds of apple, pear and quince within the isolation area all fruit trees are not eligible for certification and all related species such as wild plum and hawthorn must be absent. (4) Grown on sites where soil sampling has found no nematodes species known to be a virus vectors. The source plants (2 above) are supplied by the research stations and are from plants tested and found free from all detectable viruses. Root stock layer beds or cutting hedges or mother trees can be used for 12 years, subject to satisfactory health trees at 5 year intervals. The nursery trees may be entered for a special stock certificate. These trees must be raised from special stock rootstocks and scionwood from the nursery man's own, special stock mother trees or 'increase blocks', or rootstocks and scion wood obtained from the nuclear stock association (tree fruits).

These SS nursery trees must be grown at least 25 m from non-special stock plants. In the Elite scheme which was started in 1976, some nurseries were allowed to propagate elite root stocks which can be supplied to the SS nurseries. This was mainly because East Malling has not been able to provide source material in sufficient quantity to meet the needs of SS nurserymen. At UK, nematode transmitted diseases are major concern and schemes like Foundation strawberry, Elite and strawberry, Elite and Special Stock Cherry, Elite and Special Stock plum, Elite black current and A-plus and AA-hp were developed.

All the initial root stocks in the EMLA scheme was raised at East Malling and over 80,000 root stocks being supplied in 1969, the first year of issue and by July 1979, over 400,000 have been supplied, including 114,000 MM 106 and 75,000 M. 9 apple root stocks. Even 6073 mother trees for scion wood have also distributed. This material was supplied to the nurseries within the Nuclear Stock Association (Tree Fruits Ltd.) where it is propagated in isolation from virus-infected material and sold under Ministry of Agriculture's Special Stock (SS) Certification.

In UK, field trials were conducted during 1972, on seven farms in Kent, representing a range of soils and the observations for four cropping years, the EMLA trees of cox orange on M.9 have cropped 40 % more heavily than non-EMLA trees. In the same trials, the cox on MM-106 EMLA was slightly more vigorous than the non-EMLA trees, but the cropping was virtually the same. The EMLA material and the special stock certification scheme for fruit trees and root stocks had improved the general health of the planting stock and had allowed the nursery men to produce trees that were more consistent in quality and established more uniformity.

In the European Union, where nearly 35 countries are associated, the European plant protection organization (EPPO) plays an integral role in developing the protocols and published certification schemes for fruit crops like grapevine, rubus, strawberry, citrus species, plum, peach apricot, sweet cherry etc., in raising certified plant material. They also suggest the most reliable diagnostic tests against virus and virus-like diseases (EPPO 1991, 1992, 1994). The general procedures recommended for the production of certified virus-free plant material of deciduous fruit crops in EPPO countries are being followed.

In Australia also, the control programme for apple virus diseases involves indexing cultivars and root stocks and eventual distribution of clean stocks. Short term measures include the annual examination of bud wood sources in the main nurseries to ensure that cuttings are taken from symptomless trees (Shea 1964). More information on the production of virus-free deciduous fruit crops plant-material by different schemes and techniques is available in different review articles of Moniau et al. (1977); Barba (1998); Mink (1998); Rowhani et al. (2005).

2.7.5 Strawberry Certification Programme

Strawberry plants are propagated through runners and also by micropropagation. Depending on the variety and environmental factors, strawberry plants do not induce distinct symptoms on foliage but considerable reduction is noticed both in productivity and fruit quality. In almost all strawberry growing countries virus and virus-like diseases are systemically infected and hence the pathogen is paused from one vegetative generation to next. So to raise the disease-free stock plant, virus elimination is a pre-requisite for successful strawberry production. The certification schemes for production of virus free planting material are usually organized and administered by government (National, State or Province) or government officiated bodies similar to other crops.

In the early years, the programme for the certification of strawberry plants was based on virus testing, grafting/mechanical sap inoculation, indexing for viruses by ELISA or other serological tests, virus-free plants multiplication in isolation and net houses and distribution to the farmers as certified stock plants. Strawberry certification scheme was developed by the Arkansas Agricultural Experiment Station in cooperation with the Arkansas State Plant Board (Fulton and Seymour 1957). Plants were raised in screen houses and multiplied through runners. During the winter months the mother plants growing in the crocks of sand were indexed for the presence of viruses. After no additional 'daughter plants' were to be collected from a 'mother plant', use is made of runner grafts or runner to petiole grafts (Goheen 1955). In indexing each 'mother plant' at least 10 and often 25 or more grafts are made and multiplied in screen houses. If obvious virus symptoms appear in any of the indicator plants, the mother plant from which the grafted tissue is obtained is discarded. The first year's field increase of strawberry plants from the screen house plants was produced by the Arkansas Agricultural Experiment Station at one or more of its sub-stations. The stations are isolated from the commercial strawberry production. These field plantings are subjected to inspections by inspectors. These plants are sold to commercial growers as 'Arkansas certified purple tag plants'. Similarly 'blue tag' and 'green tag' plants are produced by growers. However, the difference is that, in the case of certified blue tag plants', the starting stock plants used were from certified 'purple tag stock', and in 'blue tag' plants were used in field production of 'green tag' plants. Strawberry plants produced under these certification programmes are planted throughout Arkansas in a variety of situations.

Similar strawberry runner distribution scheme is operating in Oregon State. Strawberry certification standards in this state are set up by a certification board at Oregon State University which also operates a strawberry plant propagation Centre. Indexing of each mother plant is accomplished by making three-leaf petiole grafts, one known to be infected with mild mottle, one infected with vein banding and the third one virus-free. This method reveals not only virus-free infection but also the components of virus complexes (Dobie et al. 1958). Plants passing this test are propagated and runners distributed to growers and co-operating institutions. No virus disease tolerance is allowed in the propagation center.

However, for nurseries growing certified plants with extremely low tolerance up to 0.5–1 % infection is allowed.

In U.K., schemes are developed for strawberry which is supervised by the Ministry of Agriculture, Fisheries and Food (MAFF), virus-tested stocks are propagated by commercial growers, with the Agricultural Development and Advisory Service (ADAS) providing the technical support. In October 1953, the Nuclear stock Association Ltd. (N.S.A.) was formed and the aim of the N.S.A. was to take over from the National Agricultural Advisory Service the practical responsibility of receiving virus-tested runners from East Malling Research Station and multiplying them in isolation on a regular basis to meet the requirements of plant raisers and fruit growers for replacement stocks. In this scheme, before planting the crop the sites and the ground below adjacent hedges are sampled to 23 cm depth by ADAs advisors for the nematode vectors and the lands free from

Propagation of special stock strawberry plants tested plants from E.M. glass house	
Foundation stock	NSA gauze houses
Elite stock	NSA isolated fields
Special stock	Commercial propagators with one mile isolation Fruit growers

the nematode vectors only are utilized for the propagation of certified material.

East Malling Research Station retains the responsibility for maintaining the virus-indexed mother plants of strawberry (the nucleus) and annually release about 12 cultivars, 50–200 plants of each cultivar, is made to N.S.A. for propagation as foundation on stock by growers, who are appointed by N.S.A. and different stages are represented in the table, and each stage is a single years propagation. The schemes stipulate the conditions of production at each stage and periodic inspections are made by MAFF Offices during the growing season.

(a) **Meristem-tip culture for production of virus-free plantlets of strawberry**

Meristem tip culture is unique technique to produce pathogen free planting material and virus elimination is most important factor for successful strawberry production. In this technique the runner tips of strawberry were used as explants and cultured in liquid MS medium having required components. In liquid MS medium, the meristem developed tiny shoots after 4–5 weeks. Before mass micro propagation, the meristem derive plantlets were tested for virus detection by serological techniques. The mass micro propagation of virus free plants was done in MS medium containing BA and Kin. The micro shoots developed were isolated and cultured on half strength of MS for root induction. Well-developed rooted plantlets were planted in net house. The plantlets developed from meristem tip culture were symptomless and fruit yield of meristem derived plants was more than that of runner derived plants. More details for this procedure can be had from articles of Spiegel (1998); Khanam et al. (1998); Biswas et al. (2007). The details of micropropagation technique for production of virus-free strawberry are presented in Fig. 2.3.

Fig. 2.3 Virus-free plant production in strawberry by micropropagation technique. *Source* Laurence Bourrain, Ctifl

Laurence Bourrain and his associates in 2009 from Ctifl *in vitro* laboratory - Bellegarde, France, have tested the strawberry mother plants (initial material) against 5 viruses by ELISA and successfully produced virus-free strawberries plants by micro propagation technique and have produced 400,000 plants by year for nurseries.

(b) **The future of fruit propagative certification schemes**

The fruit tree 'virus' research over the past several years, is widely accepted by the fruit industry and by scientists that 'virus' certification programs for deciduous fruit tree in the US and Canada have been remarkably successful.

The combination of state and national plant quarantine laws, state certification programs, grower education efforts, and continuing research and development by a dwindling group of fruit tree researchers have combined to greatly reduce the impact of 'virus' diseases of fruit trees. Because of changing consumer preferences and the competition for new international markets, US and Canadian fruit growers and nurseries will continue to find, test, and distribute many new fruit cultivars and rootstocks. If quarantine and certification programs are not continued in combination with supportive research programs, one can except that future changes in the fruit industry will create an environment for the resurgence of some of the 'virus' diseases that are currently viewed as unimportant.

2.7.6 Virus-Free Banana Production Certification

Some of the viruses that have been identified as pathogens in banana (*Musa* spp.) are *Abaca mosaic virus,* a synonym for *Sugarcane mosaic virus* (SCMV), *Banana bract mosaic virus* (BBMV), *Banana bunchy top virus* (BBTV), *Cucumber mosaic virus* (CMV), and *Banana streak virus* (BSV) (Jones 1991; Jeger et al. 1995; Selvarajan and Balasubramanian 2008). Of these five viruses, SCMV is reported to infect abaca (*Musa textilis*), which yields the Manila hemp of commerce in the Philippines. The other four cause yield losses of considerable economic importance in various banana cultivars (Jones 2001a, b). Techniques like ELISA, PCR and other molecular methods have been applied for detection of viruses in different banana-growing countries (Shamloul et al. 1995; Geering and Thomas 1996; Kiranmai et al. 1996; Wanitchakorn et al. 1997; Thomas et al. 1997; Delanoy et al. 2003; Selvarajan and Balasubramanian, 2008). Although propagation through the use of sword suckers is practiced in certain countries, this process encourages contamination by soil-borne pests and other pathogens. Hence plants raised through tissue culture (TC) are being used in different countries. The stages involved in shoot tip culture for banana micropropagation is presented in Fig. 2.4. There are many added benefits in using TC plants: (1) they are more vigorous, allowing for faster and superior yields; (2) more uniformity, allowing for better marketing; and (3) can be produced in huge quantities in short periods of time, allowing for faster and better distribution of existing and new cultivars, including genetically modified banana. In other words, the TC technology can help banana farmers to make the transition from subsistence farming to farming with income generation.

Three techniques used to generate disease-free banana plants are by suckers, macropropagation and micropropagation. Excellent information on this topic has been provided by Singh et al. (2011) as practiced in the south Pacific zone and is presented here.

The procedure for commercial macropropagation and micropropagation of banana plantlets consists of four stages: culture initiation, bud multiplication, plantlet regeneration and acclimatization in the nursery. A production program producing millions of plantlets annually needs hundreds of thousands of suckers for culture initiation each year. If suckers are obtained directly from the field, they should be subject to virus indexing. Only those shown to be free of virus should be used for culture initiation. However, virus indexing of such a large number of samples is time-consuming and costly. The solution was to establish virus-free foundation stock which could supply virus-free material for culture initiation. The establishment and management of the virus-free foundation stock in different Asian countries are as follows.

(a) **Sucker propagation**: There are two types of suckers: *sword suckers* with a well-developed base, pointed tip and narrow leaf blades, and *water suckers* with a smaller, less vigorous base, broad leaved and emerge in clumps.

Natural regeneration of cultivated bananas through suckers is very slow due to hormone-mediated apical dominance of the mother plant. A plant produces only 5–20 suckers during its life time of 12–14 months. For accelerating the propagation rate, suckers with growing buds or cut rhizomes called ‘bits’ and ‘peepers’ are used. Several good bits, each with a centrally placed germinating eye can be

Fig. 2.4 Shoot tip culture for banana micropropagation: **a** sword sucker and explant; **b** shooting after apical disabling; **c** proliferation; **d** multiple shooting; **e** rooting; **f** nursery hardening; **g** young tissue culture plantation; **h** field planted tissue culture banana at fruiting. *Courtesy* H. P. Singh and R. Selvarajan

cut from an unbunched rhizome after trimming the roots. Selection of appropriate mother plant for raising new propagules either through *in vivo* or *in vitro* methods is important. About 1 kg uniformly sized rhizomes or bits, well-trimmed around the growing sprout are the best starting material. Although immediate planting is preferred, in some parts of India rhizomes are sun dried for 2–3 days after paring and pralinage treatment (trimmed of all roots, dipped in mud slurry and sprinkled with a nematicide) and stored in shade for a week before planting.

Mother nursery block and selection of mother plants:

- Mother plant should be healthy, true to type and free from diseases and pests.
- Mother plants should be raised under roofless insect-proof shade netting of sufficient height.
- Mother nursery must be located 500 m away from other banana plantations to maintain purity and to avoid spread of virus diseases.
- Once indexed, the mother suckers can be maintained in a field or in concrete rings with frequent decapitation to facilitate production of more axillary buds. They also serve as explants for culture initiation.

(b) **Macropropagation**: Macropropagation is an excellent option for producing low cost quality planting material. This is a simple method because of the ease of multiplication, by saving the cost of producing planting material and has the potential of producing 50–60 shoots/sucker in 4–5 months. Macropropagation is achieved by two methods and could be adopted either in the field conditions (*in situ*) or in the nursery (*ex situ*). It involves decapitation, decortication and hardening.

(c) **Micropropagation**: Micropropagation is the practice of rapidly multiplying stock plant material to produce a large number of progeny plants under aseptic conditions using modern plant tissue culture methods. Application of micropropagation in banana has the following advantage of rapid multiplication, the need for limited mother stock, product uniformity, season independent production, agronomic advantages, production of disease-free planting material, and high financial returns.

(d) **Decapitation**: In this procedure, 4–6 month old plant is headed back, the pseudostem is cut down and cross cuts incisions are made on the growing meristem so as to stimulate the production of lateral buds. This method results in the production of 9–15 uniform shoots/plant in a short span of time and is highly suitable for small and marginal farmers whose requirements of planting material are relatively low. Suckers of choice cultivars can be maintained in a nursery either in a sawdust bed or in a big, bottomless concrete pot. The initial planting material should preferably be certified as virus-free and multiplied at farm level under an insect-proof net house.

(e) **Decortication**: The pseudostem of the mother corm or sword sucker is cut transversely 2 cm above the collar region and then the apical meristem is removed leaving a cavity of 2 cm diameter and 4 cm depth. This is done to overcome apical dominance. Decapitation and decortication activate the lateral buds to give rise to more side shoots. Generally, the corms that have already flowered give better results than corms that have not yet flowered. Hence, healthy corms left in the field after harvesting are also a potential explant both for *in situ* and *ex situ* mass multiplication.

(f) ***Ex situ* mass multiplication**: Sword suckers are pared partially (trimmed of all roots and the outer surface scraped) and the growing points are excised out with a sharp knife. These corms are surface sterilized by dipping in 0.3 % Bavistin (carbendazim) for 15 min, allowed to dry for a day and then planted in the initiation medium, usually comprised of rice husks or sawdust, though the latter is much preferred.
(g) **Secondary decortication**: After 30–35 days from the first decortication, 3–4 buds emerge from the mother corm, depending on the cultivar. When the side shoots attain a height of 15–20 cm with 3–4 leaves, secondary phase decortication is done by heading back with a sharp knife followed by 3–4 transverse cuts. This encourages production of multiple shoots. Third phase decortication is again attempted after 20–25 days but with greater care as the shoot buds are located nearby. Thus by the end of 140–150 days, a total of 50–60 shoots are produced from a single sucker.
(h) **Hardening**: The lateral sprouts of 8–10 cm length are shifted to pro-trays containing equal parts of cocopeat and vermiculite and after sufficient watering left in a shade net facility (70 % shade) at 80–90 % relative humidity. High humidity is achieved by intermittent misting. Sprouts are usually maintained in the pro-trays for a period of 15–20 days and then shifted to polythene bags of size $6' \times 4'$ and thickness of 120 gauges for secondary hardening. At this stage, the plants are maintained at 50 % shade and 40–50 % humidity. Watering is done on alternate days and the plants are ready for field planting in 30–45 days.

Once the plantlets are ready for shifting outside the laboratory, they are carefully acclimatized to adapt to the greenhouse and later to less protected field conditions. During hardening, the plantlets undergo physiological adaptation to changing external factors like water, temperature, relative humidity and nutrient supply.

The plantlets from culture vessels/bottles are moved from the laboratory to a room at ambient temperature and are kept open for 4–6 days. Later they are shifted to the greenhouse for primary hardening where they are first gently washed free of agar medium. This is important as sucrose in agar encourages microbial growth. Selected 8 cm shoots with 3–4 ramified roots are planted in individual micropots in a portray. In places where weather is conducive (24–26 °C temperature and more than 80 % humidity), the plantlets are hardened for 4–6 weeks in mini-sand beds. During this period, 90–95 % humidity is maintained for the initial 6–8 days under diffused light. The humidity is slowly reduced to 70 %, light intensity raised to normal, and temperatures brought to 26 °C by the end of 6 weeks.

Structures used for primary hardening vary with the climatic conditions. These can be highly sophisticated with UV-sterilized polysheet covering, multiple misting options, thermal shade netting and auto-monitoring for light intensity, temperature and humidity. On the other hand, the structures can be simple with polycarbonate roofing, shade net on all sides with fogger capabilities. Temperature, RH and light intensities are monitored manually using thermometer, hygrometer and lux meter, respectively.

Planting media for primary hardening range from sieved sand augmented with nutrients to mixtures of cocopeat and soilrite with fine sand in equal proportions. NPK nutrients are provided in liquid form on a weekly basis.

(i) **Secondary hardening**: After primary hardening for 5–6 weeks, the plantlets are transferred from microplots to polybags. Base substrate is generally soil and sand along with low cost materials like coir pith, sawdust or rice husks. Organic manure is either in the form of farm yard manure or poultry manure. In Maharashtra, India, press mud, a byproduct of sugar factories, has been found to provide the best substrate for secondary hardening along with soil (Vasane et al. 2006).

Plantlets from micropots are dipped in fungicide solution (0.1 % Bavistin) and planted in polybags containing suitable substrate. Initially, these are maintained in low light intensity shade nets and 70 % RH. The plants are hardened by gradually increasing the light intensity and reducing RH (40 %). After 5–6 weeks the plants are ready for field planting with 3–5 well-developed leaves and a good mass of fibrous roots.

During both primary and secondary hardening, the stocks should be rogued for variants at weekly intervals. These could include vegetative deformities like dwarfism, leaf variegation, rosette foliage and leaf crinkliness. Other precautions to be followed are:

- The rooting media should be completely free from pathogens.
- Water used for irrigating the plants should be free from pests and pathogens.
- Sample plants from each batch should be randomly virus indexed (at least 10 plants from each batch/explant).
- While shifting primary hardened plantlets, two longitudinal cuts should be given to the micropots to facilitate further corm growth.

(j) ***In situ* mass multiplication**: *In situ* production of suckers is induced chemically by pouring 4 ml of 40 ppm benzyl aminopurine (BAP) into the decorticated cavity and covering the individual mats with a mixture containing equal parts of sandy loam and poultry manure to 5 cm above the ground level. Such chemical induction of lateral buds could be done on the first generation suckers and continued up to third generation suckers. This method leads to the production of 45–50 shoots in a short span of months. The suckers are separated from the mother corm and subsequently rooted in sterile soil medium under intermittent misting.

(k) **Shoot tip culture**: The earliest reports of *in vitro* culture of bananas came from Taiwan in the 1970s (Ma and Shii 1974; Ma et al. 1978). Protocols have been standardized for *in vitro* propagation of a wide range of *Musa* species and cultivars belonging to various ploidies and genomes (Sathiamoorthy et al. 1998).

Shoot tips can be extracted from the pseudostem, suckers, peepers, lateral buds or even small eyes which contain a shoot meristem (Jarret et al. 1985; Vuylsteke and De Langhe 1985). Although all of them behave similarly under *in vitro* conditions, peepers and sword suckers are preferred because of their ease of handling and the minimum damage caused to the parent stool during their removal (Singh et al. 2011). It is always better to collect the explants from flowering plants so as to ascertain their trueness to type.

The steps followed for production of micropropagation-based banana planting material are:

- Selection of the mother plant and establishment of a mother block nursery.
- Virus indexing of mother plant nursery.
- Culture initiation.
- Culture proliferation.
- Roguing at various stages of proliferation.
- Rooting and primary hardening accompanied by roguing.
- Secondary hardening accompanied by roguing.
- Fidelity testing and virus indexing at various stages of mass multiplication.

2.7.7 Citrus Bud Wood Certification

In almost all the citrus growing countries, certification schemes are followed for the production of disease-free plant material, as almost all viruses and viroids infecting *Citrus* species can increase and perpetuate in orchard plantings only through the use of virus-infected buds. In California, the citrus variety improvement program (CVIP) is part of an ambitious programme to maintain California's position of world leadership in research and development. Since 1958, the objective of the project has been the "Development and maintenance of primary sources of virus-free, true-to-name propagative material", as implemented by virologists of the University of California, the USDA, and the Nursery and Seed Services of California Department of Food and Agriculture. The foundation bud wood block at the University's Lindcove Field Station (Davis), provides a reservoir of virus-free, true to-name citrus propagative material unmatched anywhere else in the world.

Nauer et al. (1975) have included citrus selections and hybrids of nearly 354 into this programme. Since January of 1974, Lindcove Foundation block bud wood has been available to the California citrus industry at a cost of 25 cents/bud and has distributed over 34,000 buds. Until recently, any selection found to be virus infected was eliminated from the programme. Since the advent of thermotherapy (Calavan et al. 1972), tissue culture (Bitters et al. 1972) and shoot tip grafting (Roistacher et al. 1976) the elimination of the diseases from the infected plants made possible that a promising, but virus-infected selection, could be treated and re-entered into the program when virus-free (Navarro 1993).

Lee et al. (1999) and Lee (2004) have published the review articles on nursery practices and certification programs for citrus bud wood and root stocks based on their long standing experiments with the citrus crop.

In Florida, the citrus bud wood programme was initiated in 1951 and was made a reality following recommendations of a standing committee of the Florida State Horticultural Society in January 1953. The programme was administered under the State Plant Board (SPB), later to become the Florida Department of Agriculture and Consumer Services. In 25 years of operation, participating nursery men and growers have produced 31,421,700 registered citrus trees, which represent 45 % of the total grove trees of Florida. In a 1976 survey that included 40 Florida citrus nurseries producing 95 % of the citrus trees in Florida, only two nurseries were not associated with the bud wood programme (Youtsey 1978). A similar type of scheme has been operated in Australia since 1928 under the New South Wales Co-operative Bud Selection Society. In 1975, a new society (the N.S.W. Horticultural propagation Society Ltd.) was constituted to replace the Co-operative Bud Selection Society Ltd., to assume responsibility for the selection, maintenance and distribution of horticultural propagation material to meet the needs of the citrus industry (Fraser and Broadbent 1979). Since then this society has played a prime role in the maintenance and supply of the virus-free citrus bud-wood material to Australian growers.

In India, a bud wood certification scheme was launched by the Indian Council of Agricultural Research (ICAR) at Poona, Ludhiana, Tirupathi and Kahi Kuchi for different cultivars of citrus fruits i.e. Mosambi, Blood red, Jaffa, Sathgudi oranges and Mandarins. Now at each center, certified bud wood is being produced. Plants which are without visual disease symptoms and with high and regular yielding characters are selected as candidate trees and indexed for important virus and virus-like diseases. The presence of these diseases is tested by short term and long term indexing. For the short term indexing procedure, the indicator hosts express symptoms in a relatively short time frame and concludes in 10–12 months after a parent (candidate) tree source is entered in to the programme. Long term indexing is done for bud transmissible diseases (ie., *Psorosis*, *Xylopsorosis* and *Exocortis*) on their respective indicator hosts and this takes about 3–4 years. Once the candidate trees are found to be free from important diseases by indexing, efforts are made to propagate them in a place reasonably isolated from the citrus-growing belt and where the insect vectors of citrus viruses are not active. In the first instance, the foundation block should be established by propagating the budwood from the registered trees on suitable root stocks. Bud wood from the foundation block is given to reliable nursery men to establish nurseries under the direct supervision of the technical staff, and frequent indexing the trees in both the foundation block and the mother block for important viruses. By implementing this scheme, a constant supply of certified virus-free citrus plants for the growers is made available on a commercial scale and this scheme is an extremely valuable asset to the citrus industry both for production and research.

Mandarins (*Citrus reticulata*) in the Darjeeling Hills of India, were 16–90 % infected with *Citrus tristeza* virus; subsequently Biswas et al. (2009) developed a certification scheme in which plants on rangpur lime and rough lemon root stocks were produced in an insect proof greenhouse after indexing with molecular tests. The certification clean stock program has helped in establishing healthy orchards of mandarins in Darjeeling Hill area and 3–4 year old-grafted mandarin plants were found to yield normal sized fruits.

2.7.8 Certification of Grapevine Planting Material

For grapes, programmes to produce clean stock, are based on aspects like vine selection, disease recognition, indexing, therapy, registration and certification. Selection is a basic step in the program for clean stock production. It determines that a specific clone possesses the desirable growth and yield characteristics of the cultivar and that the cultivar is correctly identified. Selection is the responsibility of viticulturists cooperating in the clean stock programme. The programme is effective only when all serious graft-transmitted diseases can be recognized. Indexing tests must be able to establish that the mother vines are free from recognized diseases, even at the latent stage. Disease recognition, indexing and therapy are the responsibility of plant virologists.

Calavan et al. (1970) described a rapid propagation technique for clean foundation stocks of grapes by following heat treatment and indexing which takes about 18–30 months. After this a single 2-bud cutting from a clean stock plant will root in sand at 28 °C within 1 month, and within 2 months it should have new growth adequate for 10 green-bud cuttings consisting of a single bud each. The cuttings will root in vermiculite under intermittent mist at 28 °C within 2 weeks. Secondary green-bud cuttings can be taken from each daughter plant as it develops.

Grape cuttings are rooted in 15 × 21-inch flats (180 cuttings capacity). It is important to keep the temperature of the vermiculite at 28 °C for best results. After rooting, each cutting is transplanted to soil in a pot and kept in plastic, glass or greenhouses at temperatures ranging from 21 to 35 °C to produce more green bud cuttings. A new crop of green bud cuttings can be harvested from potted grape plants at 2-month intervals provided optimal conditions for vegetative growth are maintained. The potential of this technique is about 1,600,000 cuttings from one foundation cutting per year.

Registration which is a step in the programme, assures maintenance of clean mother plants and that provides for the increase of the disease-free nursery materials. It regulates the planting site to assure freedom from disease and it monitors the increase of stock by cooperating nursery men to assure propagations only from disease-free mother plants. In California, the foundation seed and plant material service of the University of California and the California Department of Food and Agriculture manage the registration scheme. Certification is the last step in the programme. Nursery stocks produced from registered sources by prescribed procedures and certified by the Californian programme was first offered in 1960, and since then 65 million certified plants have been grown and sold. According to

California Department of Food and Agriculture, Nursery and Seed Service, 95 % of all grapes sold in California in 1976 were of certified stocks (Goheen 1980). A grapevine certification scheme in the European Union was well presented by Martelli and Walter (1998) in their review article and also by Rowhani et al. (2005).

Based on the method developed at the University of California (Davis), in which healthy plants were raised from shoot tips taken from virus-infected plants grown in a heat chamber at 100°F (Goheen et al. 1966). From New Zealand, Over de Linden and Chamberlain (1970) treated 18 grape varieties ranging from 56 to 141 days and the virus-free plants obtained were indexed negative in two-growing seasons. The virus-free selected vines were increased rapidly by propagating from hard and soft wood cuttings and when once sufficient material was available, it was released for distribution. When a cultivar of particular value to the industry was identified, vines were also sent to the Horticultural Department, Massey University (New Zealand) which undertook to raise virus-free plants of the most important cultivars for sale to vine growers. In 1968, 'Pinot Meunier', which had good commercial value was made virus-free and released. In Romania, a scheme for the production of certified grapevine propagating material was developed by Pop (1995).

Even from Hungary, Lazar et al. (2002) have developed certification programme for production of grapevine virus-free propagation material. By following the above cited certification schemes, research organizations of each country have to develop mother blocks of virus-free scion and root stock cultivars that have to be established in larger areas based on the national requirement.

2.7.9 Cassava Seed Certification Scheme

Cassava (*Manihot esculenta*), also referred to as manioc or tapioca, is propagated routinely from hard-wood stem cuttings of the plant. The use of infected cuttings for propagation is responsible for the high virus incidence in this crop. In Africa, cassava is routinely infected with *African cassava mosaic virus* (ACMV), *East African cassava mosaic virus* (EACMV), and *South African cassava mosaic virus* (SACMV) which are distinct species of circular-single stranded DNA viruses that are whitefly-transmitted. In India and Sri Lanka *Indian cassava mosaic virus* (ICMV) is prevalent in cassava. From many South American countries the prevailing virus is *Cassava common mosaic virus* (CsCMV).

The basic approach for disease control that has been widely adopted with many vegetatively propagated plants is reviewed by Hollings (1965). Attempts have been made to promote use of virus free-stocks of cassava other than to meet the requirements of plant health authorities responsible for regulating the movement of vegetative materials between countries and continents. Initially virus-free material of cassava has been produced by meristem tip culture and/or heat therapy and "foundation" plants obtained in this way could then be subjected to rigorous tests for virus detection, and such virus-free cassava plants can be maintained in insect-proof conditions to prevent contamination by vectors. There are also those who advocate setting up expensive schemes for the production and distribution of virus-free planting material of cassava similar to those used in developed countries for potato and many horticultural crops. In India, Malathi et al. (1987) have recommended the painting of the *Cassava mosaic disease* (CMD) affected plants so that one can avoid the collection of virus-infected plants when cuttings are collected from the mature plants which are usually leafless and seriously affected.

Nevertheless, there is abundant evidence from experience in Kenya, Cote d'Ivoire and Uganda that a large improvement in the health status of the planting material made available to farmers can be achieved by simply selecting cuttings from symptomless plants and raising them at sites where there is little or no spread by vectors and where roguing is practiced routinely. This was the approach adopted in the 1980s in Kenya (Bock 1983) and Cote d'Ivoire (Fauquet et al. 1987), where substantial quantities of ACMV-free material were raised for use in epidemiological trials. However, arrangements were not developed for more extensive propagation and distribution to farmers on a large scale.

Virus-free planting material production schemes have been developed in Uganda where there is a huge demand for planting material of improved virus-resistant cultivars to replace the local, mainly susceptible ones that have been severely affected during the epidemics (Otim-Nape et al. 1994; Thresh et al. 1994). About 17,500 ha of virus-resistant material have been made available at various propagation sites in different parts of the country and these were used to supply ACMV-free cuttings to farmers. Three main approaches have been adopted in building up this system of multiplication, involving institutions, groups of farmers operating together and individuals (Thresh et al. 1998).

The protocol for the production of virus-free cassava planting material by using tissue culture technique has been detailed in the FAO Plant Production and Protection Paper-195 (FAO 2010). Stems (about 1–2 m long) have to be stored as they have been harvested from the field. Otherwise they can be cut to the proper size for planting (about 20 cm long). To prevent dehydration during storage, it is recommended that the stems be cut into planting stakes just prior to planting. Bunches of long stems should be placed vertically on the ground, in shade (Fig. 2.5a), with the apical portion of the stem up. Sometimes, farmers cover the stem bunches with foliage remaining from the crop to further reduce dehydration of the stems. Ideally stakes have from 5–7 nodes and are about 20 cm long (Fig. 2.5b). The storage area should be shaded and offer high but not excessive relative humidity (about 80 %) and moderate temperatures (20–30 °C). Stems should not be stored longer than 3 months under optimal conditions, preferably less than 1 month.

In certain cassava growing countries a virus-free cassava material was generated by using meristem tip culture with or without heat therapy. Virus-free plants have been produced through meristem tip culture (Kaiser and Teemba 1979; Adejare and Coutts 1981; Ng et al. 1992). Since meristematic regions in some genotypes are not always invaded by plant viruses, it was possible to obtain a small portion of virus-free plantlets from 0.2 to 0.5 mm meristem tips when cultured *in vitro*. Heat therapy before meristem tip culture led to a greater proportion of

Fig. 2.5 **a** Storage of selected stocks to be used for multiplication. **b** Selected cassava stakes with adequate length and width and 5–7 nodes. *Courtesy* Ceballos, CIAT (2006), FAO (2010)

virus-free plants than meristem culture alone. At Central Tuber crops Research Institute (CTRI), Trivandrum, India, a strategy involving testing the planting material with ELISA and PCR for detection of virus pathogens followed by meristem culture has proven to be a very successful method for mass scale production of disease-free cassava planting material (Fig. 2.6).

As the application of tissue culture technology is constrained by high costs making seedlings unaffordable, from Kenya (Ogero et al. 2012b), have achieved success by using locally available fertilizers as alternative nutrient sources for cassava micropropagation. A Low Cost Medium (LCM) whereby the conventional sources of four Murashige and Skoog (MS) macronutrients have been replaced with locally available fertilizers which were developed by Stanes Iodized Microfood® from Osho Chemical Industries in Nairobi, as the alternative source of micronutrients and modified conventional MS medium was used as the control. Both media were supplemented with 30 g/l of table sugar and 3 g/l of gelrite. Two cassava varieties, Muchericheri and KME 1 were regenerated on the two media. Node, leaf and root formation patterns plus plant height have resulted in reduction of 95.50 % in nutrient cost. The two cassava cultivars had a significantly ($p < 0.05$) higher number of nodes on the conventional medium compared to LCM. There were no differences in node formation by the two cultivars on the low cost medium during both initiation and multiplication. KME 1 produced significantly more leaves on the LCM compared to Muchericheri during both initiation and multiplication. Acclimatization of plantlets was successful; hence, this protocol can be adopted in cassava regeneration (Ogero et al. 2012b). This technology will help in virus-free production of planting materials in almost all cassava growing countries.

At IITA, Ibadan, Nigeria, the Genetic Resources Center (GRC) conserves and distributes cassava germplasm free of the viruses. Cassava germplasm was

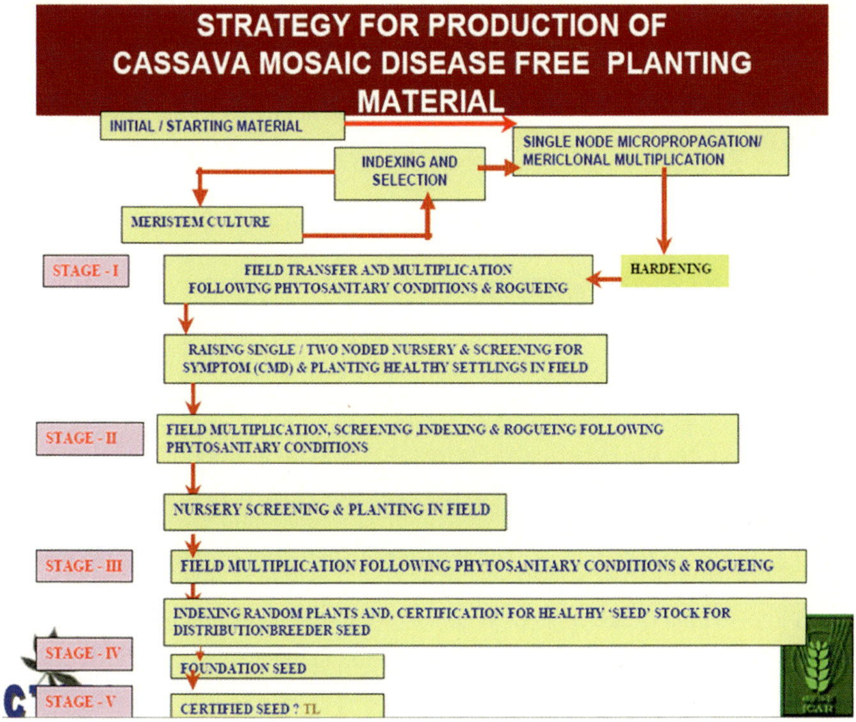

Fig. 2.6 Stages in the production of virus-free cassava planting materials. *Courtesy* Edison S, CTRI

introduced or reintroduced into the *in vitro* system following a procedure that combines heat treatment regeneration through meristem culture, virus indexing, and selection of material that are negative to viruses. Virus-tested germplasm is distributed to partners and stake holders following an internationally accepted phytosanitary procedures. Information on cassava germplasm accessions and the procedure to acquire germplasm from IITA is available at http://genebank.iita.org/ (Badara Gueve et al. 2013).

Even in Philippines, National Seed industry council of the Philippines has recommended certain cassava varieties and Acedo and Labana (2008) have developed rapid propagation technique for the released Philippines cassava cultivars through tissue culture. In their field trails the meristem-derived cassava plants produced higher yield with more and heavier storage roots than the field-sourced planting materials. In addition, the former exhibited healthier plant strand which resulted in more stakes for use in the succeeding planting season.

At Tanzania, the IITA institution is attempting to manage the spread of cassava viruses (CMGs and CBSV) through clean seed system. This objective is currently being vigorously pursued within the frame work of two Bill & Melinda Gates Foundation-funded projects in Tanzania. One of these is focusing and development of the overall 'clean' seed system model, certification system and the establishment of pre-basic 'seed' sites, whilst the second is exploring several strategies for the sustainable commercialization of cassava 'seed' systems. If these approaches are successful, they will offer great potential for application elsewhere in sub-Saharan Africa, and are likely to have a significant and lasting impact in controlling the effects of CMGs and CBSVs (Hermence et al. 2013).

2.7.10 Potato Seed Certification

Similar to other vegetatively propagated crops, potato has also received considerable attention because of the importance of planting virus-free seed pieces, and extensive and long-lived certification effort has been followed in many countries. Potato seed certification was begun in Europe as early as 1900s, largely through the efforts of Dr. Otto Appel in Germany (Appel 1934). Widely used methods in building up healthy seed stocks of potato are tuber unit planting and tuber indexing. In the former method, all seed pieces cut from each tuber are planted consecutively in a row in the field. Periodic inspections are made during a season and all infected plants are rogued. When one or more plants in a unit exhibit virus symptoms, the entire unit is removed. In the latter system, a suitable piece bearing one eye is removed from each tuber and planted in the greenhouse in the winter. If the plant produced from the seed piece develops virus symptoms, then the remaining second piece of the same tuber is discarded. In a comparative study of the two methods, Richardson and Racicot (1951) did not find any marked difference in accuracy.

Slack and Singh (1998) have reviewed the seed potato certification programs that are carried out in the US and Canada by applying integration technology for production of virus-free potato cultivars in number of EPPO countries. EPPO certification scheme for seed potatoes is intended to be used by National plant protection organization. The Scheme compliments the existing UN/ECE standard on the production and marketing of seed potatoes. The scheme has the aim of providing seed potatoes that are free from important pests and diseases to many countries and whose health status is attested by an official certificate. At all stages of the scheme, care should be taken to maintain the characters of the originally selected plants. Material imported from outside the EPPO region should also be inspected and tested under quarantine for all EPPO quarantine pests of potato according to the relevant EPPO phytosanitary procedures (EPPO/CABI 1996). EPPO standards are distributed by the EPPO secretariat to all EPPO member-governments.

In Great Britain, the potato certification scheme has led to a two to threefold increase in yield. Tested foundation stocks which are virtually free of viruses are grown in isolation in parts of Scotland that are unfavorable for aphid migration and colonization (Todd 1961). High grade stocks are grown from this seed elsewhere in Britain, in areas selected because of the low incidence of aphid vectors. The

Agricultural Development and Advisory Service (ADAS) have undertaken the responsibility of tuber-virus testing. This test estimates the level of severe viruses (PVY and PLRV) in random samples of 100 tubers from potential seed stocks by growing excised eyes in a glasshouse with supplementary light and heat during winter months. Visual assessment of the resultant plant lets after 6 weeks of growth gives a measure of the severe virus level in the stock (Hill 1977). In U.S., these certification programs are the responsibility of a state Department of Agriculture, a land grant university or a grower association with, in frequent cases, cooperative agreements among the three agencies.

India is the largest potato producer in South west Asia and potatoes are grown in diverse agro ecologies. Potato is grown both in the hills and in plains in summer and autumn/winters, respectively. Nearly 82 % potato area is in the subtropical plains where it is cultivated in the winter under short-day conditions. Nearly 10 % potato area lies in the plateau region of south peninsular India and the remaining 8 % in the mountains. Potatoes in plateau and mountains are grown under long-day conditions during summer and spring seasons, respectively. To effectively use this natural phenomenon, indexing for virus-freedom is done in hills for the crop to be raised in the plains and vice-versa. In the preliminary surveys it was observed that areas above 2000 m in the northern temperate hills in a few pockets were the most suitable for production of quality seed potatoes. During summer months in these areas when the potato crops are generally grown, low temperature, high velocity winds and frequent rains are unfavorable for the buildup of aphid populations that are vectors for potato viral diseases. In view of this observation, procedures were developed at Central Potato Research Institute (CPRI), Simla for the production of disease-free stocks in the high hills. CPRI is the back bone of potato seed production in India. Nearly 2500 tons of breeder seed are supplied every year to different organizations for further multiplication for foundation and certified seeds.

Based on the epidemiological studies on aphid vectors of potato viruses in North-eastern hills of India, Shahid Ali et al. (2013) have reported the possible ideal locations for potato seed production at the altitudes between 2000 and 2700 m viz., the Shillong Peak/Laitkor Mawri and Mawkriah West in Meghalaya; Hilley Seed Potato Farm, Okhrey, Ribdy and Rawangla Seed Potato Farm in Sikkim; Upper Wanghoo and Warjung villages in Arunachal Pradesh; Upper Ukule Kigwema and Lower Ukhule Kigwema villages in Nagaland were identified as ideal/suitable sites for quality potato seed production where aphid population was recorded below the threshold level during the main cropping season.

(a) **Maintenance of seed stocks of potato in India**: Since a number of stages are involved in the production of foundation seed, the degree of freedom from viruses/phytoplasma diseases is regularly assessed by serological testing/visual inspection in each stage during every year. The procedure adopted in the maintenance of virus-tested foundation seed at Fagu/Jallundur/Mukteswar (India) is as follows:

Stage I: The healthy counter parts of a single eye plant lots showing freedom from all the viruses/MLOs are subsequently planted in the field at the CPRI's nuclear seed station, Fagu (2700 m above sea level), where only materials having passed through glasshouse tests are grown. Virus-tested tubers of each single hill unit are separately planted in a row. Each plant is serologically tested for viruses X and S, by taking a composite sample of one leaf from each stem. The entire clone is destroyed even if a single plant is infected. The plants are visually examined for other viruses and the diseased plants are rogued along with the tuber formed.

Stage II: In the second stage, all the produce of a hill unit or a clone is separately planted in a plot. For testing against potato virus X and S, two leaves from each individual plant are taken and six leaves constitute a sample. Positive serological reaction or visual symptom in any of the samples makes it obligatory to rogue out the entire clone.

Stage III: The seed material of each clone selected under the Stage II is planted in bulk virus infection in the third stage is to be determined serologically by sample testing. Freedom from diseases and genetic purity is ensured by three periodical field inspections and plant protection measures.

The virus-free nucleus stocks produced from Stage III are subsequently utilized for the production of breeder's seed at the Institute's seed production centers at Kufri, Mukteshwar in the hills and at the regional stations at Jullundur, Daurala, Patna in the sub-tropical plains through 'seed plot technique'. The basic seed is supplied to the National Seeds Corporation and the State Agricultural Departments of India for seed multiplication programme.

The seed potatoes produced from hill stations is insufficient for sowing in the plains. So Pushkarnath and his colleagues by their intensive work from 1952–1960, after studying the epidemiology of the insect vectors in the sub-tropical plains, developed a method called 'Seed Plot technique' (Pushkarnath 1959; Nagaich et al. 1969). Methods for briefing about the necessary changes, through adaptation of seed plot technique are:

(1) For seed purposes, plant the crop (using disease-free seed) in the autumn as early as possible, preferably by the first week of October;
(2) Plant at a close spacing and into a not too rich soil. This practice ensures a larger percentage of seed size tubers;
(3) During the growing season, examine the seed plots at least twice and rogue diseased plants;
(4) Towards mid-December, when the crop has tuberized, restrict irrigation and later withhold it completely so that haulms wilt and fall over before the aphids build up. If the crop is still green remove or kill the haulms;
(5) Harvest the stocks, store in a cool situation and later transfer the stocks to cold storage to be used as seed in the subsequent autumn season.

In India, the seed potatoes produced both in low temperature hills regions and in sub-tropical plains during aphid-free period, are now being supplied to farmers through seed agencies. Periodical post-harvest checking of stocks which are introduced into the seed certification system is carried out both in the hills and plains, so that disease-free stocks are supplied. During 2007, Kadian et al. have reviewed the status of potato seed systems in Southwest Asia.

(b) **Rapid potato seed multiplication**: Many countries are now using different rapid multiplication techniques in the potato seed multiplication like tissue culture alone/micropropagation or in combination with different methods of generating cuttings are being used successfully (Chandra and Dhingra 1990). Bryan (1988) reported use of different multiplication techniques by different countries. In most of the countries the *in vitro* material was being used to produce basic nucleus disease-free material and further *in vitro* methods of multiplication were employed to multiply the basic disease-free material. The countries using these techniques are Colombia, Ecuador, Peru, Venezuela, Argentina, Brazil, Sri Lanka, Thailand, Vietnam, Philippines, Costa Rica, Mexico, Cuba, Korea, Bangladesh, Bhutan, India, Nepal, Pakistan, Indonesia, Taiwan, New Columbia, Kenya, Madagascar, Bolivia, Chile, Uruguay, Dominican Republic, Jamaica, Panama, Guatemala, Honduras, Burundi, Rwanda, etc. However, by following some of the techniques discussed here, majority of the potato growing countries have become self-sufficient in seed potato production for their country or depending on some other country. The rapid multiplication techniques are fast becoming important in developing self-sufficiency in seed production.

The commercial application of tissue culture and micropropagation techniques to the production and sale of seed potato products has been a successful propagation (Addy 1988). More details of micropropagation as the alternative to conventional propagation of potatoes is available (Chandra and Birhman 1994; Naik and Chandra 1994). The same technique can be extended to other vegetable crops propagated through vegetative propagules. Using the tissue culture techniques for micropropagation, it is now possible not only to reduce the number of field exposures but also to increase the rate of multiplication several fold. With micropropagation, the propagation takes place in the laboratory under aseptic, controlled conditions where *in vitro* potato plantlets are produced all year round in a limited space and in completely disease-free laboratory environment. Also vegetatively reproduced material can often be stored over a long period but on a limited scale. *In vitro* plantlets need no attention between subcultures and therefore, there is no labor or material requirements for watering, weeding, and spraying. Presently countries with commercial micropropagation are mainly employing *in vitro* plantlets for the large scale production of basic starting material. On the other hand, several problems are associated with the *in vitro* plants. Plantlets have to be supplied seasonally as they cannot be stored for long and are very labor intensive to produce, and are thus quite expensive.

(c) **Mini/micro tuber production**: To obtain cheaper and cleaner propagules, a compromise of the technology i.e., minituber technology was developed in U.S. and later adopted in many other countries, *in vitro* plantlets are planted in beds in a screen house (screens for aphids with sterilized soil), the plantlets produce 3–6 tubers. These tubers, called minitubers, though are not as clean as the laboratory *in vitro* plantlets yet are cleaner than the field produced tubers. Thus one would start with 3–6 tubers instead of 1. This means enough savings on the starting material to allow for 3–4 generations of multiplication instead of 8–11 traditional generations (Najjar 1993).

Another way of overcoming such problems is by producing microtubers. Almost all aspects of microtubers have been reviewed in detail (Chandra et al. 1988). Microtubers are small, *in vitro* produced tubers that are usually about the size of a pea and are produced in sterile culture under controlled pathogen-free conditions, thus they are disease-free (Chandra et al. 1992c; Randhawa and Chandra 1990). Water, air and nutrients that come in contact with the potatoes are sterile, and the test tube physically isolates the potatoes from all insects, pathogens, bacteria, fungi and viruses. Microtubers as being *in vitro* products are generation '1' or 'pre-nuclear'.

The term microtuber refers to the *in vitro* status of the tubers and not to their relative size. The range of microtuber size varies from 3 to 12 mm and the range of microtuber weight varies from 40 to 140 mg depending on the concentration of triadimefon, uniconazole, benzyl aminopurine used (Chandra et al. 1992c). Microtubers, which are physiologically and morphologically identical to conventional tubers although much smaller in size, integrate well with the current practices as they store well and are easily handled and planted. Until recently, the high cost of production prevented the commercial exploitation of microtubers in the potato industry. The cost of production priced is out of reach for seed potato growers and confined to their use for research. Some cheap methods for the production have been developed (Chandra et al. 1992a, b, 1994), but even then microtubers could not get a direct place in potato seed production system anywhere in the world. The established technology to mass produce microtubers and at competitive prices was started at the University of Wisconsin-Madison when Prof. Brent McCown and his co-worker Peter Joyce evolved the Microtuber Multiplication System of potatoes (MMS) (Najjar 1993).

The MMS is a proprietary and patented system (USA Patent No. 5,047,343). The principal concept of the process is in the production of microtubers with a high multiplication rate in novel advanced bioreactors. Several different working bioreactor designs have been developed and improvements are made regularly to optimize their cost effectiveness. In the early stages of development, the two different bioreactor systems were: System I, which utilizes small low cost bioreactor and System II, which utilizes large, automated, reusable bioreactor.

The MMS has three interconnected stages:

(1) the formation of a microtuber-shoot complex
(2) the elongation of the shoot axis; and
(3) the tuberization of the mass of shoot axis.

In the first stage, the bioreactor is charged with "seed" microtubers. These are placed under conditions which stimulate the formation of several initial shoot axes. The initial source shoots, which emerge from the microtuber explants, are then induced to extensively branch and form a mass of short shoot "initials" which will eventually produce microtubers. In the second stage, shoot elongation stage, each of the shoots of the seed microtuber is allowed to develop into an independent axis. To accomplish this, the microtuber shoot complex I is transferred to a medium and environment which promotes the rapid growth of the potato shoots. This transfer may be accomplished by physically transferring the microtuber shoot complex to a different culture vessel or by changing the medium in the vessel. In the third, tuberization stage, tuberization is achieved when numerous shoot axis have developed to the point where they become physiologically independent of each other while still being continuously sank in a growth medium. The elongated shoots emerging from the shoot complex are tuberized by changing the medium and the environment. The microtubers reach convenient harvest size within 2 months.

Presently, the result is that the original "Seed" microtuber is multiplied by this process resulting in up to thirty or more daughter microtubers. The process is well adapted for automation. Labor input is minimal and the process allows for the manipulations to be done in one vessel with no labor beyond that of inserting the original microtuber and harvesting the produced microtubers. Furthermore, costs are very low as the system is simple and requires only a low cost growth environment.

The firms dealing with potato micropropagation should give top most importance to : (i) delivery in time, (ii) prices should be reasonable and competitive, (iii) consistently high quality plantlets, (iv) homogeneity and uniformity of plants, (v) no mutations, and (vi) easy to acclimatize and to bring into production.

(d) **Through meristem-tip culture**: Through meristem-tip culture large number of plants of different potato cultivars were developed throughout the world and farmers are much benefited by this technique (Faccioli 2001; Zaman et al. 2001; Nagib et al. 2003; Nascimento et al. 2003). Even the technology of commercial micropropagation has not changed much since the eighties because of lack of automation of existing procedures. The flow chart of the use of meristem culture and micropropagation in production of virus-free tubers/ mini tubers is given in Fig. 2.7.

The most reliable method of micropropagation is still by the division of axillary shoots which requires a high input of manual labor at all stages. If commercialization

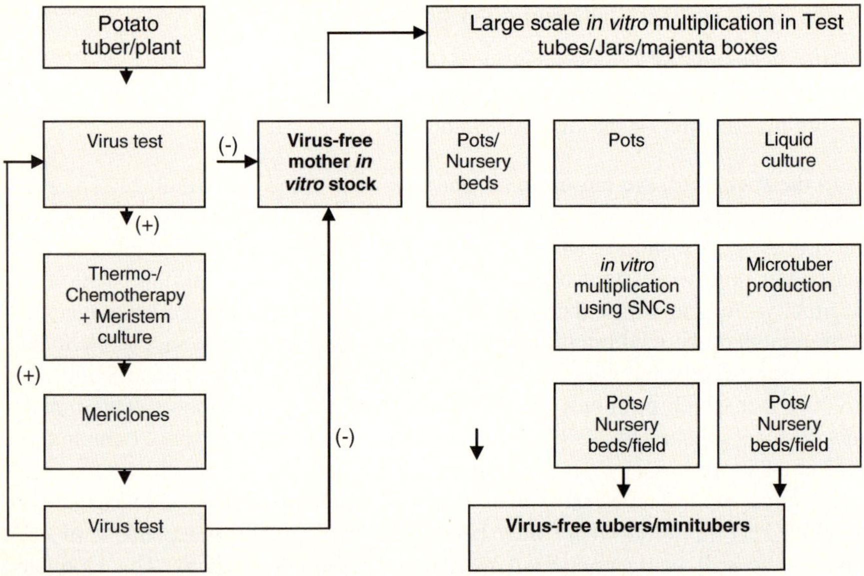

Fig. 2.7 Use of meristem culture and micropropagation in production of virus-free tubers/minitubers. *Source* CPRI (ICAR), India

of potato micropropagation has to be made, it is vital that production cost should be decreased. One estimate suggests that a 50 % reduction in average cost would allow the market to expanded to more than 10 times its current size, and that by decreasing production costs by 90 %, the potential market would become 1000 times larger than at present (George 1996). Micropropagation techniques have been introduced in clonal selection systems in most countries in Europe to provide supposedly better control of virus and bacterial diseases in the seed programme. It is also used in some countries to try to get rid of quarantine diseases although, obviously, this will be successful only if the chances of re-infection during multiplication in the field are extremely small.

The replacement of basic plants by pathogen-tested *in vitro* plantlets or mini-or microtubers is an attempt to improve the standard of health of the starting material and there by the overall quality of the seed. Clearly, the success of this replacement depends on the standard of health of the basic plants and on the chances of re-infection with tuber-borne pathogens during the field multiplication. A valid reason or reasons to combine traditional clonal selection with micropropagated material must include : (i) improvement of the quality of the pre-basic seed, (ii) ridding the crop of quarantine disease, (iii) rapid multiplication of given cultivar for which a large quantity of seed is needed in a short time period, and (iv) reducing the production cost of pre-basic seed.

2.7.11 Sweet Potato Production of Clean Plant Material

Virus diseases are also one of the limiting factors for obtaining higher yields and in almost all countries, healthy sweet potato plant material was produced by following chemo-, thermotherapy and tissue culture. Some of the successful virus-free production methods are discussed herein. As early as 1989, Love et al. have developed meristem-tip culture and virus indexing method for producing health sweet potato plants.

Introduction and availability of clean sweet potato planting materials of improved sweet potato cultivars is the perquisite to realizing high root yield. Accessibility of these materials to farmers is dependent on multiplication and distribution of clean planting materials in larger quantities. Supplying planting materials from research has not met the demand by farmers because sweet potato like other vegetative plants has a low multiplication ratio. The objective of basic seed production was to establish sweet potato seed nurseries to act as source of nuclear planting materials for secondary multiplication by farmers, NGOs and CBOs.

Different procedures that are followed for the production of virus-free sweet potato planting material in some of the countries are described here in detail. Sweet potato vines for planting should come from actively growing and disease-free plants. The vines can be:

- **Shoot tip cuttings**: These are the best parts for propagation as the meristematic cells are still actively dividing and hence grow faster and more vigorously. The tips should be 10–15 cm long. These should be planted vertically with two-thirds of the vines inserted in to the soil.
- **Three to four node cuttings**: These are mostly used when there is inadequate planting material from the shoot tips. These are also planted vertically, with two-thirds of the cutting inserted into the soil.
- **Two-node cuttings**: Two-node cuttings usually have 1 or 2 leaves intact. The leaves initiate photosynthetic activity before other leaves are formed. Two-node cuttings are also planted vertically, with one node in the soil. However, these cuttings require high humidity and this is provided by covering the beds with plastic sheeting suspended 80 cm over the beds. The sheeting is removed once 80 % of the cuttings have sprouted.

Water the beds before planting and plant the cuttings vertically, at 10 cm × 10 cm, with leaves outside the soil. After planting make sure to water the plants.

Although tubers can also be planted they are not commonly used as seed as they can also be consumed and take longer to sprout than vine cuttings. Hence, the use of node cuttings is preferred.

During 2004/2005 seasons, in anticipation of food shortages due to drought, the Tanzania government directed that each household had to grow drought tolerant crops particularly cassava, sweet potato and sorghum. The Lake Zone Agricultural Research and Development Institute (LZARDI) responded by multiplying and

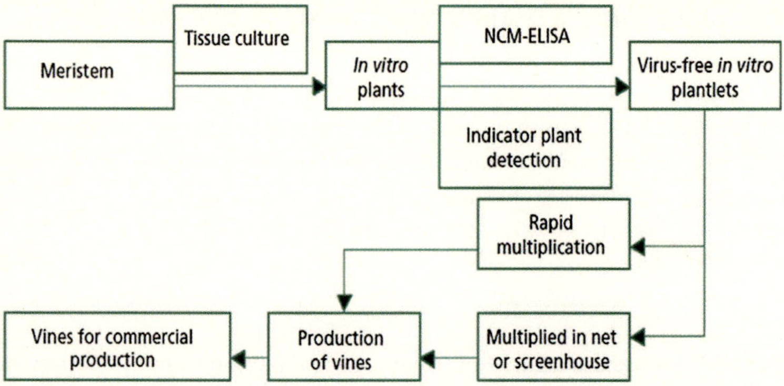

Fig. 2.8 Protocol for virus-free sweet potato planting material production. *Courtesy* Robert Mwanga and Segundo Fuentes; FAO (2010)

distributing sweet potato planting materials to farmers for secondary multiplication. This activity was done in collaboration with district councils in areas where the sweet potato project is operating (McKnight project), and the areas included Misungwi, Meatu, Sengerema and partially the Kwimba districts.

Robert Mwanga and Segundo Fuentes have provided the protocol for the rapid multiplication of healthy sweet potato seed material (FAO 2010) and the information is provided in the Fig. 2.8.

Officially released cultivars that were multiplied and distributed from sweet potato seed nurseries at LZARDI, Ukiriguru were 'Simama', 'Jitihada', 'Vumilia', 'Mavuno' and 'Sinia'. Other orange-flesh cultivars that have been "on farm tested" for acceptability between 2002 and 2004 include 'Kakamega', 'Carrot Dar' and 'Zapallo'. All the planting materials that were distributed were further multiplied in communal nurseries in order to increase the number of beneficiaries. A total of 225,450 vines were distributed.

Xiansong (2010) has developed a technique which helps in rapid production of virus-free sweet potato planting materials. Even from Kenya, Ogero et al. (2012a) have developed a low cost tissue culture technology in which there was 96.9 % reduction in the cost of the nutrients used in the media preparation. This technique may boost the production of disease-free seedlings.

2.7.12 Sugar Beet Certification Scheme

Among the viruses affecting the sugar beet crop are *Beet curly top virus*, *Beet mosaic virus*, Rhizomania (*Beet necrotic yellow vein virus*), *Beet soil-borne mosaic virus* and *Beet western yellows virus* are important. Collectively they are important and are responsible for heavy yield losses when plants are affected at an early stage.

Sugar beet is cross-pollinated by wind or insects. Pollen normally sheds over a period of weeks; therefore the chances of contamination due to out-crossing are prolonged. The sugar beet seed fields must be isolated from fields of other cultivars, are same cultivar to purify requirements of certification, and fields of swiss chard, spinach beet and garden beet at least by 5 m at mother root production stage; and by 1600 and 1000 m, at seed production stage for foundation seed and certified seed stage, respectively.

(a) **Methods of Seed Production**:

There are two methods of producing sugar beet seed.

1. *In situ* methods of producing sugar beet seed.
2. Transplanting method (Steckling method).

In the over-wintering method the crop is over-wintered in the field and the crop flowers *in situ* after the necessary thermal induction. In the steckling method, the selected stecklings (roots) are lifted at the end of the first season and stored in shallow trenches for over-wintering. These stecklings are replanted to produce seed in the following spring. Before replanting, one-third of the root tip is chopped and the top is also removed so as to retain about 10 cm of top leaves.

In Great Britain this scheme is most effectively implemented. Farmers are provided with stecklings grown in suitable isolation, and thoroughly inspected throughout the season. An evaluation is made at the end of October. The plant population is estimated by counting the plants in one meter lengths of row at twenty random positions, counts of diseased plants are made from twenty 10-m lengths of row and the percentage of infection is calculated. More than 10 % results in condemnation; less than 1 % for certification. Cases falling between 1 and 10 % are subject to arbitration (Hull 1952).

2.7.13 Virus-Free Yam Planting Material

The principal viruses infecting yam (*Dioscorea* spp.) are *Cucumber mosaic virus* (CMV), *Dioscorea bacilliform virus* (DBV), *Dioscorea latent virus* (DLV), and *Yam mosaic virus* (YMV). These viruses are widely distributed and cause considerable yield losses. The systemic nature of yam *potyviruses* and *badnaviruses* makes it unlikely that virus-free planting material could be derived from nodal cuttings or from any other vegetative part of the plant. Since the perpetuation of desirable agronomic characteristics is of prime importance, it is necessary to obtain virus-free material by apomictic means. These considerations led to the development of a methodology for apical meristem tip culture in yam (Mantell et al. 1979).

The methodology, developed in the Yam Virus Project (Mantell et al. 1978; Mantell 1980), is outlined below:

1. Source plants-grown for 4–6 months under 16-h day length
2. Tissue selected-nodal segments 5–15 mm of vine containing 1 or 2 axillary buds and 2–8 mm of petiole
3. Growth medium-based on Murashige and Skoog basal medium plus mineral salts
4. Implant nodal segments in the medium with axillary buds uppermost; incubate under lights (1200–1400 lx) at 27 ± or ± 2 °C for 20–30 days
5. Repeat 4 as needed to give required number of plantlets
6. Allow plantlets to form 6–10 nodes
7. Transfer plantlets to 9-cm plastic pots and incubate for 30 days
8. Transfer plants to 11 × 18 cm black plastic bags
9. Plant out in protected area in the field-screened gauzehouse

For further details of the techniques utilized, refer to Malaurie (1998).

Towards the end of the project when virus-free planting material became available, a brief study showed that the use of the new "seed" material could lead to yield increases of at least 30 %.

FAO/IBPGR has released the technical bulletin covering guidelines for the safe movement and transfer of yam germplasm for both seed and vegetatively propagative material (Brunt et al. 1989).

2.7.14 Seed Programmes for Sugarcane

In sugarcane a seed programme based on pathogen elimination has been developed by Indian Institute of Sugarcane Research, Lucknow, India (Singh 1977). The process of virus elimination is gradual and stalks of disease-free seed canes are built-up and available to meet the seed requirements of the cultivars to the farming community.

Integrated disease management through a three-tier seed nursery programme is being followed in many sugar factories in India. This programme involves production of disease-free seed material through heat therapy. The programme consists of three stages: (1) foundation seed, (2) certified seed and (3) commercial seed (Singh 1977). The seed material employed for raising foundation seed is obtained from a nucleus seed with 100 % genetic purity. The seed materials are subjected to aerated steam therapy (AST) or moist hot air treatment (MHAT) and a foundation nursery is raised. The crop raised from the foundation seed will provide seed for raising certified seed. Good quality certified seed is used for raising commercial seed. No heat treatment is given while raising certified/commercial seed. The nursery fields are inspected at regular intervals and diseased clumps are rogued out to keep the nurseries free from disease. The permitted limits of grassy shoot (phytoplasma) and other virus diseases incidence is 0 % at the foundation nursery and 0.5 % in certified and commercial seed.

Rigorous seed selection to eliminate even incipient infections, seed treatments, field sanitation, control of vectors and certification before distribution, are the main tools proposed to be employed. The planting material will be given heat treatment for nucleus seed plots only. It will ensure freedom for ratoon stunting and grassy shoot. Adequate precautions are to be taken to check for secondary infection with these diseases in multiplication plots. Seed plots will remain under constant surveillance for potential secondary infection.

This programme envisages the establishment of 10–15 nucleus seed plots of 1–2 ha each in each factory zone. Crops from these nucleus plots will be used to plant first seed multiplication plots which in turn will provide planting material for the second seed multiplication plots. From here it will be distributed to progressive cultivators (who can be local seed men). The program will be continuous.

The commercial seed thus raised is distributed to farmers for commercial planting. This system ensures a continuous supply of disease-free seed material to farmers in the sugar factory zone, and this programme is successful in managing the disease in different states.

In recent years, some of the government and private agencies are mass multiplying the sugarcane seedlings through tissue culture and supplying to the formers for cultivation in many countries including India.

2.7.15 EPPO's Certification Schemes for Ornamental Plants

The European and Mediterranean plant protection organization (EPPO) is a regional organization established according to the international plant protection convention. More details about EPPO have been included in the earlier chapter. EPPO is very much involved in the global efforts to harmonize phytosanitary measures. Several EPPO member governments have developed certification schemes for ornamental crops for raising the healthy (disease and pest free) planting material and to promote the trade. Since 1985, the EPPO panel on "Certification of ornamental" has developed the schemes for the five ornamental crops viz., (1) Carnation (2) Pelargonium (3) Lily (4) Narcissus and (5) Chrysanthemum. In each crop, different virus and virus-like diseases along with other pests were identified by different techniques and healthy planting material was produced by chemo, thermo and tissue culture techniques depending on the requirement of the crop. Some of the definitions used in the production of vegetatively propagated plants proposed by EPPO are as follows.

(a) **Basic material**

Propagation-stock material from all but the last stage of propagation stock, that satisfy the recommended certification standards, are certified for sale. According to the number of stages of propagation stock, there may be several grades of basic material.

(b) **Candidate nuclear stock**

Any plant that may become or may be propagated to produce nuclear stock is included in this grouping. Testing for specified pests is required before the plant can be accepted as nuclear stock. Until testing is complete and negative, the plant remains candidate nuclear stock.

(c) **Certification scheme**

System for the production of vegetatively propagated plants for planting, intended for further propagation or for sale, obtained from nuclear stock after several propagation stages under conditions ensuring that stated health standards are met. The filiation of the material is recorded throughout the scheme.

(d) **Certified material**

Propagating material from the last stage of propagation stock and satisfying the recommended certification standards and are certified for sale is addressed here. In the case of plants that are sold grafted onto rootstocks, the rootstocks must also be at least of the last stage of propagation stock, and the plants must be held under approved conditions between grafting and sale. Certified material may, according to the plant concerned, be referred to more specifically as, for example, certified plants, certified cuttings, certified bulbs, etc.

(e) **Classification scheme**

System for the production of vegetatively propagated plants for planting intended for further propagation or for sale, obtained from selected candidate material after one or several propagation stages under conditions ensuring that stated health standards are met. Different classes may be defined according to the inspections and tests used, the tolerance levels applied and the precautions taken. The filiation of classified material is not considered.

(f) **Filiation**

The line of descent by vegetative propagation from a defined parent plant applies here.

(g) **Nuclear stock**

Plants individually tested by the most rigorous procedure in a certification scheme and found free from specified pests. All such plants must be maintained at all times under strict conditions ensuring freedom from infection. According to the crop concerned, plants propagated from nuclear-stock material may remain nuclear stock provided that they do not leave the nuclear-stock conditions. In the case of plants that are maintained by grafting onto rootstocks, the rootstocks must also be nuclear stock.

(h) **Nuclear-stock material**

Propagating material derived from nuclear stock, which may be further propagated without change of ownership, or certified for sale as pre-basic material.

(i) **Pre-basic material**

Nuclear-stock material which satisfies the recommended certification standards and are certified for sale is covered here.

(j) **Propagation stock**

Plants derived from nuclear stock, propagated and maintained under conditions ensuring freedom from infection. Pathogen freedom is checked by appropriate procedures. Propagation may be done in a number of successive stages under different approved conditions. The plants are then known as propagation stock I, propagation stock II, etc. There may be several generations within each of these stages, provided that the plants do not leave the approved conditions. The number of stages and/or generations allowed within propagation stock is generally limited and will depend on the crop concerned. In the case of propagating material which is maintained by grafting on a rootstock, the rootstock should be at least of the corresponding stage of propagation stock.

(k) **Propagation-stock material**

Propagating material derived from propagation stock, which may be further propagated without change of ownership, or certified for sale as basic or certified material, according to the stage of propagation stock concerned.

(l) **Outline of requirements**

EPPO Certification Schemes describe the steps to be followed for the production of vegetatively propagated planting material of a particular cultivated plant, whose health status is attested by an official certificate. Certification and classification represent distinct alternative approaches to the production of healthy planting material. In a typical certification scheme, the certified material is descended by not more than a fixed number of steps from individual plants each of which is tested and found free from pests, and is then maintained and propagated under rigorous conditions excluding recontamination. In a classification scheme, the classified material is descended by one or more steps from material which, as a population, meets certain health standards and is maintained and propagated under conditions minimizing recontamination. In both cases, however, health status is attested by an official certificate. Which of the approaches is appropriate for a given cultivated plant depends on considerations of cost and resources, health status required, practical possibilities for testing, rate of recontamination, value of the final material.

EPPO Certification Schemes give details on the selection, growth and maintenance of the candidate material and on the propagation of the material in several stages under conditions ensuring that stated health standards are met. Appropriate checks on specified pests are specified throughout the scheme. Information is provided, as necessary, on relevant pests, cultural practices, inspection and testing methods, recommended certification standards.

Krczal (1998) has provided information on the important virus and virus-like diseases of different ornamental plants and their diagnostic techniques in his review article on virus certification of ornamental plants and the European strategy. More details on certification schemes on ornamental crops can be obtained from OEPP/EPPO Bulletin, Vol. 23 of 1993.

2.7.16 Bulb Inspection Service Scheme

About 8 % of the flower bulb production of the crops like lily, dahlia, gladiolus, iris, hyacinth, crocus, tulips, daffodil etc., from the Netherlands is exported to the countries like West Germany, Sweden, France, England and USA. On the basis of available knowledge and practical experience, a provisional certification scheme has been set up by the bulb inspection service of the Netherlands, similar to the one in operation for seed potatoes. Phytosanitary regulations of the importing countries require both a field inspection and an export inspection; and in addition to satisfying these phytosanitary regulations, bulbs intended for export must also meet the quality requirements of the Danish Government. The strict field and shed inspection for which the growers have their own fully qualified Bulb Inspection Service working in close cooperation with national plant protection service. The field inspection of the crop is done jointly, which will help for locating and roguing out the virus inspected plants which are spotted based on symptoms on flowers and leaves. The inspection will also be done at the exporters shipping rooms. The accuracy and expertness, with which field and export inspections are performed, constitute the determining factors for the risk taken in the exportation of these flower bulbs. The plant protection service of the Netherlands is conscious of this responsibility. Van Os (1980) has reviewed the different aspects of production testing and certification of virus-free plant materials of bulbous crops in the Netherlands.

2.7.17 Performance of Virus-Free Plants

In general the performance of almost all crops raised from virus-free plants is quite satisfactory. Ever since meristem tip culture method is developed, several workers have obtained virus-free specimens from virus-infected clones of a number of vegetatively propagated species valuable in horticulture and agriculture. It is of greater importance as the cultured plantlets usually retain the genetical characteristics of the parent clones. In general plants, regenerated from meristem tips show no more genetical variation than those produced by conventional methods of vegetative propagation.

In Northern Ireland potato cultivars "King Edward" and "Arran Victory" which were completely virus infected were made virus-free through tissue culture by Kassanis (1957). Through field trials, it was found that the average yield of healthy stock was 10 % more than the old stock and, in 1965, it was calculated that the annual benefit to UK agriculture was £2 million and the tubers were more uniform in size. Gregorimi and Lorenzi (1974) also freed the Italian cultivar San Michele from unknown viruses and obtained 60 % more yield. Gippert and Schmelzer (1973) observed that virus-free *Pelargonium zonale* plants cultured from meristem tips were more vigorous than untreated plants and produced 20–30 % more cuttings. In addition to the increased rooting capacity of these cuttings, there was also about 35 % increase in total production. In England, all the commercial varieties of Rhabarber (*Rheum rhaponticum*) were widely infected with viruses. In 1968, Walkey could isolate virus-free stocks by means of meristem-tip culture. A 60–90 % increase in yield of petioles was observed as compared with virus-infected plants. Walkey et al. (1974) produced virus-free cauliflower clones with higher seed yielding quality.

Wang JiXuan et al. (2000) have studied the growth and fruit production of virus-free apple cv. Fuji, Gold Spur, Tian Huangku and Ralls were investigated for 8 years. Compared with control trees, the height, crown and trunk diameter of virus-free trees increased by 6.4–16 %, 9.8–15.9 % and 8.9–17.4 %, respectively, while the number of new branches and total new branch length per tree increased by 12.7–24.4 % and 23.1–40.6 %, respectively. The fruit bearing of biennial virus-free Gold Spur, triennial virus-free Fuji, and triennial and 4-year-old virus-free Ralls reached 60, 48, 62.5 and 97.5 % respectively, while the corresponding virus-infected trees reached 16.7, 16, 0 and 12.5 %, respectively. The number of fruits in the virus-free Tian Huangkui, Gold Spur and Fuji was higher by 41.6, 54.7 and 45 %, respectively, than those of the virus-infected trees. The fruits of virus-free Fuji, Gold Spur and Ralls were 8.8–22.3 % harder than those of the virus-infected fruits. The ascorbic acid, total sugar and acid, and anthocyanins in fruits of virus-free trees were higher than those of the virus-infected trees. In India, a number of private companies are involved in production of virus-free plants of banana, sugarcane, papaya, etc. and some of the farming community is using them for planting. At the present time, the extensive use of meristem-tip culture techniques throughout the world for eliminating virus pathogens and are achieving higher yields with virus-free plant material of many cultivars of vegetables, fruits and ornamentals.

Conclusions

In seed certification schemes, precautions should be taken by restricting the regions for raising the seed lots along with the protective measures during the cropping period. If these precautions are not taken according to seed health certificate requirements, then (1) proper crop stand as expected by seed analysis will not be achieved; and (2) the disease will develop and lead to total devastation of the crop resulting in deterioration in the quality of seeds and consequently, their market value. If the diseased seed is distributed within the country or sent to other countries, these seed-borne virus diseases are further disseminated. The seed after harvest should be carefully analyzed by applying some of the advanced virus diagnostic techniques as described in Chapter-5 of Volume-I, to decide whether the seed samples in question should be released for use or rejected. For additional source of information for seed certification the reader can refer to Maury et al. (1998b); Albrechtsen (2006) and Sastry (2013a).

2.8 Production and Use of Virus-Free Transplants

The production season should begin with the virus–free transplants. This can be accomplished by purchasing or producing the required crop transplants in isolated areas away from production fields or in poly houses. The greater the distance between production areas and nurseries, the lower the incidence of virus-infected transplants will be. The systemic insecticide application to the soil at the time of seed sowing followed by foliarly applied insecticidal sprays will further reduce the insect vector population, so that virus-free nurseries can be maintained. These nurseries should be raised in the net houses or where net-house facilities are not available, then within muslin cloth cages.

Mukhopadhyay (1984) has reviewed the role of rice nursery beds in the spread of virus diseases. Seed beds raised by the cultivators in plots before the land preparation operations has resulted in high virus incidence as the source of virus inoculum was from the stubbles and cereal weed hosts. In some countries due to availability of irrigation source, rice is grown three times in a year. Because of the extension technology availability, in almost all countries, the rice nursery is raised in isolation or in poly houses/net houses along with granular and foliar insecticidal applications. In a number of developed and under developed countries the farmers, research scientists, seed companies and plant nurseries are raising their required seedlings (tomato, brinjal, capsicum, onion, tobacco, papaya, ornamentals, etc.) in net houses/ poly houses with frequent insecticidal applications. Through this technology involves slightly higher expenses, the advantage is that the crops can be raised to maturity in open fields and supplemented with suitable plant protection measures.

2.9 Need for Managing the Virus Diseases

In almost all tropical countries, virus and viroid diseases are responsible for heavy yield losses due to death of plants at a very early stage. Depending on the virus inoculum load and vector population, the infection may be at the early to late

Peanut bud necrosis virus in tomato
Courtesy: Dr. RA Naidu

Papaya ringspot virus in papaya
Courtesy: Dr. RA Naidu

Citrus tristeza virus in sweet orange
Courtesy: Dr. L. Navarro

Cadang cadang viroid in coconut
Courtesy: Dr. John Randles

Fig. 2.9 Field view of virus and viroid diseases of vegetable and fruit crops

stages of the crop development. The impact of virus and viroid diseases under field conditions in some of the crops like tomato, papaya, sweet orange, and coconut are shown in the Fig. 2.9.

Various techniques of producing the virus-free crop plants were discussed in the early part of this chapter. The primary goal of any agriculturist will be planting/sowing of virus-free seed/vegetative propagules in the fields for achieving the higher yields to feed the hungry population. We know that even though healthy planting material is used, under field conditions, the spread of the virus takes place through insect and non-insect vectors as well as by other means. Under favorable conditions, the extent of disease incidence in major crops can be as high as 80–90 %, and this has disturbed the economy of some countries (Table 2.2).

Perusal of Table 2.2 clearly indicates that the virus diseases have disturbed the livelihood of residents in a number of tropical countries. Research has been conducted throughout the world to manage virus diseases that are spread by vectors, and control measures have addressed improved cultural practices, vector management, using resistant plant material and other aspects, including quarantine. It is a well established fact that healthy planting materials and subsequent IPM application in

Table 2.2 Economic crop losses due to certain viruses

Virus	Crop	Countries	Loss
African cassava mosaic virus	Cassava	Africa	$2000 millions
Barley stripe mosaic virus	Barley	Montana	$30 million
Barley yellow dwarf virus	Barley	UK	£6 millions
	Wheat	UK	£5 millions
Beet yellows virus	Sugar beet	UK	£5–50 millions
Cacao swollen shoot virus	Cacao	Ghana	£3,650000 millions
Cassava brown streak virus	Cassava	Malawi	US $5–7 millions
Cassava mosaic virus	Cassava	Sub-Saharan Africa	2×10^9 US $
	Cassava	Uganda	US $60 million
Citrus tristeza virus	Citrus	World-wide	£9–24 millions
Cocoa Swollen shoot virus	Cocoa	Sub-Saharan Africa	1.9×10^8 (number of trees eradiated)
Cotton leaf curl virus	Cotton	Pakistan	$5 billion
Groundnut bud necrosis virus	Groundnut	Asia	US $89 millions
Groundnut rosette virus	Groundnut	Sub-Saharan Africa	1.56×10^8 US $
Potato leaf roll virus	Potato	UK	£30–50 millions
Rice hoja blanca virus	Rice	S. America	$9 millions
Rice ragged stunt virus	Rice	SE Asia	$140 millions
Rice tungro viruses	Rice	SE Asia	$1500 millions
Rice yellow mottle virus	Rice	West Africa	3.3×10^6 tons
Tomato yellow leaf curl virus	Tomato	Dominican Republic	$10 million
Yellow mosaic virus	Legumes	India	$30 million

Source Sastry (2013b)

majority of the crops in almost all countries showed that virus-free planting materials yielded two to three times more than those infected with the virus.

Generally the entire range of methods employed to manage plant virus and viroid diseases are divided into the following distinct categories: (a) Phytosanitation in managing the virus and viroid diseases; (b) Variation of the crop cultural practices; (c) Cross protection; (d) Control of vectors; (e) Interferring with vector transmission; (f) Breeding for disease resistance and development of transgenic plants; (g) Quarantines; (h) Integrated approaches.

2.9.1 Phytosanitation in Managing the Virus and Viroid Diseases

Phytosanitation approach helps in decreasing the number of *foci* of infection from which further virus spread can occur. Some of the methods which are widely used are avoidance, roguing/eradication of virus reservoirs, plant remains, field sanitation, soil management and others.

(a) **Avoidance**

Planting site selection is an important first step in establishing a production area, and can be an important tactic for disease avoidance. In general, the conditions under which disease development is favored or hosts are predisposed to disease development should be considered. Likewise, production areas in which hosts might be predisposed to disease development should be avoided.

The importance of using disease-free planting materials cannot be overstated, and any measure that facilitates their production and utilization would be useful. For example, clean nursery stock and budwood schemes are hallmarks of successful citrus programs. Vegetative propagules can harbor bacteria, fungi, nematodes, viruses and viroids, and it is with them that many economically important pathogens are moved and established. Tissue-culture plantlets should be used whenever possible, as they are free of virus and viroid pathogens when they are produced from indexed mother plants. Only in rare cases are tissue-culture plantlets are not safe. In India, some farmers are regularly using tissue culture generated banana and sugarcane for their cultivation and getting higher yields. Many diseases of perennial crops originate in propagation nurseries, and soil-borne diseases whose symptoms are not readily apparent can be most problematic.

The new plantings should not be located near old plantings. For example, new tomato, capsicum, leafy crops or cucurbits or any other vegetables should not be grown nearby old fields having the same or another susceptible crop harboring virus and viroid diseases. Removal of the earlier crops of either tomato, capsicum, cotton, tobacco, cereals, brassicas, cucurbits or any other crop, should be enforced immediately after the harvest. Since the older plants are not ideally suited for insect vector multiplication, insects will leave in search of better quality hosts and enhance the chances for virus movement to adjacent or nearby young susceptible plantings.

(b) **Eradication of virus reservoirs**

This principles aims at eliminating a pathogen after it is introduced into an area, but before it has become well established or widely spread. It can be applied to individual plants, seed lots, fields or regions, but generally is not effective over large geographic areas. If pathogen exclusion has failed or if this is not possible, other strategies are needed. These measures are diverse, always more expensive than pathogen exclusion, and seldom entirely effective. Although the goal of eradication is to reduce occurrence to zero, in practice these measures are most often rate limiting. When there are significant reservoirs of inoculum, alternative hosts are removed from plantations and destroyed. Disease pressure is usually reduced in such cases, especially when pathogen host ranges are limited. The effectiveness of removing alternative hosts depends on their size and the pathogen's mobility. For example in Cyprus in a relatively isolated regions, the spread of *Tomato yellow leaf curl virus* was successfully controlled by eradicating in the spring the primary inoculum sources in over wintered tomato, before emergence of whiteflies. In three consecutive years when this measure was under taken, primary virus spread to spring plantings was almost completely prevented, while further secondary spread to summer plantings was below 5 % as compared with 40–50 % in previous years (Ioannou 1987). In Southern region of Western Australia, Mackie et al. (2002) have eradicated *Potato spindle tuber viroid*, which was noticed on tomato in one of the isolated hydroponic green house. The action involved was destroying the tomato infected crop and decontamination of green houses and surrounding areas by which action in the other hydroponic tomato crops were completely free from PSTVd infection.

(c) **Roguing**

Roguing infected plants is a key strategy which is widely used especially if the crop plant is the primary or sole source of inoculum. For example, bunchy top of banana caused by *Banana bunchy top virus* can be managed only if affected plants are identified frequently, removed from plantations, and destroyed.

Sometimes it may be worthwhile to remove infected plants from a crop. If the spread is occurring rapidly from sources outside the crop, roguing the crop will have no beneficial effect. If virus spread is relatively slow and mainly from within the crop, then roguing may be worthwhile, especially early in the season. Even with a perennial crop, if a disease spreads slowly, roguing and replanting with healthy plants may maintain a relatively productive stand. A study of the distribution of infected plants within the field using the formula developed by Vander plank (1963) will give an indication as to whether spread within the crop is taking place.

In certain situations, roguing may increase disease incidence by disturbing vectors on infected plants (Rose 1974). In many crops, newly infected plants may be acting as sources of virus for further vector infection before virus symptoms can be seen.

Roguing or removing diseased plants based on the symptoms from the host population in glasshouses or fields is the oldest method of control and usually the first practice to be tried. The removal of infection sources by roguing out infected plants is another important line of defense against air-borne virus diseases. For viruses which cause few or no noticeable symptoms, roguing based on external symptoms can be difficult. In such cases it is desirable to support inspection and roguing by laboratory tests. This however, can be effectively applied only with some virus diseases and especially where plants are young and the vector population and the disease incidence is not too high. It is futile for viruses with fast and efficient mode of transmission like TMV in tomato and tobacco (Wolf 1933). However, with some other contact transmitted viruses, such as *Barley stripe mosaic virus,* the spread in two barley cultivars was reduced from 85 to 12 % and 46 to 0 % by early roguing (Inouye 1962). In most countries commercial growers rogue out the infected plants in seed-beds in crops like lettuce and brassicas. Zink et al. (1957) found that roguing *Lettuce mosaic virus* infected plants just once soon after thinning had no effect on disease incidence at harvest (81.7 %), presumably

Fig. 2.10 Roguing of *Lettuce mosaic virus* infected plants in lettuce fields. *Courtesy* T. A. Zitter

because much spread has already occurred, but roguing twice or three times reduced the incidence i.e. from 73 to 45 and 36 %; and 17 to 9 and 6 %, respectively in two field trials and three hoeings were very effective (Fig. 2.10). In the past, roguing was usually done by laboriously digging or pulling out infected plants and removing them from the field.

Since hand roguing involves financial commitment, Barnes (1959) demonstrated that spraying a mixture of diesel oil and growth regulating compounds like dichloropropane-dichloropropane (D-D) on the diseased seed-potato plants would rapidly kill the foliage and also resulted in the least amount of regrowth from the seed pieces or tubers. This method was most efficient and less costly than hand roguing. Zink et al. (1957) used diesel oil mixture with parathion to rogue lettuce crops, and it was slightly more effective than hoeing, probably because aphids were killed, whereas aphids on hoed plants might move to nearby healthy ones. Regarding cost, roguing an acre three times with oil required approximately 6.5 man hours and roguing three times by hoeing required approximately 5 man hours. Zink et al. calculated that labor wages at the rate of 82 cents/h, the cost of labor would be approximately $5.33 to rogue an acre three times with oil and $4.10 to rogue three times by hoeing. The former method would cost somewhat more because of the added cost of the oil.

Roguing is practiced generally in perennial fruit crops; however, encouraging results were also obtained in the annual vegetable crops like tomato and okra, among tuber crops like cassava, potato, beet, sweet potato, and ornamentals like tulips and carnations. At Uganda, in sweet potato improved plant growth and higher tuber yields were achieved by early roguing the *Sweet potato virus disease* (SPVD) infected sweet potato vines (Bua et al. 2009). In rice also roguing was recommended as a tactical control measure for tungro virus disease (Tiongco et al. 1998).

From Uganda, Gibson et al. (2004) have reported that roguing of infected sweet potato cuttings infected with virus diseases within 1 month of planting has decreased the virus incidence. It was observed that sweet potato virus diseases (SPVD) spreads more rapidly in the unrogued plots than the plots rogued. Roguing 1 month after planting did not provide any additional decrease in SPVD spread. Regular roguing of infected plants has been effective in the control of *African cassava mosaic virus* in cassava in trials carried out in tropical Africa (Robertson 1987), but the method is not widely used. In Tanzania, roguing of *Cassava brown streak virus* disease has been followed by 46 % of the farmers, by which the virus incidence was reduced (Thresh et al. 1998). Also in Western Kenya Obura et al. (2003) reduced the spread of *East African cassava mosaic virus* spread by rouging infected plants. The role of phytosanitation in *African cassava mosaic virus* control has been reviewed by Thresh et al. (1998).

Most of the successful eradication schemes have been in tree crops which are perennial in nature. Among the factors that dictate success are: (1) relatively small numbers of infected trees and infection foci; (2) low rate of natural spread; (3) good data on extent and of distribution of infection; (4) rapid, reliable and inexpensive diagnostic procedures for the virus and virus-like disease detection, and (5) resources for rapid and extensive surveys and tree removal.

One of the most successful examples of disease control by roguing of infected crop plants has been the reduction in incidence of the *Banana bunchy top virus* (BBTV) in bananas in eastern Australia (Cann 1952). Over 90 % of banana production had ceased in New South Wales between 1922 and 1925 due to BBTV and in 1927 the Department of Agriculture destroyed all plantations and permitted their replanting with virus-free suckers in 1928. Strict regulations were made with regard to movement of plants in and between all banana-growing areas and frequent inspections were made to locate the diseased plants. If infected plants were spotted, then ½ pint of kerosene was poured into the central leaf and allowed it to trickle down round the leaf bases. After few hours, the plants were dug up and chopped into pieces (Simmonds 1934). Herbicides such as 2, 4-D when syringed into the base of the pseudostem, were shown to be a quick and effective means of killing unwanted bananas. The incidence was reduced to one plant in 2000 in 1948 or only 1/12 the number found in 1937. In 1952, only 15 infected plants were noticed per 100 acres. The cost of the control was about £ A 28,000/year, *approx.* $62,000 (Cann 1952). Within about 10 years, the campaign was effective to the point where BBTV was no longer a limiting factor in production. Dale (1987) and Dale and Harding (1998) attributed the success of the scheme to the following main factors: (1) absence of virus reservoirs other than bananas, together with a small number of wild bananas; (2) knowledge that the primary source of virus was planting material, and that spread was by aphids; (3) cultivation of the crop in small, discrete plantations, rather than as a scattered subsistence crop; (4) strict enforcement of strong government legislation; and (5) co-operation of most farmers. Dale (1987) and Sharma (1988) have reviewed the various aspects of bunchy top disease of banana.

In Honduras, the combination of insecticidal treatment and roguing was recommended against another important disease of banana, namely banana mosaic caused by CMV (Adam 1962). Each farm was inspected three times a year and infected plants were sprayed with nicotine along the plants with in 25 acres adjoining it. Then the infected plants and the 80 plants around the diseased plants were dug out, chopped up and sprayed with Malathion. The 25 acres were sprayed again 20 and 40 days later. The re-growth of these cleared areas was checked and no replanting was done for at least 6 months. Cucurbits and some solanaceous crops were abandoned to grow either in between or nearby the banana crop. This ambitions program had decreased the incidence of mosaic on a 14,000 acre plantation, from about 21 infected plants/1000 acres in 1956 to 0.1 in 1960, and prevented its introduction on to any new farms.

In Egypt, BBTV was eradicated by treating infected suckers with 0.12 % Chitosan (El-Dougdoug et al. 2009). Another example where the impact of roguing was noticed is the occurrence of 'Foorkey' disease of large cardamom (*Amomum subulatum*). This viral disease which is transmitted by the aphid *Pentalonia nigronervosa*, can affect nearly 50 % of the crop. The disease has been controlled in three valleys of Darjeeling district (India) by injecting diseased rhizomes with Agroxonone (liquid 2, 4-D) herbicide. The leaves turn brown and withered in 10–15 days and clumps died in 4–5 weeks, and these were then rouged out and replaced with 1 year-old healthy seedlings. Simultaneously the plantations were sprayed with 0.04 % Folidol E605 at regular intervals of 3–4 weeks to control the aphid vector (Chattopadhyay and Bhowmik 1965). Similar procedure was followed even for 'Katte' disease of small cardamom (*Elettaria cardamomum*) which is also transmitted by *P. nigronervosa.*

Roguing of diseased sugarcane plants is still important for reducing the incidence of mosaic in the field in the early stages, but the usual practice of digging the entire seed piece of planted stock is not practical in sugarcane fields. However, Koike and Tippett (1972) have recommended roguing with 2 applications of herbicides either Erbon or Glytoc at an interval of 5½ weeks.

In Ghana and Nigeria, roguing programme was started to control *Cacao swollen shoot virus* (CSSV) of cocoa, which is transmitted by mealy bugs. Experience has shown that when only those trees, found with symptoms during the routine inspections were removed, control was unsatisfactory and uneconomic, resulting in eradication of only smaller out breaks. In larger outbreaks where infected trees tend to be more numerous which were scattered were often missed and remain as source of infection. Considerable efforts were made to persuade farmers to follow the control measures. Massive publicity campaigns were mounted and compensation was paid for the removal of trees and for satisfactory replanting. Money was paid based on the area of trees destroyed and replanted or on a per tree basis. Initially, replanting did not need to be on the treated area, but this led to the migration of farmers to previous unexploited new areas. In an attempt to conserve the industry in the eastern region of Ghana, the replanting grants were increased provided that the treatment area was replanted and farmed to a prescribed standard.

The eradication campaign started in 1946 was interrupted for a while in 1948 because of political disturbances and also due to the repercussions from the farmers. However; since 1949 it continued at a steady pace. Up to June 1957, 63 million trees had been removed. In the eastern province of Ghana, there has been evidence of a leveling off at about 215,000 trees a month. Approximately 20,000 men were involved and the total cost of the operation was huge. Due to intensive researches, techniques have improved over the years and the education of the farmers had its effect on growers' cooperation. The efficacy of control measures is improved by removing both the obviously infected trees and the adjacent apparently healthy ones, i.e. an average of ten trees were destroyed for every one obviously infected.

In Ghana, during 1962, the organization responsible for a campaign was disbanded and cutting out which was left to individual farmers, virtually stopped. Again in 1967, cocoa production division was established and was responsible for the eradication programme. To avoid complication of paying compensation to many small farmers, a scheme was introduced for the treated farms to be replanted and maintained for 3 years before being handed back to the farmer. This policy has been supplemented by two large schemes supported by the World Bank for replanting cocoa in affected areas. Legg (1979) reported the achievement of cutting-out diseased plants in minimizing the disease spread. Only 4.2 % of the trees were removed from 3 ha of the cultivar Amelonado in 6 years and only 1.1 % of the trees were removed in 9 years from 13 ha of Amazon cocoa.

In Nigeria, until 1950, only diseased trees of cocoa were removed. Thresh (1959) recommended removal of all trees within 4.5, 9 or 13.5 m of out breaks of fewer than 6, 6–50 or 50–200 trees respectively, followed by regular inspection of the peripheral trees. Farmers were persuaded to remove the ring of trees around infected trees as well. For a period of 2 years, all the outbreak areas were inspected every 2 months until no further infected trees were found during a period of 2 years. Thresh and Lister (1960) used 'coppicing' technique to determine the distribution of latent and unrecognized infection around naturally infecting outbreaks of cocoa. Chopping techniques involve the cutting down the trees to a foot or so from the ground. The new flush will express the symptoms if the tree is infected. All the apparently healthy trees within 30 yards of eighty outbreaks of various sizes were coppiced. With this method relatively few healthy trees are being destroyed compared with the farmer system of treating all the out breaks to a minimum distance of 30 yards. However, coppicing needs to be done at the advent of the dry season, if regeneration is to be successful, so its use is limited. Though in certain text books, roguing method is mentioned as example for effective control of CSSV, but control based only on "cutting out" campaigns has not been achieved due to several factors including political and socio-economic problems.

A model has shown that extensive eradication will control *Citrus tristeza virus* (CTV). An analytical model of the virus disease dynamics with roguing only in the post-infectious category confers no advantage. At low contact rates, roguing only when plants became infectious is sufficient to eradicate the disease but at high contact rates, roguing latently infected as well as infectious plants is advisable.

Fig. 2.11 Roguing as a management tactic for PBNV control demonstrated in Tamil Nadu, India. *Courtesy* IPM CRSP: Sue Tolin; R. A. Naidu; G. Karthikeyan

The model indicates that eradication is achievable with realistic roguing intensities. In Israel, an eradication programme was carried out in citrus against CTV using the combined "indicator-EM" method. About 160,000 tests were made and 134,159 and 140 infected trees out of 40,000, 60,000 and 58,000 test trees were found in the year 1971, 1972 and 1973, respectively (Bar-Joseph et al. 1974). Annual indexing around infection centers has shown a substantial reduction in the proportion of trees infected since the programme began.

Field trials were conducted under IPM CRSP: Global theme on plant virus diseases, against *Peanut bud necrosis virus* (PBNV) management in tomato crop by roguing in Tamil Nadu have clearly indicated that roguing helps in decreasing the PBNV incidence with increased yields Fig. 2.11.

A model of plant virus disease dynamics in perennial tree crops was developed (Chan and Jeger 1994) and analyzed to evaluate roguing (removal of diseased trees) and replanting with healthy trees as a disease control strategy. When applied to the published data for three important virus diseases-citrus tristeza, banana bunchy top and plum pox-it was concluded that roguing and replanting could contribute to the eradication of a disease if economically viable, and that continuing operations could certainly maintain disease below economically damaging levels (Jeger and Chan 1995).

(d) **Plant remains**

Plant remains in the soil, or attached to structures such as greenhouses, may harbor a mechanically transmitted virus and act as a source of infection for the next crop. With a very stable virus such as TMV, general hygiene is very important for control, particularly where susceptible crops (e.g., tobacco, tomato, eggplant, chilli pepper) are grown in the same area every year. ToMV may be very difficult to eliminate completely from greenhouse soil using commercially practicable methods of partial soil sterilization (Broadbent et al. 1982). A major development has been the replacement of soil by sand/peat substrates that are renewed every year.

(e) **Field sanitation**

When a crop is grown in the same or adjacent fields in consecutive years, the infected volunteer plants serve as one of the major sources of disease inoculum as well as for the vector. The role of volunteer plants virus epidemiology is well established in crops like potato, sugar beet, groundnut, wheat and onion. Diseased potato tubers and sugar beet roots escaped while harvesting or dropped from lorries carrying beet and potato to the processing factories during transit, put out new growth early in season and provide inoculum for the viruses like *Potato leaf roll, Potato virus Y* and *Sugar beet yellows*. It was suggested that it would take 5–6 years to eliminate volunteer potatoes from a field in which potatoes had been grown (Doncaster and Gregory 1948). The volunteer potatoes in the cereal crops like wheat were rarely aphid-infested and the incidence of the virus disease remained steady, but when another crop of potatoes was grown, the volunteers that were infested by aphids served as an important source of virus. In Britain, all the cereal crops, which usually follow potato crops, were treated with herbicides which destroy ground creepers, together with the weeds and there will be no longer a danger of virus spread (Heathcote 1973). Lutman (1979) reported the control of volunteer potatoes in the autumn in cereal stubbles. After cereal harvest many of the volunteer potato plants present in two winter barley crops continued to grow even after cereal harvest. Aminotriazole (200 g/l with 16.5 % ammonium thiocyanate) applied to this regrowth resulted in 50–69 % fewer potato plants emerging in the year following treatment. Glyphosate (360 g/l) gave better control averaging 72 % in one experiment and 88 % in another. Recommendations for the optimum control of ground creepers are that the fields should be timely cultivated, not ploughed and planted with winter barley; and treatment with glyphosate or aminotriazole herbicides in cereal stubbles.

The small roots of sugar beet which escape the harvesting or fall from the root cleaning machinery may survive mild winters, particularly if buried in the soil (Wallis 1967). Wild beets or beets escaped from cultivation and growing on the edges of arable fields, on clamp sites, along fence rows, ditch banks and in waste areas, served as sources of infection of beet viruses in Europe (Broadbent et al. 1960) and Washington (Wallis 1967; Howell and Mink 1971). The beets found in refuse (tare) piles and infected volunteer beets in second year beet fields are important overwintering sources for *Beet western yellows virus* (BWYV) (Wallis 1967). Further studies also revealed the role of volunteer sugar beets in most years are the primary source of early season inoculum for both *Beet mosaic* and *Beet western yellows viruses* (Howell and Mink 1971). In England, volunteer carrot plants from roots left in the field or carrot crops left in the field during winter, harbor *Carrot motley dwarf virus* (CMDV) during mild winters (Dunn and Kirkley 1966). Carrot crops grown in central Washington also grown near diseased volunteer carrots had high incidence of *Carrot thin leaf virus* (CTLV) and CMDV (Howell and Mink 1977). Roguing of volunteer onion plants infected with *Onion yellow dwarf virus* in the spring before seedling emergence was cited as a practical programme that growers could follow to reduce the incidence of this disease

(Louie 1968). Roguing of volunteer plants of turnip infected with *Turnip mosaic virus* was recommended for successful turnip growing (Laird and Dickson 1972).

Self-sown plants of wheat in North America help in completing the disease cycle of *Wheat streak mosaic virus* transmitted by a mite vector, *Aceria tulipae*. These volunteer wheat plants carried the virus and also serve as the breeding place for the vector population. Effective control was achieved by destroying early volunteer wheat and weeds present in and around the fields and also by avoiding overlapping crops (Slykhuis 1955; Staples and Allington 1956).

Another example where the virus survives in the root remnants of diseased plants is of *Grapevine fanleaf virus* in grapes. The transmission tests with its nematode vector, *Xiphinema index,* indicated that this virus survives up to 4½ years in the diseased root remnants, and because of this factor, the replanting of healthy plant material will also become infected. A break of at least 5 years before replanting vines was needed to decrease the virus incidence (Thorne and Allen 1950; Hewitt et al. 1962).

From Sudan, Tarr (1951) reported that during the early years of cotton growing, the plants were cut at soil level after harvest. Summer rains or watering resulted in heavy ratoon growth and most of the ratoon growth was found to be infected with *Cotton leaf curl virus*. The ratoon growth was reduced by cutting the plants below the soil level but hand pulling the plants soon after cotton harvest is also very effective in reducing the disease. Even leaving the ground fallow or at least not irrigated for a minimum of one season after a cotton crop also allowed substantial control of the disease. Mechanical cutting or pulling will be equally effective and less laborious.

Klesser and Le Roux (1957) recommended the removal of all volunteer peanut plants whether diseased or not, to check the epidemics of *Peanut mosaic virus* in some districts of South Africa. From eastern Africa, Bock (1973) noticed a 20 % *Peanut mottle virus* incidence from the seeds of diseased plants left in the soil. The roguing of these diseased plants should be followed for 2 months after planting. Demski (1975) reported that in Georgia, the source of *Peanut mottle virus* for the soybean crop was infected volunteer peanut crop and the disease incidence was reduced by destroying the volunteer plants.

From Australia, Garrett and McLean (1983) illustrated an unusual case of high incidence of insect-transmitted viruses in brassica fields, which was due to the presence of a nearby previously infected crop that had not been removed. Broadbent et al. (1951) have recommended that for successful management of *Lettuce mosaic virus*, in addition to the use of virus-free seed, the growers should cultivate lettuce fields in discrete blocks rather than allowing them to be scattered throughout the farm. They also recommended that lettuce crop residues should be ploughed in as soon as possible after harvest to eliminate the source of virus inoculum to the newly planted crops. In the tropics, crop residues of legumes, tomato, and okra, where the geminiviruses cause economic losses, would benefit from the early season removal of crop residues to reduce the virus inoculum source for subsequently planted crops, that are grown continuously throughout year under irrigated or nonirrigated conditions.

(f) **Soil management**

Managing the air-borne viruses is easy when compared with soil-borne nematode/fungus transmitted viruses. Generally fallowing measures are recommended for the control of soil-borne virus diseases. But as some of the nematodes live for long periods and also some fungal vectors survive as resting spores, a short period of fallow will not necessarily decrease the vector populations. Fallowing of the land is not economical and small land holders do not implement this practice. Encouraging results are available, wherein the soil-borne diseases are minimized by crop rotation of non-susceptible hosts of virus/vector. For example, French and Wilson (1976) demonstrated that the incidence of *Tobacco rattle virus* (TRV) in potatoes, which is transmitted by *Trichodorus* spp. tend to decrease as the increase in the number of barley crops grown in the 3 years prior to potato sowing. Similarly the outbreaks of *Wheat mosaic virus*, transmitted by soil-borne fungal vector *Polymyxa graminis* can be minimized when four crops of maize, oat or soybean were grown between wheat crops (Koehler et al. 1952).

Crop rotation has also given encouraging results in grapes against *Grapevine fan leaf virus* and its vector *Xiphinema index.* As this nematode vector has a narrow host range, Vuittenez (1970) reported that a fallowing period of 10 years before replanting of grapes in the infested soil was necessary. However, such a long period without a grapevine crop is not practicable. In the south of France, the population of *X. index* disappeared almost entirely within 2 years after thoroughly pulling out the old vines followed by cultivation of lupine, lucerne and cereals for 7 years before replanting (Dalmasso 1971; Dalmasso et al. 1972a, b). In California, Hewitt et al. (1962) and Raski et al. (1965) have shown that both root pieces and *X. index* could be found living up to 4–5 years after the vines were removed and concluded that a successful grapevine replanting in infested soil is possible only after a gap of 5 years. At Delaware, even in case of soybean crop, *Soybean severe stunt virus* (SSSV) transmitted by nematode vector, *Xiphinema americanum,* was reduced by crop rotation for 2 years of continuous corn or grain sorghum, wheat followed by soybean or 2-year fallow reduced both nematode vector density in the soil and SSSV severity (Evans et al. 2007).

Lengthy crop rotation does not always work well against all nematode vectors. Calvert and Harrison (1963) noticed that *Tomato black ring spot virus* and its nematode vector *Longidorus elongates* persisted through a 4 year rotation of non-susceptible plants and affected susceptible leeks and onions in the following year. Harrison and Hopper (1963) also extracted *L. elongates* nematodes even after 29 months incubation at room temperature, in dark, from a light loam soil which did not had any living roots.

Throughout the world, the use of plastic sheets for solarization to control the soil-borne fungal vectors has been attempted. In India, at ICRISAT (Hyderabad), Reddy and his associates during 1991 used layers of transparent polythene sheeting for solarization. The well cultivated soils which were profusely irrigated prior to the treatment were covered with polythene sheeting for at least 70 days during the summer months. They noted reduced incidence of *Indian peanut clump virus* which is transmitted by the soil-borne fungus *Polymyxa graminis*. The economics of this practice was not established (Reddy 1991).

Even by following special cultivation methods, the soil-borne vector may be effectively reduced. For example, Oostenbrink (1964) demonstrated that the population of *Trichodorus flevensis* was reduced from 1210 to 160/1 cc of soil by mixing soil with a rotary cultivator.

(g) **Hygienic measures**

Hygiene deals with measures directed towards decontamination of plant material, storage rooms, packing material and tools, and established rules of conduct to prevent virus/viroid spread. Diseases for which no definite vectors are known, *Tobacco mosaic virus* (TMV), *Potato virus X* (PVX), *Potato spindle tuber viroid* (PSTVd) and *Citrus exocortis viroid* (CEVd), can be spread while handling, pruning or by contact with infected plant debris or contaminated implements. The maximum contact spread occurs in crops like tomato, tobacco, potato, chilli/capsicum, cucumber, and several ornamental crops which are grown under glass houses, which are frequently handled during normal cultural operations. The following measures will help to reduce the disease spread and it is generally practiced by commercial growers and their employees in some countries.

For some mechanically transmitted viruses, and particularly for TMV or ToMV, human activities during cultivation of a crop are a major means by which the virus is spread. Once TMV or ToMV enters a crop such as tobacco or tomato, it is very difficult to prevent its spread during cultivation and particularly during the cultural operations. These viruses survive for a long periods and are very difficult to manage once the crop is infected, in either the field or in net houses. However, one of the management measures consists of treatment of implements and washing of the hands. For this, Broadbent (1963) recommended a 3 % solution of trisodium orthophosphate. Workers' clothing may become heavily contaminated with TMV and thus spread the virus by contact. TMV persisted for over 3 years on clothing stored in a dark enclosed space (Broadbent and Fletcher 1963), but was inactivated in a few weeks in daylight. Clothing was largely decontaminated by dry-cleaning or washing in detergents with hot water.

While TMV is the most stable of the mechanically transmitted viruses, others can be transmitted more or less readily on cutting knives and other horticultural tools. These include CymMV, PVX and PSTVd. These mechanically transmitted agents may be a particular problem in glasshouse crops, where lush growth, close contact between plants, high temperatures and frequent handling are important factors in facilitating virus transmission.

Since mechanical transmission is an important means whereby viroids are spread in the field, decontamination of tools and hands is an important control measure. However, this presents difficulties because of the stability of viroids to heat and many decontaminating agents as normally used. Brief exposure to 0.25 % sodium or calcium hypochlorite is probably the best procedure (Garnsey and Randles 1987). One of the earliest and most effective methods for the control of viroids has been the avoidance of sources of infection. This method has been particularly successful for vegetatively propagated crops susceptible to viroids, such as potatoes and chrysanthemums. This success is probably due to the absence of vectors in the field apart from humans.

It is found difficult to free sap-engrained hands with TMV. The best recommendation is washing hands in 3 % solution of trisodium orthophosphate and then scrubbing well with soap and water, but this does not always eliminate the virus present under the nails (Broadbent 1963). Even wetting the worker's hands with skimmed milk is also recommended for reducing the TMV contamination in tomato fields (Hare and Lucas 1959). Similarly, very high TMV incidence was noticed in experimental plots of bell pepper at Indian Institute of Horticultural Research (I.I.H.R.), Bangalore (India) due to indiscriminate handling of healthy and infected plants during breeding work by using the contaminated forceps, needles and brushes (Sastry, unpublished).

Besides heat sterilization, various chemicals and detergents have also given encouraging results in preventing the contamination through farm implements. In an experiment conducted at Victoria (Australia), it was observed that transmission of TMV, through implements was reduced from 95 to 4 % when they were washed with 10 % trisodium orthophosphate and 12 % with reconstituted skimmed milk (Zitter 1991). Mulholland (1962) recommended that TMV contaminated cutting knifes can be sterilized by a 10 % trisodium orthophosphate solution with a little detergent by filling the hollow plastic handles with this solution and allowing to trickle slowly over the blades. Most of the viruses infecting orchids and gladioli spread through the contaminated knives while removing the flower spikes. Franklin (1966) sterilized the contaminated knife blades with viruses infecting orchids, by drawing them across a moist sponge in the neck of a bottle containing 2 % formaldehyde and 2 % sodium hydroxide.

Although many new products are available for the decontamination of cutting tools used to prevent the spread of TMV in test petunia plants, some surprising results were found for establishing the most effective treatments (Lewandowski et al. 2010). Two "old standbys", a 20 % (wt./vol.) solution of nonfat dry milk plus 0.1 % Tween 20 or a 1:10 dilution of household bleach (0.6 % sodium hypochlorite) completely eliminated TMV-transmission in petunias.

It is also established that TMV spreads from one crop to the next on the contaminated trellis wires. Nitzany (1960) reported that trellis wires from the infected fields stored in a shaded store, retained infectivity for 1½ to 4 months. He found that the virus is inactivated by dripping contaminated trellis wires in 1 % formalin for 5 min or 5 % of soap solution for 10 min or 0.1 % caustic soda solution. Detergent or washing-soap solutions were only satisfactory if the wires were immersed in high concentrations of the solutions for a long time.

Winter-Nielsen (1972) observed that a tractor which passed from a PVX-infected potato field into one free was capable of infecting the plants the first 150 m that it passed through this field. Donald and Bonde (1954) also observed similar transmission of PSTVd and percentage of transmission ranged from 4–13 in three cultivars. Washing the machinery thoroughly with the earlier mentioned virus-inactivating detergents before it is taken into the healthy crop is recommended.

Even under protected cropping where the crops are continuously raised inside glass or wire mesh houses and in virus experimental glasshouses, occasionally high incidence of virus diseases is noticed as the glasshouse structures are commonly contaminated. Even small fragments of infected leaf debris either on the benches or floor can serve as a source of virus inoculum. Through washing of glass or wire mesh houses with a disinfectant when they are empty between crops will help to get rid of the contamination.

Like some of the viruses which are highly contagious, the majority of viroid diseases are also spread by contaminated hands and implements. Roistacher et al. (1969) reported the transmission of CEVd by hands. Fingers rubbing over freshly cut stems of various exocortis-affected citrus cultivars and then rubbing over young emerging sprouts on 20 healthy Etrog citron stems, resulted in 14 infected plants. Care should be taken in pulling young sprouts by hand rather than rubbing.

Similar to plant viruses, even contamination of the viroid diseases from implements were eliminated by using disinfectants and heat treatments. In Germany, Timmerman et al. (2001) have tested the efficiency of several disinfectants to eliminate the contamination of viroid diseases. They reported the efficient disinfection by MENNO-Florades (Menno-Chemie-GmbH, Norderstedt, Germany) by selection of different concentrations of the disinfectant and various times of incubation. In these tests, PSTVd was eliminated by using the effective combination: 2 % of disinfectant while incubating for 1 min or alternative 3 % for 30 s. They have effectively controlled the contamination of viroid disease from contaminated-tools and tables by using the above discussed chemicals (Timmerman et al. 2001).

By using heat treatment some of the viroid diseases were eliminated from contaminated farm implements. Freytag (1965) sterilized the knives by keeping them hot by an attached propane torch. CEVd disease of citrus, is also readily transmitted by contaminated clippers, knives and pruners and tools remain contaminated for as long as 8 days. Similarly, *Peach latent mosaic viroid* in stone fruits also transmitted through contaminated blades (Hadidi et al. 1991). Garnsey and Jones (1968); Garnsey and Weathers (1972) found that a solution of 2 % formaldehyde plus 2 % sodium hydroxide effectively decontaminates the grafting tools. Subsequently, Roistacher et al. (1969) recommended that transmission on contaminated knife blades was prevented by 1 s dipping in a solution of household bleach (containing 5.25 % hypochloride) with enough water to make 1 gal. The corrosive action of the bleach on tools was avoided by dipping in a mixture of 1 part of vinegar, 3 parts of water and 2 tea spoons of emulsifiable oil per pint of mix. Heating contaminated blades for 4–6 s in a flame of a propane torch or flaming blades dipped in 95 % alcohol failed to inactivate the exocortis viroid. Garnsey and Whidden (1971) also found that house hold bleach (5 %) or formaldehyde plus 2 % sodium hydroxide decontaminates the knife blades. They also noticed that exocortis viroid transmission is prevented by 10 % bromodine for 5 min in 2 % sodium hydroxide or 10 % trisodium orthophosphate and also by dipping in fresh milk.

2.10 Avoidance of Sources of Infection/Inoculum

It is obvious that there will be no virus problem if the crop is free of virus when planted and when there is no source of infection in the field, or none near enough to allow it to spread into the crop.

The extent to which it will be worthwhile to attempt to eliminate sources of infection in the field can be decided only on the basis of the detailed information of such sources and of the ways in which the virus is spreading from them into a crop. Eradication as a control measure has been reviewed by Thresh (1988) and eradication schemes directed against CSSV and CTV are described by Ollennu et al. (1989) and Garnsey et al. (1998).

(a) **Living hosts for the virus**

Living hosts as sources of infection may include: (1) perennial weed hosts, annual weed hosts in which the virus is seed transmitted, or annual weed hosts that have several overlapping generations throughout the year; (2) perennial ornamental plants that often harbor infection in a mild form. For instance, gladioli are often infected with BYMV that can spread to adjacent annual legume crops (Hull 1965); (3) unrelated crops; (4) plants of the same species remaining from a previous crop (these may be ground keepers, as with potatoes, or seedling volunteers); and (5) seed crops of biennial plants that may be approaching maturity about the time the annual crop is emerging.

In theory it should be possible to eliminate most of such sources of infection. In practice it is usually difficult and often impossible, particularly in cropping areas that also contain private gardens. Private household gardens and adjacent farms in temperate and subtropical regions often contain a diverse collection of plants, many of which can carry economically important viruses. It is usually difficult or impossible to control such gardens effectively. In tropical countries, the diversity of cropping makes control of living sources of infection very difficult.

The extent to which attempts to remove other hosts of a virus from an area will depend greatly on how large the host range of the virus is in nature. It may be practicable to control alternative hosts where the virus has a narrow host range (OYDV infecting onion), but with others, such as CMV and TSWV that have very wide host ranges, the task is usually impossible.

The natural host range of *Sugarcane mosaic virus* (SCMV) is restricted to members of *Poaceae*. A number of cultivated and wild grasses which serve as natural reservoir of SCMV have been identified. Of these, corn and sorghum are of most importance for SCMV spread to sugarcane as they are planted in areas adjacent to sugarcane. Eradication of these hosts at frequent intervals during the cropping season yields good results.

(b) **Role of weeds in virus incidence**

The importance of weeds in crop cultivation is well known, as they not only compete with the crop plants for nutrients, moisture and solar energy, but also harbor pests and diseases. The weed host's role in the occurrence and spread of plant virus, viroid, or phytoplasma diseases is an integral part of the ecological aspect of virus transmission, and has been understood throughout the world. The majority of plant viruses that infect economic crops have weed hosts which are widely distributed along irrigation channels, ditches, road sides, in plantations, orchards, by the side of railway tracks, in fallow lands and in neglected crops. The viruses present in these weed hosts serve as the primary source of inoculum and they spread to economic crops by vectors. Infected perennial weeds are more dangerous than annuals as they serve as virus source for long periods. Virus diseases are primarily introduced in to a crop by man planting infected seed/plant material or by vectors who acquire virus from other infected crop plants/weeds/ wild hosts. Wherever the planting material is healthy, the virus inoculum source will usually be from the weed hosts or from other infected economically important crops. The epidemiology of these diseases in a given area is a highly complex phenomenon, involving a number of weed hosts, commercial crops, different vectors and agro-climatic conditions. The chances for the outbreak of the disease epidemics will be more in an area where abundant virus-infected weeds harbor and support the fast multiplication of the particular vector.

Wild plants are often infected with different viruses and act as a bridge for further spread into cultivated crops. For some virus diseases the wild plants may belong to the same family or completely unrelated family. For example, in Africa the virus causing *Cacao swollen shoot* disease infects a species belonging to a different genera in the case of *Cola acuminata* (the cola nut) of the same family as the cacao (*Sterculiaceae*), but can also infect a genus from a completely different family such as the silk cotton tree (*Ceiba pentandra*), belonging to the family *Malvaceae*. Similarly, Konate et al. (1995) have reported that *Okra leaf curl virus* occurs naturally on wild hosts like *Sida acuta* in Burkina Faso.

In almost all rice growing countries in South East Asia, tungro caused by *Rice tungro spherical virus* (RTSV) and *Rice tungro bacilliform virus* (RTBV) is a major threatening problem which is transmitted a leafhopper in a semi-persistent manner. The virus cannot persist in the leafhopper vector for more than 5 days and it is not transmitted through seeds or soil (Ling 1972). Rice again does not usually remain in the field through all the months of a year. In rice-rice rotation, there is always a crop free period between the two cropping seasons. How then does tungro survive? Does it positively interact with weeds and other allied plants? Wathankul (1964) in Philippines observed natural infection of RTV in *Eleusine indica*, *Hemartharia compressa, Echinochloa crusgalli* and *E. colonum, Polypogon monspeliensis, Sorghum halepense* and *Sporobolus tremulus.* Mukhopadhyay (1980), however, found natural infection in *Echinochloa colonum,* when it was collected from a field where tungro chronically occurred. He did not find any infection in *E. colonum* collected from the fields where disease occurred occasionally.

Rivera et al. (1969) observed RTV symptoms in *Dactyloctenium aegypticum, Eragrostis tenella, Ischinum gogossum, Leersia hexandra, Oryza barthii, O. officinalis, O. ridley, O. rufipogon, Paspalum scroboculatum, Setaria glauca, Sorghum vulgare* and *Triticum aestivum*. They however, could not find easy recovery of the virus from these plants and they also could not confirm the infection of *E. colonum*. The role of weed hosts in RTV epidemiology was also studied by Mishra et al. (1973); Rao and Anjaneyulu (1978); and Tarafder and Mukhopadhyay (1979, 1980).

The epidemiological studies suggest that a few weeds and wild rice have the potentiality to carry the virus under the forced situation of the repeated appearance of the disease and under improperly managed field conditions.

In many regions, investigation of the begomovirus(es) involved and their biological properties have revealed that these diseases are caused by locally adapted viruses with narrow host ranges, often with the susceptible crop plant being the primary host. This has led to development and implementation of regional crop or host-free periods as a management strategy. This practice involves not growing the susceptible crop plant, usually for a period of 2–3 months, thereby disrupting continuous cropping patterns. This can result in a dramatic reduction in virus inoculum in the agroecosystem and, in some cases, a reduction in the insect vector population. Gilbertson (2011) has recommended two successful applications of a host-free period for managing tomato-infecting begomoviruses in developing countries with tropical environments.

It is clear from the above mentioned facts that weeds serving as hosts for both virus and vectors, and pose a great threat to the economy of important crops grown nearby. Therefore, it is necessary to establish the biological relationship between weed, virus/phytoplasma and vector for a clear understanding the manner of the spread of the disease. This will form a sound basis for devising control measures against weed hosts and vectors, thereby helping to minimize the incidence of virus diseases and reduce economic losses to the bare minimum.

(c) **Reducing virus incidence by weed management**

Successful elimination of the weeds by various means has helped in reducing the epidemics of many economically important virus diseases. Manual eradication of weeds can be followed if it is in a limited population and when labor is cheap, and is generally done while raising nurseries. In small plots, weeds are pulled and destroyed or occasionally ploughed; however, it is impracticable to follow this procedure when the crops are grown extensively. Under field conditions there have been some spectacular achievements in limiting the spread of viruses from weed hosts to economically important crops around the world. For example, in Florida, southern celery mosaic virus (CMV strain) was eliminated by eradicating *Commelina nudiflora*, which is a perennial weed host for this virus (Townsend 1947). In Australia, the incidence of *Lettuce necrotic yellow virus* on lettuce was reduced from 75.6 to 6 % by eliminating *Sonchus oleraceus,* which is a host for both virus and vector by spraying the herbicide diquat around the field (Stubbs et al. 1963). Herbicidal sprays 1–2 months before planting helps to maintain the crops free of weeds. In the Yakima valley of Northwest of U.S., increase in the yield of sugar beet crops by about 3.8 tones/ha was the result of decreasing the *Beet western yellows virus* incidence by destroying the weeds in the drainage ditches (Wallis and Turner 1969). In some sugarcane growing countries, volunteer plants like sorghum, maize and other graminaceous hosts are destroyed to check the *Sugarcane mosaic virus* source (Louie 1980; Shukla and Teakle 1989).

Crop rotation measures also help in decreasing the disease spread, if weed hosts responsible are annuals and do not transmit virus through seed. Crop rotation will not prevent the survival of viruses in seed transmitted viruses since the infected weed seeds often remain dormant along with infectious virus for long periods.

In Simla (India) Nagaich et al. (1972) found the application of herbicides like Stam F either alone or in combination with Eptam, to be effective in checking the infection and spread of PVX in potato seed plots provided initial stocks are of acceptable health standards.

Chemical control of weeds hinges on the property of a given herbicide to affect only a given weed or group of weeds. Some herbicides can affect healthy and virus-infected plants differently. Adsuar (1961) reported that sprays of Simazine intensifies the chlorosis and other symptoms of *Sugarcane mosaic virus*-infected plants without affecting the healthy sugarcane crop. Ulrychova and Blattny (1961) also observed similar results in TMV-infected tobacco plants. The use of herbicides to bring out the masked or latent infections for easier detection can be of help in roguing programmes, if percentage of infection is low. Herbicidal applications have both merits as well as inadequacies. Predominant factors that limit herbicidal use are injury to non-target crops resulting from spray drifts and persistence of herbicidal residues in soil.

From the above information it is understood that many viruses have weed or wild hosts that act as foci of infection from which there is spread of viruses into or within crops and also breeding point for vectors. For example, the initial distribution of *Maize dwarf mosaic virus* is often associated with patches of the perennial grass weed, *Sorghum halepense,* that occur commonly within and around crop stands (Damsteegt 1976). Similarly, the common compositae weed *Tridax procumbens,* is the host of the aphid-borne virus or viruses that cause ringspot and leafcurl diseases of sunflower in eastern and southern Africa (Theuri et al. 1987). These examples emphasize the importance of adopting effective weed control measures, not only to avoid competition, but also to eliminate sources of virus inoculum and vector breeding foci. The advantages to be gained are also apparent from experience with *Cacao swollen shoot virus* in the western region of Ghana, where many of the outbreaks in cocoa are associated with the understory forest tree *Cola chlamydantha* (Attafuah 1965). These trees are indigenous hosts of the virus from which spread occurs to nearby cacao and these wild hosts are removed when outbreaks are treated during the eradication campaign. There have been also restrictions on cultivation of okra and malvaceous ornamentals in the Gezira irrigation scheme in the Sudan, as these plants are known to be hosts of *Cotton leaf curl virus* and the whitefly vector *Bemisia tabaci* (Tarr 1951). Such measures are likely to be more widely applicable, but they are difficult to enforce in sub-Saharan Africa where little attempt is made to control virus diseases by removing weed or alternative hosts.

(d) **Inter-seasonal carryover of virus**

A few weeds of wild rice and the intact stubbles of susceptible cultivars have the potentiality for providing the virus in the field for the inter-seasonal carryover of RTV.

Field studies revealed that the nursery beds may perform a key role in the perpetuation of the virus. The virus may remain in the infected intact stubbles. Seed beds are usually raised by the cultivators in nearby plots before the land preparation operations for transplanting. As soon as seedlings grow, viruliferous leafhoppers may move from the stubbles to the seedlings. Upon transplanting these seedlings, the field may receive the initial inoculum. The studies made by Mukhopadhyay (1980) have suggested that the pathway of the carryover of the virus is from infected stubbles/standing crop to nursery beds then to transplanted crop. *Rice yellow mottle virus* is another important virus disease of rice in Africa. The primary infection source is the rice seed beds and Traore et al. (2006) have recommended raising rice nurseries in protected conditions as the prerequisite. In Nigeria also, *Rice yellow mottle virus* (RYMV) survives between seasons in rice stubbles. Even the virus was also transmitted when rice roots from infected plants became intertwined with non-infected plants. Hence the presence of virus infected rice roots increases the incidence of RYMV. Leaf contact from closely spaced plants aids in virus spread under field conditions (Abo et al. 2013). Similarly in Sudan, *Cotton leaf curl virus,* which is transmitted by whiteflies, will spread from infected cotton plants that regenerate from stumps of the previous year's crop (Tarr 1951).

2.11 Variation of the Crop Cultural Practices

Manipulations and variations of cultural practices which are simplest, that an agriculturist can employ, offer a wide measure for decreasing the incidence of virus diseases, but have limitations imposed by different growing seasons and crop requirements. Cultural management of virus and viroid diseases includes any modification in the production of the crop that lowers the disease incidence by reducing or eliminating the vector population or inoculum sources. In this section of discussing cultural management, the production practices that act directly on the vector/agent responsible for the disease spread, like insecticide application, biological control and genetic control are excluded. Cultural management includes a diverse set of practices including the following: sanitation which leads to the destruction of alternate habitats acting as sources of infection, plant density, crop rotation and fallow, crop planting date, trap cropping, fertilizer use and harvesting time. Sanitation is the removal and destruction of crop residues which act as source of inoculum. Destruction of alternate habitats and hosts is usually aimed to eliminate the weed, wild hosts etc., crop rotation entails changing the crop in subsequencing planting and crop fallow involves suppressing all plant growth on a field for a production season. Both practices can disturb the normal life cycle of the virus/viroid disease, thereby reducing its vector population or inoculum source and the infection. These measures are aimed in eliminating the sources of the inoculum within and outside the field and also in reducing the vector population or by changing their feeding behavior. Use of virus-free seeds and/or propagative organs results in minimizing primary infection. This can be complemented by removal of sources of infection in and around the crop, removal of plant remains from a former season and, if necessary, creation of a time gap between crops and/or space gap between plots. These operations will reduce the numbers of viruliferous insects that reach the crop. When addressing these two approaches, cultural practices can be grouped into two categories: (1) elimination of sources of inoculum, and (2) controlling the disease in the field. More information on the role of cultural practices in plant virus disease management has been reviewed by number of researchers (Zitter and Simons 1980; Cohen and Berlinger 1986; Hilje et al. 2001; Thresh 2003; Rao et al. 2006; Hooks and Fereres 2006).

In this section, attempts are made to reduce the disease spread by avoiding the vectors or by creating an atmosphere which is not congenial for their feeding or multiplication. Based on the epidemiological studies, attempts are also made to break the disease cycle for minimizing the disease incidence. Depending on the vector activity, it is also possible to select the safe periods for raising healthy crops. This aspect is discussed under the following sub-heads:

(a) **Isolation by distance**

Overwintered biennial and perennial crops are generally harbored with many virus and viroid diseases and they may spread to nearby first-year fields of the same crop or susceptible crop. When successive crops are grown adjacent to old ones during high vector periods, these diseases spread to the new crops and reach epidemic proportions. The spread of these diseases tend to decrease with increasing distance from foci of infection. The minimum distance for adequate isolation is difficult to determine because much depends on the type of virus and vectors involved, wind direction and vector population and other factors. It is known that aphids and leaf and planthoppers can be blown hundreds of miles, sometimes from one continent to another, even across oceans. Effective control is possible especially against stylet-borne viruses by this isolation technique, as they would cease to be viruliferous during prolonged flights and sometimes they will settle on non-susceptible hosts and lose the virus after probing; whereas for persistent viruses, as the viruliferous vectors can transmit throughout their life after acquiring the virus, role of isolation plays more minor role. Around outbreaks of many virus diseases, there will be a distance beyond which spread will be so infrequent or so unimportant that it can be disregarded. This distance has been termed as the "horizon of infection" (Vander plank 1949). Where land availability and other factors permit, isolation of planting from a large source of aphid-borne infection might give a useful reduction in disease incidence. As early as 1921, Robbins has observed the rare occurrence of beet mosaic in fields which were 1 mile away from the infected beets.

In California, Shepherd and Hills (1970) successfully minimized the aphid-borne *Beet mosaic virus* and *Beet yellows virus* by planting the new sugar beet fields at different distances, and the amount of mosaic disease decreased parabolically with distance. The incidence of early disease in each case was very much reduced when 10 miles distance was maintained from the inoculum sources. At greater distances, yellows infection continued to fall off at a constant rate (1.5 %/mile) whereas mosaic virus decreases much more rapidly reaching essentially zero at 13 miles from the inoculum source. A minimum of 15–20 miles between beet crops was necessary to prevent the spread of *Beet yellows virus*. Thus, isolation of beet fields from a large source of beets infected with two viruses markedly reduced infection. Similarly, in majority of countries production of virus-free seed potatoes is frequently carried out in areas that are well separated from crops being grown for food. Distances may be controlled by legislation, and the planting of home garden potatoes forbidden within prescribed area. Some of the crops in which the diseases were reduced by isolation technique are discussed below.

Broadbent et al. (1951) demonstrated an effective control measure against *Lettuce mosaic virus* by growing early lettuce crops in one area and the next series in the other areas, about half a mile away, and not returning to the first area until all the plants had been harvested or destroyed. Only 0.3 % disease incidence was noticed in these crops as compared to 60 % incidence in the neighboring farms. Successful control of *Southern celery mosaic virus* (strain of CMV) was achieved by isolating the celery crop by 240 ft away from the weed source in which case the crop was 100 % virus-free. Even the crops at 75 and 120 ft distance, disease from the inoculum source has 12 and 4 % virus incidence, while the plants at 3–30 ft away from the weed source had 85–95 % incidence (Wellman 1932).

In Louisiana, Martin and Kantack (1960) found that internal cork of sweet potato, an aphid transmitted virus disease could be managed by isolating the plants at least 100 yards from infected crops. Incidence in isolated crops averaged 1 % at harvest and 3–9 % after storage, compared with 11–23 % incidence in non-isolated areas and 18–58 % after these had been stored. Even at Uganda, from NAARI, Gibson et al. (2004) have observed that sowing sweet potato crop from early sown infected crop given by such short distances as 15 m will considerably decreases the spread of *Sweet potato feathery mottle virus* in susceptible sweet potato cultivars. From Washington, Hampton (1967) observed maximum incidence of *Bean yellow mosaic virus* (BYMV) up to 31 % in bean crop nearby virus infected red clover fields. BYMV infection levels among bean plants decreased to less than 1 % at distances of 100–200 ft on the upwind side and 750–900 ft. on the downwind side of the red clover. From central Washington, Howell and Mink (1977) reported that carrot seed fields less than 2 miles from root fields the incidence of *Carrot thin leaf virus* was 20–52 %, while the incidence ranged 0–2 % in fields grown more than 2 miles away from root crops.

Similar type of decreased virus incidence due to growing in isolation were also noticed in crops like potato against viruses and phytoplasmas (Rykova 1977); Papaya against papaya viruses (Wolfenbarger 1966) and chillies against *Pepper vein banding mosaic virus*. In each instance, diseases tend to decrease with increasing distance from the source and accounted for gradients of infection that are a very common feature in many crops (Thresh 1976). In Zambia, it has been reported that the incidence of *Potato virus Y* (PVY) in green pepper was low when the crop was grown in fields isolated from other solanaceous crops (Ndunguru and Kapooria 1999). From these examples it is clear that there are great differences between diseases in the amount and distance of spread, but many gradients have the same curvilinear shape. The incidence of disease usually decreases steeply near the source and less steeply at greater distances, to reach zero or a low background level. However, gradients may be flattened near the source due to multiple infections (Gregory 1948). Gradients also tend to become flatter with time as secondary spread occurs, or following an increase in the number of particularly active vectors.

The definition of isolation also covers the meaning of separation of the crop plants from the infected source. Uozumi and Nishimura (1971) reported that TMV infection persisted in soil even after 20 months after removal of the diseased plants and also in dried debris virus may survive higher temperatures (Bartels 1955; Broadbent et al. 1965). Due to continuous growing of tomatoes or other susceptible crops in the glasshouses, disease incidence will increase yearly due to contaminated soil. It can be isolated with polyethylene sheeting and the plants can be grown in a suitable compost above this (Wheeler 1961).

But the isolation technique cannot always be generalized against all crops as many factors can influence virus spread. Some insect vectors are strong and active and will be viruliferous over a long period or even lifelong. The resulting gradients of infection tend to be shallow, diseased plants being scattered widely over a large area. The spectacular wind-assisted flights of the leafhopper vector of *Sugar beet curly top virus* result in very shallow disease gradients over many miles to and from ephemeral weed hosts in the desert foothills of mountain ranges in California and neighboring south western states (Bennett 1971). Flights over low ranges of hills and for several miles along valleys are referred to as 'local'. Migrations of 30–60 miles are common place and spread of curly top virus and/or its vector have been traced beyond 398 miles (Bennett 1967; Fulton and Romney 1940). Wind-associated flight is a regular feature of many aphid species and other insect species (Johnson 1969).

Virus and viroid diseases in woody perennials are difficult to infect and their spread is very slow when compared to annual crops. For example, in a typical orchard in California, *Citrus tristeza virus* spread to an average of two citrus trees a year for each infected one already present (Dickson et al. 1956). In the case of cocoa, young trees support few mealy bugs and there was little spread of *Cocoa swollen shoot virus* until the branches form a continuous interlocking canopy (Strickland 1951a, b; Cornwell 1958). In general, reasonable wide spacing between the plants and isolation between the orchards will hamper the movement of vectors and impedes virus spread. Thresh (1964) observed that blackcurrant plants will be free from mites and reversion when planted at least 100 m in isolation from the old plantings and the new plantations remained healthy for many years.

The above-cited examples clearly indicate that the spread of non-persistently transmitted viruses tend to be very localized around primary foci of infection, whether these are within or alongside crops. *Sugar beet mosaic virus*, for example, was mainly restricted to plantings within 100 yards of seed crops (Watson et al. 1951). Nearby wild or garden sources of infection are also important in the epidemiology of virus diseases of celery (Wellman 1935, 1937; Walkey and Cooper 1971); pepper (Simons 1956; Simons et al. 1956); cucurbits (Nelson and Tuttle 1969; Adlerz 1972a, b); lettuce (Tomlinson and Carter 1970; Tomlinson and Walker 1973) and many other crops (Duffus 1971). Growing the crops in isolated areas away from the old infected crops or from foci of infection will help to minimize disease spread of several non-persistently transmitted viruses to a negligible extent.

(b) **Cages, barriers, border and cover crops**

Physical barriers such as fine-mesh screens have been used in Mediterranean countries since 1990 to protect crops from virus diseases (Berlinger and Mordechi 1996; Berlinger et al. 2002; Cohen and Antignus 1994). For example, in tomato whitefly transmitted TYLCV was successfully reduced by using houses covered with 50-mesh screens which have decreased dramatically the number of invading whiteflies into net or greenhouses. When combined with a few insecticide sprays, the incidence of TYLCV in the covered structures was only 1 %, while in control (unscreened greenhouse), infection was up to 100 % (Berlinger and Mordechi 1996). Even the trials conducted by Diez et al. (1999) to manage the TSWV in tomato by using cultivars ['RDD', carrier of the Sw5 gene, which confers resistance to tomato spotted wilt virus (TSWV); 'Pitihue', tolerant to the virus; and the susceptible cultivar Rutgers]. The pre-inoculated three tomato cultivar plants with viruliferous thrips were subsequently transplanted either into the open field or in tunnels 10 × 5 m, protected with a mesh of 14 × 10 threads/cm. 'Rutgers' exhibited severe systemic symptoms regardless of treatment and a high number of plants died. The level of infected plants remained low when protective measures were applied to seedlings of 'Pitihue' and acceptable yields were obtained. In open air cultivation, where seedling infection was severe, <20 % of 'RDD' plants became infected and high yields were obtained; protected cultivation did not reduce yield. Although the percentage of infected plants was higher when cultivated under mesh, the yield of all three cultivars was greater than in the open field. The environment created under mesh stimulated growth, neutralizing the effect of the infection. The results were positive in reducing the disease incidence when the clean TSWV resistant transplants were raised under thrips proof mesh houses. On the other hand when susceptible tomato cultivars were grown under mesh, the TSWV infection was enhanced and observed reduced yields.

In recent years, insect-exclusion screening (IES) (screen barrier) has been successfully used against fruit crops like banana, grapes and papaya to manage the leaf hopper-transmitted phytoplasma diseases. For example, in Australia Walsh et al. (2006) controlled the leaf hopper vector of papaya phytoplasma disease by covering the papaya plants with IES. When the effectiveness of IES was compared with systemic insecticide (imidacloprid) treatments, and the non-treated control, the net-covered trees had significantly fewer phytoplasma symptoms. Even Minnini (2007) stated that insect exclusion screening is the only method to maintain phytoplasma-free vine yards. Production of clonal or mother plants have to be covered continuously with IES to maintain the plants free from leaf hopper vectors. Another example is that, in Taiwan papaya is grown was protected against aphid vectors of *Papaya ring spot virus* by netting and was effective in producing marketable fruits, because late infection which occurs when the net was removed after fruiting, resulted in little damage to the fruit yield (Kiritani and Su 1999). A similar approach can be extended to exclude the other insect vectors of viruses infecting other crops in the tropics. Since fruit crops are long duration and also use of nets is costly, not much information is available on this aspect.

From Israel, Cohen and Marco (1979) achieved considerable reduction in the incidence of PVY and CMV in peppers by covering the plant rows with nets of different colors. Complete prevention of the spread of these two viruses was achieved, whereas in unprotected control plots virus incidence reached 35–100 %. Zimmerman (1979) has also tested the use of nets by covering the potato plots and observed very low percentage (0–3 %) of *Potato leafroll virus* incidence as compared to 12–48 % in the untreated plots. Hanafi et al. (2003) have demonstrated that 10×20 threads/cm screens were quite effective in capturing the whitefly vectors and in turn reducing the TYLCV incidence in tomatoes. A combination of net covering and insecticidal sprays almost completely prevented the leaf roll spread. At Yemen, reduction in *Bemisia tabaci*-transmitted *Tomato leaf curl virus* to 3.4 % was observed by covering the tomato crop for a period of 65 days (Alshami et al. 2008). Lower incidence of stylet-borne viruses was observed when a potato crop was surrounded by 10 rows of cabbage or mustard, than when no border crop was used or when corn, wheat or rice was used, as a border crop (Duriat et al. 1990). The mustard and cabbage crops trapped significantly more aphids (*M. persicae*) than the other crops.

Raising the crops in greenhouses/net houses has been attempted to reduce the vector population and to raise the healthy crop. Krishnakumar and Eswara Reddy (2006) reported that incidence of ToLCV was 86.72 and 42.50 % in open field during September–March and June–December respectively; whereas in greenhouse, virus incidence was 30.68 and 16.67 %. ToLCV has also contributed to 20.05–22.50 % yield loss under protected cultivation as compared to 26.51–33.50 % in open field depending on time of infection. There was least incidence of PVY and CMV when bell pepper was grown under greenhouse conditions (unpublished data).

Due to the high cost involved and as the nets could not be used for many times, attempts were made to use tall plants to protect the crops from the insect vectors. Fast growing taller species which are economical and were tried against many diseases. One should raise the barrier crops which are not alternate host plants for virus and vector multiplication; hence monocot plants were tried as barriers for dicot crops. The barrier or border crops should be selected carefully so that no adverse effects like competition by the protection crop or dangerous pest and disease development should result.

Several kinds of crops like wheat, maize, sorghum and pearl millet were tried as barrier crops when the main crop was of dicot family against non-persistent aphid-borne viruses in different crops like chilli, pepper, potato, French beans, soybean and lupines and were effective in reducing the virus incidence under field conditions. More information can be had from the review article of Hooks and Fereres (2006).

A tall cover crop will sometimes protect an under sown crop from insect-borne viruses. For example, cucurbits are sometimes grown intermixed with maize. Broadbent (1957) found that surrounding cauliflower seedbeds with quite narrow strips of barley (about three rows 0.3 m apart) could reduce virus incidence in seedlings to about one-fifth. Barley is not susceptible to crucifer viruses. Many incoming aphids were assumed to land on the barrier crops, feed briefly, and either stay there or fly off. If they then land on the *Brassica* crop they may have lost the non-persistent virus they were carrying during probes on the barrier crop. Under field conditions, main spread of the virus and virus-like diseases takes place through vectors, which bring the inoculum from outside sources. More than 80 years earlier, the farmers have made many attempts to restrict the entry of vectors into the crops by using barriers of either plants or muslin cloth screens. Generally valuable crop materials are maintained or multiplied in insect proof glass or net house, which is a expensive way of growing plants; however it excludes birds and mammals, including men who are not conversant with the precautions needed. In recent years, crops like tomato, capsicum, brassicas, cucurbits, strawberries and flower crops are grown in glass or net houses in several countries. Sufficient care is taken to check the insects, mites, glasshouse whiteflies and aphids generally infest the crops grown and are being controlled by insecticidal sprays. Frequent sprays are required to make the cages insect proof. In the earlier years, wire mesh or plastic sheets or cloth screens are used as barriers to protect the crops from the incoming air-borne vectors. Linn (1940) found that the introduction of mosaic disease in lettuce from weed sources was reduced to 12 % from 35 % by using 1.8 m high cloth screens. In Argentina, mosaic-free seed of lettuce was obtained by raising the crop under insect proof cage houses (Crnko 1969). Large cages prevented *Tulip breaking virus* spread (35.2 % reduction). Further, roguing of infected tulips reduced the incidence to a very low level. Gibson and Plumb (1976) reported that the eriophyid mite vectors of *Rye grass mosaic virus* were excluded from ryegrass by covering plots with polyethylene film house ventilated by filtered air. The Korean scientists found that tunnels of No. 200 mesh (13.5 lines/mm) blue plastic screen, placed over rows of Chinese cabbage greatly reduced *Turnip mosaic virus* symptoms and dramatically increased yields (Paul Daly and Bruce Tomkins 1995).

These barrier crops are more effective for protecting the crops from the stylet-borne viruses rather than persistent viruses. As in the case of the former group for most of the time, the aphids loose the virus after probing the barrier crops for a brief period. Normally these will fly over the crop plots when they leave the barrier crop because they fly upwards towards the brightest light, but if they subsequently land once again on barriers, the chances of these becoming non-viruliferous is increased. The same principle applies to cover crops also. For example, by alternating 3 rows of barley with 12 rows of brassica seedlings, *Cauliflower mosaic virus* incidence was reduced by 80 % (Broadbent 1957). The barriers when effective were sown at the same time as the brassica seed between every 12 rows of seed bed, with single cross rows of cereal at intervals in long seed beds. Hull (1952) reported that French sugar beet growers were using 1-m wide barriers of maize, oats, hemp or sunflower around 20 × 5 m beds of beet stecklings. The incidence of beet yellows under barley cover was 12 %, when compared with 100 % in open bed. Bonnemaison (1961) also recommended that plots with 2 rows of maize and 3 rows of sunflower, curtailed to about 21 % of infected plants, while unscreened plots had 56 %. In another experiment on plot size, unscreened areas had 54 % yellows affected plants and those with barriers for every 5, 10, and 15 m distance, had 28, 31 and 36 % disease incidence, respectively. Heathcote (1968) found that two rows of mustard or barely acted as cover crop for sugar beet stecklings against yellows virus. The disease incidence was 1.3 and 0.1 % in barley and mustard cover crops, respectively while the open beds had 7.2 % incidence. However, the use of mustard as cover crop is discouraged because it enhances the risk of attack by downy mildews. But Hansen (1950) recommended barley barrier rows at more than 7 yards distance to isolate the clamped beet from the other inoculum sources, for successful beet cultivation. In *Capsicum* spp. the virus (PVY, TSWV) spread was effectively decreased by growing the barriers of 3 rows maize and 2 rows of cowpea. The barriers were sprayed every 7 days and the plots every 15 days with 1.5 % Folidol. Yield was considerably increased and virus incidence was generally less than the untreated plots. In Florida, Simons (1957, 1958, 1960) demonstrated that barrier crops of sunflower and other barriers (beans, corn or cucumber) were effective in reducing the spread of aphid-borne viruses viz. *Pepper vein banding mosaic virus* and PVY in pepper, celery and tomato from the infected weed host, *Solanum nigrum.* Barriers around pepper plots 45 m from a weed source did not prevent but decreased and delayed infection by about 10 days. Mohammad Roff and Ho (1991) reported that maize as a border crop was found effective in reducing the vector population in chilli. Sunflower barriers around pepper plots, 150 ft from the source recorded 63 % infection in the fifth week, compared to 95 % in plots without barriers. A low single barrier of 38 cm high was nearly as effective as single 3 m wide strips of non-susceptible crops such as beans, corn or cucumber. In number of countries CMV infection in chilli has been reduced by sowing barrier crops like maize, sunflower, sesame, sorghum or *Pennisetum americanum* around chilli plot (Deol and Rataul 1978; Avilla et al. 1996; Fereres 2000). From south Florida, Simons (1960) has reported that the primary spread of PVY from the weed host to pepper was decreased to 50 % by a row of sunflower around three sides of the plots, 70 % by a 50 ft barrier of beans and by 85 % when beans were sprayed weekly with parathion. After 6 weeks, all the plants within sunflower barriers were infected, as before, but those with bean barriers had one third less diseased and those within sprayed barriers had only half as many as unprotected ones. Gay et al. (1973) obtained low incidence of cowpea viruses by pearl millet border crop, sprayed with insecticides like disulfoton, ethoprop etc. *Cauliflower mosaic virus* incidence in the seed beds of broccoli was reduced by surrounding with a barrier crop of kale or barley even when the seed beds were only 5 yards away from infected plants (Jenkinson 1955). The same virus incidence in the seed beds of brassica was reduced by growing 3-row band of barley between every 12 rows of seed beds, with single cross rows of cereal at intervals in long seed beds and both barley and brassica crops could be sown simultaneously (Broadbent 1957). From Japan, Hidaka (1960) recommended growing of sunflower (*Helianthus annus*) between the source of inoculum and tobacco, which acts as a barrier against aphid flights, for reducing the PVY spread. Inter-planting barley in between tobacco has reduced the virus spread.

The spread of yellows viruses in sugar beet was reduced by growing the beet stecklings under barley in their first year (Hull 1965). Toba et al. (1977) also reported that the incidence of aphid-borne viruses in muskmelon was reduced by the barrier crop of wheat, but crops like radish and swiss chard as barriers for muskmelon were unsuitable as they provided undesirable competition and served as hosts for aphids. Keener (1956) also reported that barrier crops of maize or alfalfa around tomato crop has reduced the *Beet curly top virus* incidence by preventing the leaf hopper vector *Circulifer tenellus* into the tomato crop. As these barrier crops contribute in isolating the crops from the old infected crops/ weeds, is being practiced in seed certification schemes. For example in Victoria (Australia) in the seed certification scheme of Phaseolus bean, crops like maize or other tall barriers, 30 m wide are used to subdivide bean seed crops over 2 ha in any area submitted to inspection for certification (Bartha 1963). There is a little evidence to support or deny the merits of inter-cropping as a means of avoiding virus disease problems because epidemiology trials are usually done with single crops and varieties. During 2006 Hooks and Fereres, have reviewed the use of barrier crops to guard crops from aphid borne non persistent viruses. For example in Kenya, Ramkat et al. (2008) have reported negative results on the role of inter-cropping in reducing the *Tomato spotted wilt virus* in tomato. Tomato–maize inter-crop produced the lowest fruit weights and marketable yield in 2004 and 2006, while yield of onion with kale and maize as inter crops were not significantly different in their mono crops and there was no impact on TSWV incidence in the treatments.

(c) **Effect of strip cropping**

A candidate strip crop plant has qualities such as serving as cover crop or improving soil fertility and more importantly emit semiochemicals, some of which (allelochemicals) could act as attractant, repellent or toxins to targeted insects or the pathogens they vector. This strip cropping technology helps in attracting pests away from the target crops and adding valuable biodiversity to the agro-ecosystem. Thus strip crops are plant stands that are setup to attract, divert, intercept and, or retain targeted insects or the pathogens they vector so as to reduce damage to the main crop (Shelton and Badenes-Perez 2006). This technique is less expensive and if economic crops are grown the yields, also will generate revenue.

Keeping this in view Ewusie et al. (2010) have conducted field trials at Ghana by using cassava as the main crop and strips of cotton hybrid and *Jatropha* were grown in five rows each on all sites of cassava. They have studied *Bemisia tabaci,* whitefly population which is the major vector for majority of the *Geminiviruses*. They have recorded significantly fewer number of whitefly both adults and nymphs in *Jatropha* strip crop, on the other hand on cotton strip crop and also on cassava, the whitefly population were a higher level. Similar trend was noticed with whitefly population in sticky trap counts. The objective was to identify the potentiality of selected strip crops in managing whitefly (*Bemisia tabaci*) population and it clearly indicates that strip cropping of jatropha can be a viable management strategy for the control of *Bemisia tabaci* populations in cassava production. The studies on different strip croppings with various virus-host combinations may results an effective management strategy against virus disease spread.

(d) **Effect of mixed cropping on virus incidence**

Magee (1967) reported that by cultivating bananas under coconuts, the incidence of bunchy top disease can be reduced as this treatment reduces the aphid vector, *Pentalonia nigronervosa* populations, when compared to banana plants grown in nearby open fields. Even Ahohuendo and Sarkar (1995) also showed that intercropping cassava (*Manihot esculenta*) with maize, cowpea, and peanut reduced the population size of the whitefly, *Bemisia tabaci* Gennadius and the associated *African cassava mosaic virus* (ACMV) disease. Bean grown with maize, the incidence of bean common mosaic decreased (van Rheenen et al. 1981) similarly, *Cassava mosaic viruses* when cassava was intercropped with sorghum or cowpea, the virus incidence was reduced (Page et al. 1999; Fondong et al. 2002).

(e) **Use of bait crops**

In some of the field crops, certain crops are used as bait for the vectors to check the virus spread. For example, Delfosse et al. (2002) reported that *Peanut clump virus* incidence can be minimized by using pearl millet as bait crop. These are planted soon after the onset of monsoon preferably under irrigation and uprooted in 3 weeks after germination. This permits the infection by *Polymyxa graminis* but not the development into sporosori by which virus inoculum load is reduced.

(f) **Effect of barrier cropping on viruses**

The exact mechanisms that reduce the number of virus infected plants in crops with protector plants are little understood. It has been suggested that insects flying over areas with several plant species will have several inappropriate landings on the wrong host plants. The tendency is then to leave the general area completely. Likewise, it has been reported that during their host-recognition phase, if aphids determine they have alighted on an unsuitable host, they immediately resume flight. This flight may take an aphid out of the vicinity of a crop field. Further, because their ability to transmit non-persistent viruses is lost soon after acquiring a virus, aphids may lose the ability to transmit a virus while searching for suitable host plants.

When aphids search for a host plant, they commonly lose their ability to transmit a virus after making a few brief probes on a healthy or non-susceptible protector plant. If aphids then alight and feed on a susceptible cash crop, there will

Table 2.3 Effect of barrier plants on certain insect transmitted plant viruses

Crop protected	Virus targeted	Barrier plant	Country	Reference
Chilli	CMV	Maize, sorghum, sunflower	India	Anandam and Doraiswamy (2002)
	CMV	Sunflower, sorghum sesame, pearl-millet	India	Deol and Rataul (1978)
	CVMV	Maize, brinjal	Malaysia	Hussein and Samad (1993)
Cowpea	CpBMV	Pearl millet, maize	India	Sharma and Varma (1984)
French beans	BCMV	Sorghum, maize, okra, sunflower, amaranthus	India, Netherlands	Van Rheenen et al. (1981), Dhanju et al. (1995)
Green gram	GBNV	Pearl millet	India	Sreekanth et al. (2004)
Lupine	BYMV	Cereal	Australia	Jones (1993)
Okra	YVMV	Maize	India	Singh et al. (1979b), Pun et al. (2005)
Potato	PLRV	Wheat, mustard	Bangladesh	Mannan (2003)
Radish	PVY	Wheat, oat, barley	India	Usha Sharma et al. (2005)
Tomato	TLCV	Maize	India	Sastry et al. (1974)

be no opportunity for virus transmission, because the virus particles might have been removed from their mouthpart while probing the protector plant.

There are some more examples from U.S., Germany, Spain, India and Australia where barrier crops of cereals were tried against viruses like *Watermelon mosaic virus 1* (strain of PRSV), *Cucumber mosaic virus, Potato virus Y, Bean common mosaic virus*, *Bean yellow mosaic virus* and *Soybean mosaic virus,* and details are given in the Table 2.3.

The examples cited above clearly indicate that barrier or border crops are quite effective against stylet-borne viruses. Researchers have also conducted experiments with persistently transmitted viruses by aphids, whiteflies and leafhoppers, in different crops as discussed below.

Several field experiments have shown that barrier cropping can be successfully used to lessen the occurrence of insect transmitted virus diseases and resulting in increased crop yields. In addition to their potentiality to reduce the severity of aphid-borne non-persistently transmitted viruses, barrier plants have also helped to reduce the diseases of persistently transmitted viruses. For example, Heathcote (1968) found that rows of mustard (*Brassica juncea*) or barley intercropped with sugar beet (*Beta vulgaris*) could reduce the occurrence of *Beet mild yellowing virus* (BMYV), a virus transmitted in a persistent manner. In Bangladesh, Mannan (2003) has reduced the incidence of *Potato leaf roll virus* in potato fields by using wheat as border crop. Page et al. (1999) found that intercrops of beans and millets reduced the occurrence of *Maize streak virus* (MSV) transmitted by leafhoppers, *Cicadulina* spp., in maize. During 2004, Coutts et al. have found that a 15 m-wide barrier of cabbage could delay infection and reduce spread of *Tomato spotted wilt virus* in lettuce (*Lactuca sativa* L.). In India, growing tall barrier crops viz., pearl millet, sorghum or maize around the field of groundnut or cowpea, the incidence of *Tobacco streak virus*, whose transmission by thrips in a persistent manner was reduced as the barrier crops have protected against the alighting of thrips on the main crops (Prasada Rao et al. 2003a, b).

In India, border cropping of okra plants with maize has reduced the incidence of *Okra yellow vein mosaic virus* transmitted by the whitefly *B. tabaci* (Singh et al. 1979b; Pun et al. 2005). Similarly, the incidence of *Tomato leaf curl virus* transmitted by *B. tabaci* was reduced by border cropping with maize (Sastry et al. 1974).

There are several aspects of aphid behavior that makes them manageable by the barrier cropping strategy, much of which centers on their flight activity while searching for a suitable plant for colonization, feeding, and reproduction.

(1) During flight, aphids respond strongly to visual stimuli and locate host plants by contrasting the soil background with the green color of plant foliage. Therefore, the greater the percentage of plant coverage in a crop field, the lower the probability an aphid will land in that field.
(2) Plant infection with non-persistently transmitted viruses usually starts at the crop edges, because aphids entering a field tend to land on the field perimeter first. Thus, if protector plants are grown around the perimeter of a cash crop, aphids may initially probe the protector plants instead of the cash crop and clean the virus off their mouthpart while probing the protector plants.
(3) Aphids cannot distinguish hosts from non-host plants until after landing on a leaf surface and examining it with their mouthparts. Their initial behavior after landing on a plant is to walk over the surface of the leaf while testing it. During the test phase, aphids make brief, shallow exploratory probes with their mouthparts. Thus any virus particle on their mouthparts can be released into a protector plant.
(4) This behavior, whereby aphids probe and/or feed on non-host plants, has important implications in designing disease management strategies. During host-seeking, aphids may spend a significant amount of time and energy assessing unacceptable host plants in habitats of plant mixtures, and they would therefore allocate less energy to colonizing and feeding on the host crop. Thus, several behavioral aspects of aphids suggest that they may be managed by using protector plants.

In some of the crops, depending on the vector involved, monoculture has resulted increased virus incidence as compared to intercropping. For example, Van Rheenen et al. (1981) found that growing beans with maize resulted in less incidence of *Bean common mosaic*, when compared with monoculture stands. Even Fargette and Fauquet (1988) and Fondong et al. (2002) have observed reduction in *African cassava mosaic virus* when cassava was grown as an intercropping with maize/or cowpea.

(g) **Role of plant density on virus incidence**

Plant density also affects virus incidence and very interesting observations on the effect of plant spacing on the spread of virus diseases in crops have been made by several authors. It has been observed that closer spacing generally reduces the number of infections per unit area; this method requires a higher seeding rate. The planting rate should be such that it should cover the soil without reducing the yield due to competition. The mathematical relationships of size of plants and spread of systemic diseases have been discussed by Vander plank (1947). He observed that as an approximation, in cases where disease is brought from outside the crop, the proportion of infected plants is proportional to their size, provided the proportion is small. Size is defined as the catchment zone of the invading insects which is considered to be inversely proportional to the number of plants per unit area. Even in cases where a systemic disease spreads within a crop, the rate of infection is proportional to the size of the healthy plants. When the percentage of infection is low, systemic diseases could be controlled by close planting, which reduces the size of plants and increases the number per unit area. When the percentage of infection is high, control is more difficult, but is still possible if infection strikes when plants are small. Young rapidly growing plants are highly susceptible to virus infection than older plants. As the young plants are generally light yellow in color, the vectors are attracted and also young plants provide a more suitable diet for developing vectors. Examples where decreased infection occurs with increased plant age exist in peppers against CMV (Simons 1957), potato against *Potato leaf roll virus* (Knutson and Bishop 1964), sugarcane infected with *Sugarcane mosaic virus* (Abbott 1961), soybean infected with *Peanut mottle virus* (Demski and Kuhn 1975); groundnut infected with *Groundnut rosette virus* (Adipala et al. 1998); barley infected with *Barley yellow dwarf virus* (Slykhuis et al. 1959); Pineapple infected with TSWV (Linford 1943) and *Pineapple wilt virus* (Singh and Sastry 1974); *Sugar beet curly top virus* in tomato (Shapovalov et al. 1941); *Pigeonpea sterility mosaic virus* in pigeonpea (Ramakrishnan 1963) and *Cowpea vein banding mosaic virus* in cowpea (Sharma 1975).

In addition to above cited examples, increased plant density/close spacing has been found to reduce losses in other crops like groundnuts, sugar beets, field beans, cowpea, tomato, bell pepper, okra etc. (Clarke 1937; Linford 1943; Shapovalov et al. 1941; Blencowe and Tinsley 1951; Jameson and Thomas 1954; A'Brook 1964; Way and Heathcote 1966; Heathcote 1970, Ghanekar 1980; Gill et al. 1982; Chowfla and Parmar 1995; Aliyu and Balogun 2011). Losses from air-borne (vector) viruses have been shown to be less if the plants are densely spaced than if widely spaced. As early as 1928, Harper observed the phenomena of close spacing aiding in decreased disease incidence in groundnut against rosette disease (as quoted by Hull 1964). The more plants present in unit area, the smaller will be the proportion of infected plants by the same number of viruliferous aphids (*Aphis craccivora*). Generally, the subsequent virus spread within the crop will be very little, after introduction of *Groundnut rosette virus* by the winged aphids. In South Africa peanut cultivators delay the weeding operations which help in covering the soil and also helps in increasing the plant density, which is a favorable factor for reducing the rosette disease incidence (Storey 1935). A'Brook (1964) by his several experiments confirmed the earlier work and observed that the dense plantings had few or no aphid-infested plants after 80 % ground cover was achieved, and more aphid colonies were noticed in thinned plantings. For this reasons suggested by Kennedy et al. (1961) were checkerboard effect i.e. contrast between bare earth and crop, which resulted in a differential alighting response, alatae tending to settle on plots with a more 'broken pattern' such as encountered at lower plant densities. Aphids are thought to be attracted by the brown and green checkerboard of bare soil and widely spaced plants, more than by the uniform green of a more closely sown crop. Other possible cause of aphids becoming fewer may be due to the density of host plants which increases moisture and cooler air which is unfavorable for aphid movements, and their predators are being more effective in dense crops. Hull (1964), Kousalya et al. (1971) and Davies (1976) also obtained the less rosette disease incidence with increased plant population. In Malawi, Farrell (1976) compared the results of plant density and inter-sowing of beans in groundnut crop against rosette disease. In two trails during 1966–1967, the spread of virus in groundnut inter-sown crop with beans (*Phaseolus vulgaris* L.) was less than the crop with high plant density, comprising plant populations equal to the number of groundnut plants and to the total number of crops in the inter-sown crop. Davies (1976) stated that high plant density with 134 kg seed/ha instead of 67 kg or 100 kg seed/ha followed by Menazon spray at a rate of 294 g a.i./ha four times at 10-day intervals further reduced the rosette incidence and the yields were highest at the highest plant densities.

Hull (1969) also observed that *Beet mild yellowing virus* (BMYV) incidence was minimized by higher sugar beet plant populations. The BMYV was more prevalent in sparsely planted sugar beet crops than when the ground was well covered. The percentage of plants with BMYV was inversely related to the plant population during the years 1965–1968, but the number of infected plants per unit area differed little between treatments except that they were fewest where the plant stand was dense. This is mainly because as the denser foliage provides an environment unsuitable for aphids to multiply and on a percentage basis, a given number of viruliferous aphids will infect similar percentage in denser planting.

Heathcote (1972) demonstrated that with increase in beet plant density ranging from 7,000 to 18,000 plants/acre, the percentage of plants with yellows disease usually decreased from 30.8 to 9 %. Considerable decrease in the incidence of virus diseases with higher seed rates within the row and by decreasing the row width was observed in the sugar beet crops and brassica seed beds, wherein the viruses were introduced from outside by the vectors (Blencowe and Tinsley 1951; Broadbent 1957).

In addition to increased plant density, treatments like use of resistant/tolerant cultivar, insecticidal sprays, planting 2 plants/hill, close planting and later thinning, weeding etc., have also reduced the virus incidence in certain crop plants. For example, by close planting and later thinning out the diseased plants will help in reducing disease incidence. This method has been suggested as a practical measure of control for the *Tomato spotted wilt virus* of tomato and tobacco in South Africa (Vander plank and Anderssen 1944; Vander plank 1947). They calculated that if the incidence of the disease was 50 % with single plant spacing, it would decrease to 9 % by transplanting 2 plants/hill and 1 % by three, if the distance from the hills as not altered. Broadbent (1948) and Reestman (1960) also advocated the dense planting and thinning later to reduce virus infection in potatoes. Way and Heathcote (1966) also observed that by increasing the density of planting of field beans, the *Pea leaf roll virus* incidence was greatly reduced. They recorded that the number of aphids per plant and per acre was inversely related to plant density. In Uganda, highplant density of groundnut cultivar 'Igola-1' significantly reduced groundnut rosette virus and aphid vector infestation (Adipala et al. 1998).

In certain fruit crops the higher plant density factor is also used to accommodate more plants per unit area in crops where in the viruses do not affect the fruit size and quality. Cohen (1968) proposed the use of exocortis viroid as a possible agent to produce dwarf citrus trees. Exocortis diseased trees were stunted but often continue to produce fruit with no apparent deterioration of the trees. Cohen believes that a close replanting of exocortis viroid infected trees may produce more fruits than standard size tree plantings. As originally proposed by Cohen (1968) and Mendel (1968), the use of viroids as a means of producing dwarfed citrus trees has been under continuous investigation (Long et al. 1972; Stannard et al. 1975; Bevington and Bacon 1977; Castle 1978; Duncan et al. 1978; Bacon 1980; Broadbent et al. 1992; Perez et al. 1992; Semancik et al. 1997, 2002; Stuchi et al. 2007; Vidalakis et al. 2007, 2010), and reviewed by Hutton et al. (2000) and Semancik (2003). More details on cross protection are discussed later in this chapter.

High-density planting is not a new concept for citrus. Older groves in Italy, Spain, Algeria and Israel had free densities ranging up to 988 and in some cases 1483 trees/ha.

The use of transmissible small nuclear ribonucleic acids (i.e. TsnRNA-IIIb *syn. Citrus dwarfing* viroid) reduces tree volume, enhances fruit size and increases the yield per canopy volume (CV) of sweet orange (*Citrus sinensis*) on trifoliate (*Poncirus trifoliata*) root stock in the absence of any other symptoms (Polizzi et al. 1991; Albanese et al. 1996; Semancik et al. 1997, 2002). Reduced height and CV induced by TsnRNA-IIIb for Clementine trees (*C. clementina*) on trifoliate rootstock have also been reported (Verniere et al. 2002, 2004).

Stannard et al. (1975) estimated the reduction of canopy surface area (CSA) and yield of Parent Washington navel orange on *P. trifoliata* rootstock treated with the Australian dwarfing budline '3538' (later found to contain the viroid species CDVD-IIIa) at 36.5 and 22.4 %, respectively while the yield per CSA increased by 6.5 % (Broadbent et al. 1992). Albanese et al. (1996) in Italy estimated the reduction of CV and yield of Valencia orange trees on *P. trifoliata* rootstock treated with TsnRNA-IIIb at 54 and 37 %, respectively while the yield per CV increased by 32.5 %.

The term 'transmissible small nuclear ribonucleic acids' (TsnRNAs) describes well-characterized viroid RNA species that do not induce any pronounced disease syndromes in specific citrus hosts, but rather act as regulatory genetic elements modifying tree performance. The CV of 13-year-old navel orange trees (*Citrus sinensis*) on *Poncirus trifoliata* rootstock treated with TsnRNA-IIIb was reduced by 45 and 53.5 % in standard density (6 m × 6.7 m) or high-density (3 m × 6.7 m) plantings, respectively. The total yield of eight consecutive harvests was not affected significantly by the TsnRNA-IIIb treatments or the two planting densities. However, the yield per land surface unit (Y/LSU) was almost doubled (increased by 97.5 %) for the high density plantings over the standard density plantings of the untreated controls. The Y/LSU of the TsnRNA-IIIb-treated navel orange trees in the standard-density planting was reduced by 32.7 %. The TsnRNA-IIIb-treatment in both planting densities concentrated significantly more fruit production (approximately 60 %) in the economically advantageous middle canopy height zone (0.6–2.4 m) in comparison with the untreated controls (35 %). Fruit grade, size, appearance, organoleptic characteristics or time of maturation of the TsnRNA-IIIb-dwarfed navel trees were not significantly different between the two planting densities and the controls. Fruit with higher commercial value was produced in the TsnRNA-IIIb-dwarfed navel trees in the high-density planting by 3.9 and 4.6 % over the TsnRNA-IIIb or controls in standard-density planting, respectively. The increase in Y/LSU and fruit value for the TsnRNA-IIIb-treated navel trees in the high-density plantings in combination with the reduced management cost of dwarfed trees could result in substantial higher profits for a commercial grove despite the higher establishment cost of high-density plantings.

The reduction of tree height and the accompanying alternation in the profile of fruit production could have important economic and practical benefits. For example, orange harvesting rate is dependent upon tree height, yield and fruit size. For the same yield and fruit size, the predicted harvesting rate was increased per picker of 1 box/h for an approximate decrease of 1.8 m in tree height (Whitney et al. 1995). It has also been estimated that while harvesting oranges 25.7 % of the pickers' time was unproductive. The majority of the unproductive time (62.7 %) was spent in positioning and climbing the ladders (Jutras and Coppock 1958). In the same study, it was also estimated that the picking rate was 67.2 % faster when picking oranges from the ground than when picking from a ladder. Having fruit production optimized on a restricted canopy zone of a compact citrus tree also has potential benefits for mechanical harvesting (Jutras and Coppock 1958; Fridley 1977; Whitney et al. 1995).

(h) **Breaking the disease cycle**

Continuous cultivation of particular crop in a locality will lead to disease epidemics after some years. Breaking of disease cycle is possible by having a crop-free period prior to the main crop against viruses and vectors which have restricted host range. In India, *Pigeonpea sterility mosaic virus* spreads through mite, *Aceria cajani* and the host range of both virus and vector is restricted to pigeonpea only. But the virus survives due to volunteers and ratoon cropping and this disease can be successfully eradicated by removal of volunteers at least for one season (Ramakrishnan 1963). For a successful cotton crop, okra was forbidden in Sudan for at least 2 months before or 1 month after cotton was sown, as the *Cotton leaf curl virus* perpetuates on this host (Tarr 1951).

In Philippines, a 2 month break in rice planting has significantly reduced the tungro disease incidence (Chancellor et al. 2006). Some other field experiments have indicated that staggered planting contributes to increased tungro incidence (Cabunagan et al. 2001; Loevinsohn 1984). In south Sulawesi, Indonesia, staggered planting was thought to be a major cause of tungro problems and regulations were introduced to synchronize planting during periods when vector populations were relatively low (Sama et al. 1991).

In the Marshlands district near Christchurch (New Zealand), onion growing was hampered due to *Onion yellow dwarf virus* (OYDV). Seeding of onions was prohibited and removal of all unharvested bulbs and shallots was enforced. This resulted in virtual eradication of (OYDV) from that area (Chamberlain and Baylis 1948).

In south Florida, celery is said to grow 13 months of the year because spring seedbeds are already established to produce transplants for use in the fall, before the spring celery crop is completely harvested. Generally the isolation of seed beds and the mature crop is not maintained. Besides this, celery transplants grown by one grower often sold to other area growers or transplanted to different farming operations in central Florida, where virus later occurred. These factors during 1972 were responsible for the appearance of *Celery mosaic virus* in all ten farms in south Florida, which accounted for nearly 10,000 acreages of celery. Due to these enormous losses the celery growers voluntarily imposed a celery-free period, so that the maturing celery crop would not overlap with the new seed beds. After 1975 season, only two farms sustained losses which occurred late in the season and no incidence was noticed in the next year (Zitter 1977).

In most of the tomato growing countries TMV is one of the major constraints for successful cultivation as the old infected root debris is infectious even 2 years after the soil is covered. Crop rotation is one of the ways of freeing soil from TMV satisfactorily, as the root debris and virus is eventually destroyed by fungi and bacteria in cultivated soil. Jones and Burnett (1935) observed that tomatoes planted in glasshouses within 5 weeks of clearing diseased plants became infected, while those planted after rotation with other crops for a 6 month period between tomato crops remained healthy. Reports from the East Malling Research Stations, Kent indicated that better and more permanent success could be achieved by cultural means involving the provision of hop-free intervals of 2 years against *Arabis mosaic virus* (ArMV-H). This could be done either by bare-fallowing or by planting virus-immune hosts ('break crops'). During the hop-free interval the vectors gradually lose their infection either by moulting or presumably, by erosion or inactivation or the virus particles adhering to the cuticular esophageal lining. If conditions allow the grower to apply hop-free break periods for minimum 19–20 months, a permanent control would result, provided that care was taken to ensure that ArMV-H was not reintroduced in the field. Further research may allow the hop-free interval to be reduced to little over a year, virtually halving the loss of crop resulting from a 2-year break. Fallowing techniques are less likely to succeed with other strains of ArMV which have a wider weed host range and would therefore demand a much more rigorous fallow regime (Crosse 1976).

(i) **Effect of priming on virus disease incidence**

Poor crop establishment is a major problem in many areas of the world, particularly for subsistence farmers in rainfed and poorly-irrigated environments. 'On farm' seed priming-soaking seeds, usually overnight, before sowing-is a simple technology that farmers can use to improve crop establishment and increase yield. It has been tested, developed and refined in a range of crops, countries and agro-environments using a combination of *in vitro*, on-station and farmer-participatory research and regularly increasing yields. In addition, priming has been associated with increased disease resistance in some crops and can also be used as a vehicle to alleviate some micronutrient deficiencies. Since 1990, on-farm priming of seeds of a range of tropical and sub-tropical crops has been tested as a means to promote rapid germination and emergence and to increase seedling vigor and hence yield (Harris 2004).

In Pakistan, Rashid et al. (2004a, b) have reported that soaking (priming) of mungbean seed for 8 h before sowing has enhanced the germination and vigor of the plants. The effect of priming seed before sowing on the incidence and severity of infection of mungbean (*Vigna radiata* (L.) Wilczek) with *Mungbean yellow mosaic virus* (MYMV) was investigated in a replicated, large-plot trial in 2002. They have recorded that primed seeds germinated and emerged faster and more completely, resulting in the establishment of 45 % more plants per unit area than non-primed seeds. However, subsequent mortality during the growth cycle in both treatments reduced this to a small, non-significant difference between treatments at final harvest. Symptoms of MYMV infection were assessed using a visual scoring index (VSI). More than 70 % of the non-primed plants had exhibited severe viral symptoms whereas only 14 % of the primed plants were similarly affected. These marked differences between priming treatments in the incidence and severity of disease were reflected in the components of yield. Primed crops produced 80 % more above-ground biomass (3.3 vs. 1.9 t/ha), 264 % more pod yield (1.0 vs. 0.28 t/ha) and 415 % more grain (0.36 vs. 0.07 t/ha) than did non-primed crops. From Bangladesh Musa et al. (2001) have reported the effect of priming in chickpea where significant reduction in collar rot incidence caused by *Sclerotium rolfsii*. The information on the priming treatment in disease control is very meager. Hence further studies have to be carried out to test the role of priming in managing of plant pathogens in different pathogen-host combinations, as the results so far are quite encouraging.

(j) **Effect of sowing and harvesting dates**

Plants are often more susceptible at certain stages of development and they are susceptible to virus and virus-like diseases in the seedling stage when high vector population exists and crops become resistant with advanced age, resulting in less susceptibility to the virus and the vector. The correct choice for time of planting has been a major concern of farmers throughout history. However, based on long standing experience, farmers themselves have identified the period to avoid the virus or other diseases.

Adopting planting dates so as to avoid exposing young plants to infection at times when the largest numbers of viruliferous insectvectors are likely to occur is often practiced. Seeding time is important for reducing yield losses, manipulation of sowing and harvesting dates also affect the virus diseases which are carried by air-borne vectors. Adjusting the planting time based on the vector migration i.e. avoidance of vectors by sowing late if vector migrates early and *vice-versa*. The manipulation of planting or harvesting dates depends mainly on the environmental factors which have direct effect on vector population and also the growth of the plants. As the climatic factors and season vary from one country to another, the vector population and disease incidence also vary similarly. Epidemiological studies conducted at each place will help to fix the date of planting or harvesting based on vector flights. In many crops early plantings decreased the disease incidence as the plants become older and more resistant when the insect vector invaded the crop. The period of maximum vegetative growth should not coincide with the period of maximum vector activity because plants infected when flowering or fruiting suffer less than those infected when younger. Some of the positive virus management results are seen for groundnut rosette (Wallace and Murphy 1938; Hansen 1950; Smartt 1961); *Tomato spotted wilt virus* in peanut (Culbreath et al. 2003); *Barley yellow dwarf virus* in wheat (McKirdy and Jones 1997); *Sugar beet curly top* in sugar beet (Hull 1952); *Sugar beet yellows virus* in sugar beet (Bennett 1960); *Bean mosaic virus* in common bean (Fajardo 1930); and *Lettuce mosaic virus* in lettuce (Kemper 1962).

Available information regarding sowing and harvesting dates for virus disease management in different crops is furnished below.

In groundnut, early sowing decreased the incidence of *Groundnut rosette virus* in several African countries (Reddy 1991; Sreenivasulu et al. 2008). Under Tasmanian conditions, autumn sowing of broad beans (*Vicia faba*) not only maximized potential yield, but it also greatly reduced the incidence of *Subterranean clover red leaf virus* (SCRLV). The yield losses in broad beans due to SCRLV in 1972/1973 were 31, 30, 61 and 8 % in plots sown in May, July, September and November, respectively (Johnstone and Rapley 1979), Whereas under Sudanese conditions, October sowing of broad bean is effective against *Sudanese broad bean mosaic virus*. When sowed in October, November and December yields were 1387; 494 and 50 kg, respectively (Abu Salih et al. 1973).

In India, thrips-transmitted *Groundnut bud necrosis virus* (GBNV) and *Tobacco streak virus* (TSV) in peanut and sunflower, respectively are economically important. The experimental results indicated that in southern India early sowing in Kharif (rainy) season, GBNV would be reduced in peanut (Reddy 1991; Reddy et al. 1983). At Solan (northern India) the incidence of *Pepper mosaic virus* complex was less (21–28 %) in the bell pepper plants when planted between 25th April and 10th May, while it was 88 % when the crop was planted on 9th July (Chowfla and Parmar 1995). The same virus incidence was less in mungbean sown in second 2 weeks of May to first half of June in summer and late sowing in spring has reduced GBNV and its thrips vector (Reddy et al. 1983; Sreenivasulu et al. 2008). In northern India, GBNV on the other hand, incidence in groundnut was high when sown on 10th May (Thira et al. 2004). In Central India, in potato crop high temperature (30–35 °C) and dry weather during September–October favor thrips activity to the maximum extent and thus potentially cause higher incidence of the GBNV incidence. In the crops planted after October 15th, low activity of thrips and thus low GBNV incidence occurred. Therefore, potato planting after the end of October is helpful in reducing the disease incidence by avoiding crop exposure to thrips vectors (Somani et al. 2007).

Based on the studies carried out at NBPGR regional center Hyderabad, India, by Sreekanth et al. (2002) it could be inferred that early or timely (second week of May to first 2 weeks of June) sowing of green gram in *Kharif* and late October sowing in *Rabi* would help in containing the thrips infestation and PBNV incidence. The usual sowing dates in rice fallows and summer sowings need not be disturbed. However, irrespective of sowing date, the maximum *T. palmi* population build-up coincided with the beginning of flowering stage of the crop phenology.

Sunflower crop sowing in post-rainy season (September onwards) was found to be beneficial for minimizing the necrosis disease caused by TSV (Shirshikar 2003). In India, the incidence of the whitefly-transmitted YVMV was less in okra crop sown in June when compared to July sowings (Gill et al. 1982).

In both Syria and Egypt, faba bean crops planted early in September are often severely attacked leading to 100 % fava bean FBNYV infection. In such circumstances, farmers plough the crop under and replant another crop. Delaying sowing until October or November often resulted in lower levels due to virus infection, most probably due to the reduced vector populations and activity. More studies are still required to establish the most suitable sowing date.

Broadbent et al. (1957) demonstrated that early planting and harvesting of seed potatoes will reduce the leafroll, mosaic and spindle tuber infections, but tuber yields were low and uneconomic. Earlier Hansing (1943) and Bagnall (1953) also found that early harvesting was not an economical control measures against yellow dwarf and PVY, respectively. Whereas Bankina (1976) reported that harvesting 7–14 days after the beginning of the mass aphid flight protected plants from repeated infection by *Potato virus X* and *M* more effectively than harvesting at the beginning of leaf drying with early harvest yields decreased by 13–30 %, but in the following year there was an increase of 8–11 %. Early harvest increased the number of seed potatoes (25–125 g wt.) and the content of starch and dry matter. In the potato cv. Majestic planting in mid-August had reduced the leaf roll virus incidence to 6 %, instead of late June during which the disease incidence was 80 % (Cadman and Chambers 1960). In India, based on the population of the green peach aphid (*Myzus persicae)* potato planting and harvesting dates were manipulated. Aphids normally appear July-August in the northern hills and during December–January in the plains. The crops raised during summer in the hills and during autumn in the plains approach maturity before the appearance of the high aphid population and consequently produce was healthier. Growth of virus-free potatoes in the Indian Gangetic plains during aphid-free period (October-December) is demonstrated by

Pushkarnath (1959) and this is known as "Seed plot technique". Towards the mid-December, when the crop has tuberized well, the haulms should be destroyed before viruliferous aphids colonise the healthy crop (Pushkarnath et al. 1967; Nagaich et al. 1969). In many parts of Europe and America, potato seed production is based partly on early haulm destruction, before virus brought into healthy crops by aphids dispersing in late summer from diseased plants (Salzmann 1953; Munster 1958; Kabierson 1962; Wright and Hughes 1964). In the Netherlands, based on the aphid flight, the warning will be given to the farmers to destroy the haulms within 3–5 days so that healthy seed potatoes were multiplied.

In Australia, carrot crops free from *Carrot motley dwarf virus* were produced by delaying sowing until spring dispersal of its vector *Cavariella aegopodii* was over (Stubbs 1948). Lettuce crop was successfully grown in Germany without severe losses with LMV by sowing in March, which comes to maturity by June, before the large populations of *M. persicae* developed, while the June to August grown crop was severely infected due to maximum vector population (Kemper 1962). In Taiwan, Chen (1980) has observed that brown planthopper (*Nilaparvata lugens*), the vector of *Rice wilted stunt virus*, populations were not found during December through May and no rice plants were found to be infected with this virus and the viruliferous insects were detected first in June and July.

In Bangladesh, Rashid et al. (2008) studied the effect of different dates of planting time on prevalence of TYLCV and whitefly vector in tomato crop. They have observed that the percentage of TYLCV incidence in different dates of planting time (from mid October 2000 to mid-September 2001) of tomato cv. BARI Tomato 4. They have also studied the epidemiological aspects at 45, 60 and 75 DAP on virus incidence and vector population in tomato. The highest TYLCV incidence (%) was observed at 75 DAP during the period of March and April, 2001 planting followed by May, 2001 planting, but the lowest TYLCV incidence (%) was found in November, 2000 planting followed by December, 2000 planting. A strong correlation was obtained between the incidence of TYLCV and number of whitefly in tomato plants. Even in India similar type of results were reported by Sastry et al. (1978) and Saikia and Muniyappa (1989) in TLCV infection in tomato.

Based on the activity of the soil-borne vectors like nematodes and fungi, the sowing and harvesting dates were recommended for certain crops. In Nebraska, the spread of *Wheat streak mosaic virus* (WSMV) was decreased by delaying sowing of winter wheat until early September when the previous wheat crop has been harvested and the destruction of volunteer wheat plants carrying the vector, *Aceria tulipae* at least 3 weeks before wheat is planted (Slykhuis 1955; Staples and Allington 1956 and Slykhuis et al. 1957). A direct correlation between sowing times and infection by WSMV in wheat has been observed by Dimov and Stefanov (1975). The incidence was highest in early sowings (5–25 %) and the lowest in those made between 20th October and 5th November.

Maize rough dwarf virus (MRDV) is another economically important virus disease transmitted by planthoppers like *Javesella pellucida, Loadelphax straitellus, Calligypona marginata, Peregrinus maidis* and *Sogatella vibix* and occurs in most of the maize growing countries. In Milan (Italy), Zaragoza (Spain), Tel Aviv (Israel) and Prague (Czechoslovakia), the mere delaying of maize planting dates until such time as the seedlings will emerge from the soil when the mean air temperature exceeds 24 °C, resulted in reduction of MRDV incidence to negligible proportions. In Israel, maize trial fields sown during the last week of May 1959, and 1960, suffered not more than 3 % infection as compared to 45 % infection in adjacent fields of the same cvs., but planted in mid-April (Harpaz 1961). It is mainly because the vector of this disease does not breed on maize crop and remain in low numbers during the hot weather from May onwards. It has since become common practice in Israel not to plant maize before the last week of May instead of previous custom of planting early in season as possible, normally during April. Also in Italy, it has long been known that early planted maize fields are likely to be affected by the disease to a much greater extent than late sown ones. Thus, Grancini (1958) stated that in northern Italy maximum infection usually occurs in maize fields planted during second half of May and the first 2 weeks of June. Maize sown at the end of June and later will as a rule evade infection and end up with only very low percentage of MRDV incidence. In keeping with the same principle of evading infection by proper adjustment of the planting date, Blattny et al. (1965) in Czechoslovakia suggested advancing the sowing of maize to the month of May instead of the normal practice of planting in June. In this manner the emergent maize seedlings will precede and thus avoid the massive migration flights of the plant hopper vectors of this disease, which begins in June and goes on to the end of August.

In Africa, *African cassava mosaic virus* (ACMV), transmitted by whitefly *Bemisia tabaci* is the primary threat to the successful cultivation. From the Ivory Coast, Fargette et al. (1994) reported that maximum spread of ACMV occurred between November to June when the maximum temperature was present; small increase in disease was noted between July to October when the minimum temperature occurred. In Israel, processing tomatoes are transplanted in the field in the early spring, usually the end of March or the beginning of April and are harvested 3 months later. *Tomato yellow leaf curl virus* (TYLCV) normally spreads during the late summer and autumn and corresponds to the whitefly population which peak from September to November. Processing tomatoes are harvested prior to the built up of large whitefly populations when incidence of TYLCV-infected plants is lower and impact on yields is minimum (Polston and Lapidot 2007).

The above cited examples clearly indicate the possibilities of minimizing the disease incidence in different crop plants by adjusting the planting and harvesting times (Table 2.4). These dates will vary from place to place as the vector behavior depends on the environmental factors like temperature, humidity, rainfall etc. Manipulating the planting and harvesting dates based on vector flights coupled with insecticidal application will greatly help to raise healthy crops successfully.

Table 2.4 Effect of changing the sowing and planting dates in decreasing the plant virus diseases

Crop	Virus	Vector	Recommendations	Country
Broad bean	*Sudanese broad bean mosaic virus*	*Aphis craccivora, Acryrthosiphon sesbaniae*	Vector population peak in January, sow in October to produce adult plants in January	Sudan
	Subterranean clover red leaf virus	*Aulacorthum solani*	Autumn sowing; vector population peak in September; sow in May to July or November	Tasmania
Carrot	*Carrot motley leaf virus*	*Cavariella aegopodii*	Sow after spring when dispersal of vectors is over	Australia
Common beans	*Bean golden mosaic virus*	*Bemisia tabaci*	Cooler temperatures reduce vector population; grow in the late dry season	Brazil
Groundnut	*Peanut bud necrosis virus*	*Frankliniella* spp, *Thrips tabaci*	Early sowing in kharif	India
	Groundnut rosette virus	*Aphis craccivora*	Early sowing, sow in June rather than in July when vector population are lower	Nigeria
Lettuce	*Lettuce mosaic virus*	*Myzus persicae*	Sowing in March is recommended	Germany
Maize	*Maize rough dwarf virus*	*Laodelphax striatella*	Vector population declines after late May, sow in late May	Israel
Mungbean	*Peanut bud necrosis virus*	*Frankliniella schultzei, Thrips tabaci*	Second fortnight of May to first half of June; late sowing in spring	India
Potato	*Potato virus*	*Myzus persicae*	Sow between October–December in Gangetic plains	India
Potato	*Peanut bud necrosis virus*	*Frankliniella* spp., *T. tabaci*	Sowing at the end of end of October	India
Rice	*Rice wilted stunt virus*	*Nilparvata lugens*	Sowing in December to May with no plant hoppers	Taiwan
Sugar beet	*Sugar beet yellows virus*	*Myzus persicae*	Sow early for plants to achieve less susceptible stage as vectors increase	World wide
Sunflower	*Tobacco streak virus*	*Thrips tabaci*	Post rainy season, sowing after September	India
Tomato	*Tomato spotted wilt virus*	*Thrips tabaci, Frankliniella schultzei*	Vector population peak in warm humid seasons, and avoid such climate	World wide
	Tomato yellow leaf curl virus	*Bemisia tabaci*	Avoid sowing in September–November; and also vector population peak in summer, sow in spring	Israel
Wheat	*Wheat streak mosaic virus*	*Aceria tulipae*	Relaying sowing until early September	Nebraska
Winter wheat	*Barley yellow dwarf virus*	*Rhopalosiphum padi*	Sow after September 10th by which time vector population are in decline	Montana
	Wheat streak mosaic virus	*Aceria tulipae*	Spring sown wheat crop acts as virus and vector source for autumn sown wheat crop; sow in September	Alberta

2.12 Cross-Protection in Crop Plants

Cross-protection offers a promising strategy for biological control of plant viruses. It is well known that attenuated strains of viruses are being successfully used as vaccines for preventing various human and animal diseases. In plants also, cross-protection by a milder strain against a virulent strain of the same virus offers some opportunity for virus disease control. In nature most of the viruses and viroids exist in various strain forms and they can be isolated and differentiated based on serology and reactions on differential hosts. It is known that if a plant is already infected with one virus strain, it resists infection by different strains of the same virus. This phenomenon was first demonstrated by McKinney (1929) and he observed that tobacco plants infected with a strain that produced light green mosaic, showed no change in symptoms when challenge inoculated with a yellow mosaic strain.

Cross-protection is the "deliberate use of a mild or attenuated virus strain to protect against economic loss by the severe strain of the same virus". In recent years, there has been some interest to use this phenomenon as means of protection against severe strains to minimize yield losses under field conditions. For cross-protection as a method of control, the knowledge of strain interrelationships is of particular interest. Viruses vary widely with respect to the number of strains that occur under natural conditions.

Different strains of viruses were isolated in crops like citrus, apple, banana, papaya, cocoa, tomato, sugarcane, turnip, etc. and were tested for cross protection, which have given either partial or incomplete protection. It is generally accepted that either partial or complete protection occurs between closely related strains, provided that the interfering strain is systemically established in the plant before challenge inoculation with severe strain. Partial or complete cross protection may be seen in plants in which the first virus is incompletely invaded and also depends on the order and method by which the two viruses are introduced and on the relatedness of the virus isolates.

2.12.1 Theories to Explain Cross-Protection

In the absence of well-defined immune system in plants as compared with that found in animals, the basic resistance towards cross protection and its use in control may originate in our total ignorance of its mechanism. The proposed mechanisms explaining the phenomenon of cross protection are: (1) precursor exhaustion theory, (2) competition for sites of infection in the cell, (3) antibody theory, (4) adsorption theory, and (5) genetic recombination theory. More details are furnished in reviews of Fulton (1986), Sherwood (1987), Gonsalves and Garnsey (1989), Gonsalves (1989), Fuchs et al. (1997), Lecoq (1998), Fraser (1998), Gal-on and Shiboleth (2006), Tripathi et al. (2008), Ziebell and Carr (2010) and Zhou and Zhou (2012).

2.12.2 Mechanisms of Cross Protection in Plants

The mechanisms underlying the cross-protection are not fully understood, however, posttranscriptional gene silencing (PTGS) is considered to be functionally involved in the cross-protection (Ratcliff et al. 1997, 1999). PTGS as a defense mechanism against viruses in plants is mediated by virus-derived small interfering RNAs from 21 to 24 nucleotides long (Baulcombe 2005; Meister and Tsuchl 2004; Voinnet 2005).

Recently, a computational estimation system to predict potential targets for PTGS was developed (Xu et al. 2006). Nakazono-Nagaoka et al. (2009) using this estimation system, have predicted the presence of putative effective small interfering RNA segments in the M11 genome (both negative and positive senses) and analyzed whether corresponding interfering RNA segments exist in the challenge virus genome. In fact, numerous interfering RNA segments commonly recognized in the genome of M11 and in each of the challenge viruses were generally consistent with the cross-protection effects observed in their studies. With the estimation system for potential RNA segments for PTGS, 876 putative interference RNA segments were recognized in the M11 genome (positive senses). Of the 876 RNA segments, 866 were found in the genome of the parental isolate IbG. When M11 and the gladiolus isolates Gla and G1 that M11 completely cross-protected against were compared, 447 and 411 common interference RNA segments were found, respectively, even though G1 was molecularly and biologically the most distantly related to M11 among the three gladiolus isolates. In contrast, less than 100 common interference RNA segments were found between M11 and non-gladiolus isolates, against which M11 provided partially cross-protection. The CS isolate, which is the most distantly related to M11, shared only 10 RNA segments. ClYVV-No30, a virus distinct from BYMV but was partially protected against by M11, shared two RNA segments with M11. WMV-Pk, against which M11 did not cross-protect, had no target segments with M11. Similar results were obtained from the analysis for negative sense RNA. These results support the idea that PTGS is involved in cross-protection (Nakazono-Nagaoka et al. 2009).

2.12.3 Practical Use of Cross-Protection

The effectiveness of cross protection in crops like citrus, banana, papaya, cocao, passion fruit, peach, apple, tomato etc. were tried in different countries under field conditions with varying degrees of success. The following is a brief review of this phenomenon in different crops.

(a) **Citrus**

Among the virus diseases infecting *Citrus* spp. *Citrus tristeza virus* (CTV) is prevalent and economically very important. This is not a very complex disease, since it is caused by a single virus, but exists in nature in the form of numerous strains. There are three general types of tristeza which can cause tree decline. McClean (1977) described the three types of decline induced by tristeza: (1) the necrosis of the phloem cells in the sour orange rootstock just below the bud union causing girdle and death of trees; (2) seedling yellows tristeza-strains of tristeza which induce a seedling yellows reaction in certain seedlings such as lemon, grapefruit or sour orange; and (3) stem pitting tristeza which induces stem pitting in the scion or rootstock of lime, grapefruit or sweet orange.

Strains of tristeza: The existence of strains of tristeza virus, some of which cause mild symptoms and others severe symptoms has been demonstrated (Grant and Costa 1951; Grant 1959; Capoor and Rao 1967; Balaraman 1981, 1987; Tanaka et al. 1968). In most instances strainal differences have been recognized by differences in the behavior of lime seedlings. Grant (1959) distinguished three strains as mild (T1), strong (T2) and severe (T3) of CTV. Similarly, Capoor and Rao (1967) from India, and Omori and Matsumoto (1972) from Japan were able to identify three different strains based on key lime reactions. In India Balaraman and Ramakrishnan (1977, 1979a) isolated six strains denoted as very mild (S1), mild (S2), moderate (S3), moderately severe (S4), severe (S5) and severe with corky veins (HD) based on parameters like height, vein clearing, stem and root weight and girth.

Cross protection experiments: These experiments with mild strains of tristeza were reported from several laboratories on potted glasshouse plants (Grant and Costa 1951; Olson 1958; Stubbs 1961, 1964; Giacometti and Araujo 1965; Capoor and Rao 1967; Tanaka et al. 1968). However, the cross protection trials under the field conditions were limited. During 1946, U.S. Department of Agriculture and Institute Agronomic, Brazil started a co-operative research project. In the State of Sao Paulo, the cultivation of citrus cultivars like ‘Pera’ sweet orange, Galego lime and grape fruit have become increasingly difficult because of severe tristeza strains. Grant and Costa (1951) presented the initial evidence that the presence of mild tristeza strain offered some degree of protection against severe strains. Costa et al. (1954) working with several stionic citrus combinations inoculated with mild and severe tristeza strains as well as with comparable non-inoculated plants, observed that after 3 years of exposure in the field, grape fruit and West Indian lime scions on sweet orange root stocks infected with the mild strains grew slightly better than when infected with the severe strain. It was also noticed that the symptoms shown by naturally infected plants tended to be severe, indicating that severe strains were present in the natural tristeza virus complex. Muller and Costa (1968, 1972, 1977, 1979) reported the protective effect of 45 mild isolates in ‘Pera’ sweet orange, ‘Galego’ lime and ‘Ruby red’ grape fruit scions on three different root stocks i.e. Rangpur lime, Caipira sweet orange and Cleopatra mandarin. Strong or intermediate reactions in ‘Galego’ lime scions were noticed with 40 isolates and they were inadequate for preimmunization. The other 5 isolates, all from ‘Galege’ lime sources, induced mild symptoms and conferred good protection in preimmunized scions, compared with the non-inoculated controls exposed to natural infection. In the 1972 observations revealed that 2 of these 5 isolates provided outstanding protection. Trees immunized with them made apparently normal growth and with the highest yield. 'Pera' sweet orange plants, preimmunized with each of 3 mild strains out of the 45 selected originally, were able to grow satisfactorily in the area where they have been exposed to infection with severe CTV strains present in infected surrounding trees. Plants cross protected with many other selected mild isolates also performed better than the controls, but not as well as the five best. As the immunized plants of 'Pera' sweet orange and Galezo lime trees are performing well in field tests, in Brazil beginning in 1977, about 5 million cross-protected trees had been planted with the sweet orange predominating. By 1980 the number of cross protected 'Pera' sweet orange trees exceeded eight millions. Greenhouse and field tests have shown that protection given to citrus scions by inoculation with mild CTV isolates was not broken down by super inoculations with the vector or by field exposure to natural populations of the vector for long periods. Large scale propagation of cross protected 'Para' sweet orange by budding has revealed almost no breakdown in protection in successive clonal generations. Currently, most of the 'Para' budwood used in the State of Sao Paulo is derived from the original preimmunized material produced in the virus department of the Institute Agronomico deCampinas. In India, cross protection of tristeza on acid lime field trials were also tried (Balaraman 1987; Balaraman and Ramakrishnan 1979a, b, 1980; Sharma 1989).

Citrus exocortis viroid (CEVd) has been shown to exist in a number of strains based on differences in the severity, type of scaling, period of incubation and degree of stunting in root stock plantings of rangpur lime and in citrus trees on *P. trifoliate* root stock (Calavan and Weathers 1959; Fraser and Levitt 1959; Salibe and Moreira 1965; Rodriguez et al. 1974). Fraser and Levitt (1959) have reported dwarfness in citrus trees on trifoliate orange root stock without the association of scaling symptoms. Similar types of symptoms were also most prevalent in Washington navel orange and Marsh grape fruit trees and were reproduced in progeny trees (Fraser et al. 1961). The possible commercial use of citrus trees dwarfed by some strains of exocortis has been developed from the experimental plantings initiated by the Citrus Improvement Committee (CIC) of the New South Wales, Department of Agriculture, Australia.

Bevington and Bacon (1978) and Duncan et al. (1978) have reported the effect of time of inoculation with exocortis mild strain on both Navel and Valencia oranges and also on other root stocks. Trees inoculated in the nursery stage were the most dwarfed, while in trees inoculated in the field 1, 2, 3 or 5 years later were less dwarfed. At 10 years of age, trees on *P. trifoliate* inoculated in the nursery showed a 50 % reduction in canopy surface area. The effect on Carrizo citrange, Troyer Citrange and Rangpur (*C. reticulate* var. *austere* hybrid) was not as pronounced. With the citranges, a 25 % reduction in canopy surface area was observed at 10 years of age, whereas with Rangpur lime an 18 % reduction had been observed (Bevington and Bacon 1978). Where trees on *P. trifoliate* and citrange were inoculated with a dwarfing budline that causes butt scaling (exocortis mild strain) on *P. trifoliate,* canopy surface area was reduced by approximately 70 and 50 %, respectively (Stannard et al. 1975; Bevington and Bacon 1978).

In Brazil, Rodriguez et al. (1974) have studied the effect of six mild and two severe strains of exocortis on the yield of 5 year old Hamilton orange on Rangpur lime over a 4 year period. The average production of the trees affected with the mild strains reached 68.4 kg/tree during this period, which was essentially the same as the healthy controls (68.3 kg). On the other hand, trees affected by severe exocortis produced fruits 59.1/tree, 13.1 % less than the protected plants.

The experimental results of Duncan et al. (1978) with dwarfed (immunized) and uninoculated navel and Valencia orange trees on *P. trifoliate*, confirms the high production potential of dwarf trees when planted at high density. Because of the reduction in tree size, the yield per tree was also reduced. However, there was a loss of production efficiency as the reduction in yield was proportional to the reduction of the canopy surface area. Inoculation had no adverse effects on fruit size and quality. Encouraging results were also obtained with exocortis strains in the field trials at Lake Alfred and Fort Pierce (Florida) involving *C. sinensis* 'Pine apple' and 'Valencia', *C. reticulate* 'Dancy' and *C. paradise* "Marsh' and 'Ruby' trees on the three susceptible root stocks plus rough lemon and orange root stocks for comparison (Castle 1987).

The mild strain virus induced dwarf trees have obvious advantages as compared to larger vigorous ones in terms of greater production per hectare, especially in the early years following establishment. The other advantages are reduced harvesting costs, because of little or no ladder work is necessary; better pest and disease control through more efficient spraying and lower spraying costs because less expensive equipment can be used. High density plantings make more efficient use of inputs such as irrigation, fertilizers and herbicide, and are capable of high early production and economic returns. Planting of dwarf trees might be a desirable alternative to plantings of vigorous trees that require annual hedging. In large scale field trials conducted in New South Wales since 1961 to commercialize the technology of using exocortis mild strains, a good response was received from engaged citrus growers.

(b) **Passion fruit**

Passion fruit woodiness virus (PWV) causes woodiness, pimpling and pocking of the fruit. In Queensland (Australia) during 1932 the commercial life of a passion fruit plantation had dropped to a low production level in just 2 years due to woodiness virus (Simmonds 1959). Before the disease became well established vines had been productive for up to 8 years. McKnight (1953) put forward a strain complex hypothesis for this virus and suggested the possibility of obtaining of mild strain immunity. Simmonds (1959) conducted cross protection experiments in field with two mild strains (10 and 15) and a virulent strain (16). In second and third year of vines carrying a mild strain, they showed marked superiority over those carrying a severe strain as accessed in terms of number, weight and size of fruits free from woodiness symptoms. He had demonstrated the possibility of inoculating the passionfruit vines with mild strains of Queensland woodiness virus which would protect them from severe strain infection under natural conditions.

The mild strains apparently remained mild through successive transfers over a 5 year period. He observed only 4 % of the immunized plants were infected with severe strain, in contrast to 88 % of severe strain infection in the control treatment. Similar work on passion fruit woodiness virus was reported by Greber (1966) who worked on commercial plantations in central Queensland. In this area, the disease caused terminal necrosis of shoots, severe stunting and severe commercial loss. Greber used *P. edulis* seedlings which were infected by mild strains of the virus following a petiole graft technique. After symptoms appeared these plants were challenged by virulent tip blight isolate which was grafted to the mild strain-inoculated seedlings. From this work there was a strong indication of protection of inoculated plants by the mild strain. However this method does not seem to have been widely adopted.

(c) **Banana**

The existence of at least two distinct strains of *Banana bunchy top virus* has been demonstrated by Magee (1948). While investigating sporadic causes of apparent recovery from the disease in the "Viemana" (local name for Cavendish banana in Fiji) in Australia and Fiji, Magee found that the plants in question had not recovered but contained a strain of the virus so mild in its effect on the plant that symptoms were hard to discern; however, the characteristic green streaking could be detected and it was shown that the plants were highly resistant to infection by the typical form of the virus. The recovered plants were more difficult to infect by the transfer of viruliferous aphids than healthy plants i.e. as if they had acquired resistance to the virus, but if large numbers of viruliferous aphids were used, the partially recovered plants develop normal severe symptoms within the usual incubation time; in these plants, however, after the emergence of five or six yellowed leaves, normal green leaves again began to appear, indicating that cross-immunity was incomplete in banana. By suitable feeding tests it was shown that the plants did not carry the severe (normal) strain of the virus and under natural conditions in the field, transmitted no disease at all to their neighbors. Their growth was virtually normal and they appeared to bear normal bunches, but contributed a poor yield. For reasons which are discussed here, it would be highly desirable to have information on a detailed and rigorous comparison of yield of healthy and infected plants. The origin of plants so affected is not general known, but Magee observed a stool in Fiji of which one vigorous sucker showed mild symptoms, while all others were severely infected. This suggests that the mild strain in a non-infective mutant which confers a high (though not absolute) resistance against infection by the normal type virus. Presently, no resistant cultivars are available and no effective control exists other than phytosanitary measures. Experiments on the cross protection deserves proper attention as a possible control measure.

(d) **Cocoa**

In Ghana, Ivory Coast, Nigeria and Sierra Leone, *Cocoa swollen shoot virus* (CSSV) is a menace to the economic stability. There are many diverse strains of this virus (i.e., IA, IB and IC). The virulent strain (IA) predominates and causes various types of leaf chlorosis, root necrosis, root and stem swellings and die-back in Amelonado trees, which are usually killed within 2–3 years. The avirulent strain (IB) is not lethal and rarely causes leaf symptoms, although swellings are often pronounced, but they do not cause appreciable yield reduction. Strain IC caused no swellings and shows motley leaves. This strain has a pronounced debilitating effect on most trees, the leaf canopy becomes thinner and yield is reduced. The existence of mild form of strain IB was also found in apparently healthy trees found surviving on cocoa farms devastated by swollen shoot disease. Clones developed from such trees did not suffer when attempts were made to infect them with virulent strain IA by grafting (Posnette 1945, 1947). Crowdy and Posnette (1947) in their immunity studies between the three strains found a variable response from tree to tree and temporary remission when graft inoculation with IA. The protection appeared to be more effective against the insect inoculation than against graft transmission. It seems likely that there are two factors to consider in interpreting this delaying effect; the distribution of the virus in the plant tissues and the concentration of the virus. The results of the yield and immunization field trial involving three acres showed a greater protective effect by 1B against virus 1A infection, when the interval between the respective inoculations was 6 months than when it was 1 year, and this may indicate a fall in the concentration of strain 1B after 6 months. This decrease in concentration was suggested also by the leaf symptoms of strain 1B, which occurred only on one flush of leaves, while subsequent flushes were symptomless (Posnette 1947). Strain 1C conferred no protection against strain 1A and it may be unrelated virus. Regarding yield performance, strain 1A reduced yield by 50 % in the first year after inoculation and killed the trees in the second year, whereas strain 1B had no appreciable effect on yield.

Posnette and Todd (1955) carried out field trials to test the ability of mild strain to protect mature trees against infection with virulent strains allowed to spread naturally by mealy bugs and to ascertain the effect of the mild strain on yield of mature trees. During a 3 year field trial period in which 273 trees out of 387 (c70 %) previously uninfected trees became severely diseased, only 35 out of 416 (c8 %) trees initially infected by the mild strain showed severe symptoms. Five years after infection with the mild strain, trees were yielding 1 pod/tree more than in the year they were infected, whereas the decrease in yield on trees infected with the virulent strain was 16 pods/tree.

Investigations into mild strain cross-protection therefore started with this conviction. Earlier observations by Posnette and Todd (1955) indicated that mild strain of CSSV could protect cocoa against the effect of the severer strains, was thus reviewed and pursued further. Ollennu et al. (1989, 1999) have studied the cross protection of CSSV with CSSV N1 and CSSV 365B isolates, which are capable of protecting cocoa plants from the adverse effects of the virulent 1A type isolates (Ollennu et al. 2003). Current data from mild-strain cross-protection studies has shown two progenies of cocoa, T85/799 × T79/501 and T85/799 × Pa7/808, to be suitable for use in the mild strain cross-protection work (Domfeh et al. 2009).

This result reaffirmed the reported high yielding nature of T85/799 × Pa7/808 in the presence of CSSV N1 mild strain (Ollennu and Owusu 2003). It is envisaged that this mode of control would be most suitable only in the "areas of mass infection" in the eastern and western regions of Ghana, where the disease is so widespread (Dzanini-Obiatey et al. 2010).

From Ghana, Ollennu et al. (1996) have reported that mild strains of CSSV confer some protection against the severe strains available when tested in the field (Hughes and Ollennu 1994). However, the degree of protection offered so far is not economically feasible and further research is needed.

(e) **Sweet potatoes**

Sweet potato feathery mottle virus (SPFMV) exists wherever the crop is grown and causes russet crack and skin distortion of the sweet potato roots and yield losses of 20–50 %. Yamasaki et al. (2009) have tested the SPFMV mild strain 10–01 which has given the highest yields. When sweet potato cuttings were inoculated with mild strain the russet crack symptoms were much reduced.

(f) **Potatoes**

As early as 1933, Salaman demonstrated that a mild strain of *Potato virus X* (PVX) provided protection against super infection with a virulent strain of PVX in potatoes. Even Webb et al. (1952) showed that cross protection against the phloem limited virus *Potato leaf roll virus* (PLRV) could be achieved by infection with the aphid vectors.

(g) **Papaya**

In Taiwan, mild strains of *Papaya ringspot virus,* PRSV HA 5-1 and PRSV-HA 6-1, were first tried in large-scale field plots (Wang et al. 1987; Yeh et al. 1988; Yeh and Gonsalves 1994). In preliminary results, plants pre-inoculated with the mild strains did not show complete protection against the severe strain in the field, but did show a delay in the severe effects of the challenge virus. Although this degree of protection was insufficient for plants under severe disease pressure, economical returns could be obtained by isolating orchards as much as possible and by roguing out severely affected trees until the flowering period. About 100–200 ha of papaya were protected by PRSV HA 5-1 from 1985 to 1991 (Yeh and Gonsalves 1994) (Fig. 2.12). Currently, the mild strain is sparsely used, mainly because it does not provide consistent economic returns to the farmers. The failure of PRSV HA 5-1 to completely protect against PRSV in Taiwan is due to differences between the mild strain and the wild-type virus, as shown by green house experiments (Tennant et al. 1994). Similarly, these mild strains (PRSV HA 5-1 and 6-1) did not provide good field protection against PRSV strains in northeast Thailand (Gonsalves 1998). In India Badak et al. (2006) have tried cross protection principle against PRSV in Maharashtra area, but did not achieve satisfactory results.

Fig. 2.12 Cross protection in Papaya against PRSV (*Left* Control, *Right* Cross protected papaya plants). *Source* The American Phytopathological Society (APS)

The mild strains performed much better in Hawaii, presumably because the strains had been derived from a Hawaiian PRSV isolate. A series of field experiments were conducted with the Hawaiian Solo cultivars line 8, Kamiya, and Sunrise on the island of Oahu (Ferreira et al. 1992; Mau et al. 1989; Pitz et al. 1994). These studies showed that mild strain PRSV HA 5-1 gave good protection against the local strains, although PRSV HA 5-1 produced noticeable symptoms on leaves and fruit, with the degree of symptom severity markedly dependent on the cultivar. Cultivars line 8 and Kamiya were the least affected and could be grown economically, but Sunrise was too severely affected in fruit appearance. Protection of line 8 has been the most successful (Mau et al. 1989). Field trials were conducted with cross protected papaya plants at Oahu, Hawaii during 1993 and partial success was achieved. By using cross protected papaya plants, the growers have obtained lower yields, but more consistent and predictable crops for a known period of time. Thus the growers avoid the wild swings in production often associated with crop failures due to high virus levels. The last-resort approach of cross protection is its adopted to produce a crop economically with acceptable yields of grade A quality. It was also observed that the mild protecting strain does not reduce fruit quality as measured by brix or sugar content for the important cultivars grown in Hawaii ('Line 8', 'X-77', 'Kamiya', and 'Sunrise'). However, cross protection has not been widely adopted on Oahu, Hawaii, for several reasons: (a) the adverse effects of the mild strain on Sunrise and to a lesser extent on Kamiya, (b) cross protection requires extra cultural management and care, and (c) the reluctance of farmers to infect their trees with a virus. The positive and negative aspects of cross protection against PRSV have been reviewed by Gonsalves (1998) and Tripathi et al. (2008).

(h) **Grapevine**

In France, Komar et al. (2008), have tried the efficacy of cross-protection for mitigating the impact of *Grapevine fanleaf virus* (GFLV) on grapevines (*Vitis vinifera*) by using mild strains. Test vines consisted of scions grafted onto rootstocks that were healthy or infected by mild protective strains GFLV-GHu or *Arabis mosaic virus* (ArMV)-Ta. Challenge GFLV infection via the nematode *Xiphinema index* was monitored over 9 consecutive years in control and ArMV-Ta cross-protected vines. Results were consistent with a significantly reduced challenge infection rate in cross-protected vines compared with control vines, more so in those protected with GFLV-GHu (19 vs. 90 %) than with ArMV-Ta (40 vs. 65 % in field A and 63 vs. 90 % in field B, respectively). However, the two mild strains significantly reduced fruit yield by 9 % (ArMV-Ta) and 17 % (GFLV-GHu) over 8 years and had a limited effect on fruit quality. Therefore, in spite of offering a great potential at reducing the incidence of challenge field isolates, cross-protection with natural mild protective strains GFLV-GHu and ArMV-Ta are not attractive for the control of GFLV because of the negative impact on yield.

(i) **Apple**

The use of mild strains to protect apple trees against the effects of virulent strains of *Apple mosaic virus* (ApMV) and *Apple chlorotic leaf spot virus* (ACLSV) were well worked out in different countries. From Kent, Posnette and Cropley (1952, 1956) showed that mild strains of ApMV would protect trees of different apple cultivars against the effects of severe strains. They also studied the yield differences due to mild and severe strains during 1950–1956 and obtained much higher yields in mild strain-infected plants when compared with severe strain inoculated plants. In New Zealand, Chamberlain et al. (1964) have also demonstrated the usefulness of mild strains. They found that trees infected with severe strains of mosaic could be regenerated by top working the trees with graft wood carrying a mild strain. Such trees grew much more vigorously, cropped earlier and more heavily than comparable trees that were top-worked with either healthy graft-wood, or with graft wood infected with a severe strain. In Denmark, Thomsen (1975) tried cross protection experiment in 12 cultivars of apple against ApMV. The symptom development depended on the apple cultivar and the virus strain and varied from year to year and among trees of the same cultivar. Results from the cross-protection tests with a mild strain showed that varieties 'Graasten' and 'Spartan' were completely protected from the virulent strain of the virus.

Work at Summerland by Welsh (1964) has shown that certain clones of some apple cultivars are tolerant to viruses causing fruit symptoms and that this tolerance can be graft-transmitted to sensitive clones of the same cultivar. Buds from a clone of 'McIntosh', tolerant to leaf pucker virus, would induce tolerance in sensitive clones of this cultivar if grafted into the sensitive clones before the infector buds were grafted into the sensitive clone. Fruit and leaf symptoms in 'Spartan' apple causes by leaf pucker virus were eliminated by top-working with a 'Golden Delicious' clone, and symptoms caused by ring russeting virus in 'Spartan' were eliminated by grafts of a 'Winesap' clone. Welsh concluded that this suppression of symptoms was caused by an interfering virus or viruses in the tolerant clones.

In France, Marenaud et al. (1976) isolated mild strains of ACLSV and tried their utility in cross protection. They isolated M1 and GR2 strains from apple and

1n 21 and P 328 strains from prunus and plum. They tried cross protection tests on ACLSV and line pattern symptoms were noticed on *Malus platycarpa* and bark split on *Prunus domestica* P 707. The results of the cross protection greatly depended on the host plant and the nature of the strains. On *M. platycarpa* no cross protection was observed and line pattern symptoms did develop with whatever the strain combination was tested. On the contrary, inoculation of *P. domestica* P 707 with a mild strain (unable to induce bark split symptoms) always prevented the plant from developing symptoms when super infected with a strain which normally induces severe bark split symptoms. Protection was only observed with the mild strains CR2 and 1n 21. In their studies total protection appeared when the mild strain was inoculated 1 year before the severe strain treatment. Two years after the infection more than 60 % of the trees inoculated with CR2 and P 863 expressed severe symptoms but they disappeared the following year. When both mild and severe strains were simultaneously inoculated, the results were more irregular and the degree of protection was low.

(j) **Soybean**

Kosaka and Fukunishi (1993, 1994) have developed attenuated virus strains of *Soybean mosaic virus* (SMV) derived from low temperature treatments of infected plants. They have observed protection against SMV infection in soybeans when immunized with attenuated strain.

(k) **Tomato**

Tomato mosaic, a disease caused by *Tobacco mosaic virus* (TMV) is economically important in almost all tomato growing countries. Yield losses can be as high as 23 % and fruit quality is also often affected (Broadbent 1964). Although the use of a mild strains to protect the plants against more virulent strains was suggested by Kunkel (1934), the technique had not been widely adopted by tomato growers until Rast (1972, 1975) from the Netherlands, demonstrated the commercial benefits of inoculating crops with the avirulent TMV strain MII-16. Growers in the Isle of Wight (UK) were the first to adopt cross protection and a normal fairly severe strain of *Tomato mosaic virus* was used for inoculation in the seedling stage. The proportion of unsalable and poor quality fruit was reduced from about 30 to 3 % in 1965. These results lead to the idea that the benefit gained from seedling infection would be greater if a mild strain of TMV could be used, that would provide protection from more severe strains. Rast (1972, 1975), isolated an almost symptomless mutant using the nitrous acid mutagenic technique, which gave sufficient protection to five strains of TMV, when challenge inoculated. One of the effects observed on tomato plants following early infection with MII-16, was a temporary check of growth which delayed flowering and fruit set, and so it was necessary to advance the sowing date by about a week to compensate for this. He found that plants inoculated at the seedling stage with MII-16 yielded better than those inoculated at the same growth stage with the parent strain. Upstone (1974) reported that in 27 trials in the UK, plants inoculated with the Rast's MII-16 strain, on an average yielded 5 % more than the un-inoculated plants. The value of cross protection as a control measure against TMV infection in glasshouse tomatoes has been well established in different countries like Denmark (Paludan 1975); France (Migliori et al. 1972); U.K. (Evans 1972; Fletcher and Rowe 1975; Channon et al. 1978); Israel (Zimmerman and Pilowsky 1975); Belgium (Vanderveken and Coutisse 1975); The Netherlands (Rast 1972, 1975; Buitelaar 1974); Japan (Oshima et al. 1965; Goto et al. 1966; Nagai 1977; Oshima 1981); New Zealand (Mossop and Procter 1975); German Democratic Republic (Schmelzer and Wolf 1975) and the U.S. (Ahoonmanesh and Shalla 1981).

Pepino mosaic virus (PepMV) is another economically important virus in solanaceous crops like tomato, chillies, and capsicum, especially when grown under glass house/net house conditions. The large scale application of attenuated strains of PepMV has been attempted by a number of workers (Schenk et al. 2010). In the field trials, the yield losses were reduced by the challenge inoculation of attenuated strain of PepMV and the virus accumulation levels and also the symptom severity were also reduced. In Japan Sayama et al. (1993) have tested satellite-containing attenuated strain of *Cucumber mosaic virus* for tomato protection under field conditions.

(l) **Cucurbits**

Lecoq et al. (1991) have isolated a mild strain which is a variant of *Zucchini yellow mosaic virus* (ZYMV-MK) which induced mild foliar mottle without symptoms on fruits and is a poor aphid transmitted isolate. ZYMV-WK reduced the marketable fruit production of zucchini squash only slightly when plants were infected at the seedling stage. In field trials conducted in France in 1988 and 1989, ZYMV-WK proved to be very efficient in protecting two cultivars of zucchini squash from severe ZYMV under intense disease pressure. Increase in weight of marketable fruit in cross-protected plots was up to 14.7 times that in unprotected plots.

In Taiwan also Wang et al. (1991) have observed effective cross protection in cucumber, melon and squash crops against severe strain of *Zucchini yellow mosaic virus* (ZYMV) by using the mild strain of ZYMV-WK from France. Yields of marketable fruit were 2.2 and 40 times greater than those of control plants under moderate and high disease pressure conditions, respectively. In subsequent studies Wang et al. (1991) have used four different mild strains of ZYMV and under greenhouse conditions, 50 days after challenge inoculations, full protection was observed against mechanical inoculation in melon for ZYMV-CT and ZYMV-FR and 80 % for ZYMV-FL and ZYMV-TW. In cucumber and squash, the effectiveness of cross protection against the four severe strains varied from 50 to 90 %. In all three crops, cross protection was most effective against ZYMV-FR, from which ZYMV-WK was derived. Conversely, the lowest effectiveness of cross protection was for the local strain, ZYMV-TW. In conclusion, mild strains selected from local severe strains in the area may provide better cross protection. In cucumbers, Kosaka et al. (2006) observed the effectiveness of an attenuated ZYMV on protecting severe infection with ZYMV in cucumbers. Even in cantaloupe in the US, Perring et al. (1995) have effectively achieved higher yields and effective ZYMV control by using ZYMV-WK strain against severe strain. At Hawaii, Cho et al. (1992) have commercialized ZYMV cross protection for zucchini production and achieved higher yields.

(m) **Broad bean**

In Japan, protection was observed in broad bean against severe strain of *Bean yellow mosaic virus* (BYMV). Cross protection was complete against BYMV isolates from gladiolus but incomplete against BYMV isolates from other hosts (Nakazono-Nagaoka et al. 2009).

(n) **Yam**

Fuji et al. (2000) have selected an attenuated strain of *Japanese yam mosaic virus* (JYMV) and have proven the effectiveness of attenuated strains in protecting yam plants against severe strains of JYMV.

Conclusion

The commercial use of mild strains to protect crops under field conditions is not widely practiced except in few crops like citrus, tomato, passion fruit etc (Muller and Rezende 2004). The reluctance of scientists and growers in using this technique is due to their fear that either the primary mild virus may regain potency and virulence or subsequent field infection by one or more viruses unlike the primary virus might result in a virus-complex diseases increase as the number of viruses in the complex increases. Such effects have already been demonstrated with virus combinations in fruit trees such as *Prune dwarf virus* and *Raspberry ring spot virus*, which cause severe raspberry leaf diseases. Another drawback is that even inoculation with a mild strain will cause reduction in yield by about 5–10 %. It would be premature to attempt the use of mild strains in commercial gardens until much more data is available on yield losses and unless the virus is slow spreading.

Sometimes a mild strain isolated from a particular cultivars of a crop may not behave in the same manner in other economically important cultivars. Hence the crop protected with the mild strain may provide virus inoculum for another crop in which the same mild strain might be severe. There are also reports that all the mild strains may not give protection against severe strains of the same virus. Examples exist for some PVY strains in tomato (Silberschmidt 1957), a Tobacco veinal necrosis strain in *Physalis floridana* (Richardson 1958), a *Beet western yellows virus* strain in *Capsella* (Duffus 1964) etc. Even Muller and Costa (1968) have reported that mild strains of CTV collected from para sweet orange and grape fruit cultivars, induced severe reactions on the Galego lime test plants. Some isolates that were mild for ruby red grape fruit scion induced strong stem pitting symptoms on the para sweet orange. As the prevalence of the strains varies from place to place and also with different host varieties, distributing the immunized plants to different parts of the country is not advisable. It is better to use the strains isolated from that particular locality.

In India in 1974 field trials, when experiments using immunized Kagzi lime seedlings with mild strains of tristeza were conducted, there was no change in the strain behavior even after 7 years (Balaraman 1981; Balaraman and Ramakrishnan 1979a, b). Nearly 6 lakhs of immunized seedlings were distributed to six states in India. Even though the mildest strains available may diminish the crop yield, this must be weighed against the probability of death or crippling from virulent strains. Nevertheless, when crops are virtually certain to become infected with virulent strains because of the activity of vectors soon after planting, and when effective resistance is unknown in commercially acceptable cultivars, the value of mild strain protection seems to be irrefutable.

2.13 Vector Control

There are number of aerial insect vectors like aphids, leafhoppers, planthoppers, thrips, mealybugs, beetles, and mites which spread the viruses very fast because of their flight advantage and lighter body weight. In a majority of these vectors, they are being carried out by wind to shorter and longer distances and hence very high incidence of virus diseases are caused by these vectors. On the other hand, the spread of the viruses which are transmitted by soil-borne vectors like nematodes and fungi is very slow. Hence the management measures against insect-borne viruses and soil-borne viruses are different.

Before management of virus spread by vectors is attempted, it is necessary to identify and have available information about the population dynamics of the vector, for which basic brief information was provided in Chapter-4 of Volume-I. The vector may be an occasional visitor or a regular colonizer on the main crop. Furthermore, some aphid species are more efficient vectors than others. For CMV, *Aphis gossypii* is proven to be efficient vector, even though more than 10 other aphid species are known to transmit. Similarly, *Myzus persicae* is the most efficient vector for potato viruses, though there are other aphid vectors available in the field. In the case of CTV, the brown citrus aphid (*Toxoptera citricida*) is a much more efficient vector of CTV than is the melon aphid (*A. gossypii*).

Since the viruses are causing heavy yield and economic losses and the vectors are playing pivotal role in spread of the viruses, maintaining the healthy crops for higher yields, demands vector control. The various measures followed in vector control are being discussed.

As far as spread of virus within the crop is concerned, favorable ecological factors need to exist for maximum spread to occur. Generally the spread of a virus transmitted non-persistently are difficult to control with an insecticide treatment as would occur with virus diseases spread in a semipersistent or a persistent manner. In the latter cases the insect requires a fairly long feeding period on an infected plant, thus allowing time for the insecticide to be effective. Hence in number of field trials conducted in different places and countries, encouraging results were not achieved in reducing the spread of viruses like PVY, LMV and CMV in different vegetable crops. Some examples of where encouraging results were achieved in virus management by insecticidal or other chemical applications are discussed in detail. The spread of the persistently-transmitted PLRV in potato crops was substantially reduced with the appropriate application of insecticides, but reduction in the spread of PVY was not (Burt et al. 1964; Webley and Stone 1972).

Their results emphasized the importance of early spread in the season by winged aphids and the need for plants to be made lethal for aphids as early as possible. For example, in okra the incidence of the whitefly-transmitted *Okra yellow vein mosaic virus* was reduced to 57.97 % by spraying the pearl millet barrier crop with Achook (a neem product) (Kalita 2003).

The effectiveness of systemic insecticide treatment in controlling the non-persistently transmitted PVY in tobacco varied by season, and was probably dependent on climatic factors that affected the numbers and movement of the aphid vectors (Pirone et al. 1988).

The disease forecasting data is an important factor in the economic use of insecticides. Sometimes a long-term program of insecticide use aimed primarily at one group of insect vectored viruses will help in the control of another virus. Thus, the well-timed use of insecticides in beet crops in England, aimed mainly at reducing or delaying the incidence of yellows diseases (*Beet yellows virus* and *Beet mild yellowing virus*), has also been a major factor in the decline in the importance of *Beet mosaic virus,* a potyvirus infecting the same crop (Heathcote 1978). A warning scheme to spray against the vectors of beet yellows viruses was initiated in the UK in 1959 and is based on monitoring populations of aphids in crops from May until early July (Dewar and Smith 1999).

2.13.1 Insecticides

Insecticides are often effective against the spread of persistently aphid-transmitted viruses like PLRV but not against the spread of non-persistently aphid-transmitted viruses. Sometimes, use of conventional insecticides even increases the incidence of the virus within the crop, presumably due to increased probing's by the agitated vector. Non-persistently aphid-transmitted viruses are transmitted quickly even by the short duration probings of less than a minute by the vector. Consequently, only the newer classes of insecticides, called synthetic pyrethroids, which can quickly knock down the vector, hold some promise. The pyrethroid, λ cyhalothrin (Karate, ICI) application gave maximum control of the aphid vectors *Macrosiphum euphorbiae* and *Myzus persicae* and halved the incidence of PVY (Loebenstein et al. 2001).

Since the natural spread of CTV in citrus orchards is attributed to aphids, aphid control is sometimes suggested as an approach to reduce CTV problems. Aphid vector suppression has not been practiced in commercial citrus, and is not considered an effective strategy except, perhaps, in a citrus nursery. Unlike direct aphid feeding damage which requires large infestations to cause damage, even a small number of winged aphids may transmit virus infections. However, under specific circumstances vector control may have more potential than previously recognized.

If aphid control is based on insecticides applied to protect healthy trees, these must be extremely potent and kill the aphid quickly enough to prevent transmission. Long-term chemical control is difficult because citrus is a perennial crop and any control must continue year after year. In temperate regions, CTV spread is seasonal, and is concomitant with flushes of new growth in spring and fall. Under subtropical or tropical regions, new short growth is continuous as long as rainfall or irrigation is sufficient. Therefore, maintaining aphid control with insecticides is expensive, requires repeated applications, and has deleterious effects on the environment and non-target organisms. Control of the brown citrus aphid may actually be easier to achieve since the aphid reservoir is confined to citrus.

While the use of insecticides for general long-term control of aphids to prevent CTV spread does not appear realistic, there are situations where chemical control could be effective for the short term. For example, chemical suppression of a new and limited infestation of the brown citrus aphid may be useful to contain it until its distribution is known and the potential to eradicate or contain it are evaluated and quarantines can be imposed. While protecting healthy plants from inoculation with CTV by aphids is difficult, application of an insecticide to inoculum source trees could help to prevent aphid spread of CTV to nearby healthy plantings. This strategy may be useful in conjunction with eradication or suppression programs to limit further spread until infected trees can be removed, especially if the infected trees are localized. If it can be determined that spread of CTV in some regions is sporadic and only occurs during short periods which can be defined then a single properly-timed insecticide application may be useful to slow further spread.

Pineapple mealybug wilt associated-virus can be managed by controlling mealybug vector and ants. Dipping of planting material is recommended rather than by spraying the growing plant with an organo-phosphorus insecticide (Vilardebo 1955).

2.13.1.1 Role of Insecticides in Vector Control

A wide range of insecticides are available for the control of insect pests on plants. To prevent an insect from causing direct damage to a crop, it is necessary only to reduce the population below a damaging level. Further, insects like aphids, thrips, leafhoppers, beetles, whiteflies etc., also act as virus transmitting agents or vectors and depending on favourable situation heavy crop losses were also encountered in different virus-host combinations. Management of insect vectors to prevent infection by viruses is a difficult problem, as relatively few winged individuals may cause substantial spread of virus. Contact insecticides would be expected to be of little use unless they were applied very frequently. Persistent insecticides, especially those that move systemically throughout the plant, offer more hope for virus control. Viruses are often brought into crops by winged aphids, and these may infect a plant during their first feeding, before any insecticide can kill them. When the virus is non-persistent, the incoming aphid while feeding, can rapidly loses infectivity anyway, so that killing it with insecticide will not make much of a difference to infection of the crop from outside. On the other hand, an aphid bringing in a persistent virus is normally able to infect many plants, so that killing it on the first plant will reduce spread. Persistently transmitted virus whose spread depends predominantly on colonizing aphid species which require long acquisition and inoculation feeding periods for becoming efficient vectors, and their spread should theoretically be controlled efficiently by using systemic aphicides against the virus vector. The efficacy of vector control by chemical means is governed by the persistence of the aphicide in the plant and the time intervals at which they are applied. Systemic insecticides have been shown to reduce the incidence of persistently-transmitted viruses.

In the majority of the tropical countries, use of insecticidal sprays are an effective means for controlling aphid populations but they do not act fast enough to prevent inoculation of those viruses that, like CMV and PVY, are transmitted during the brief probing phase on the host plant. However, properly applied treatments can reduce disease incidence by killing the aphid before it has time to move to another plant. In practice, the efficacy of insecticide sprays is greatly reduced because: (i) viruliferous aphids are dispersed and survive long enough to transmit the virus, and (ii) the low aphid specificity of CMV transmission allows aphids either to acquire from or to transmit the virus to plant species on which they probe but do not feed. Effective aphid control and consequential low disease incidence can be achieved if winged aphid forms are kept at low levels. In several instances it has been observed that CMV and PVY epidemics correlate positively with extraordinary high aphid movement. In temperate regions like those of the Mediterranean basin, this usually occurs after mild winters or after high spring rainfall followed by rapid temperature increase. The conditions greatly favour reproduction within aphid colonies that rapidly reach a crowded stage leading to emergence of numerous winged forms. Such colonies often develop on weeds that can be virus reservoirs for CMV, PVY and other viruses, and therefore their destruction is a valid means for controlling and preventing virus outbreaks. More details on vector transmission can be obtained from Chapter-4 of Volume-I (Sastry 2013b).

The use of systemic insecticides such as imidacloprid (Gaucho 480) as a seed-dressing was effective at experimental level in reducing the incidence of BYDV in wheat (Gray et al. 1996); *Tomato yellow leafcurl virus* in tomato (Ahmed et al. 2001) and *Sugarcane mosaic virus* in sorghum (Harvey et al. 1996). Field trials in soybean conducted by Magalhaes et al. (2009) with seed treatment of imidacloprid at the rate of 62.5 g a.i./kg seed offered reasonable protection from mosaic virus disease for 2–3 months after sowing and this treatment has proved to be economical and less hazardous to the environment. Imidacloprid has both contact and systemic properties and affects various aphid species and had greatest effect on the fecundity of aphids rather than adult mortality; a reduction in the overall population of aphids developing on the crop would be beneficial in reducing secondary spread of the virus from infection foci. Even Castle et al. (2009) have reduced the incidence of *Cucurbit yellow stunt disorder virus* (CYSDV), a crinivirus transmitted by whitefly vector *Bemisia tabaci,* in cantaloupes (*Cucumis melo*) by the use of imidacloprid alone and in combination with thiamethoxam to reduce high densities of eggs and nymphs of the whitefly vector and CYSDV incidence. Even *Watermelon bud necrosis virus* (WBNV) incidence in watermelon was effectively minimized by seed dressing and foliar spray of imidacloprid (Kamanna et al. 2010; Rajasekharam 2010). Krishnakumar et al. (2006a) observed that imidacloprid seed treatment was significantly superior only up to 25 DAS to control thrips carrying WBNV. Even tuber treatment with imidacloprid was effective for control of potato stem necrosis disease which is caused by thrips-transmitted *Peanut bud necrosis virus* (Khurana et al. 2000).

In India, the sunflower necrosis disease caused by pollen-borne *Tobacco streak virus* (TSV), whose transmission is assisted by thrips was reduced by the seed treatment with imidacloprid (5 g/kg seed) followed by three sprays with Confidor 200 SL (0.05 %) at 15, 30 and 45 days after sowing (Shirshikar 2008).

Since TSV is major factor limiting sunflower production, extensive trials with more treatment combinations were tried. For example, at Hyderabad (India) during 2009–2011 field trials were conducted revealed that management of TSV in sunflower was effective by integrated approach involving seed treatment with thiamethoxam and three sprays of thiamethoxam at 15, 30 and 45 days after sowing along with 3 rows of border crop of sorghum. The next effective treatment was seed treatment with imidacloprid and three sprays of imidacloprid along with 3 rows of border crop of sorghum. Both of the treatments have reduced the incidence of TSV to 3.8 and 4.7 %, respectively as compared to 21.2 % in the control. Even the thrips population was reduced viz., 0.56 and 0.70 thrips/plant compared to 4.06 thrips/plant in the control. Due to the effective treatments the increase in sunflower seed yield over the control ranged from 20.5 to 41.2 %. These most effective treatments not only reduced TSV infection and thrips population, but also increased the yields (Bhat Bharati et al. 2012).

The feasibility and applicability of carbofuran (systemic insecticide) as seedlings dips to protect crops like tomato from *Tomato leaf curl virus* (TLCV) has been found to be effective. Sastry (1989) has observed reduction in whitefly transmitted *Tomato leaf curl virus* with seedling dip in carbofuran liquid alone and also in combination with oil sprays has reduced TLCV under field conditions. In Greece, Chatzivassiliou (2008) has reported that TSWV incidence in tobacco was reduced by the soil drench application of carbofuran (Furadan 35 SC) @ 400 ml/10 ha through irrigation water and recorded reduced thrips vector population and in turn disease incidence. The additional application of Malathion as foliar sprays @ 75–100 ml/100 l of water further reduced virus incidence. The final incidence of tobacco plants with TSWV infection fluctuated from 10–20 % in Virginia tobacco crops, whereas in the control crop it was 60 and 85 % in 2004 and 2005 respectively. In Africa, *Maize streak virus* (MSV) transmitted by leafhopper *Cicadulina mbila* was significantly reduced by application of carbofuran granules to the furrows @ 0.2 g a.i./m at the time of sowing (Drinkwater et al. 1979). Even carbofuran applied as a seed dressing at 0.80 and 1.04 g a.i./kg seed suppressed *Cicadulina* populations. Carbofuran pesticide was also reportedly better than conventional spray treatments with endosulfan at 31 and 45 days after emergence of maize plants or aldicarb in granules at 0.3 g/m in controlling MSV vectors (Drinkwater et al. 1979). Mzira (1984) reported that application of carbofuran effectively reduced the rate of MSV disease by 10 times in the treated plots.

Around the globe a number of researchers and farmers have used an innumerable number of different liquid insecticides (either contact or systematic) against different virus-host combinations under field conditions and encouraging results were achieved. In most countries, there are no crops which have not been sprayed with insecticide and in some instances the frequency of insecticidal sprays was every 5–7 days. Some of the examples where virus incidence and vector populations were reduced by insecticidal spray are: YVMV in okra (Sastry and Singh 1973a, b; Dahal et al. 1992). *Okra mosaic virus* and *Okra leaf curl virus* in okra (Lana 1976), *Mesta yellow vein mosaic virus* in Mesta (*Hibiscus cannabinus*) (Seetharam et al. 2011); *Mungbean yellow mosaic virus* in mungbean (greengram) (Agarwal et al. 1979; Borah 1995). *Indian cassava mosaic virus* in cassava (Saraswathi et al. 2002); *Peanut bud necrosis virus* in mungbean (Sreekanth et al. 2003) and potato (Somani et al. 2007); *Mungbean yellow mosaic virus* in mungbean (Ghosh et al. 2009); *Pigeonpea sterility mosaic virus* in pigeonpea (Rathi 1979) and *Subterranean clover red leaf virus* in broad bean crops (Johnstone and Rapley 1981). More information on role of insecticides in managing the plant virus diseases can be obtained from the review articles of Perring et al. (1999), Castle et al. (2009).

Carbofuran in granular form @ 1.5 kg a.i./ha was effective in reducing the *Tomato leaf curl virus* incidence in chillies and tomatoes (Singh et al. 1979a; Sastry et al. 1976). Rice tungro virus disease was minimized by the application of carbofuran (Shukla and Anjaneyulu 1980). In glass house studies, Palaniswamy et al. (1973) have observed that besides carbofuran, even aldicarb granules were proved to be quite effective and accounted for 90 % whitefly mortality within 2 days of caging. Soil application of methyl phosphoro-dithioate (Furatox 10G) @ 1.5 kg/ha was shown to be superior in reducing YVMV incidence and whitefly incidence in okra (Khan and Mukhopadhyay 1985). *Pigeonpea sterility mosaic virus* (PPSMV) was minimized by organophosphorus chemicals and seed dressers (carbofuran, aldicarb) (Rathi 1979; Reddy et al. 1990; Nene 1995). Soil application of carbofuran and aldicarb (Temik 10G) or as foliar sprays (oxythoquinox, kelthane, dinocap, monocrotophos, metasystox) (Ghanekar et al. 1992; Rathi 1979) were also successfully used to manage mite-transmitted PPSMV of redgram. In a number of vegetables and ornamentals the insecticidal application was also proved to be effective. For example, insecticide application either directly to the soil or as seedling drenches just before transplanting, potentially offers promise in reducing thrips vector populations and suppressing TSWV spread (Seaton et al. 1997). Even in southwest Australia, Coutts and Jones (2005) have reported that drenching plants with two systemically active neo-nicotinoid insecticides (thiamethoxam and imidacloprid) just before transplanting or to soil around plants reduced the spread of TSWV by thrips. Coutts and Jones (2005) demonstrated that imidacloprid drench treatments applied to lettuce seedlings suppressed the incidence of TSWV by 80–92 %. Csinos et al. (2001a) have also managed TSWV in tobacco with imidacloprid application. In recent years, spinetoram, acetamiprid, flonicamid, spirotetramat etc., are widely used for controlling the major vector of TSWV, the western flower thrips (*F. occidentalis*) in a number of countries (Culbreath et al. 2003; Pappu et al. 2009).

2.13.1.2 Role of Pyrethroids in Vector Control

Pyrethroids, the new type of insecticides quickly knock down the insect vectors like whiteflies, aphids, leafhoppers, thrips etc. For example, Perrin and Gibson (1985) have recorded reduction in PVY incidence by controlling the aphid vectors *Macrosiphum euphorbiae* and *Myzus persicae* by use of pyrethroid (Karate: ICI, PP 321 ICI). The field trials conducted at Australia, in barley and wheat crops against *Barley yellow dwarf virus* (BYDV) by application of synthetic pyrethroid insecticides, alpha-cypermethrin or beta-cyfluthrin before flag leaf emergence at rates as low as 12.5 g a.i./ha, has decreased spread of BYDV by 75 % and increased grain yields by up to 41 %. Effective reduction of aphid vector *Rhopalosiphum padi* was also recorded. Seed dressing with imidacloprid (70 g a.i./ha) followed up by foliar sprays of alpha-cypermethrin (25 g a.i./ha) gave excellent control of BYDV (McKirdy and Jones 1996). Even Perrin and Gibson (1985) have reduced *Barley yellow dwarf virus* with the same pyrethroid. In Kenya, virus incidence in seed potato was reduced by spraying the pyrethroid bifenthrin (2.4 l a.i./ha) which suppressed the aphid populations to very low level (Olubayo et al. 2010). Raccah (1985) has used the combination of mineral oils and pyrethroids for the control of non-persistently transmitted viruses and obtained effective reduction in virus incidence.

While working on rice tungro virus disease management, Satapathy and Anjaneyulu (1984) have reported that cypermethrin was proved to be very effective in controlling the leafhopper vector, *Nephotettix virescens* which was killed within 7 min after releasing an insecticide-treated plant. Subsequently, Bhaktavatsalam and Anjaneyulu (1984) have noticed decamethrin to be effective in having excellent insecticidal activity. In subsequent studies Anjaneyulu et al. (1994) have concluded that decamethrin plus bufrofaziu and tralomethrin were equally effective as that of cypermethrin and decamethrin and vector population and tungro disease incidence in rice were possibly correlated. Besides, effectively reducing the aphid-borne and leafhopper transmitted viruses, pyrethroids were also proved to be effective against thrips vectors. Pappu et al. (2000) have provided the evidence by managing the spread of *Tomato spotted wilt virus* in flue-cured tobacco by the application of acibenzolar-S-methyl and imidacloprid.

2.13.1.3 Control of Nematode Vectors

Nematode genera viz., *Xiphinema, Longidorus, Paratrichodorus,* and *Trichodorus* transmits virus diseases like *Grape fanleaf, Raspberry ringspot, Pea early browning* and *Tobacco rattle viruses*. The two major groups of nematode-vectored viruses are nepoviruses with isometric particles and tobraviruses with straight tubular particles. Six species of *Xiphinema* Cobb and five of *Longidorus* (Micoletzky) Filipjev (Nematoda, Longidoridae) are implicated in transmission of the 15 known nepoviruses, and five species of *Trichodorus* Cobb and seven of *Paratrichodorus* Siddiqi (Nematoda, Trichoridae) transmit two known tobraviruses (Lamberti 1981a). Nematode transmitted viruses cause heavy yield losses in a number of fruit and vegetable crops, and viruses like *Grapevine fan leaf*, *Cherry leaf roll*, and *Tobacco rattle*, are very important in many temperate regions of the world. Others such as *Tomato black ring, Raspberry ringspot, Arabis mosaic, Strawberry latent ringspot* and *Pea early browning viruses* are confined largely to Europe. Still others like *Peach rosette mosaic, Tobacco ringspot,* and *Tomato ringspot viruses* are seen only in North America. Viruses like *Pea early browning, Pepper ringspot,* and *Tobacco ring spot* infect a large number of vegetable crops.

The chemical treatments for nematode control are most effective when the soil preparation, timing, and method of application are done meticulously according to label requirements. Soil to be treated should be cultivated deeply to break up layers that prevent vertical diffusion of the product, residues of the previous crop should be removed, and soil clods that may be protecting nematodes should be broken up.

Chemicals have apparently given satisfactory control of soil-borne viruses of grapevines in the shallow soils of France (Vuittenez 1957, 1958, 1960). A number of interacting abiotic and biotic factors affect efficiency of nematode control. All nematode vectors of plant viruses are ectoparasites and therefore are readily exposed to chemicals in the soil. Soil disinfection with fumigant nematicides provides a quick method of eliminating the nematode vectors. One of the early experiments was done in Burgundy by Vuittenez (1958, 1960, 1961). Laboratory tests had shown that *X. index* is less sensitive to 1,2-dibromo-3-chloropropane (DBCP). Application of 1,200–2000 l/ha of fumigants containing 1,3-dichloropropene (DD) (Raski and Lear 1962) eradicated *Meloidogyne incognita* to a depth of 2 m in the field. Laboratory tests indicated that a 100 % kill of second stage larvae of *M. incognita* would require a 3-day exposure to 2.5 μg of DD/g of dry soil (Abdalla et al. 1974). The same doses should be enough to control 99.9 % of all stages of *X. index* (McKenry and Thomason 1974). Larvae of *X. index* seem to be less sensitive than adults to methyl bromide; 27 h exposure to 600 ppm is required to kill larvae, compared with 43 h exposure for adults (Brown 1978). All stages of *X. index* were totally killed with 530, 350, and 50 ppm of methyl bromide and exposure times of 24 h, 3 days, and 21 days, respectively (Abdalla and Lear 1975). Carbon disulfide and methyl bromide in shallow applications gave no satisfactory results. Soil fumigation with DD stimulates the growth of the newly planted grapevines (Vuittenez 1960; Bruckbauer 1969a, b) but the real cause of this stimulation is not known. Application of DD at 800 l/ha reduces the nematode population up to a depth of 180 cm (Sharma and Swarup 1982).

Nonfumigant nematicides such as aldicarb and oxamyl seem to have nematostatic rather than nematicidal action. At certain doses, these agents temporarily affect nematode behavior by delaying egg hatching, reducing motility, or prevent feeding and therefore are effective only on crops with a short period of susceptibility to a virus (Lamberti 1981b).

The efficiency of a nematicide is greatly influenced by its movement and persistence in the soil. The toxicant must reach the nematode and be exposed long enough to absorb a lethal dose. Generally, the organophosphates such as fenamiphos are most effective at temperatures above 25 °C, while the carbamates, such as aldicarb and oxamyl, at temperatures below 20 °C. Fumigants, on the other hand, move through the soil primarily in the gaseous phase, when diffusion is at least 1,000 times faster than in the water phase. Clay soils are the most difficult soil type to treat because pores are very small and cracking occurs during drought. Fumigants tend to diffuse upward, especially when the deep soil layers are compacted or water-saturated and are not retained in clay or other coarse soils long enough to be effective.

A nematicide must be thoroughly diffused to control nematode vectors of plant viruses. Viruliferous *X. index* have been found to a depth of 2 m in old vineyards in southern Italy and 3.6 m in California. The largest populations however are generally located at depths between 0.3 and 0.6–0.8 m.

High doses of DD completely killed plant parasitic nematodes to a depth of 2 m in California. McKenry (1978) suggested that lethal doses are possible to a depth of 0.75–1.5 m when the rate of application is selected on the basis of soil structure, moisture, and temperature, which are the factors affecting diffusion of fumigants into the soil. Methyl bromide moves first by mass flow and, therefore, disperses faster in the soil, especially downward; injection to 0.75–0.8 m was lethal to nematodes to a depth of 2.44 m (Abdalla et al. 1974).

2.13.2 Chemical Control of Fungal Vectors

Soil inhabiting fungal vectors belonging to chytridiomiomycota members (*Olpidium virulentus [a noncrucifer strain of O. brassicae sensu lato], O. bornovanus*) transmits *Mirafiori lettuce big-vein virus* (MLBVV), *Tobacco necrosis virus, Tobacco necrosis satellite virus* and *Cucumber necrosis virus*; and Plasmodiophorid members (*Polymyxa graminis, P. betae, Spongospora subterranea*) transmits *Barley mild mosaic virus, Barley yellow mosaic virus, Peanut clump virus, Soil-borne wheat mosaic virus, Beet necrotic yellow vein virus* (BNYVV) and *Potato mop-top virus* (PMTV). Thick walled resting spores are formed inside the roots and young tubers. Depending on the occurrence of favourable conditions, the resting spores germinate and release the primary zoospores and the zoospores penetrate the fresh root hairs and begin the life cycle. For more details of fungal vector transmission refer to Campbell (1996) and Chapter-4 of Volume-I. (Sastry 2013b)

The first record of chemical control of a fungal vector of a virus is established in 1956 and since then sufficient information is available for controlling soil-borne vectors of plant viruses by chemical application. In a number of countries, horticultural crops are now cultivated in various soilless substrates irrigated with nutrient solution, hydroponic systems, which may or may not be recirculated. This is an ideal environment for many zoosporic fungi and most of the viruses transmitted by *Opidium virulentus, O. brassicae* and *O. radicale* have been found causing damage in such environments. For example in the UK, MLBVV started to cause major problems in lettuce grown by the re-circulated nutrient film technique in 1978 and on some holdings 100 % virus infection of crops was observed. Laboratory experiments following these outbreaks identified chemicals controlling the zoospore stage of *O. virulentus* (*O. brassicae*) (Tomlinson and Faithfull 1979). Chemicals like copper (4 μg/ml), zinc (10 μg/ml), and the fungicide carbendazim (as Bavistin WP, a wettable powder containing 50 % (w/w) carbendazim; BASF) have been shown to be effective in vector control. Further tests showed that a "blank" formulation of Bavistin containing only surface active agents (surfactants) was equally effective in killing zoospores. This led to tests on a range of surfactants of which Agral, Centrimide, Deciquam 222, Ethylan CPX, Hyamine 1622, Monoxol/OT and sodium lauryl sulphate were found to be toxic to zoospores at concentrations of 1–10 μg/ml. A commercial trial was subsequently carried out with Agral (non-ionic liquid containing 90 % (v/v) alkyl phenol ethylene oxide condensate: ICI Plant Protection Division) added at a concentration of 20 μg/ml at 4-day-intervals to the circulating nutrient solution in a greenhouse containing c. 140,000 lettuce plants, where the two preceding lettuce crops had shown 100 % MLBVV symptoms. After 6–7 weeks, when the plants had reached maturity none of them showed MLBVV symptoms and no *O. virulentus* was observed in their roots (Tomlinson and Faithfull 1980). In an adjacent area of lettuce irrigated with untreated nutrient solution >90 % of plants had developed severe MLBVV symptoms and their roots were infected by *O. virulentus*. In the UK, cucumber crop is generally affected with *Melon necrotic spot virus* (MNSV), transmitted by *O. radicale* in field trails. Agral has also been shown to be lethal to zoospores of *O. radicale* and when added twice weekly at 20 μg/ml to the hydroponic nutrient solution supplying a cucumber crop gave good control of *O. radicale* and *Melon necrotic spot virus* (MNSV), increasing cucumber yields two to threefold (Tomlinson and Thomas 1986).

A number of fungicides have been tested for their ability to control viruses in hydroponic systems. Although carbendazim precipitates in solutions at pH values higher than pH 3.0 a slurry of Bavistin (0.06 g product/l) gave good control of MLBVV symptoms and *O. virulentus* (*O. brassicae*) in a recirculated hydroponic system (Tomlinson and Faithfull 1979). Although not tested alone, thiophanate-methyl as a carbendazim regenerating product (25 mg a.i./l) gave good control of lettuce ring necrosis symptoms and its vector *O. virulentus* (*O. brassicae*) when used in conjunction with zinc (>0.05 mg/l) in a recirculated hydroponic systems (Vanachter 1995).

(a) **Soil amendments and fungicides for vector control**

Various amendments have been tested for their ability to reduce the transmission of fungal vectored viruses. Decreasing the pH of infective soil to 5.0 by applying sulphur greatly decreased the infection of potatoes with PMTV and *S. subterranean* in field experiments (Cooper et al. 1976). This treatment did not eliminate either disease and when the pH of treated soil was raised, the transmission of PMTV resumed. Soil pH was also shown to affect development of rhizomania of beet caused by BNYVV and again the application of sulphur to lower soil pH to 5.5 controlled the disease (Abe 1987). Acid soils in West Africa are considered to be deleterious to *P. graminis* and *Peanut clump virus* (PCV) (Thouvenel et al. 1988). A disadvantage of such a low soil pH has an adverse effect on plant growth. Soil applications of chemical or organic fertilizers containing high levels of nitrogen were reported to reduce the incidence of *Wheat spindle streak mosaic virus* (WSSMV) (Slykhuis 1970).

One of the first attempts to control a fungal virus vector by using a fungicide was in 1960 when quintozene at 50 ppm was shown to give good control of *O. brassicae* on lettuce in sand (Rich 1960). Further tests showed that this chemical also suppressed the incidence of LBVV in field crops (Thomas 1973). Captan controlled TNV which is also transmitted by *O. brassicae* (Harrison 1977). A range of fungicides were tested for their ability to control PMTV in the field, but only mercurious chloride showed promise (Cooper et al. 1976). However, its use commercially was not considered as practical because of its cost and toxicity to humans. Mercuric chloride, captan, and 2,4,5-trichlorophenol as soil drenches have been shown to give good control of WSSMV in laboratory experiments (Slykhuis 1970). Dibromochloropropane, carbofuran, and aldicarb have been shown to control *Indian peanut clump virus* (IPCV) at nematicidal dose rates (Reddy et al. 1988). The first report of a systemic fungicide giving control was prothicarb which was reported to give substantial control of *P. betae* and *O. brassicae* when incorporated into the coating of sugarbeet pills at the rate of 1 mg/seed (Horak and Schlosser 1978). This was followed by various reports of Bavistin and benomyl, incorporated into peat blocks of transplants or drenched prior to transplanting, reducing the incidence of MLBVV (Campbell et al. 1980; Tomlinson and Faithfull 1980). Benomyl is degraded to methyl benzimidazole-2-yl-carbamate (carbendazim) which is the active ingredient of Bavistin, so it is considered that their modes of action in soil applications are identical (Campbell et al. 1980). Fosetyl-aluminum has exhibited activity against *P. betae* in a polytunnel experiment (Walsh 1992). Thiabendazole has been shown to give some control of LBVV in pot tests and field trials but is phytotoxic to lettuce. Many of these fungicides often give variable results under field conditions.

2.13.3 Soil Sterilants and Disinfectants for Reduction of Vector Populations

Many different soil partial sterilants have been tested for their ability to control many of the viruses like *Cucumber necrosis virus* transmitted by *Olipidium radicale, Tobacco stunt virus* by *Olpidium brassicae, Beet necrotic yellow vein virus* by *Polymyxa betae, Potato mop-top virus* by *Spongospora subterranean,* and *Rice necrosis mosaic virus* by *Polymyxa graminis* fungal vectors in different crops like cucumber, tobacco, potato, beet, rice, etc., and have been reduced by using different herbicides, surfactants fungicides and other fumigants (Thomas 1973; Tomlinson and Faithfull 1980; White 1983; Schlosser 1988; Reddy et al. 1988). Chloropicrin, pentachloronitrobenzene (PCNB), methyl bromide, and 1,3-dichloroporopene with 1,2-dichloropropane (D-D), have been shown to be partially or completely effective in controlling LBVV when injected into soil, although some findings were inconclusive or contradictory. Methyl bromide killed *O. brassicae* and prevented transmission of TNV to cucumbers and beans in greenhouses in New Zealand and methyl bromide and steam were shown to control *O. brassicae* transmitting *Freesia leaf necrosis virus*. Methyl bromide, metham sodium, chloropicrin, dazomet, dichloropropene and D-D have been shown to control a number of fungal-transmitted plant viruses in different crops and has been reviewed (Walsh 1998). In California, Telone (dichloropropene) was used routinely to reduce the damage caused by *Beet necrotic yellow vein virus* (BNYVY) disease (Schlosser 1988). Various treatments have been also been found to control SBWMV, WSSMV and PCV which are all transmitted by *P. graminis* (McKinney et al. 1957; Slykhuis 1970; Thouvenel et al. 1988).

Even through the above cited examples have shown apparent successes, the expense of such control methods is often prohibitive for field application. In a number of field trials, methyl bromide proved to be the most effective, but it is most toxic and expensive. Many fumigants are also hazardous to the environment, methyl bromide is listed as an ozone-depleting substance and the Environmental Protection Agency of the U.S. has strictly prohibited this fumigant. In lettuce, problems have also been experienced with excessive bromine residues some of the chemicals discussed here are usually only partially effective as they only kill resting spores of vectors in the upper layers of the soil and re-infestation from lower layers occurs during the growing season and treatments must be repeated each time whenever a susceptible crop is grown.

As the fumigants and herbicides were proved to be having high toxicity, residual effect and high cost, subsequent studies showed benomyl and ribavirin to be effective in reducing *Lettuce big vein virus* vector and its spread (Campbell 1980). The active ingredient of Bavistin (carbendazim) is also found to be effective in reducing the MLBVV. Similarly, the fungicide fluazinam (Omega) was shown to have activity against fungal species that can vector a number of virus diseases, but need to be demonstrated in field trials. In recent years, different species of *Trichoderma* and *Pseudomonas fluorescence* are being applied to the soil in controlling some of the soil-borne diseases including certain species that can vector plant viruses. The effectiveness of insecticides in controlling different vectors including insects, fungi, and nematodes has been reviewed by number of workers (Perring et al. 1999; Palaniswami et al. 2005; Castle et al. 2009).

Conclusions

In a number of countries farmers often spray insecticides to control insects and the virus diseases they transmit, as for example in faba beans where insects are a pest. However, the application of 3–5 sprays is mostly uneconomical and can have adverse environmental effects. Therefore, proper timing of a single spray or a minimum number of sprays to prevent the build-up of high insect vector population and the movement of viruliferous vectors within a crop should be economical and given priority.

Despite the wide range of insecticides, use of insecticides is not the preferred solution to prevent vector activity. Many viruses are introduced into crops by visiting insects that inoculate during their first probing activities. Vectors for nonpersistent (and partly semipersistent) viruses need relatively short inoculation times - much shorter than the time needed for insecticides to kill. In addition, insecticides can induce restlessness in insects, with the result that they make more inoculation attempts than do calm insects. Exceptions are vectors that colonize the crop and transmit persistent viruses, for which insecticide control may result in reduced spread of virus. However, a number of research articles have mentioned the success story of reducing the vector population in different crops by application of systemic and non systemic insecticides (Perring et al. 1999; Castle et al. 2009). There is the necessity of insecticidal application to reduce vector populations along with other integrated approach methods like cultural and genetic factors (resistance) for virus disease management is to be worked out depending on the virus, vector and environment.

As the whitefly transmitted begomoviruses are occurring to a greater extent in number of crops and in the majority of the cases the insecticides have not given encouraging results. The reason is that the whitefly feeds on under surface of leaves and the waxy covering of immature insect stages provides protection from insecticides. However, the introduction of neonicotinoids group of insecticides into the market helps to increase whitefly control. This group of insecticides was found efficient in controlling sucking insects such as whiteflies, aphids and leafhoppers (Horowitz and Isshaaya 2004)

Spraying insecticides in farming operation has certain disadvantages. It is an extra operation to be performed, tractor damage to the crop occurs, and spraying may not be practicable at the time required. Drifting spray can also lead to damage in other crops. Persistent systemic insecticides applied in granular from at time of planting overcome many of these difficulties. With a crop such as potatoes, granules can be metered into the furrow through applicators attached to the planting machines. For example, various systemic granular pesticides such as aldicarb and thiofanox have given control of aphids and PLRV in potato crops, applied in-furrow at planting, or as side bands some weeks later (Woodford et al. 1988).

The side-band treatments were generally less effective than those applied in-furrow.

Entomological research should also include development of chemicals which can deter or repel feeding by aphid vectors to reduce plant virus disease spread. Use of inappropriate insecticides may increase virus infection either by disturbing the aphids present in crop or disturbing the predators. Hence the knowledge of virus, vector and its relationship is very much required before starting the insecticidal spray. Exotic parasitoids are to be evaluated for their ability to reduce populations of the insect vectors for plant virus transmission.

2.14 Role of Botanicals in Plant Virus Management

From different countries efforts were made to test the efficacy of certain plant products (botanicals), to induce systemic resistance for managing the incidence of plant virus diseases. For example; Srivastava et al. (2006) have identified the systemic resistance inducing protein from *Clerodendrum aculeatum* (CA-SRIP) against *Papaya ring spot virus* (PRSV) infection in papaya. Following challenging inoculation by PRSV, to the CA-SRIP treated papaya plants, no virus symptoms were noticed up to 6 months after inoculation. There was also a significant difference in the vegetative growth of the treated plants as compared to control set of plants. Even, Singh and Awasthi (2006) have applied 8 sprays of leaf extracts of *Clerodendrum aculeatum* for the control of *Mungbean yellow mosaic virus* in mungbean and urdbean and the reduced virus incidence was 66.7 and 63.6 % respectively. Even root extract of *Boerhaavia diffusa* and *Azadirachta indica* leaf extracts were effective in reducing virus incidence. Parveen et al. (2001) have isolated and characterized an inducer protein (Crip-31) from *Clerodendrum inerme* leaves that induces systemic resistance against CMV, PVY and ToMV in *Nicotiana tabacum* cv. White Burley. Prasad et al. (1995) have isolated two basic proteins from *Clerodendrum inerme* which are inducers of systemic antiviral resistance in susceptible plants.

Chen et al. (1991) have proved that pokeweed antiviral protein (PAP) from *Phytolacca americana* inhibited the infection of TMV, AMV, PVX, PVY, *Cauliflower mosaic virus* and *African cassava mosaic virus.* Antiviral activity is dependent on the concentration of PAP and not of the virus. Picard et al. (2005) have also observed the inhibition of *Brome mosaic virus* replication in plant cells with pokeweed antiviral protein. Sastry and Singh (1982) have proved that tuber extract of *Dioscorea floribunda* was quite effective against TMV infection in *Chenopodium amaranticolor,* a local lesion host.

Ragetli and Weintraub (1962) and Plobner and Leiser (1990) have showed that the leaf extracts of *Dianthus caryophyllus* inhibited 100 % local lesion production of TMV in *N. glutinosa* and the inhibitory factors were dianthin 30 and 32 which were glycoproteins containing mannose. During 2008, Choudhary et al. have isolated an antiviral proteins (AVPs)/ribosome-inactivating proteins (RIPs) from

Bougainvillea xbuttiana which have been suggested to be multifunctional proteins capable of inducing systemic resistance, and their virus inhibitory activity may be directly correlated to their enzymatic activities, such as Nglycosidase, RNase and DNase.

The botanical insecticides composed of essential oils may be an alternative to the more persistent synthetic pesticides for the management of vectors responsible for disease spread. A proprietary emusifiable concentrate containing 25 % essential oil extract of *Chenopodium ambrosoides* (EOCA) as the active ingredient at 0.5 % caused mortality percentage (43.6) and insecticide soap (55.2 %) and were more effective against *Myzus persicae* than the control. The extract of *C. ambrosoides* at the concentration of 0.5 % gave excellent control of thrips, *Frankliniella schultezi* 95.7 % and insecticidal soap 83.6 % than neem oil (17.7 %) and water (10.8 %). EOCA proved to be more effective than commercial product in controlling the major virus vectors such as *M. persicae* and *F. schultezi,* than neem extract, insecticidal soap endosulfan and abamectin (Rajapakse and Janaki 2006). For more details on role of botanicals in plant virus inhibition, one can refer the review articles by Baranwal and Verma (2000).

2.15 Role of Oils in Arthropod Vector Control

Intensive researches of virus disease management have also led to the discovery of the application of oils for the control of virus diseases. The finding that vectors carrying viruses seldom transmit after probing a leaf surface which is lightly coated with oil offers a novel way of minimizing the disease spread. Even in *in vitro* studies, oils exhibited strongest inhibition of many viruses. The attraction of using these non-toxic materials at recommended concentrations are great because they are less likely to cause the environmental pollution than the chemical pesticides, have excellent spreading and sticking properties, are not subject to resistance development, and are economical to use. The pioneering work in this field was done by Bradley (1956). He reported that when viruliferous aphids probed into a membrane of low melting paraffin wax, the wax greatly reduced the subsequent transmission of *Potato virus Y* (PVY). It was shown later (1962) that this phenomenum was caused by the oil present in the wax and not by wax itself, as believed earlier. Bradley (1963) showed that transmission of PVY by *M. persicae* was impeded by coating either the source plants or the test plants with mineral oil (liquid paraffin). These observations have led to considerable work aimed at developing oil sprays to protect plants from infection by aphids carrying especially non-persistently transmitted viruses. Studies on the possible role of oils in preventing the viruses both in glasshouse as well as in field conditions in countries like U.S., Israel, U.K., India and Germany have given encouraging results in crops like pepper, cucumber, beet, potato, tomato, lilies, tulips, gladiolus etc. (Zitter and Ozaki 1978; Sharma and Varma 1982; Sastry 1984; Singh 1992).

2.15.1 Oil, Virus-Vector Relationship and Virus Particle Morphology

Oil spray impedes essentially aphid-borne non-persistently transmitted viruses (Bradley 1963; Loebenstein et al. 1964, 1966, 1970; Allen 1965; Bradley et al. 1966; Nitzany 1966; Hakkaart 1968; Wenzl 1970; Converse 1970; Lu 1970; Zschiegner et al. 1971; Asjes 1972, 1975; Slusarek 1973; Genereux 1973; Verhoyen 1973; Wenzl and Foschum 1973; Esparza-Duque and Verhoyen 1975a, b; Hein 1975; Proeseler et al. 1975).

Even semipersistently transmitted viruses like *Beet yellows virus* were also curtailed by oil application (Vanderveken 1968; Vanderveken and Semal 1966; Crane and Calpouzos 1969; Russell 1970). Besides the aphid-transmitted viruses, transmission of semipersistently whitefly-transmitted (*B. tabaci*) viruses like *Tomato leaf curl virus, Mungbean yellow mosaic India virus, Chilli leaf curl virus* and yellow mosaic of French bean were also effectively reduced by oil application (Butter and Rataul 1973; Nene 1973; Singh et al. 1979a).

Earlier studies indicated that circulative viruses like *Pea enation* and *Potato leaf roll* were not inhibited by oil application (Hein 1971; Peters and Lebbink 1973). But in subsequent years, good success was achieved against *Tomato yellow top virus* which is probably a strain of *Potato leaf roll virus* (Zitter and Everett 1979).

Several studies have also indicated that there is no influence of virus particle structure on the oil action in virus inhibition. Viruses with particle morphology of flexuous rods of 530–1250 nm, icosahedral particles of 25–30 nm and bacilliform particles were inhibited with oil sprays (Vanderveken 1977).

It was also established that the action of oil was not specific against any virus or vector. It inhibits the transmission of many viruses like *Cucumber mosaic virus - Aphis gossypii* (Loebenstein et al. 1964, 1966; Nitzany 1966; Zitter 1978; Proeseler et al. 1976), *Beet yellows virus - Aphis fabae* (Vanderveken and Ohn 1968), *Bean yellow mosaic virus - Acyrthosiphon onobrychis* (Vanderveken 1968), *Pea mosaic virus - Acyrthosiphon pisi* (Cousin and Grison 1969), and *Pumpkin mosaic virus - Aphis gossypii* (Singh 1981), which indicates that oil has a wide spectrum on many viruses and aphid vectors, which transmit the virus either in a non-persistent or semi-persistent manner.

Studies also indicated that the mechanism of virus transmission inhibition was not dependent on the nature of the treated foliage. Bradley (1963) impeded the transmission of PVY by using the vector *M. persicae* on test plants like tobacco, datura, potato, pepper and rape seed which were coated with oil before inoculation.

2.15.2 Types of Oils

The oils presently used are mineral oils, silicone oils and plant oils. Mineral oils are from a petroleum source which are highly refined, whereas silicone oils are synthetic ones composed of polymers containing silicon, carbon, hydrogen and oxygen. The oils of plant origin used for spraying are composed of triglycerides. Mineral oils with different trade names like paraffin oil, Bayol 50, Albolinum oil, Sunspray oil (7E and 11 E), Esso 2214, ESSO N 61, Krishi oil, Blancol oil, E WOS SGE 2967, Sehering 11 E, WZ-4, JMS Stylet-Oil®, Citol oil etc. are found to be effective against different viruses (Allen 1965; Bradley et al. 1966; Calpouzos 1966; Asjes 1975; Peters 1977; Vanderveken 1977; Singh 1981). Silicon oils like Bayer M10 and Bayer M100 are found to be inhibitory to *Beet mosaic virus* (Vanderveken 1977). Horticultural oils and insecticidal soap have been shown to interfere with both mechanical and insect transmission of viruses (Peters and Lebbink 1975; Zinnen and Vachris 1990) as well as controlling aphid, leaf miners, mealybugs, mites, scales, and whiteflies (Bethke and Parrella 1990; Larew 1988). Allen et al. (1993) reported that Wilt-Pruf, oil and Dow Corning 36 products reduced *Tomato spotted wilt virus* transmission by 73, 57, and 46 %; and feeding activity of western flower thrips vector by 40, 4, and 41 %, respectively. Only Dow Corning 36 significantly reduced (66 %) reproduction.

Lipids and milk spray were found to be effective against non-persistently transmitted aphid-borne viruses (Bradley et al. 1962; Hein 1964, 1965; Jaeger 1966; Bhargava and Khurana 1969). Vanderveken (1972) compared the efficacy of some mineral oils, lipids and silicone oils on the transmission of *Beet mosaic virus* and *Beet yellows virus*, with beet as test plant and *M. persicae* as vector. His experimental results indicated that inhibition by mineral oils seems to be higher than inhibition by lipids or silicone oils. Similar results were also obtained with other virus-host combinations (Bradley et al. 1962; Hein 1965; Vanderveken 1977).

Certain plant oils have also given effective control of certain plant viruses. For example in Germany, 3–4 % maize oil (Hein 1965) and 1–3 % groundnut oil (Jaeger 1966) when applied as 4–6 weekly applications have been used against *M. persicae* to manage *Lettuce mosaic virus spread.* Bradley et al. (1962) reduced PVY spread by the application of corn oil and olive oil. Joshi and Gupta (1974) tested six vegetable oils and only coconut oil at 0.5 and 1 % prevented *Soybean mosaic virus* spread in soybeans. Weekly spraying up to 5 weeks checked the natural spread of this virus. Even Fenigstein et al. (2001) have reported that out of eight vegetable oils evaluated against sweet potato whitefly *B. tabaci* on cotton, cotton oil spray was found to be best followed by corn oil in reducing whitefly activity. In India, Pun et al. (2005) while working with YVMV in okra have observed that spraying with neem oil and neem kernel extracts were effective in reducing the virus disease incidence. Neem oil was found to be effective in reducing the survival of the leaf hopper, *Nephotettix virescens* and its transmission of RTV (Mariappan and Saxena 1983). Also the incidence of *Rice ragged stunt virus* disease transmitted by the planthopper, *Nilaparvata lugens* was reported to be significantly reduced in rice fields when sprayed periodically with neem oil emulsion (Saxena et al. 1981). Coundriet et al. (1985) reported that neem seed oil extract repelled whitefly from alighting on sweetpotato. Thus anti-feedent and repellent action of insect vectors by rendering the treated plants unattractive or unpalatable, offers a novel approach for the management of virus diseases of different crop plants. Even other oils like groundnut oil, castor oil, clove oil, mustard oil and neem oil were tested against different viruses with various degrees of success (Bhargava and Khurana 1969; Nene 1972; Khatri and Sekhon 1973; Xiuhua et al. 1976; Singh and Varma 1977; Singh 1981).

2.15.3 Characteristics of Mineral Oils

Literature on the physico-chemical properties responsible for the activity of mineral oils is scarce. It is difficult to compare the characteristics of various oils, as nearly all researchers used different oils in their studies without providing much information on their properties like viscosity-gravity constant, the paraffine-pour point, the unsulfonated residue (USR), the distribution of carbon atoms, specific weight (Sp.W), the refractive index, molecular weight, boiling range etc. A mineral oil as an inhibitor of virus transmission, can be evaluated by the following four characteristics - (1) the classification of oil based on the viscosity gradient constant (VGC); (2) the viscosity; (3) the paraffin pour-point, and (4) the unsulfonated residue (USR). Some of the properties of oils are discussed herein.

(a) **Viscosity**

The virus transmission inhibitory activity of paraffinic and slightly naphthenic oils depends on their viscosity. High viscosity is a problem at the level of the spraying process and may prevent homogenous covering of the leaf surface. The effect of viscosity on drop retention varies with drop size (Furmidge 1968). The retention of large drops (500 μm dia.) increases with viscosity, although the drops tend to rest as spheres on the leaves. Changes of viscosity have less effect on the retention of smaller drops (100 μm dia.) which spread after impaction, covering a surface area dependent on their constant angles. The most active oils have viscosities between 12 and 30 C St (60–150 SVS). Wijs (1980) and Wijs et al. (1979) while working with mineral oils like Bayol 50, Sunoco 7N and Sunoco 7E against PVY on red pepper plants showed that the transmission inhibitory activity was rapidly lost with decreasing viscosities below 12 C St, but diminished only slowly with increasing viscosities about 30 C St. Wijs et al. (1979) recommended a viscosity of at least 12 C St at 37 °C for the inhibition of PVY transmission. Apparently the activity was not totally lost, that low viscosity oil (45 SUS = 65C St at 37 °C) was more effective than three oils with higher viscosity (80 V; 140 V and 310 V = 16, 30 and 67 C St at 37 °C) in the greenhouse against *Beet mosaic virus*. In the field however, the 45 V oil proved to be ineffective, whereas the other three oils reduced the virus incidence significantly. Walkey and Dance (1979) showed that low viscosity oil fractions (6.1 C St) are less effective than high viscosity oils (18.7 C St). Zitter and Simons (1980) tested oils of 60, 70 and 110 s SUS viscosities (viscosity expressed in Saybold Universal Seconds) and found that the 70 s oil was superior.

The oils sprayed on the plants do not immediately spread into a film, but cover the leaf surface with small droplets, which stretch out and join together between 24 and 36 h after treatment, thus imbibing the cuticle (Kulps 1969). The spreading of oil drops on leaves is not only a function of physical properties of the liquid, but is also governed by the nature of the leaf surface. The influence of each type of surface roughness on drop spread will depend on drop size. Boize et al. (1976) showed that the small oil drops and portions of larger drops tend to spread rapidly on dicotyledonary leaves following the reticulate pattern of epidermal grooves, whereas on monocotyledon leaf, spreading tended to follow the uninterrupted grooves between longitudinally aligned rectangular epidermal cells. Kulps (1969, 1971) found that leaves with wax coating are easily wetted by oil than glabrous or pubescent leaves. However, Vanderveken (1968) could not observe any difference in inhibition of *Bean yellow mosaic virus* in broad bean and French bean which have glabrous and pubrescent leaves, respectively.

(b) **Surface tension**

No inhibition of virus transmission can be obtained with highly volatile essential oils because they do not persist at the surface or inside the foliar tissues (Calpouzos 1966; Fraselle 1968). Surface tension of oils conditions the coating and impregnation of plant organs; coating of a leaf surface by oils is favoured by affinities between cuticular waxes and oils (Hassall 1969).

Mineral oils have an optimal surface tension and Kulps (1969) showed by means of micro autoradiography of the stylets that a mineral oil rapidly and easily covers the mouth parts of the aphids. The height of oil creeping up in the labial groove might depend on its surface tension. In the case of *Beet yellows virus* (a semi-persistently transmitted virus) the surface tension of lipids might be too high to allow access to the attachment sites of the virus on the vector mouth parts. Bradley (1963) reported that the virus attachment sites at the stylet tip might be reached by compounds with a high surface tension including lipids.

Aromatic compounds are of no importance in virus transmission inhibition properties of mineral oils. Besides this, aromatic oils are more phytotoxic than naphthenic or paraffinic oils. Therefore, mineral oils known to be recommended as adjuvants to insecticides and fungicides and as inhibitors of virus transmission by aphids must be paraffinic or naphthenic in nature with low unsulfonated residues (USR) of 95–100.

(c) **Oil concentration and size of the droplets**

The concentration and the quality of the oil used as spray are important to get maximum inhibition and it varies with the virus and also with host. The transmission of PYV was reduced by more than 60 %, using 0.03 mg of oil/10 cm^2 of leaf surface (Bradley 1963), whereas the amount required to impede 80 % transmission of *Beet yellows virus* was about 0.07 mg of oil/10 cm^2 (Dutrecq and Vanderveken 1969). They attribute this difference in results to the difference in persistence of the viruses used. At higher concentrations the sprays cause severe scorching and it was necessary to find the lowest concentration of oil that would suppress the virus without damaging the plants. For example, Loebenstein et al. (1964) found that the summer oil at 10, 5 and 2 % caused leaf scorching, even though it prevented infection with CMV. It also prevented infection significantly at 1 % concentration until 7 days after spraying, whereas the lower concentrations (0.5 %) prevented infection only slightly from the third day onwards. Infectivity of aphids was very much reduced by both concentrations, but only when they fed on freshly-sprayed plants immediately or after 1 day. Dubey and Nene (1974) and Singh (1981) also noticed less effectiveness of the oil sprays at lower concentrations. This less effectiveness may be due to the thinness of the film and its denaturation or evaporation or, it may be that when the leaf continues to grow, the film from the low concentration is ruptured, leaving leaf tissue exposed. With the higher concentration the film stretches without rupturing.

The relationship between insecticide drop size and velocity has been studied by Himel (1969) who found that the impaction of small drops (60 μm dia.) generally increases as their velocities increase the shape, size and surface characteristics of the target which influence deposition. Courshee (1960) showed that such drops tend to land preferentially on sharply curved objects of small diameter, such as leaf hairs, but to a lesser extent on leaves. The orientation of leaves on plants may also enhance or decrease deposition depending on whether the surfaces are horizontal or vertical.

Apparently phytotoxicity is related also to the size of the spray droplets. The phytotoxicity will increase when the spray droplets are larger. Zitter and Simons (1980) showed the effectiveness of spray pressure and nozzle orifice with Sunoco 7 E and Sunoco 11 E oils at 0.75 % concentration with PVY on pepper. They found that efficacy was generally improved as spray pressure was increased, with a pressure of 400 psi required for optimal effectiveness in most instances. They noticed the evidence that the oil droplets obtained by using recommended spraying technology, were very small measuring 0.2 μm in diameter, whereas the emulsion droplets had an average diameter of 40 μm. These small oil droplets do coalesce readily and probably have considerable mobility on the leaf surface, accumulating in the grooves around the borders of the epidermal cells. They also studied the interactions between the spray nozzle orifice and leaves of different plant species. They found that Teejet Tx-4 nozzles provided better protection on smooth-surface leaves such as on pepper. For use on pubescent leaves such as on tomato, cucumber and squash the Teejet Tx-5 nozzles were found to give good protection. Differences in effectiveness ranged from 10 to 20 % and were consistent, again indicating some subtle distributional effects.

From Israel, Loebenstein et al. (1966) tested the effect of low-volume and high-volume spray applications of Blancol oil on the incidence of CMV in cucumbers. Low volume applications were markedly more effective and reduced infection by 80 % or more. The plots sprayed with 5 or 10 % Blancol oil in a low volume spray had only 14 % CMV infection as compared to 80 % in the control plots. An average increase of 50 % in the yield of marketable cucumbers was obtained in the low volume-sprayed plots; from 84 kg/plot in the control to 126 and 120 kg in the

5 and 10 % sprayed plots, respectively. The average yield of the high-volume-sprayed plots was intermediate, i.e. 106 kg. Yields of the low-volume-sprayed plots were highest and infection rates were the lowest. A good coverage of both sides of the leaves was obtained with the low-volume application. No phytotoxic effects were observed with high volume or with low volume applications of high oil concentrations. Wenzl and Foschum (1973) recorded 58–78 % inhibition of PVY spread, according to the different application modalities of oil emulsions and maximum inhibition was noticed when the whole aerial parts of the plants were perfectly sprayed.

Addition of 1 % detergent to the oil will help for the uniform spread and distribution of oil. Asjes (1974) obtained best control of brown ring formation virus in lily by the addition of 0.1 % Agral L.N. to the 2.5 % albolinium oil. Zitter and Simons (1980) observed that adding small amount of solvents to oil with a viscosity of 70 s. SUS can significantly lower the viscosity and effectiveness of the oil.

Simons and Zitter (1980) tested a number of commercially available emulsifiers, using various concentrations and oils of different viscosities. Their results indicated that significant effects related to both the kind and quantity of emulsifier used and that optimal concentrations of emulsifier exist. Increasing the amount of emulsifier improves performance, but once the optimal concentration was reached there was at plateau in performance; any further increase in the emulsifier will reduce the efficacy of the formulation. The reasons were not known, but it seems likely that distribution of the oil on the leaf surface was affected somehow.

(d) **Longevity of the oil**

The longevity of the oil on the leaf varies with the type of oil, concentration and also on the environmental factors. Bradley (1963) observed that oil-coated mature tobacco leaf was protected against virus transmission by aphids for more than 60 days. Experiments by Kulps (1968, 1969) showed that young lettuce leaves once coated with oil remainded protected even after they had grown to 11-times the size of the leaf when treatment started. The intervals between sprays differ with the crop and will be determined on the longevity of the oil film, and how fast the plant grows. For example, peppers are generally sprayed once a week, while the faster growing cucumber and squash plantings are sprayed twice weekly. Simons et al. (1977) reported that the oil persists on sprayed leaves for at least 10–14 days. Since mineral oils are protectant sprays, any new leaf area that is not protected with a regular spray schedule is vulnerable for new infections to occur.

Singh (1981) found that Sunspray oil at 2 % concentration showed 100 % inhibition of PMV transmission up to 24 h. The effectiveness decreased beyond 24 h and as much as 60 % infection was noticed after 168 h, while the controls showed 100 % transmission. Similar results have also been reported earlier by Dutrecq and Vanderveken (1969) and Khatri and Sekhon (1973). The loss in the effectiveness of the oils may be due to the rupture of the oil film during leaf surface expansion, thus expositing small portions of the leaf tissues for the aphids to feed.

Another reason may be due to the oil itself, thereby making the leaves susceptible to virus transmission. As the inhibitory activity of the oils depend also on the growing rhythm of the plants, the application dates and number of sprays should be manipulated based on type of the oil and the host plant.

(e) **Translocation of the oil inside the host**

Kulps (1971) and Kulps and Hein (1972) noted that oils not only penetrated the cuticle, but also the cell walls of the underlying tissues. Stomata are considered to be one of the important routes of oil penetration into the leaf (Gudin et al. 1976). Stomatal penetration of small quantities of oil over a leaf depends largely whether the stomata are sunken or raised. There is no definite conclusion at present whether the oil is a contact type or systemic type, as some reports are published for each category. When once the leaves are sprayed, the oil moves in the intercellular system from the upper surface to the lower one and vice-versa. There are reports which indicate that oils move systemically and prevents the virus infection by aphids, when the lower surface of the leaves are coated with oil and viruliferous aphids were fed on the upper epidermis. However, the percentage was very less. When both the surfaces of the leaf were sprayed, the aphids could not transmit the virus. Singh (1981) also reported that *Pumpkin mosaic virus* (PMV) transmission was completely inhibited when both the leaf surfaces were sprayed. When the oils were applied on only one side of the leaf and inoculation of the virus on the opposite side, there was little transmission of PMV that varied from 20–40 %. Spraying the upper surface of the leaf was comparatively more effective in inhibiting the transmission of the virus than the lower surface. Earlier Kulps (1972) also demonstrated the systemic action of mineral oil and noticed the translocation of oil into the roots and younger parts of the plant. He observed the inhibition of virus infection in lettuce leaves even after an 11-fold enlargement of the leaf surface, though the plants were treated when they were young. On the other hand, Crane and Calpouzos (1967) could not observe systemic translocation of oil inside the host plants. In their experiments oil was applied only to one leaf which 24 h later was probed by viruliferous aphids. They noticed the symptoms on the other unsprayed leaves, whereas the oil sprayed (probed leaf) leaf remained symptomless, suggesting that the oil film on the leaf surface did not inhibit the viral movement and induces symptoms in the remote unsprayed leaves. These studies indicate that for the effective inhibition of virus spread, the whole plants should be sprayed at shorter intervals, perhaps right from the nursery stage.

(f) **Oil sprays and vector behavior**

Hein (1971, 1972) and Vanderveken (1973) did not observe any change in the feeding behavior of *M. persicae* when released on oil sprayed plants. Peters and Lebbink (1973) also observed similar results with *Acyrtosiphon pisum.* Simons et al. (1977) found that *M. persicae* required a greater preprobe time on oil treated plants, although the delay was less than a minute. Russell (1970) reported that direct spraying of aphids with oil decreased the larviposition on the test plants which has diminished the population of potential vectors present on them.

(g) **Interference with the transmission process**

Mineral oils are hydrophobic liquids that interfere with virus acquisition and retention by aphids. Mineral oils of an appropriate viscosity and unsulfonated residues were found to be effective in reducing the efficiency of transmission by vectors. This measure is still popular for protection against nonpersistent viruses, particularly in nurseries. The mode of action seems to be by interference with virus binding by probing aphids. The leaf surface must be fully covered: full coverage demands frequent applications (up to twice a week) of large volume at high pressure. Combination of oil with pyrethroids (insecticides that have insect-repelling qualities) was tested successfully in Israel and in England (Raccah 1986).

2.15.4 Mechanism of Virus Inhibition by Oil Sprays

There is no clear understanding of the mechanism involved in the inhibition of virus transmission due to oil sprays. The possible assumptions and explanations are presented herein.

2.15.4.1 Oils and Role of Stylets

Bradley (1963) proposed 'bipartite hypothesis' considering that oil may remove the virus particles from the stylets, which prevents the transfer of virus from one plant to another. It is also possible that virus particles adhere more firmly on the stylets, thus preventing their introduction into a test plant. Autoradiographic studies indicated that stylets are easily over sprayed by oil along their whole length after probing other oil-sprayed leaves. The oil is carried at the tip of the labium and a brief contact of even 1 second between labium and treated leaves cause an 80 % reduction in transmission. Simons et al. (1977) treated the labium with Huber's clay which absorbs oil and showed that removal of oil restored the transmission to the level found with similar untreated control aphids. These experiments indicate that the tips of the labium carries oil which interferes with virus transmission.

Peters and Lebbink (1975) were of the opinion that the action of oil may even occur during the early stage of infection process. Vanderveken and Semal (1966) earlier demonstrated that oil strongly inhibited acquisition of *Beet yellows virus* when aphids (*M. persicae*) came in contact with oil, either before or after they probed the virus infected source plant.

A 'mechanical surface adherence' hypothesis was suggested by Vander Want (1954) in which differences in the surface structure of the stylet accounted for differential adsorption of virus to, and elution from the stylets. When applying this concept to the inhibition of aphid transmission of stylet-borne viruses by oil would lead to the following proposition. Oil modifies the surface structure (or charge) of the stylets thus impeding adsorption or elution of virus particles. It may be that mineral oils have an optimal surface tension and inhibit aphid transmission by altering the virus-vector relationship.

2.15.4.2 Role of Electrical Properties of Oil

Kassanis and Govier (1971) felt that electric properties of oil have to be taken into consideration, especially their dielectric nature. According to these authors, *Potato virus C* (PVC) and *Potato aucuba mosaic virus* (PAMV) are not adsorbed to the stylets, but PVY can be adsorbed. When PVY is adsorbed, it modifies the surface structure, thus allowing PAMV or PVC to be adsorbed either on the modified stylet surface or on to the PVY particles.

The effectiveness of oils might result from their electric insulating properties, which would impede the exchanges of charges between virus particles, aphid mouth parts and plant cells. These exchanges of charges might account for "the components that aid or inhibit adsorption or infectivity", "the specific attachment substance" or "the prosthetic groups for adsorption and the reciprocal sites" to which Watson and Plumb (1972) had made reference.

2.15.4.3 Interference of Oil in the Infection Process

There are indications that oil affects a stage in the infection process after uncoating of the virus particle. For example, Peters and Lebbink (1975) obtained reduced number of local lesions on *N. glutinosa* plants when the intact virus or isolated virus nucleic acid was treated with oil. Complete inhibition of local lesion development was obtained with repeated oil treatments at 15 min interval after inoculation. They concluded that oil seems to interfere in the development of infection at an early stage, but later than the uncoating. The possibility of oil inactivating the virus is excluded as the virus-oil emulsion could be separated when layered on sucrose gradient and ultracentrifuged and the infectivity at higher percentage could be restored. It is also possible that when the stylet carrying the virus comes in contact with oil, the virus will be sheathed by oil, so there is no chance of initiation of virus replication. Earlier results of Bradley (1963) did not agree with these results as the application of oil after aphid inoculation did not result in a decrease in a number of plants infected.

There are also reports which support that oil application inhibit virus replication or some times because of very slow virus replication, the movement of the virus in the treated plant may slow down and also in some cases, the symptoms would not appear. For example, Crane and Calpouzos (1967, 1969) and Crane (1968) observed the suppression of symptoms in oil treated mature sugar beet plants infected with *Beet yellows virus*. Symptoms of virus yellows are suppressed for at least 3 weeks by a single spray of paraffin oil on sugar beet foliage either before or after the viruliferous aphids probed the plants. The oil film did not inhibit aphid transmission and furthermore, the virus passed readily through the oil treated leaves and induced symptoms in remote leaves that were unsprayed. Oil is also effective even if it is sprayed 24 h after viruliferous aphids have feed and significant suppression of symptoms were noticed. Even Jadot and Vanderveken (1973) found that spraying oil at short intervals (7 times in 50 days) to the beet plants delayed the symptoms of *Beet yellows virus*, a semi-persistent virus, and that symptoms appeared once the sprays were discontinued. On the other hand, Vanderveken and Dutrecq (1970) did not agree with the above results, as they did not observe the hampering of transmission of *Beet mosaic virus* or *Beet yellows virus*, even when the test plants were coated immediately after inoculation.

It is also possible that in certain oil-sprayed plants, the aphid's stylet could easily break through the oil films, losing some virus in the process. Plants infected with a lower amount of virus naturally take longer time to show observable symptoms. This explains to some extent the reason for the increase in the incubation period of the virus in the treated plants when compared with the control. These possible explanations, however, need experimental support.

2.15.4.4 Oil Plus Virus Complex Formation

There are number of examples where oils were found to be inhibitory *in vitro* studies, indicating that oils might have exerted some inhibitory effect directly on the virus. For example, Loebenstein et al. (1964) reported that the number of lesions were only 3 for 12 half leaves inoculated with TMV plus oil mixture, compared to 289 from the control inoculum. Similar type of virus inhibition was also obtained in different virus host-oil combinations (Loebenstein et al. 1964; Dubey and Nene 1974; Peters and Lebbink 1975; Singh 1981).

Zschiegner et al. (1974) in his electronmicroscopical studies on the effect of oil on *Brome mosaic virus*, he found that in the presence of mineral and sunflower oils, *Brome mosaic virus* remained intact. The reduction of the electrophoretic mobility by oils especially on sunflower, indicates the possible formation of virus-oil complex with rather low stability. Whereas Hein (1972) reported that the local lesion production of CMV on *Chenopodium quinoa* was not decreased by the addition of mineral oils to the plant sap. This means that oil affected a stage in the infection process after uncoating of the virus particles. Repeated oil treatments at 15 min intervals after inoculation resulted in a nearly complete inhibition of local lesion development.

2.15.4.5 Factors Responsible for the Breakdown of Oil Efficacy

(a) **Oil-number of aphids**

Some evidence exists which indicate that with increase in aphid population on the oil treated plant, the virus inhibition/protection was reduced. Vanderveken et al. (1966) and Vanderveken and Semal (1966), Vanderveken and Ohn (1968) reported that with 4–5 aphids on each plant, 3 of 32 beets became infected in oil coated series and 27 of 32 in non-sprayed control series. By increasing the aphid number to 10 and 15 on each plant, the results were 29 of 52 and 40 of 52, respectively. In these experiments the aphids were transferred to plants 2–3 h after spraying. It was observed that the oil spray did not spread into film immediately, and remained as distinct oil drop lets. The increase in virus transmission due to an increase in aphid number may be due to probing of aphids in between the oil droplets. From India, Singh (1981) also observed similar results while working with *Pumpkin mosaic virus* on pumpkin with its vector, *A. gossypii.* This failure of the oil film can be restored by increasing the oil concentration (Vanderveken and Ohn 1968).

(b) **Oil concentration**

Oil sprayed at lower concentrations will lose its efficacy very soon, as the oil film breaks apart as the leaf expands and the vectors feed in between the exposed areas. This can be avoided by using higher concentration which is not phytotoxic, so that the oil film also expands when the leaf expands. Vanderveken and Ohn (1968) reported that the virus inhibitory effect was restored by increasing the oil concentration and was effective even when large numbers of aphids were used per plant. When the crop is within its rapid growth period, the spray intervals should be shortened.

(c) **Incompatibility effect**

When oil is used in combination with other pesticides as part of the normal production practices, serious problems of incompatibility can arise. Most incompatibility will be with involvement of fungicides including sulfur, chlorothalonil (Bravo) and dichlone (Phygon). Aerially-applied chemicals can be more problematic than when dilute materials are applied with ground sprayers, because aerial applications use much lower volumes of water resulting in the spattering of rather large deposits of chemicals on foliage. Consequently, the efficacies of oil diminishes when the oil sprays are followed by other fungicidal or insecticidal sprays. The oil sprays should be done separately, and no pesticides applied for at least 1 or 2 days after the oil application.

(d) **Environmental factors**

If the air temperatures are above 30 °C during oil sprays applications, plant injury will be reduced, especially when sprays are followed by rains.

2.15.5 Application of Oil Under Field Conditions

Oil sprays under field conditions have shown varying degrees of promise as a practical method of controlling the spread of virus diseases. Further spread of the diseases inside the field takes place through the secondary spread by vectors and oil sprays are effective in limiting the plant-to-plant spread. The success of virus control in different crops is discussed herein.

(a) **Capsicum and chillies**

Among the most important viruses infecting capsicum and chillies are PVY and CMV, which are aphid-transmitted. Experiments conducted by Loebenstein et al. (1970) at Israel, indicated that oil (Blancol) spray increased the average height of pepper plants (51.5 cm) as compared to control plants (26.5 cm) and virus infection was delayed for at least 2 months. Significant increases in yields were noticed with oil spray and the yield in green pepper were 2800, 6000 and 5300 kg/ 1000 m^2 in plots sprayed with 0, 2.5 and 5.0 % oil, respectively. Grade "A" yields were 3–3.5 times higher as a result of oil sprayings. Earlier, Nitzany (1966) also reported that spray of 1.25 and 2.85 % concentrations of light oil emulsion at intervals of 5 or 6 days reduced the field spread of PVY and CMV and high pepper yields. Lange and Hammi (1977) recommended weekly sprays of summer oil at 1 to 2 % concentration to pepper in the seed beds and also after transplantation in the field, to reduce the spread of CMV and PVY. From Argentina, Ramallo et al. (1975) recorded the low incidence of PVY in capsicum with YPF No. 1 and Hoescht oil emulsion sprays at 0.5–2 % concentration at 7–15 day interval. Sprays with Curafrutal C at 3 % at 25-day intervals from nursery stage to the fruiting stage were effective against mosaic diseases - (PVY, CMV) of pepper (Gracia and Boninsegna 1977).

Simons and Zitter (1980) reported that the spread of *Tobacco etch virus* can be reduced to 8 % after 10 weeks, whereas in the unsprayed control field plants became 100 % infected in about 6 weeks. In Florida, the use of oil on peppers has become so effective in suppressing inoculum potential which made feasible the double cropping from the same plants. The first pickings were made in November to December, the plants were cut back to a height of about 15 cm in late December to early January and a second crop of fruit was produced on the regrowth for harvest in March–April. During this period, weekly sprays of JMS Stylet-Oil®, was found to be very effective. This results in a roughly 50 % increase in yield, with the second crop, costing only about 15 % as much to grow as the first crop.

Zitter and Ozaki (1978) controlled the *Pepper mottle virus* in three bell pepper varieties (Early Calwonder, Florida VR-2 and Delray Bell) by the application of JMS Stylet-Oil® at 0.75 % at weekly interval. The control pepper plots averaged 73 % infection compared with 15 % in the oil plots 10 weeks after virus spread began. Significantly more infected pods were harvested from control plots and both marketable count and weight were greater in the treated plots.

In India, Khatri and Sekhon (1973) obtained complete inhibition of *Chilli mosaic virus* which spreads by *A. gossypii* by spraying 2 % emulsion of light paraffin oil, but its efficacy lasted for 48 h only.

The field trails by Lowery et al. (1997) have indicated that Neem seed oil inhibits the transmission of PVY in pepper.

(b) **Potato**

In almost all potato growing countries, PVY is the most prevalent virus and good success has been achieved in reducing virus spread with the application of various oils. Bradley et al. (1966) demonstrated that paraffin-base oil emulsion sprays could substantially reduce virus spread. Wenzl (1970), Wenzl and Foschum (1973) recommended that nine applications of emulsified mineral oil at 3 % concentration be applied at weekly interval, and this reduced the spread of PVY by 54–84 % in 1967, and 79 % in 1968, while spraying with crude oil or refined sunflower oils reduced it by only 37 and 52 %. Increasing the spray intervals from 1 week to 2 and 3 weeks reduced efficiency from 79 to 55 % and 45 %, respectively. Zschiegner et al. (1971) reduced the spread of PVY by three sprays of 2 % mineral oil emulsion. Mineral oils YPF No. 1 and Hoescht oil emulsion, each at 0.5–2 %, effectively reduced the spread of PVY (Ramallo et al. 1975). Similar encouraging results were obtained from various countries (Kostiw and Iskrzycka 1976; Singh and Nagaich 1976; Shands 1977; Kvist and Morner 1977; Simons and Zitter 1980; Olubayo et al. 2010).

Allen (1965) reported that six weekly sprays of No. 3 paraffin oil at the rate of 1 gal of oil/acre/100 gal of spray for the control of *Potato virus A* (PVA). In the first year, the oil sprays had decreased the spread of PVA by 50 % and in the next year oil sprayed potatoes had 30 % less new infections than the control.

Even though the spread of viruses infecting potato were reduced, a large number of undesirable side effects have been reported. A reduction in tuber weight has been found by several authors, whereas an increase was expected. From Wageningen, Schepers et al. (1977) found a decrease in tuber weight of 6 %. They found a reduction of virus incidence by 8.5, 10 and 22 % in plots sprayed with oil at intervals of two, one or half a week. Quemener (1976), Peters (1977), Schepers and Bus (1979) and Simons and Zitter (1980) also reported a reduction in weight of the tubers harvested. Toxic effect in the form of burning of the hulm after oil-spraying has also been reported (Bradley et al. 1966; Quemener 1976; Schepers et al. 1977). There are also many reports which support the good control of virus spread without any deleterious side effects. It appears that tuber reduction depends on the variety used for oil treatment (Kostiw and Iskrzycka 1976). It was observed that sprays of repellent neem oil at every 7–10 days after emergence until the end of the peak of aphid flight (Handizi and Legorburu 2002) helped to reduce the incidence of aphid-transmitted viruses in potato, but is labor intensive. Intensive research is required to decide the proper oil concentration, time of application and number of sprays, so that it can be used effectively in seed potato production.

(c) **Tomato**

Based on the field trails conducted at Florida, Zitter and Everett (1979) have established that stylet oils were effective against tomato yellows virus, which is aphid-transmitted in a persistent manner. They recommended the application of JMS Stylet-Oil® at weekly spray intervals at the rate equivalent to 3 qt/100 gal water by using 7 × 5-nozzles at 400 psi pressure, had greatly reduced the spread of this disease. A total of 11 sprays were applied at weekly interval and the overall spread of the virus in the oil treated plots was 9 % as compared to 21 % in the control plots. Yields/plant and fruit quality differed depending on the time of symptom expression. Yields on oil sprayed plants varied from 3.6 to 9.7 lb/plant compared with 3.2 to 5.7 lb/plant for the controls.

(d) **Okra (Ladies Finger/Bhindi/Bhendi)**

The field experiments carried out at Indian Institute of Horticultural Research (I.I.H.R) Bangalore (India) by Sastry and Singh (1973b) revealed that 1 % mineral oil sprays applied at 10-days-internals reduced the incidence of whitefly transmitted *Okra yellow vein mosaic virus* (OYVMV). Even Nene (1973) reported that spraying with an emulsifiable mineral oil called "orchard oil" at 2 % gave 100 % whitefly mortality within 30 min and it could immobilize the individual whitefly adults immediately after they settle on leaf surfaces.

(e) **Squash**

Zitter and Ozaki (1978) found that by spraying JMS Stylet-Oil® at 0.75 % has reduced the incidence of *Watermelon mosaic virus-1* (strain of PRSV) in zucchini squash. They noted a fivefold reduction in virus spread in the winter trial and nearly a fourfold reduction in spread in the spring trial. The control averaged 75 % infection compared with 15 % in oil sprayed plots during a 9 week period.

(f) **Cucumber**

From Israel, Loebenstein et al. (1964, 1966) recommended the sprays of 1 % summer oil at weekly intervals for preventing CMV spread. High-volume applications (80–120 l/1000 m^2) of a 1 % emulsion, wetting both leaf surfaces, reduced virus incidence by 50–70 % compared to controls. Low-volume air blast sprayings (10–20 l/1000 m^2) with 5 or 10 % emulsions were even more effective, reducing the infection by 80–90 % compared to unsprayed controls. At the last observations when the average rate of infection in the control plots reached 80 %, plots sprayed with 5 or 10 % oil in a low volume had only 14 % infected plants. An average increase of 50 % in the yield of marketable cucumbers was observed in the low-volume sprayed plots: from 84 kg/plot in the control to 126 and 120 kg in the 5 and 10 % oil sprayed plots, respectively. The average yields of the high-volume-sprayed plots was intermediate, i.e. 106 kg. No phytotoxic effects were observed with high volume or with low volume applications of high oil concentrations.

In California, Toscano et al. (1979) reduced the incidence of virus disease of cucurbits and their vector population to 23 % by the application of a mineral oil (citrus spray oil) at the rate of 2–4 gal/acre.

(g) **Muskmelon**

In southern California, Toba et al. (1977) sprayed a misible oil (citrus soluble oil) at 5 % concentration for protecting muskmelons against WMV-2. The first sprays were applied 4 weeks after planting, and subsequent applications were made weekly. The oil-sprayed plants appeared as green and vigorous as the check plants, but mild mottle symptoms appeared 2 weeks later in oil sprayed plants. The oil sprayed plants produced about the same number of fruits, and more of them (67 %) were marketable and had 1 % higher soluble solids.

(h) **Watermelon**

WMV-1 is a limiting factor in watermelon production in most countries. Kitaba (1979) reported that by spraying summer oil emulsion at 0.25 % concentration he encountered low disease incidence due to reduced aphid (*A. gossypii*) population. The treated plots yielded more, in comparison with the untreated plots. Witcher and Smith (1980) observed a low incidence of WMV-1 in JMS Stylet-Oil® sprayed plots. Significantly lower number of fruits with WMV symptoms was observed in sprayed plots when compared to the untreated controls. Similarly, Chalfant et al. (1977) also obtained an effective control of the disease by spraying mineral oil. A significant control of WMV-1 was achieved by the twice weekly application of JMS Stylet-Oil® at 0.75 % and the results indicated that the virus incidence was only 4 % when compared to 33 % in control. Similarly, the fruit weight in lbs was 288 in stylet oil treatment as compared to 105 lbs in the control (Zitter and Ozaki 1978).

(i) **Sugar beet**

Vanderveken and Vilain (1967) and Vanderveken and Semal (1966) reported that the transmission of *Beet mosaic virus* (BtMV) and *Beet yellows virus* (BYV) which are non-persistently and semi-persistently transmitted, respectively, were greatly inhibited with the applications of mineral oils. Crane and Calpouzos (1967, 1969) noticed that treating test plants with oil either before or after inoculation with BYV delayed the onset of symptoms. Proeseler et al. (1976) reported that the addition of mineral oil 'citorol' to the systemic insecticide exhibited greater effectiveness against BtMV. In field experiments insecticides reduced the number of infected plants by 53 %. The addition of citorol oil increased effectiveness by 25 %. The oil could be applied twice without causing leaf damage. In U.S., diesel oil sprays were used as early as 1955 to reduce *Circulifer tenellus*, the vector of *Beet curly top virus*, to reduce the disease incidence. This oil keeps the insect vectors away from the sugar beet.

(j) Soybean

From India, Joshi and Gupta (1974) tested seven different oils viz. castor, clove, coconut, groundnut, light paraffin, mustard and olive for preventing the spread of *Soybean mosaic virus* by *A. craccivora.* Except at lower concentrations (0.5 and 1 %) of coconut oil, all the other oils caused phytotoxic although they partly prevented virus spread. Coconut oil sprayed weekly at the 1 % concentration, for 5 weeks prevented virus spread to a great extent. Even Lu (1970) reduced the spread of *Soybean mosaic virus* to 27 % over unsprayed controls by spraying vegetable oil emulsions at 5-day intervals.

(k) Broad bean

Zschiegner et al. (1971) recommended the spraying of mineral oil 'W6735' for successful control of pea mosaic virus spread, with the oil spray resulting in 69 % protection.

(l) Celery

The spread of *Celery mosaic virus* (CeMV) can be successfully curtailed by spraying mineral oil on the seed beds or during the major aphid flights. Esparza-Duque and Verhoyen (1975a, b) reported that CeMV was reduced by oil sprays and that the combination of oil sprays with repellents i.e. aluminized craft paper, had a better effect than the oil treatment alone. Although there was no significant difference in the weight of the plant, the quality of plants from the oil-treated plots was improved since less plants had virus symptoms when treated. From Belgium, Matthiew and Verhoyen (1979) recommended the spraying of 20 % Albolineum oil at 30 l/ha. The tests were conducted with three treatments applied 4–5 day interval during a period of high risk of contamination. The infection rate was reduced from 63.85 (control) to 6.85, 13.49 and 15.98 %, respectively. The very low risk of phytotoxicity and of decrease in root development was largely compensated by the total yield gain. The first treatment proved best in controlling the disease and increased leaf and root yields.

(m) Strawberry

In Corvallis (Oregon), Converse (1970) tested the effect of 2 % summer oil on the spread of strawberry viruses. In plots where oil sprays were applied the levels of non-persistently transmitted mottle virus and persistently transmitted crinkle virus were reduced by 24 and 23 %, respectively compared with the control treatment (78 %). Spraying with oil weekly or biweekly significantly decreased the production and length of strawberry runners, compared with unsprayed plots. None of the oil treatments significantly influenced the average number of daughter plants per mother plant or the fruit yield, compared with unsprayed plots. There were significantly fewer weak mother plants in plots sprayed monthly with oil than in unsprayed plots.

2.15.6 Efficacy of Oils in Combination with Insecticides and Cultural Practices

Most of insecticides, either organophosphorus or chlorinated hydrocarbons, are ineffective against stylet-borne aphid-transmitted viruses, as they do not act quickly enough to prevent virus transmission. It is well established that oil sprays are quite effective against stylet-borne virus provided they are sprayed at shorter intervals, but when applied weekly they have proved to be too expensive. However, the use of oil sprays in combination with insecticide have proven to be more effective than either treatment alone, and this applies for viruses transmitted either by aphids or whiteflies. When the insecticides are applied in combination with oil, one can expect an increase of effectiveness of the insecticide, as it penetrates well into the plant. The use of an insecticide-oil combination has been suggested by some workers on theoretical grounds (Hein 1965; Vanderveken 1968; Vanderveken et al. 1968; Cousin and Grison 1969), and a number of field trials have supported the effectiveness of this combination in crops like potato, sugar beet, pepper, broad bean, etc.,. For example, Zschiegner et al. (1971) reported that the spread of *Pea mosaic virus* (PMV) in *Vicia faba* was effectively controlled by mineral oil and dimethoate and proved to be superior to a treatment with mineral oil alone.

Zschiegner et al. (1971) also obtained encouraging results with a mineral oil-insecticide spray on virus spread in seed potato. The percentage of virus-infected plants were 9.4, 4.4, 4.9 and 2.4 in untreated control, mineral oil (2 % W 6735), dimethoate (900 ml/ha Bl 58 EC) and 2 % oil plus 600 ml/ha dimethoate, respectively. Even large scale commercial trials in East Germany, strongly limited virus spread by the combination of mineral oil with dimethoate (Kramer et al. 1978). Proeseler et al. (1976) also successfully reduced the virus diseases of beet root with combination sprays of citorol oil and a systemic insecticide. The insecticides reduced the number of virus-infected plants by 53 %. The addition of citorol oil increased effectiveness by 25 %. Nawrocka et al. (1975) also observed decreased CMV incidence in lettuce with the application of accephate + oils at 0.5 + 1 gal lb a.i./ha. Significantly higher production of up to 64 % was noticed in the treated plots as compared to control that yielded 52.9 %. Studies carried out in India, by Kalleshwaraswamy et al. (2009) revealed that mineral oil (0.1 %) with imidacloprid and deltamethrin applications as alternative fortnightly sprays were effective in reducing PRSV incidence in papaya.

2.16 Biological Control of Plant Virus Vectors

Most of the pest problems in agriculture are man-made as a result of abuse of pesticides. These agrochemicals often reduce the population of beneficial organisms which normally feed on insects pests. The elimination of beneficial biocontrol organisms like parasitoids, predators or entomopathogens favour the increase of insect-pest populations. There are three basic types of biological pest control strategies: Importation, Augmentation and Conservation. Importation involves the introduction of the pest's natural enemies to a new locale when they do not occur naturally. Augmentation involves the supplemental release of natural enemies boosting the naturally occurring populations. The third strategy is conservation of existing natural enemies in an environment.

Although some workers have attempted to control plant virus vectors by means of predators, parasites or pathogens, such measures have met with little success. Harpaz (1982) have made attempts at Chile to control the rapid spread of aphid vectors of *Barley yellow dwarf virus*, by using natural predators such as the insect *Aphidius ervi*. It was thought that the low incidence of *Barley yellow dwarf virus* in Israel was due to the presence of natural predators of the aphid vectors and that the introduction of such predators to Chile was successful (Van Den Bosch 1976). Dader et al. (2012) have studied the effect of aphid parasitoid, *Aphidius colemani* in managing CMV (non-persistent virus) and *Cucurbit aphid-borne yellows virus* (*Cucumber aphid-borne yellows virus*) (CABYV) (persistent virus) in cucumber and the *Aphis gossypii* is the vector for both virus. Their studies indicated that the parasitoid, *A. colemani* has significantly limited the spread and incidence of persistent virus CABYV at long term (Dader et al. 2012).

In Cuba, Ginarte (2010) has evaluated the parasitoid, *Paranagrus perforator* (*Hymenoptera*) and the predator, *Tytthus parviceps* (*Hymenoptera*) against the planthopper *Tagosodes orizicolus,* the vector of *Rice Hoja blanca virus*. Biological control of *Bemicia tabaci* biotype Q in Spanish greenhouse pepper production was achieved by using the parasitoid, *Eretmoceus mundus* (Stanley et al. 2005). Natarajan (1990) has proposed the biological control of whitefly, *B. tabaci* by using the parasites like *Encarsia formosa*, *Encarsia lutea* and *Eretmocerus mundus* in Mediterranean regions. Even certain microbes have been used as biocontrol agents. In India Colvin and Muniyappa (1999) reported that entomopathogens *Beauveria bassiana* and *Verticillium lecanii* proved to be effective in controlling the whitefly vector *B. tabaci*; the percentage of mortality of whitefly adults treated with *B. bassiana* and *V. lecanii* were 20 and 5.83 %, respectively at 24 h of evaluation; after 72 h the respective mortality percentage were 100 and 88.3. The control recorded a mortality of 10.8 %. The same entomopathogens *B. bassiana* and *V. lecanii* were proven to be effective in the management of onion *Thrips tabaci*. Application of *B. bassiana* at 1×10^9 spores/ml twice was found to be the best treatment with lowest mean of *thrip* population (9.53/plant) with highest per cent reduction (64.15) followed by *V. lecanii* at 1×10^9/ml at 20 and 30 days after planting of onion crop (Sudhir Kumar et al. 2012).

During 2007 Chitturi et al. have noticed interesting observation that the availability of pine pollen has been found to positively influence the host selection and fecundity of thrips species (Thysanoptera: Thripidae). Studies suggest that pine pollen which is deposited on tomato and peanut plants each spring could increase the settling and reproduction potential of thrips vectors on host plants of *Tomato Spotted wilt virus* potentially increasing its role as a vector of this virus. Effect of pine pollen (*Pinus elliottii* Engelm.) on the settling behavior and oviposition of

Frankliniella occidentalis (western flower thrips) and *Frankliniella fusca* (tobacco thrips) was established. Thrips settling behavior was increased by sevenfold with *F. occidentalis* and onefold with *F. fusca* with the addition of pine pollen over both tomato and peanut. Oviposition was increased by threefold with *F. occidentalis* and onefold with *F. fusca* with pollen (Chitturi et al. 2007). Since the preliminary results are quite interesting in the spread of TSWV, requires further intensive investigations to confirm the role of pollen of pollen on thrips vectors.

A large number of field trials with different biological agents have to be tried against plant virus vectors to get definite effectiveness of the biological agents in reducing the plant virus incidence under field conditions (Napometh 1988).

Conclusions

The examples cited above clearly indicate that oils of various types can be used against a number of viruses and their vectors in different crop plants. The efficacy of oil can be enhanced by the judicious application of oil at the correct concentration and the stage of the crop to be sprayed. For example, for *Cucurbitaceae* and *Leguminosae* crops when raised by direct seedling, the spray schedule should be started from the primary leaf stage; if the crops are raised through nurseries the spray schedule should be from nursery stage. At present there are more than 231 viruses which are seed-transmitted and in some crops this remains to be the primary mode of introducting the virus disease into the field. Generally speaking, when oil sprays are applied to the crop, they will help in limiting the further secondary spread of the disease, even though some infected plants occur in the field.

Some of the oil applications are proved to be ovicidal also in many insect vectors (Nene 1972; Zwick and Westigard 1978). Nene (1972) noticed 100 % ovicidal action on *B. tubaci* with 2 % orchard oil. Even Zwick and Westigard (1978) reported that the application of petroleum oils to pear trees, prior to oviposition by overwintering pear psylla (*Psylla pyricola*) reduced oviposition by 98 % for upto 5 weeks. Studies on the ovicidal action of some of the mineral oils which are widely used for virus control are worth pursuing.

The use of mineral oil sprays has several advantages. They are relatively cheap, have a good spreading capacity, are easy to mix and are reasonably safe to animal life. An important consideration is the fact that insects have not yet developed any resistance to oils, which is the case of certain insecticide classes. It is also observed that the oil sprays boost the growth of the plant. For example, Loebenstein et al. (1966) reported that growth of the cucumber plants 5 weeks after emergence. The average width of plants in the sprayed block reached 51.5 cm compared to 26.5 cm in the control rows.

The use of oil sprays are attractive in several ways: (1) they are non-polluting and fit well into an integrated pest management scheme; (2) they are safe and economical to use; (3) they are broad spectrum and should be tried wherever aphid-borne virus diseases are an important production problems; and (4) they appear to have no detrimental effect on treated plants when used at proper concentrations.

The efficacy of the oil sprays in achieving maximum success can be increased by (1) choosing oil of proper viscosity (highly refined paraffinic petroleum oil works best); (2) adding the correct emulsifier to ensure that oils will collect in the areas where aphids are known to probe; and (3) using proper application technique to optimize the coverage of plant surfaces. Efforts should be made to test the efficacy of different mineral oils on other vectors like leafhoppers, thrips, mealybugs etc. Much work remains to be done on questions as to the type and concentration of oil, frequency of application, phytotoxicity, compatibility with conventional fungicides and insecticides, and costs of application.

2.17 Avoidance of Vectors

2.17.1 Bait Crops in Plant Virus Management

It was shown that by planting alternate rows of tomatoes and cucumbers (cucumbers planted 30 days before transplanting tomatoes) the spread of *Tomato yellow leaf curl virus* (TYLCV) in the tomatoes was significantly delayed during the first 2 months (Al-Musa 1982). The controlling mechanism of the cucumber bait plants is probably due to the combined effects of (1) the huge difference between the canopy volumes of cucumbers and tomatoes, and (2) cucumber is a much better host for whiteflies and once they land on this host they do not leave it as long as the plants remain a good host for colonization. Tests in a flight chamber showed no preference of adult whiteflies in landing on tomato or cucumber leaves, but once they landed, cucumbers were preferred over tomatoes.

The same approach was later used by Schuster (2004), who protected tomato from TYLCV by using squash as a bait plant. A greater cumulative proportion of TYLCV-infected plants were observed in plots that were surrounded by tomato than on tomato plants surrounded by squash.

(a) **Pheromone derivatives**

Pheromones are important chemicals used for signaling between members of the same insect species. There are three main uses of insect pheromones. viz., (1) insect monitoring (2) for mass trapping where insects are lured and killed (3) involves as sex pheromones and is called as mating disruption. The number of male insects is reduced and hence fewer matings, resulting in lowered insect populations in the subsequent generations.

The process of virus transmission is influenced by a variety of insect behaviors, such as leaving and alighting on plants, probing and feeding, interactions with adjacent insects and response to alarms. These behaviors are influenced by a wide range of chemicals and it is suggested that manipulating them may be a way of effecting virus control (Irwin and Nault 1996).

Derivatives prepared from the pheromone (E)-β-farnesene and related compounds interfered with the transmission of PVY by *Myzus persicae* in glasshouse experiments (Gibson et al. 1984). It is possible that these substances act in a manner similar to that of mineral oils because they possess aliphatic carbon chains. While new compounds continue to be developed (Dawson et al. 1988), this type of vector control is still very much at the experimental stage.

2.17.2 Role of Plant Volatile Organic Compounds in Vector Management

Volatile organic compounds released by plants can act as semiochemicals which have insect repellant properties. They play an important role in enabling insects to recognize host plants from a distance (Schutz et al. 1997; Bruce et al. 2005).

A large number of different plant volatiles, with numerous ecological roles, have been identified so far (Sacchettini and Poulter 1997; Pichersky et al. 2006). The largest class of plant volatiles is derived from the isoprenoid or terpenoid pathway. Solanaceous plants like tomato often make use of these terpenes for the defense against herbivores (Kennedy 2003). Some terpenes have been shown to exhibit repellency properties to insects (Birkett et al. 2004). These plant-produced semiochemicals have been potentially used as insect repellents of natural origin, thus providing an alternative to the use of pesticides (Peterson and Coats 2001).

Extensive researches was conducted to study the effect of volatile organic compound against whitefly-transmitted plant viruses. Begomoviruses are a large group of whitefly-transmitted plant viruses, which are responsible for significant yield losses in a wide variety of crops in all most all tropical countries. Intensive studies conducted by Bleeker et al. (2009) on the role of specific tomato volatiles in tomato-whitefly interaction demonstrated a clear preference of whiteflies for cultivated tomato plants over the wild tomato accessions. *Solanum pennellii* LA2560 and *Solanum habrochaites* PI127826 ranked as least preferred, whereas *Solanum peruvianum* LA1708 was only slightly less attractive than the cultivated tomato plants as indicated by the classification in seven preference groups. They further reported that the collected headspace volatiles of the two least preferred tomatoes (*S. pennellii* LA2560 and *S. habrochaites* PI127826) could reduce the attractiveness of the cultivated tomato (*S. lycopersicum* 'Moneymaker'). The headspace volatiles collected over 24 h were applied to a *S. lycopersicum* 'Moneymaker' plant on 10 filter paper cards distributed over the entire plant. Free-choice bioassays were performed with four plants, of which one received the wild tomato volatiles. In this setup, the cultivated tomato with the headspace volatiles of LA2560 or PI127826 attracted up to 60 % less whiteflies. The carrier of the headspace volatiles, pentane-diethylether, alone did not affect the choice of *B. tabaci*.

Finally Bleeker et al. (2009) concluded that two functional metabolomic approaches led to the identification of several volatile mono- and sesquiterpenes from tomato that influenced the whiteflies' choice for the host. The addition of some of these terpenes to the headspace of an otherwise attractive, cultivated tomato plant significantly reduced the number of whitefly visitations or, in other words, made the plant less preferred. Based on these observations, i.e. host selection based on odor cues instead of visual cues, together with a specific antennal response to certain terpenes, they proposed a significant role for plants volatiles in the choice behavior of whiteflies. Further studies with different plant volatile organic compounds on other insect vectors with different virus-host-combinations are needed.

2.18 Role of Repelling and Attracting Surfaces in Vector Control

Numerous studies on the role of reflective/repelling surface aluminum or plastic mulches have proven to be effective primarily against aphid, whitefly and thrips vectors transmitting different viruses. In recent years much importance has been directed to the management of TSWV incidence in tomato by use of mulches (Funderburk 2009). As the aphids respond differently to various wave lengths of light, the use of attractive colors as traps or repellents to avoid landing of the vector on susceptible crops, is advantageous in minimizing the spread of virus diseases. The visible infrared and ultraviolet lights emitted from different surfaces are responsible for vector repellency. It was first demonstrated that white surfaces reflecting ultraviolet or short wave light which was unattractive to alighting aphids and were even avoided by them (Moericke 1954). In explanation it was proposed that aphids first taking off from a plant, are strongly attracted to ultraviolet or short wave light. After flying for varying periods aphids enter the alighting or searching phase when they are repelled by the short wave light of the sky and are attracted to long wave light reflected from plants. At this point they descend, flying close to the ground. The mulches used in between the crops reflect the Sun's ultraviolet rays which the aphids 'see' instead of blue-green light (colour) of the plants. In effect, they receive a signal to "keepflying" instead of landing and they fly further. The use of this principle in virus disease control, is the outcome of the pioneering experiments conducted by Moericke (1954), Kring (1964), and Cartier (1966). Based on these principles at New Delhi (India), Vani et al. (1989) have successfully reduced the mosaic disease incidence in muskmelon by using mulches. In South Africa, aphid-transmitted mosaic virus diseases were effectively reduced in *Cucurbita pepo* by use of white reflective mulches (Cradock et al. 2001).

In the early experiments Moericke (1954) found that when yellow pans were set on bare soil or on cotton cloths of different colors, aphid response varied according to species. Although most of the aphids were caught in pans set on bare soil, 30 % were captured in pans set on black cloths, 30 % on blue cloths and 7 % of white cloths. Moericke suggested that the increased attractiveness was due to increased remission of UV light. Later Kring (1964) also found that when the aluminium pans were placed around yellow ones, repelling of aphids was noticed. Jones and Chapman (1968) tested plastic sheets of nine different colours in addition to aluminum foil to determine their attractiveness to aphids. Yellow was most attractive, followed in order by pink, green, red and black, whereas white, orange, light blue, aluminum foil and dark blue attracted the fewest aphids. The use of yellow color has been used to avert serious losses due to viruses during heavy aphid flight period over susceptible crops.

Another possible control approach is to avoid the physical contact between aphids and plants. This is relatively easy to achieve in nurseries where the plants can be grown under screen houses equipped with double doors, insect-proof nets and yellow sticky traps. Although simple in principle, in practice this control measure could cause serious problems in reducing ventilation, thus leading to a damaging increase of temperature in the screen house. Elimination of this problem by using forced ventilation and cooling may be very expensive. Reflective mulches have been used under several circumstances to reduce incidence of non-persistently transmitted aphid-borne viruses in the field. Summers et al. (2004) made a comparison of the efficiency of spray mulches, film mulches and nets - the latter being either suspended on wire hoops to form tunnels or placed on soil surface - in protecting zucchini squash from non-persistently-transmitted aphid-borne viruses. Mulching performed better than netting because it covered more than 60 % of soil surface. Silver mulches were significantly more effective in repelling *Aphis gossypii* than other tested colored mulches (white, brown, green, black). The repelling power of each color should be tested for each species of aphid before suggesting its use in large scale agriculture. In addition, silver netting used as tunnels was initially as effective as other silver mulches in repelling aphids, but after removal for the first fruit harvest, the plants became infested by aphids and showed viral symptoms soon thereafter. Also plants protected by silver mulches became infected but only in a later stage when canopy cover, significantly decreased the efficiency of mulches in repelling aphids. However, older plants can tolerate virus infection better than young ones, thus yielding marketable fruits. Finally, spray mulches appear more convenient than polyethylene film mulches as they are biodegradable, thus eliminating the problem of disposal at the end of the crop. Cucurbits posses a strong enough structure to support unwoven tissue laid directly onto the plant leaves even at the seedling stage. Tomassoli et al. (1995) showed that this early protection from incoming viruliferous aphids delayed CMV infection in a highly infested area. As for net tunnels, the infection was not totally escaped because the coverage had to be periodically removed for routine agricultural practices and fruit harvesting. More information is provided by Greer and Dole (2003) regarding the role of reflective mulches in decreasing insect vectored viral diseases of vegetables.

2.18.1 Aluminum Mulches

UV reflective aluminum mulches have been used successfully to delay and reduce the incidence of aphid-borne virus diseases in squash and other crops (Brown et al. 1993; Summers and Stapleton 1999; Stapleton and Summers 2002) and to delay colonization by *B. argentifolii* and reduces the incidence of squash silver leaf disorder (Summers and Stapleton 2002). These mulches reflect short-wave UV light, which confuses and repels incoming alate aphids and adult whiteflies, thus reducing their incidence of alighting on plants.

Many attempts were made to reduce the spread of virus diseases in crops like bell pepper, tomato, squash, cucumber, lettuce, beet root, etc. by the use of aluminum mulches. These mulches are effective at least against 12 species of aphids including 4 commonly occurring virus-vectors viz. apple grain, English grain, green peach and potato aphid (Smith et al. 1964). The effectiveness of the aluminum mulches in different crops is discussed here.

(a) Pepper (Bell and pungent): Black and Rolston (1972a, b) and Black (1980) achieved good success in reducing the spread of viruses like PVY, TEV and CMV by using aluminum mulches. The number of aphids trapped over the aluminum foil mulch plot was less than 10 % of those trapped over rows with black polyethylene or no mulch during the first 3 weeks after planting. During the next 5 weeks, the percentage gradually increased to about 50. At the first harvest, 10 % of the plants grown on aluminum foil were showing mosaic symptoms as compared to 85 % of the plants grown on black plastic and 96 % grown without mulch. In a separate experiment, *Capsicum frutescens* (Tabasco) with the same treatments due to TEV, 42 % of the plants grown on aluminum foil had died, as compared to 96 % on black plastic and 98 % without mulch. From Bet Dagan (Israel), Alper et al. (1973) and Loebenstein et al. (1975) obtained effective control of viruses by using aluminum foil or grey plastic sheets. In the experiments conducted during 1972, the virus incidence before harvest reached only 6 and 4.5 % in aluminum and gray plastic-mulched plots, respectively, compared with 44 % in the control, and the mulch treatments resulted in higher yields. Initial experiments conducted at Taiwan agricultural research institute (TARI), Taichung in the early 1980s (Chen and Chen 1980) showed that aphid population could be significantly reduced, virus infection delayed and virus incidence reduced when using reflective mulches such as silver-painted or silver-coated commercially available plastic mulches. Yield increases of more than 100 % could be obtained using Taiwan-made silver coated reflective mulch (Chen and Chen 1983). Another plastic mulch, PVC-UV400, has also been shown to reduce not only aphids but also thrips and spider mites on pepper, grown under shelter (Kuo and Wu 1989). This type of plastic, however, is not yet used commercially. Experiments conducted at TARI, Taichung showed that the use of the locally produced reflective mulch combined with weekly application of mineral oils (Orchex 796, summer oil) delayed virus infection and reduced virus incidence by 80 % or more in sweet pepper (Cheng and Tung 1987).

(b) Squash: From Florida, Moore et al. (1965) noticed low *Watermelon mosaic virus* (WMV) incidence and reduced aphid population by the use of aluminum foil and black plastic mulches. The aphid catch was reduced by 93 % in aluminum mulch plots as compared to 41 % in black plastic mulch, indicating that the latter treatment was less effective. Wolfenbarger and Moore (1967) observed that plots of squash grown in soil mulched with aluminum on paper, white on black plastic and black plastics were effective in reducing virus incidence. Low disease incidence and higher yields were obtained with aluminum foil mulch. Wolfenbarger and Adlerz (1971) noticed 90–100 % reduction in aphid number due to mulching. In two experiments WMV incidence was reduced from 29 to 4 %. The disease development was delayed by about 2 weeks.

In California, Wyman et al. (1979) and Toscano et al. (1979) tried aluminum and white plastic mulches for WMV control in summer squash plantings. Aluminum mulch reduced aphid trap catches by 96 % while white plastic mulch reduced them to 68 % over the entire season. The reduction in virus incidence was 94 and 77 %, respectively. The effectiveness of white plastic mulch was reduced during harvest, whereas aluminum foil remained effective throughout the season. Aluminum and white plastic mulches increased total fruit yield by 43 %. Yield responses were particularly evident during early production with 85 and 69 % increases resulting from the aluminum and white plastic mulches, respectively. The rate of production would increase gross income by more than $750/acre and would justify the initial mulching costs of between $150–200/acre.

(c) Vegetable marrow: Daiber and Donaldson (1976) reported that by covering the plots with aluminum mulch, the incidence of WMV was decreased. Only after 8 weeks, the first symptoms of virus were found in the mulched plots, whereas in control plots the plants were infected after 4 weeks and all the plants were affected after 8 weeks. Eulitz (1977) also noticed delay in WMV disease development for 3 weeks. The yield of marketable fruits of vegetable marrow from an aluminum foil mulched plot reached 33.5 t/ha 41 days after sowing, compared with none from an unmulched plot.

(d) Watermelon: Adlerz and Everett (1968) reported that the epidemics of WMV were delayed by 5 days with aluminum foil treatment while the white polyethylene was not effective and it attracted more aphid vectors. Mansour et al. (2000) reported that aphid vectors of WMV-1 and virus incidence could be minimized by application of aluminium foil (surface mulched aluminium foil, aluminium foil boards). Brust (2000) reported that using reflective and black polythene mulching minimized aphid population and delayed the onset of WMV-1 in watermelon, and increased yields of cucurbits in Western Australia under conditions where both insecticide and oil sprays proved ineffective (McLean et al. 1982). However, these reflective mulches are expensive, and where they have come into regular use difficulties in disposal at the end of the season may occur, at least when disposal was forbidden (Nameth et al. 1986).

(e) Cucumber: Schalk et al. (1979) noticed reduced incidence of mosaic virus diseases among aluminum-mulched plots. They have also observed similar effectiveness even in squash crop.

(f) Sugar beet: In Suffolk (UK) during 1965–1967, Heathcote (1968) used aluminum foil strips between the rows of beet stecklings and noticed low aphid population. The mean percentage of virus yellows was 7.2 % in untreated control plots and 4.4 % in aluminum-treated plots.

(g) Turnip and lettuce: Heinze (1966) found that the incidence of yellows in turnip and mosaic in lettuce was also decreased significantly. At Geneva, field tests indicated that mulches like aluminum foil and black plastic mulches used at lettuce seedling emergence reduced the spread of LMV infection in late planted lettuce.

(h) French bean: Tachibana (1981) reported that silver polythene film reduced the aphid transmission of *Bean common mosaic virus* spread.

(i) Broad bean: Reflective mulches have been used to reduce the incidence of PSbMV by as much as 78 % in broad beans in Japan (Tachibana 1981).

(j) Tomato: In tomato, aluminium surfaced plastic mulch reduced the thrips population and incidence of TSWV (Greenough et al. 1990; Diaz-Perez et al. 2003). The ultraviolet reflective mulch in the typical raised bed and plastic mulch, repels the migrating adults of western flower thrips (*Frankliniella occidentalis*) and Florida flower thrips (*Frankliniella bispinosa*) (Momol et al. 2001, 2004). This cultural tactic is most effective from early to mid-season before the plants to grow to cover the mulches. Effective management of whitefly transmitted plant viruses were noticed by Hilje and Stansly (2008) at Costa Rica. Even at Israel, Paris et al. (2005) have recommended the practice of plastic mulches and also with metalized silver mulch (82%) reflectivity and yellow mulch in the range of 32% were proved to be effective in delaying virus infection. The highest yields of watermelon and melons were observed with the application of metalized silver mulch (Paris et al. 2005).

Interference with vector landing on crops is achieved by altering the attraction of insects to colours. Insects (e.g. aphids) are repelled from reflective surfaces: this effect led to the use of metallic reflective surfaces, straw mulches or kaolin particle films. Landing can be prevented by the use of physical barriers. Insect-proof nets greatly reduced virus incidence and the need for insecticide applications against the *Tomato yellow leaf curl virus* in tomato. Camouflaging nets greatly reduced insect landing and also virus infection (Raccah and Fereres 2009). This measure is now being used commercially for the protection of papaya from *Papaya ring spot virus* in Taiwan.

2.18.1.1 Factors Responsible for the Failure of Aluminum Mulches

Failure of aluminum mulches to protect plants usually have resulted when: (a) too little aluminum was used (Dickson and Laird 1966; Hakkaart 1967; Rotham (1967); (b) too many vectors penetrate the area (Rotham 1967; George and Kring 1971); (c) the reflective surface was covered by the plant before virus spread occurs (Adlerz and Everett 1968); or (d) the plant grew taller than the protective boundary of the aluminum (Hakkaart 1967). Application of certain fungicides and other pesticides reduces the UV reflectance and hence the efficacy of the mulch. A single application of copper or mancozeb fungicide can reduce the reflectance by approximately 49 %. Repeated application can consequently lead to higher incidence of TSWV (Reitz et al. 2003).

2.18.2 Plastic Mulches

As plastic mulches are cheaper than the aluminum mulches, attempts were made in many countries to reduce the virus incidence by using different colored plastic mulches. Jones and Chapman (1968) tested plastic sheets in nine different colors in addition to aluminum foil to determine their attractiveness to aphids. Yellow color was most attractive, followed in order by pink, green, red and black. Whereas white, orange, light blue, and dark blue colours attracted the fewest aphids.

This novel and most promising approach was developed by Antignus and coworkers (as cited in Raviv and Antignus 2004). Polyethylene sheets and nets that absorb UV were found to greatly reduce virus incidence. An impressive reduction of insect landing was recorded for whiteflies, aphids or thrips. The mode of action and benefits of using this type of materials has been extensively reviewed.

Plastic mulches of different colors were tried and achieved varied degrees of success in virus disease management. Plastic UV reflective mulch (metalized mulch) and wheat straw mulch delayed colonization by *Bemisia argentifolii* (Summers et al. 2004).

Summers et al. (2004) have compared the effect of mulches with a preplant treatment of imidacloprid and an untreated, unmulched control. In 2000, yield of marketable squash in the plastic and straw mulched plots was approximately twice that from the imidacloprid plot. Yields from both mulched plots and from the imidacloprid plots ranged from 3 to 12 times higher than those from the control plots. The mulches were more effective than a preplant application of imidacloprid in reducing the incidence of both *B. argentifolii* and aphid-borne viruses. Plants growing over the plastic mulch and the straw mulch grew more rapidly and reached a larger size, as determined by plant dry weight, than did those growing over bare soil, with or without imidacloprid. The spectral quantum flux from the plastic averaged between 80 and 90 % of ambient spectral quantum flux values in the UV (300 to 400-nm) range. Spectral quantum flux values of wheat straw were similar to those of the reflective mulch and ambient near 300 nm but were virtually identical to bare soil beyond 320 nm. The metalized mulch reflected 94 % of the incoming photosynthetically active radiation (PAR) in the 400 to 700-nm range compared with ambient, whereas the straw mulch reflected 85 %. Bare soil reflected only 41 % of incoming PAR compared with ambient (Summers et al. 2004).

In the earlier studies during 1995, Summers et al. observed low incidence of aphid-borne viruses in zucchini squash when they tested sprayable and film mulches in spring seeded squash. Approximately 30 % of the plants on unmulched beds were infected with one or more viruses by the first harvest while only 10–15 % of those grown over the silver mulches showed virus symptoms. Marketable fruit yields in the spring planting were approximately 70 % higher in plots mulched with silver than the unmulched control. Even in the field trials conducted during 2000 and 2001, Summers et al. (2007) successfully managed three non-persistent aphid-transmitted viruses of squash by using plastic reflective or wheat straw much. Yields of marketable fruit increase approximately tenfold in plot trials and twofold in a commercial squash field.

Similarly, encouraging results have been achieved by using grey plastic sheets (Alper et al. 1973), PV 400 plastic mulch (Kuo and Wu 1989), black plastic mulch (Moore et al. 1965), and white plastic mulch (Wyman et al. 1979). Toscano et al. (1979) mentioned that straw mulch would be effective for a short period when compared to aluminum or plastic mulches.

In Malaysia, Ong (1975) found that the incidence of *Chilli veinal mottle virus* was reduced in chillies by using polyethylene sheet coated with aluminum paint. From Israel, Loebenstein et al. (1975) reported that plastic mulch and aluminum foil covering 57–83 % of the ground in peppers from emergence effectively protected the crop against PVY and CMV. In 1970, virus incidence in the white on black plastic-mulched plot was only 5.5 % infected at the time of harvesting compared with 45.5 % in the control. Similarly in 1972, virus incidence before harvest reached only 6 and 4.5 % in aluminum and gray plastic mulched plots, respectively, compared with 44 % in the control. Pepper yields in the plastic-mulched plots were markedly higher (40–100 %) than in the controls. Ong (1984) reported that incidence of insect-borne viruses in chilli was significantly reduced with aluminium foil mulch (aluminium-painted polyethylene sheet mulches). Black (1980) reported that the influx of winged aphid vectors and the incidence of potyviruses in chilli could be reduced by mulching the chilli plots with aluminium foil or black plastic mulch sprayed with aluminium paint. Basky (1984) found that blue and transparent plastic mulches placed between the rows and an application of Agridex EEC318 and Atplus 11 oil sprays decreased virus incidence in cucumber. Effective results were achieved in reducing the virus diseases of aphid-transmitted non-persistent virus diseases of cucumber, squash, tomato and potato from the extensive field trials conducted at ARO, Volcani Center, Bet Dagen (Israel) by the application of plastic mulches (Alper et al. 1973; Cohen and Marco 1973; Loebenstein et al. 1975). A field showing the application of plastic mulches to control *Cucumber mosaic virus* spread in cucumbers is shown in the Fig. 2.13.

Fig. 2.13 Use of plastic mulches in the control of *Cucumber mosaic virus* spread in cucumbers. *Courtesy* S. Cohen, S. Marco

2.18.3 *Sticky Yellow Polyethylene Sheets*

In Israel, Cohen and Marco (1973) reduced the spread of CMV and PVY in peppers by using sticky sheets of yellow polyethylene along the edges of the field or surrounding the plots. The sheets were stretched at a height of 0.7 m from the soil surface. The sheets were covered with "Rimi-foot" glue which remained sticky for 4 weeks. The reduction in the spread of the virus was greater in the plots surrounded by "sticky" baits than in those treated with plain polyethylene sheets or untreated controls. From the same laboratory, Zimmerman's (1979) field trials also indicated that seed potatoes can be protected by 'sticky' yellow polyethylene sheeting which showed marked reduction in the incidence of virus-infected tubers. In his trials, *Potato leaf roll virus* (PLRV) was more frequently noticed when compared with *Alfalfa mosaic virus* (AMV) and PVY. The percentage of PLRV-affected tubers from the protected plots and untreated controls were as follows: for 1974, 2 and 17.2 %, for 1975, 6 and 29 %, respectively. There was also a reduction in the spread of both AMV and PVY.

In Hong Kong, Lee and Lee (1973) reported the effective control of *Turnip mosaic virus* infection in Chinese cabbage by using polyethylene with reflective borders or reflective strips surrounding the beds from sowing until harvest and obtained increase in yields. Smith et al. (1972) recommended the use of yellow sticky board traps for the control of greenhouse whitefly.

2.18.4 *Saw Dust/Straw/Rice Husk Mulches*

In zucchini squash, aphid-borne viruses and *Bemisia argentifolii* were successfully managed by using wheat straw mulches and also by UV reflection plastic (Summers et al. 2004). In experiments in Colorado during 2005 to 2007, straw mulch application to the center of onion beds at the early to mid-bulb growth stage has reduced abundance of thrips vectors (*T. tabaci*) as much as 33% when compared to non-treated plots of transplanted onions, along with reduced *Iris yellow spot virus* incidence (Schwartz et al. 2009). Even at Geneva (US) the results of trials conducted by Larentzaki et al. (2008) have indicated that in onion the populations of *Thrips tabaci* adults and larvae have been significantly reduced by the use of straw mulch without comporomising overall onion yield.

In Brazil, nursery seedlings of some vegetables were protected from PVY and WMV by the use of rice husks, which had aphid repellency quality, and reduced the early infection rate up to 90 % by preventing the landing of the aphid vector on the nursery seedlings. Mulching soil with this material also reduced the virus infection spread by migrating, *M. persicae* landing on tomato plants (Costa 1972), and reduced the incidence of virus transmitted by *A. gossypii* to bush squash plants (Costa and Costa 1971). Costa et al. (1974) reported that for citrus plants under conditions where root stock infection by *Citrus tristeza virus* prior to budding might be a problem, that with rice husk mulching of the soil between plants in seed beds and nurseries, there was a valuable method for reducing tristeza infection. They recorded a lower number of alates of *Toxoptera citricidus*.

Attempts were also made to spray the plants with rice polish to repel the vectors. Shands et al. (1962) achieved very little success (non significant results) in reducing PVY spread in potato by spraying 0.5 or 1 % rice polish extract.

2.18.5 Effect of Whitewash on Virus Incidence

Marco (1993) has conducted field experiments over 6 years by using sprays of whitewash (Yalbin or Loven) at 10 % concentration for reducing the incidence of PVY in peppers. He observed reduction in virus incidence by 40 %. The best control was achieved when the product Yalbin was combined with bifenthrin or mineral oil, which reduced virus infection by approx. 60 %. Whitewash treatments caused slight damage to pepper seedlings at the first-true-leaf stage but did not damage older plants. In three of four experiments, whitewash-treated plots had significantly higher yields than control plots. In the remaining field experiments there was reduction in the yields of pepper due to whitewash spray. Although encouraging results were obtained with whitewash spray in reducing virus spread, only based on large number of field experiments with different virus-host combinations could one draw appropriate conclusions.

2.18.6 Effect of Silver Spray Paint on Planting Beds on Virus Incidence

For managing the virus diseases in vegetables, Aluminum foil works, but is far too expensive for commercial production. Hence alternatively silver spray paint on planting beds was tried to reduce the aphid transmitted non-persistent viruses in the vegetable crops. Silver spray paint on planting beds could deter the aphid vectors and disease problems in vegetable crops. When an aphid is flying along, looking for a place to land, it cues in on reflections coming off of plant surfaces.

The field trials conducted during 1993, with zucchini squash have indicated that the plants grown on beds treated with spray-on silver mulch repels aphids and delays the onset of virus infection by 10 days to 2 weeks and produced 3,973 pounds of good quality zucchini per acre. The untreated control beds produced 867 pounds of marketable zucchini per acre due to high virus incidence. The advantage of spraying on the silver mulch is that the plants need less water as the mulches reduce evaporation from the soil surface and the crust formed on the spray-painted surface suppresses weeds. The water-based paint can simply be disked down and incorporated into the soil with no harmful effects (Anon 1994) (Fig. 2.14).

Fig. 2.14 Silver spray paint on planting beds of Bell pepper to control plant viruses. *Source* T. A. Zitter

2.18.7 Protection of Greenhouse Crops by UV-Blocking Cladding Materials

(a) **Inhibition of insect infiltration**

Polyethylene films are used as greenhouse cladding materials to provide efficient protection against wind, rain, and hail. They can also act as radiation filters that enable growers to control light quality and intensity within the greenhouse. Originally UV-blocking polyethylene films were commercially produced to protect greenhouse-grown roses from petal blackening. However, UV-blocking cladding materials have been found useful not only for horticultural purposes, but also as a tool to combat fungal diseases (Raviv and Antignus 2004). Further studies have demonstrated the dramatic inhibitory effect of UV-blocking polyethylene films on the invasion of greenhouses by insects (Antignus et al. 1996a, b, 2001a). UV-blocking polyethylene films are produced by the introduction of a UV absorbing additive into the polyethylene raw material. The resulting modified polyethylene blocks over 95 % of light transmission in the range of 280–380 nm, while allowing 80 % transmission of radiation in the range of 380–700 nm. UV-blocking films allow the transmission of 5 % of the total UV irradiation of the sunlight compared to 13–20 % that flow through films of the ordinary type. It has been shown that UV-blocking greenhouse cladding materials can reduce the entry of insects and the spread of insect-transmitted viral diseases of plants (Antignus et al. 1996b).

In experimental "walk-in" tunnels (6×6 m) the number of whiteflies (*B. tabaci*) trapped on yellow sticky traps under a UV-blocking film was 4–10 times lower than the number trapped under a standard film. In a similar experiment, the number of aphids (*A. gossypii*) recorded under UV-blocking films was $\sim$100 times lower than that recorded under standard films and these films reduced the invasion of thrips (*Frankliniella occidentalis*) by a factor of 10 (Antignus et al. 1996a, b).

In a similar set of experiments, UV-blocking films dramatically reduced the infestation of mint with nocturnal moths (*Spodoptera littoralis, Laphygma spp.*) and of leafminer flies (*Liriomyza trifolli*) that infested the same crop heavily under standard films (Antignus et al. 1997). The reduction in pest populations under the UV-blocking films enabled growers to reduce the number of pesticide applications by 50–80 % from the usual level (Antignus et al. 1997).

(b) **Inhibition of the spread of viral diseases in greenhouses by UV-blocking films**

The protective effect of UV-blocking has a highly significant effect in reducing the spread of insect-borne viral disease of tomatoes grown in "walk-in" tunnels covered with a UV-blocking polyethylene films. Growing tomatoes with these polyethylene films (without insecticide application) had a 1 % infection rate of TYLCV compared with 80 % infection rate for tomatoes grown under standard films (Antignus et al. 1996a).

In all experiments that tested the protection of tomato from whiteflies and TYLCV, the reduction in the rate of viral infection was always greater than the reduction in whitefly population density, which may indicate that beside the reduced invasion into the protected structures, the viral transmission efficiency is lowered under UV-blocking films.

(c) **Parameters affecting the efficiency of protection by UV-blocking films**

The degree of the UV-blocking by a plastic film determines its protective efficiency. PVC films, that are highly efficient UV blockers, gave significantly better protection against insect pests than standard polyethylene films (Antignus et al. 1996a, b, 1999).

To achieve an adequate level of protection against large populations of insects, in addition to the UV-blocking roof, the greenhouse side walls should be covered with 50-mesh screens; a combination of a UV-blocking roof and 30-mesh screens failed to provide protection against whiteflies and the spread of TYLCV (Antignus et al. 1999). This is consistent with a report from California that UV-blocking roofs did not reduce the numbers of whiteflies in open-sided greenhouses, although reductions in the numbers of aphids and thrips were achieved in those greenhouses (Costa et al. 2002).

(d) **The effects of UV blocking polyethylene on plants**

No significant differences were found in growth, yield, maturing time, fresh or dry weights of tomatoes grown in greenhouses under standard and UV-blocking films. Physiological disorders were reduced by 38 % under the UV-blocking plastic compared with their incidence under the standard material. Similarly, the yield and quality of peppers and cucumbers were not affected by UV-blocking. No differences were found in pigment intensity and total soluble solids of tomato and pepper fruits grown under regular and UV-blocking films, and the percentage of viable pollen grains in these plants were not affected by the type of polyethylene film covering (Pressman et al. 1996). The firmness and shelf life of tomato fruits were also unaffected by the type of polyethylene cladding (Antignus et al. 1999) and UV-blocking had no effect on the yield and marketing quality of tomato and pepper (Antignus et al. 2001b, 2005, 2007; Antignus and Ben-Yakir 2004; Gilbertson et al. 2007)

(e) **Impact of nylon nets on virus incidence**

The general practice for almost all vegetable crops is to raise the nursery in a protected condition in order to avoid the virus transmission through aerial vectors. For this purpose, the majority of the growers are covering the nurseries with nylon mesh cages. Some of the examples are:

From India, Nilakshi Kakati and Nath (2006) have reduced the TLCV incidence by raising the tomato seedlings in the nursery bed covered with nylon net and no incidence of TLCV as against 8.0 % in the uncovered nursery bed. In the main field where the seedlings raised under nylon net covering when 2 sprays of Nimbicidine @ of 2.0 % at 20 and 35 days after transplantation showed no TLCV incidence and a high yield of tomato (318.89 q/ha). Whereas in untreated control (nursery raised in open field without spraying in the main field) recorded highest TLCV disease incidence 89.33 % with a corresponding low yield of 98.89 q/ha.

Anitacherian (1998) has reported that tomato nursery raised without nylon net covering resulted in ToLCV incidence in early weeks of transplanting which reached 100 % by the fourth and fifth week. She has also reported that tomato nursery grown under nylon net covering and sprays with mycopesticides and triazophas after transplanting at weekly intervals delayed ToLCV incidence by 2–3 weeks and increase the yield considerably.

A combined treatment of nylon net covering of tomato nursery beds and 2–3 sprays of monocrotophos or dimethoate or cypermethrin after transplanting in the field was effective in reducing the spread of ToLCV (Muniyappa and Saikia 1983). Venkatesh (2000) has reported that nylon net (152 cm height) painted with yellow colour and sprayed with deltamethrin delayed ToLCV infection by reducing the migrating population of *B. tabaci* into the tomato crop and also recorded highest crop yield (12.15 t/ha). At ARO, Volcani Centre, Bet Dagen have used nylon net covering of pepper plants even under the field conditions to restrict the spread of *Cucumber mosaic virus* as shown in the Fig. 2.15.

2.18.8 Use of Row Covers in Vegetable Production

One of the more popular techniques used by commercial growers to extend the growing season of a crop is the use of row covers. Row covers are generally made of flexible transparent to semitransparent materials and are used to enclose one or

Fig. 2.15 Nylon net covering of pepper plants to restrict the spread of *Cucumber mosaic virus*. *Courtesy* S. Cohen, S. Marco

more rows of plants in order to enhance crop growth and production by increasing both air and soil temperatures and reducing wind damage (Hochmuth et al. 2000).

Lightweight floating row covers are often used to cover some crops in the spring to protect them from insects. Edges of the cover must be covered with soil to ensure exclusion of insect vectors like aphids, leafhoppers, mealybugs etc., resulting in reduced virus incidence (Fig. 2.16). This will result in less use of insecticides and less frequency of insect-borne diseases. Floating and supported row covers can also be used to protect plants from birds.

Conclusion

The tendency of the aphid vectors to be attracted towards colors depends on the species and is not a uniform characteristic. For example, Cartier (1966) found that white was a preferred color after orange and yellow by the pea aphid *Acyrthosiphon pisum* but it was not very attractive to the potato aphid *Macrosiphum euphorbiae* in laboratory tests. Similarly Zettler et al. (1967) reported that *R. maidis* was much attracted towards yellow. Even Mound (1962) observed that whitefly (*B. tabaci*) was attracted by blue ultraviolet and yellow wave lengths of the light spectrum. Depending on the vector involved, different colors of reflective mulches should be tested in order to obtain effective control. It is also reported that the aluminized mulches are effective against thrips (*Frankliniella tritici*) (Smith et al. 1972) and attempts to control other vectors like leafhoppers, mealy bugs, and mites etc., in different crops is worthwhile to pursue.

There are a number of advantages when considering the mulching of crops. This will increase the early and total yields in most vegetables by reducing the number of several economically important pests. For example in Florida, with aluminum mulch application, the leaf miner damage was less on squash. A similar result was obtained on tomatoes with vertical aluminum strips. In three cases, the number of honey bees visiting squash was increased (Wolfenberger and Adlerz 1971). Mulching also helps in weed control. Mulches can conserve soil moisture by eliminating evaporation from the surface and they prevent the crusting of soil on the

Fig. 2.16 Removal of floating row covers in the management of plant virus spread. *Source* T. A. Zitter

soil surface. They also reduce loss of fertilizer by preventing water from percolating through the soil and leaching out soluble nutrients. Thus nutrients are available to the plants for a longer part of the growing season. Mulched soils is less compacted, has more uniform moisture content and a higher temperature. This results in a higher soil microorganism activity, a more rapid breakdown of organic matter and a release of plant nutrients. Mulching also prevents root pruning and injury due to cultivation. A more extensive root system results and this contributes to improved utilization of nutrients and moisture. Another advantage of mulching is that the fruit is kept off the ground and a higher percentage is marketable due to less fruit rot.

However, there are also certain disadvantages found with the use of mulching. Aluminum mulch may seriously inhibit watermelon seed germination and aggravate damping-off problems, since aluminum mulch can depress soil temperatures. Use of aluminum mulches are not effective in all crops against all viruses. For example, in California, Dickson and Laird (1966) did not achieve control of *Watermelon mosaic virus* in melon crop. It just delayed the infection by 5–10 days in the range of 2–50 % infection. Similarly, Mowat and Woodford (1976) could not achieve success in reducing the spread of two non-persistent aphid-borne viruses in lilies by using aluminum-coated building paper as mulch. Rotham (1967) also observed the failure of aluminum foil to protect winter oats from aphid vectors of *Barley yellow dwarf virus*. The possible factors for the failure of protection due to mulches were discussed earlier.

The maximum vector repellency will be obtained if the mulch area covered is more than 50 %. For example, Smith and Webb (1969) found that 96 % of the aphids were repelled if 50 % of the area was covered with aluminum sheets, whereas repellency was reduced to 70 % if only 30 % of the area was mulched. They also showed that aluminum sheets repelled equal number of aphids whether the sheets were placed at the ground level or suspended 45.7 cm above ground near the tops of the plants. Aluminum sheets suspended vertically over the plants were less effective than those placed in a horizontal position on the ground.

Johnson et al. (1967) also observed that the aluminum strips of 60 cm wide between the rows were more highly repellent than 30 cm strips.

For increased effectiveness of the mulches, fertilizer and herbicide should be applied before laying the mulch. As the foil prevents weeds from coming up in the covered area, the uncovered area should be treated with a suitable herbicide. Before laying out the mulch, the field should be well prepared for a seed bed or for transplanting. The mulches are usually applied 2–3 days before planting or sowing of the crop. The aluminum mulch can also be applied with some difficulty by tractor drawn machinery designed to lay plastic mulches. Soluble fertilizers can be applied to the crop during growing season through a drip irrigation system. Dry loose soil may not hold the mulch in high winds. In small-scale applications, the mulch can be held down by stones or pieces of 2.5 cm mesh chicken wire attached to stakes. However, burying the edges with soil seems to be more satisfactory. Occasionally walking on the mulch after the soil has thoroughly settled down does not ruin the mulch. However, one should not walk on the mulch more than necessary.

The initial cost of aluminum mulch is over $200/acre. As the cost of aluminum and plastic mulches are too high, attempts were made to reduce the cost of cultivation by using other reflective matters like saw dust, rice straw, wheat straw, rice husk, rice polish etc., which are reasonably priced.

Methods and materials should be developed which will have repellent activity for longer periods. While breeding crops, it may be kept in mind to incorporate the factors responsible for colors, which will repel the vectors. Muller's studies (1964) showed that many more *M. persicae* and other flying aphids alighted on green or yellow-green varieties of lettuce as on bronze varieties and that virus infection was correspondingly reduced. Intensive research is also required for wider application of mulches against different viruses infecting other agronomically important crops. Some more information on mulches and their role in plant virus control can also be obtained from review article of Cohen (1982).

2.19 Resistance

2.19.1 Introduction

Use of resistant cultivars is relatively inexpensive to deploy and has no adverse environmental consequences. The farmer will not have to provide additional production cost or decision making actions during production season. Resistant cultivars have been developed for insects, nematodes, fungal and bacterial, viral and viroid diseases. The mechanism of resistance in plants may be biochemical, physiological, morphological or anatomical.

Resistance to viruses manifests itself as absence of symptoms and /or restriction of virus multiplication and spread within the field, and is an effective measure to manage virus diseases. It may not often be practicable to adopt cultural and chemical control measures against virus diseases in time because of cost, feasibility and other considerations. However, among the available virus disease management measures, use of resistant/tolerant cultivars has been found to be the most efficient and cost effective method, provided that stable sources of resistance are available. Development of virus resistant cultivars needs awareness of the viruses confronted viz., vector involved, virus-host range and the epidemiological information associated with the host crop. Essentially, resistant cultivars are required when disease pressure is consistently high, whereas cultural methods are more appropriate where disease pressure is moderate or low.

There are basically three ways for plants to prevent pathogen infection and reproduction: (1) pre-formed plant defense compounds like structural barriers or pathotoxins restrict successful pathogen infection; (2) the plant does not supply the essential living requirements for a potential pathogen and is therefore a non-host; (3) plants are capable of defending themselves by means of a combination of constitutive and induced defenses. The latter induced resistance mechanism depends on recognition of the attacking pathogen and describes a situation known as 'gene-for-gene' infection. These interactions are mediated by plant disease resistance (R) genes. There are instances of breaking down of resistance due to evolution of new virus strains as has been seen in case of cassava with *Cassava mosaic virus,* in number of legumes with geminiviruses, and in tomato with *Tomato spotted wilt virus* and *Tomato leaf curl virus*. Hence the plant breeder's goal is to introduce disease resistance into cultivars that will provide useful disease control for a period long enough to ensure that the commercial life of a cultivar is not curtailed. The major advantage of breeding for resistance to viruses is that once a resistant cultivar is developed, farmers often do not need to take additional control measures like spraying insecticides to reduce the insect vector population. More information on virus disease resistance can be had from reviews of Fraser and Vanloon (1986), Nene (1988), Fraser (1990), Hull (1990), Foxe (1992), Hull and Davies (1992), Boulcombe (1994), Grumet (1994), Culver (1995), Khetarpal et al. (1998), Varma and Mitter (2001), Lapidot and Friedmann (2002), Varma et al. (2002), Goldbach et al. (2003), Deom (2004), Kang et al. (2005), Ritzenthaler (2005), Maule et al. (2007), Quint and Lubberstedt (2008), Madhavi Reddy et al. (2010), Palloix and Ordon (2011).

2.19.2 Types of Resistance

Vander plank (1963, 1968) has postulated that resistance to disease in plants could be placed into one of two categories, vertical or horizontal resistance. When a cultivar is more resistant to some strains of the virus than to many others, the resistance is called vertical (qualitative). The absolute definition of vertical resistance is that it involves a gene-for-gene relationship. This resistance is absolute and mostly appears as hypersensitivity or extreme resistance. Differential interactions exist between genotypes and pathotypes among virus strains. In contrast, when resistance is evenly spread against all strains of the virus, it is called horizontal resistance (quantitative). Horizontal resistance does not involve a gene-for-gene relationship (Robinson 1980) and the more commonly acknowledged horizontal resistance is usually polygenic control. It has been established that horizontal resistance is by nature more stable while vertical resistance is unstable. Usually a higher degree of resistance is obtained by vertical than by horizontal resistance. Vertical resistance can be useful in cases where the number os pathotypes/biotypes/strains are limited.

The number of genes for resistance varies greatly with the host and the virus. Disease resistance is usually conditioned by monogenic, oligogenic or polygenic factors in plants that have been locally collected, in exotic cultivars, or in wild germplasm. In the case of monogenic resistance, reports suggest that the dominance of resistance and susceptibility are equal. Fraser (1982) classified resistance mechanisms into three broad classes based on the concepts of underlying mechanisms rather than observed symptoms-(1) non-host immunity, (2) genetically controlled resistance and (3) individual resistance. In a breeding programme, any one of the resistance types, immunity, hypersensitivity, tolerance and disease escape (klendusity) are utilized depending on the ease with which the characteristics can be transferred. The complete genetic mechanism of resistance is known for only a few plant virus diseases and efforts are being made to incorporate the resistance into improved cultivars.

Besides the above discussed types of resistance, the major development has been the exploitation of pathogen (virus)-derived resistance which depends on different mechanisms of which two are widely used: (i) protection thought to be mediated by expression of native or modified viral proteins (e.g., coat protein, replicase, defective replicase), and (ii) protection mediated at the transcriptional level ('RNA-mediated resistance') which requires transcription of RNA either from full or partial sequences derived from the target virus (including genes for coat protein, replicase, defective replicase, protease, movement protein etc.) and these aspects are discussed later while dealing with transgenic approach for virus disease management.

(a) **Gene pyramiding**

The concept of transferring several specific genes into one plant is called gene pyramiding. Whenever a virus or viroid pathogens show strain variation, then the breeder may want to transfer a number of genes for conferring resistance against different strains in to a cultivar. The success of gene pyramiding depends upon several critical factors, including the number of genes to be transferred, the distance between the target genes and flanking markers, the number of genotypes selected in each breeding generation, the nature of the germplasm, etc.

Shi et al. (2009) have done pyramiding of *Rsv1, Rsv3,* and *Rsv4* for *Soybean mosaic virus* resistance using molecular markers. J05 carrying *Rsv1* and *Rsv3* and V94-5152 carrying *Rsv4* were used as the donor parents for gene pyramiding. Similarly Werner et al. (2005) have worked with yellow mosaic virus resistance in barley and identified *rym4, rym5, rym9 and rym11.*

(b) **Resistance linked molecular markers**

One of the main objectives of the plant breeder in agriculture is to improve the existing cultivars which are deficient in one or more desirable traits. A conventional breeding programme is a laborious and time consuming, involving several crosses, several generations, and careful phenotypic selection. With the advent of DNA marker technology, the development of several types of molecular markers and molecular breeding strategies has contributed to overcoming many of the problems faced during the conventional breeding. Traditional plant breeding relies upon crosses and subsequent selection of genotypes to meet desirable traits. The incorporation of marker-assisted selection into breeding strategies would result in a reduction in the number of offspring to be propagated, selected and tested. Screening by molecular markers (linked to resistance genes) is quick and accurate (Watanabe 1994; Barone and Frusciante 2007). Marker-assisted selection is useful where phenotypic selection is difficult or where it is not possible or convenient to use the virus for direct screening. It can also be useful for back cross breeding, for the introgression of resistance genes from wild species, while selecting against the undesirable characteristics of the wild parent (Young and Tanksley 1989).

Molecular markers include bio-chemical constituents (e.g., secondary metabolites in plants) and macro molecules (proteins and DNA). These macro molecules show easily detectable differences among different strains or different species. A wide range of molecular marker technologies are now available for genetic studies. Among these are RAPD, AFLP, SSR and STR marker systems that are being utilized by breeders. The marker assisted selection (MAS) breeding programme has been attempted in certain virus-host combinations. For example in pea for PSbMV, the resistance is conferred by a single recessive gene (*eIF4E*) localized on LG VI (Sbm 1-locus) (Smykal et al. 2010). The specific gene for MAS was the *bc-3* gene for introgression into Andean common bean against BCMV (Johnson et al. 1997). In cassava Lokko et al. (2005) identified SSRY28-180 to be linked with resistance to *Cassava mosaic disease* (CMD) using bulk segregant analysis. In tomato, *Tm-2* against *Tomato mosaic virus* (Shi et al. 2011), *sw-5* and *sw-7* against *Tomato spotted wilt virus* (Shi et al. 2011) are being used in tomato breeding programme. Even the commercial hybrids and wild sunflowers were screened in Argentina against *Sunflower chlorotic mottle virus* by gene markers were MS0022 (5 cM) and ORS-307 (4 cM) (Lenardon et al. 2005). More information on this aspect can be obtained from review and book chapters (Bernardo and Charcosset 2006; Prasanna and Firoz Hossain 2010; Bhat et al. 2010; Jonah et al. 2011; Setti Benali et al. 2011).

2.19.3 Sources of Resistance

Breeding for resistance to virus diseases has been successfully accomplished in certain vegetable and field crops and are presented in Table 2.5.

Table 2.5 Resistant and tolerant sources against virus diseases in different crops

Host	Virus	Reference
Beans	*Bean golden mosaic virus*	Morales and Niessen (1988)
Capsicum spp.	*Tobacco mosaic virus*	Provvidenti (1977), Rast (1982), Sowell (1982), Patezas et al. (1989)
	Tomato leaf curl Bangalore virus	Maruthi et al. (2013)
	Tomato spotted wilt virus	Boiteux et al. (1993a), Nono-Womdim et al. (1994), Roggero et al. (2001), Cebolla-Cornejo et al. (2003)
Chillies	*Cucumber mosaic virus*	Narasimha Prasad et al. (2001)
	Pepper veinal mottle virus	Caranta et al. (1996)
	Potato virus Y	Zitter (1979)
Common bean	*Bean common mosaic virus*	Temple and Morales (1986), Allavena (1989), Gupta and Chowfla (1990), Ogliari and Castarro (1992), Kelly (1997), Jones and Cowling (1995)
	Tomato yellow leaf curl virus	Lapidot (2002)
Cotton	*Cotton leaf curl virus*	Sabhiki et al. (2004), Akhtar et al. (2005), Daljeet Singh et al. (2006), Bhatoa et al. (2013)
Cowpea	*Black eye cowpea mosaic virus*	Ndiaye et al. (1993), Kamala et al. (2007)
	Cowpea aphid-borne mosaic virus	Ladipo and Allen (1979), Mali et al. (1987)
	Cowpea mosaic virus	Robertson (1966), Beier et al. (1977)
	Cowpea mottle virus	Allen et al. (1982)
	Cucumber mosaic virus	Gillaspie Jr. (2001), Mali et al. (1987)
	Tobacco ring spot virus and Cowpea mosaic virus	Mali et al. (1987), Ponz et al. (1988)
Cucumber	*Melon mosaic virus*	Cohen et al. (1971)
	Zucchini yellow mosaic virus	Pitrat and Lecoq (1984)

(continued)

Table 2.5 (continued)

Host	Virus	Reference
Faba bean	*Pea seed-borne mosaic virus*	Fagbola et al. (1996)
French bean	*Bean common mosaic virus*	Zaumeyer and Meiners (1975), Provvidenti (1977), Walkey and Innes (1979), Sastry et al. (1981)
	Bean yellow mosaic virus and pea virus II	Provvidenti and Schroeder (1973)
	Mungbean yellow mosaic virus	Aghora et al. (2010)
Lentil	*Pea seed-borne mosaic virus*	Haddad et al. (1978), Latham and Jones (2001a, b), Coutts et al. (2008)
Lettuce	*Lettuce mosaic virus*	Ryder (1970, 1976), Walkey et al. (1985)
Lupin	*Bean yellow mosaic virus*	McKirdy and Jones (1995a)
Maize	*Maize streak virus*	Rodier et al. (1995)
Mung bean	*Cucumber mosaic virus*	Sittiyos et al. (1979)
	Mungbean yellow mosaic virus	IIPR (2006), Kooner and Cheema (2007), Akhtar et al. (2009), AVRDC (2000)
Muskmelon	*Cucumber green mottle mosaic virus*	Rajamony et al. (1990a, b)
	Cucumber mosaic virus	Webb and Bohn (1962), Karchi et al. (1975), Lecoq et al. (1979), Lecoq and Pitrat (1983), Yang et al. (1986), Diaz et al. (2003)
	Cucurbit aphid-borne yellows virus	Dogimont et al. (1996)
	Watermelon mosaic virus	Gilbert et al. (1994), Diaz-Pendon et al. (2005)
	Zucchini yellow mosaic virus	Diaz et al. (2003)
Okra/Bhendi	*Yellow vein mosaic virus*	Nariani and Seth (1958), Dhankar et al. (1991), Arora et al. (1992), Bora et al. (1992), Sharma et al. (1993), Rattan and Arvind Bindal (2000), Rashid et al. (2002), Zulfeghar Ahmed and Patil (2004), Kumar and Singh (2013)

(continued)

Table 2.5 (continued)

Host	Virus	Reference
Papaya	*Papaya ring spot virus*	Conover et al. (1986)
Peanut	*Groundnut rosette virus*	Subrahmanyam et al. (1998)
	Peanut bud necrosis virus	Nagaraja et al. (2005)
	Peanut mottle virus	Kuhn et al. (1968)
	Tobacco streak virus	Kalyani et al. (2005, 2007)
	Tomato spotted wilt virus	Culbreath et al. (1997)
Pepo-cucumber	*Cucumber mosaic virus*	Zitter et al. (1991)
Pea	*Bean common mosaic virus*	Provvidenti (1991)
	Pea seed-borne mosaic virus	Hampton (1980), Hampton and Braverman (1979a, b), Muehlbauer (1983), Baggett and Kean (1988), Provvidenti and Alconero (1988), Khetarpal et al. (1990), Maury et al. (1992), Thakur et al. (1995), Dhillon et al. (1995), Kraft and Coffman (2000), Latham and Jones (2001a, b), Coutts et al. (2008)
	Peanut stripe virus	Musil and Jurik (1990)
Phasey	*Bean common mosaic virus*	Provvidenti and Braverman (1976)
Pigeonpea	*Pigeon pea sterility mosaic virus*	Sreenivas et al. (1997)
Pumpkin	*Watermelon mosaic virus*	Gilbert et al. (1994)
	Watermelon mosaic virus-2	Gilbert-Albertini et al. (1993)
Soybean	*Bean yellow mosaic virus*	Provvidenti (1975)
	Peanut mottle virus	Demski and Kuhn (1975)
	Peanut stripe virus	Warwick and Demski (1998)
	Soybean mosaic virus	Koshimizu and Iizuka (1963), Ross (1977), Kwon and Oh (1980), Cho and Goodman (1979, 1982), Bowers and Goodman (1982, 1991), Buzzell and Tu (1984), Lim (1985), Suteri (1986), Mohammad Arif and Sher Hassan (2002), Domier et al. (2007), Pedersen et al. (2007)
Subterranean clover	*Peanut stripe virus*	McKirdy and Jones (1995b)

(continued)

Table 2.5 (continued)

Host	Virus	Reference
Sweet potato	*Sweet potato virus disease*	Mwanga et al. (2001)
Tomato	*Peanut bud necrosis virus*	Venkata Ramana et al. (2006)
	Pepino mosaic virus	Soler-Aleixandre et al. (2007)
	Potato yellow mosaic virus	Rampersad and Umaharan (2003)
	Tobacco mosaic virus	Honma et al. (1968), Besedina (1985)
	Tomato leafcurl virus	Muniyappa et al. (1991, 2002), Maruthi et al. (2003, 2005), Mukhtar et al. (2009), Colvin et al. (2012)
	Tomato spotted wilt virus	Black et al. (1991), Boiteux et al. (1993b), Kumar et al. (1993), Gardner and Panthee (2012)
	Tomato yellow leaf curl virus	Vidavsky and Czosnek (1998), Hanson et al. (2000), Yang et al. (2004b), Lapidot et al. 1997, 2006, Brown (2008)
Urdbean	*Blackgram mottle virus*	Krishnareddy (1989)
	Urd bean leaf crinkle virus	Indu Sharma and Dubey (1984), Bashir et al. (2005), Ashfaq et al. (2007)
Vegetable marrow	*Cucumber mosaic virus*	Pink and Walkey (1984), Walkey and Pink (1984)
Watermelon	*Watermelon bud necrosis virus*	Pandey and Pandey (2001)

(a) **Resistance in poaceaeous crops**

Some of the crops having resistance to virus diseases include: Tungro virus in rice (Anjaneyulu 1977; Anjaneyulu et al. 1982; Mohanty et al. 1989; Srinivasulu and Jeyarajan, 1989; Tarafdar et al. 2001; Prasad et al. 2004); *Rice yellow mottle virus* in rice (Konate and Fargette 2001; Onwughalu et al. 2011); *Sugarcane mosaic virus* in sorghum (Teakle and Pritchard 1971; Henzell et al. 1982); *Barley stripe mosaic virus* in barley (Timian and Sisler 1955; Jackson and Lane 1981; Catherall 1984; Timian and Franckowiak 1987; Edwards and Steffenson 1996). *Maize dwarf mosaic virus* (Miao-Hongqin et al. 1998) and *Maize streak virus* in maize (Welz et al. 1998; Salaudeen et al. 2013); and *Wheat streak mosaic virus* in wheat (Murugan et al. 2011; Sharp et al. 2002).

The examples of durable resistance include both monogenic and polygenic resistance, whereas theoretically polygenic resistance would be more durable.

(b) **Resistance in oil seed crops**

Resistant/tolerant cultivars/lines of groundnut against rosette disease in Nigeria were identified, especially the groundnut line ICGV-15-96805 and MS 16-791, following 2 years of trials (Iwo and Olorunju 2009). Even Nutman et al. (1964), Van der Merwe and Subrahmanyan (1997), and Subrahmanyam et al. (2002) have identified resistance in early groundnut breeding lines against *Groundnut rosette virus*. Similarly in Thailand two resistant lines of groundnut were identified against *Groundnut bud necrosis virus* (Pensuk et al. 2004). Brown et al. (2005) have identified resistant sources of *Tomato spotted wilt virus* in groundnut breeding lines (Fig. 2.17). At AVRDC - Taiwan, Green and Lee (1989) have identified tolerant and immune lines of soybean against *Peanut stripe virus*. Breeding for virus resistance in peanut has recently been reviewed (Nigam et al. 2012). Cotton which is primarily grown for fiber and also oil extracted from the seed is affected

Fig. 2.17 Yellowing and general late season decline in peanut resulting from *Tomato spotted wilt virus* (TSWV) infection. Susceptible cultivar on the *left* versus a resistant breeding line on the *right*. *Courtesy* Brown et al. (2005)

with whitefly-transmitted *Cotton leaf curl virus*. Daljeet Singh et al. (2006) have screened 783 entries and identified 17 to be resistant.

(c) **Resistance in tuber crops**

Some of the virus resistant/tolerant cultivars that are among tuber crops identified through field and lab screening in different tropical countries are: in sweet potato (Mwanga et al. 1995; Carey et al. 1997; Turyamureeba et al. 1998; Aritua et al. 1998a, b; Gibson et al. 2004; Miano et al. 2008); in potato (Wilson 2001); and in cassava (Thresh et al. 1994; Mba and Dixon 1997; Mignouna and Dixon 1997; Obura et al. 2003; Vanderschuren et al. 2012; Maruthi et al. 2013). The lines/cultivars identified to be resistant have developed symptoms that are less conspicuous than those of sensitive cultivars and in some cases they contain lower virus concentrations. IITA (Ibadan Nigeria) has released high yielding mosaic-resistant cassava as part of the Tropical Manihot Series (TMS) cultivars viz., 50395, 63397, 30555, 4(2) 1425 and 30572, which were cultivated in Ghana, Uganda and Nigeria. In Uganda, the area under cassava that deployed the TMS cultivars increased from 20 to 60 % in 1996 and the virus disease declined over 90 % (Beck 1982; Mahungu et al. 1994; Thresh et al. 1998; Felix Nweke 2009). Vanderschuren et al. (2012) have identified cassava TME-7 as resistant to *Cassava brown streak* and *Cassava mosaic virus*.

(d) **Resistance in wild species and utilization in breeding programme**

Certain wild species of crop plants have been screened when the resistant source was not identified among the cultivars. A few examples are provided for field crops and selected vegetables. For example in field crops, Culver et al. (1987) have identified resistant sources to *Peanut stripe virus* in *Arachis diogio* (PI46814 and PI468142), *A. helodes* (PI468144), *Arachis* sp. (PI468345) and of the rhizomatosae section (PI468174; PI468363 and PI468366). In India, peanut accessions of the *Arachis* section *A.cardenasii* (PI11558) was absolutely resistant to PStV even after sap, aphid and graft inoculation, while *A. chacoense* (PI4983), *A. chiquitana* (PI11560), *A. cardenassi* (PI11562 and PI12168) and accessions of Erectoides section, *A. stenophylla* (PI8215), and *A. paraguariensii* (PI8973) were resistant after aphid and sap inoculation only (Prasada Rao et al. 1989, 1991). At ICRISAT (India), Reddy et al. (2000) have identified resistance source against *Peanut bud necrosis virus* in peanut, *Arachis* germplasm accessions out of 83 wild *Arachis* species screened. Even Gururaj et al. 2002 have identified field resistance against PBNV in peanut. Similarly, sources of resistance to *Peanut mottle virus* in eight peanut entries of *Rhizomatosae* and *Arachis* sections were also identified (Melouk et al. 1984). Kalyani et al. (2007) identified 8 resistant accessions against *Tobacco streak virus* (TSV) in wild *Arachis*, which are cross compatible with *A. hypogaea* for utilization in breeding programme. Jones et al. (2008) have reported that *Lupinus angustifolius* exhibited milder symptoms on sap inoculation with *Alfalfa mosaic virus* and seed transmission was only 0.8 % in this host. From Canada, Singh (1985) reported clones 1726 and 1729 of PI473340 of *Solanum berthaultii* to be resistant to PSTVd, but it had high seed transmission. For PLRV in potato, resistance source was identified in *Solanum brevidens* and *S. acuale* OCH 13823 (Barker and Solomon 1990).

Provvidenti et al. (1978) working with curcubits, identified *Cucurbita ecuadorensis* and *C. foetidissima* to be resistant to CMV. While working with tospoviruses in pepper, Roggero et al. (2001), Black et al. (1996) and Latham and Jones (1998) have reported *Capsicum chinense* PI152225 and PI159236 to be resistant to TSWV and utilized these sources in breeding program.

In tomato, *Solanum* (*Lycopersicon*) *pimpinellifolium* (Venkata Ramana et al. 2006) and *S. peruvianum* were found to be resistant to TSWV (Stevens et al. 1992, 1994; Latham and Jones 1998). In different parts of the world, *Tomato yellow leaf curl virus* is the major limiting factor and resistance genes were identified in *Solanum pimpinellifolium, S. peruvianum, S. habrochaites,* and *S. chilense* (LA 1969) etc., and used in breeding programmes (Zakay et al. 1991; Fargette et al. 1996; Vidavsky and Czosnek 1998; Pico et al. 1998; Vidavsky et al. 1998). In India, *Tomato leaf curl virus* is the major limiting factor and Saikia and Muniyappa (1989); Muniyappa et al. (1991) have identified three *Solanum* (*Lycopersicon*) species to be resistant and have developed high yielding cultivars by using the wild sources in their breeding programme. Garcia-Cano et al. (2010) have identified resistance in wild tomato species against whitefly transmitted *Tomato chlorosis virus* (ToCV) which is one of the emerging virus problems in tomato. Rampersad and Umaharan (2003) from Trinidad and Tobago have identified the resistance sources to *Potato yellow mosaic virus* in five tomato species.

In wild lettuce (*Lactuca* spp) certain lines were identified to be resistant to TSWV (O'Malley and Hartmann 1989). In India, success has been achieved, in okra in transferring YVMV resistance genes from the wild species in to cultivated species which is considered as a classical research work of Indian plant breeding. YVMV resistant okra cultivars such as P7 (*Abelmoschus esculentus* x A. manihot ssp manihot) (Thakur 1986), Parbhani Kranti (*A. esculentus* x A. *manihot* (Jambhale and Nerkar 1986), selections 4 and 10 (Arka Anamika) (*A. esculentus* x A. *manihot* ssp *tetraphyllus*) (Dutta 1978; Arumugam et al. 1975) have been released. For more details one can refer the review of Pitrat (2012).

(e) **Developing cultivars with low seed transmission**

As large number of viruses are seed transmitted, attempts have also been made to locate cultivars in which the virus transmission through seed is either low or nil, since the reduction of seed transmission has a major impact on virus spread as they form the primary source of virus inoculum for plant to plant spread in the field. Soybean lines PI86736, improved Pelican and UFV-1 were found to possess low rate of seed transmission of SMV with superior agronomical qualities (Goodman and Oard 1980; Irwin and Goodman 1981). Against the same virus, Goodman and Nene (1976) also identified twelve lines of soybean in which seed transmission was not noticed. In Columbia (Central America), the lowest incidence of BCMV (0–1 %) was recorded in French bean lines viz., Pinto-114, Imuna, Great Northern

123 and 31, while in susceptible lines the extent of seed transmission was upto 54.4 % (Morales and Castano 1987). *Peanut mottle virus* was not seed transmitted in the peanut lines EC-76446 (292) and NCAC 17133 (RF), and these were used in resistant breeding programme (Bharatan et al. 1984). At Montana (U.S.), Mobet barley germplasm (PI467884) was identified for its resistance to seed transmission of three isolates of *Barley stripe mosaic virus* (Carroll et al. 1983). Combined resistance to seed transmission of four viruses viz., BICMV, TMV, CpAMV and CMV in *Vigna unguiculata* was identified by Mali et al. (1987) in genotypes like CoPusa-3, N-2-1 and V-16.

While screening the germplasm, locating a resistant line against a number of virus strains with good agronomical characteristic is a major objective. Often the cultivars found to be resistant to one virus strain turn out to be susceptible to another virulent virus. For example, several cultivars of pea were reported to be resistant to an isolate of *Pea early browning virus* at one site in Britain whereas at another site, all of them were susceptible (Harrison 1966). Plant breeders are therefore advised not to rely on results from only one site in making selections for resistance.

(f) **Breeding for vector resistance**

There has been substantially increased interest in breeding crops for resistance to insect pests as an alternative to the use of pesticide chemicals. This has been due to various factors, including emergence of resistance to insecticides in insects, the cost of developing new pesticides, and increasing concern regarding environmental hazards and the effects on natural enemies (Painter 1951). Along with these developments, there has been increased activity in breeding for resistance to invertebrates that are virus vectors. Some virus vectors are not pests on their own hosts but others, especially leafhoppers and planthoppers, where they may be severe as pests. In this situation there is a double benefit for achieving a resistant cultivar, sometimes with striking improvements in performance. The subject has been reviewed by Jones (1987) and Barker and Waterhouse (1999).

Some decades ago genetic resistance to the vector was advocated, but not much progress has been made in this area. However, the development of insect resistance to pesticides and the public awareness of environmental hazards resulting from their heavy use have generated new interest in breeding for insect resistance in plants.

Resistance to aphid vectors has been reported in a number of legume crops. Gunasinghe et al. (1988) reported the presence of soybean genotypes resistant to the aphid species *Aphis citricola*, *Myzus persicae* and *Rhopalosiphum maidis* and that such resistance reduced the spread of *Soybean mosaic virus* in the field. Even Gray et al. (1986) have observed virus suppression and aphid resistance effects on spread of *Watermelon mosaic virus*-2 in *Cucumis melo*. Amin (1985) also reported resistance to *Groundnut bud necrosis virus* through crop resistance to the thrips vector, *Frankliniella schultzei*. In pepper, from the Netherlands, thrips resistant cultivars were reported by Maris et al. (2003). In soybean, the mechanism of resistance is directed against the vector due to leaf pubescence, whereas the mechanism of resistance in groundnut is not known. Among the virus diseases of maize, *Maize streak virus* is most important in cereals and Kairo et al. (1995) have identified the resistance against *Cicadulina mbila*, the main vector species.

Aphid resistance of faba bean to *A. craccivora, A. fabae* and *Acyrthosiphon pisum* has been reported (Clement et al. 1994). In Egypt, over 1000 lines were screened for aphid resistance, and 36 were classified as resistant. The faba bean line BPL 23 was resistant to both *A. craccivara* and *A. fabae* (Bond et al. 1994). Whether or not the use of such cultivars could reduce FBNYV spread in the field awaits further evaluation.

The basis for resistance to the vectors is not always clearly understood, but some factors have been defined. In general terms, there are two kinds of resistance relevant to the control of vectors. First, non-preference involves an adverse effect on vector behavior, resulting in decreased colonization, and, second, antibiosis reproduction and survival after colonization has occurred. These two kinds of factors may not always be readily distinguished. Some specific mechanisms for resistance are: sticky material exuded by glandular trichomes such as those in tomato (Berlinger and Dahan 1987); heavy leaf pubescence in soybean (Gunasinghe et al. 1988) a-type hairs on *Solanum berthaultii* which, when ruptured, entrap aphids with their contents, and B-type hairs on the same host, which entangle aphids making them struggle more and so rupture more A-type hairs (Tingey and Laubengayer 1981); inability of the vector to find the phloem in *Agropyron* species (Shukla et al. 1987) - although this effect was not operative with an aphid vector of BYDV in barley (Ullman et al. 1988); and interference with the ability of the vector to locate the host plant. For example, in cucurbits with silvery leaves there was a delay of several weeks in the development of 100 % infection in the field with CMV and CIYVV (Davis and Shifriss 1983). This effect was likely due to aphids visiting plants with silvery leaves less frequently because of their different light-reflecting properties.

Combining resistance to a vector with some other control measures may sometimes be useful. For example, in field trials, rice tungro disease was effectively controlled by a combination of insecticide application and moderate resistance of the rice cultivar to the leafhopper vectors (Heinrichs et al. 1986). Sprays would be unnecessary in fully resistant cultivars were available.

There may be various limitations on the use of vector-resistant cultivars. Sometimes such resistance provides no protection against viruses. For example, resistance to aphid infestation in cowpea did not provide any protection against CABMV (Atiri et al. 1984). In addition, if a particular virus has several vector species, or if the crop is subjected to infection with several viruses, breeding effective resistance against all the possibilities may not be practicable, unless a non-specific mechanism is used (e.g. tomentose leaves). Perhaps the most serious problem to arise would be that of an emerging new vector biotype following the widespread cultivation of a resistant cultivar, as often happens following the repeated use of insecticides with the same mode of action.

This difficulty is well illustrated by the recent history of the rice brown planthopper (*Nilaparvata lugens*) (BPH). With the advent of high-yielding rice cultivars in South-east Asia in the 1960s and 1970s, the rice BPH and *Rice grassy stunt virus*, which it transmits, became a serious problem. Cultivars containing a dominant gene (*Bph1*) for resistance to the planthopper emerged. A new recessive resistance gene (*bph2*) was exploited in cultivars released between 1975 and 1983. They were grown successfully for a few years until a new planthopper biotype emerged that overcame the resistance (Thresh 1988, 1989). In an experiment to study the adaptation of three colonies of *N. lugens* to rice cultivars containing different resistance genes, Alam and Cohen (1998) showed that the *bhp1* and *bhp3* resistance genes were overcome more readily by colonies that had been exposed for about 10 years to those genes. However, cultivar IR-64 which contained *bph1* and some minor resistance genes showed greater durability of resistance than other cultivars. DNA markers to BPH-resistance genes are mapped in the rice genome and are used in breeding programs (Yencho et al. 2000).

In spite of these difficulties and the problems associated with the identification of plants with resistance to vectors, it seems certain that substantial efforts have been continued to be made to improve and extend the range of crop cultivars with resistance to virus vectors. The ultimate goal was to have a combination of resistance to the vector and to the virus.

(g) **Vector-resistant cultivars**

Resistance to a virus vector is likely to exert a complex influence on virus spread. Since nonpreference, antibiosis and tolerance are often combined into a single resistant cultivar, their relative contribution to the resistance as well as the overall magnitude of the resistance will influence the effect of resistance on virus spread. In addition, observed virus spread is the result of both primary and secondary spread and the relative importance of these and the effect of the resistance on them is important. The complexities involved are such that without a thorough understanding of the ecology of the virus and vector and the biology of vector resistance, it would be impossible to predict the effect of vector resistance on virus spread. Each combination of virus, vector and host resistance must be considered separately.

The impact of developing resistant/tolerant cultivar against insect vector resistance which spreads economically important virus diseases is well worked out. For rice tungro, which is transmitted by green leafhopper *Nephotettix virescence* and causes severe yield losses in many rice growing countries in Asia, resistance to the vector provides good protection against tungro under field conditions; green leafhopper resistant rice cultivar 1R-28 treated with insecticides had a very low incidence of RTV, whereas susceptible cultivar IR-22 under similar treatment had an extremely high virus incidence. In contrast, only insecticide-treated plots of the moderately resistant cultivar IR-36 showed lower infection of RTV than untreated plots in which RTV incidence was high (Heinrichs et al. 1986). These experimental results suggest that the use of resistant/tolerant cultivars are a major labor saving and economically sound management practice. Major efforts should be made to avoid yield losses in other economically important crop plants and the viruses that attack them.

The effectiveness of a vector-resistant cultivar for the spread of seed-borne plant viruses will mostly depend on the type and effectiveness of resistance, its relative importance in primary (introduction of virus from outside the crop) and secondary (spread of virus within the crop) virus spread, and virus-vector relationship (Kennedy 1976). The cumulative interaction of these factors may result in eventual differential effect on the spread of viruses. For example, Wilcoxson and Peterson (1960) found less incidence of *Pea seed-borne mosaic virus* in fields of aphid-resistant red clover cv. Dollard than in adjacement fields of the relatively susceptible cultivar. Lecoq et al. (1979) have identified *Cucumis melo* Songwhan Charmi (PI161375) to be resistant to CMV when tested with aphids (*M. persicae* and *A. gossypii*) and the mechanism was not specific to any virus strain. Another example of vector resistance is observed in soybean genotypes which are resistant to aphid species viz., *Aphid citricola*, *Myzus persicae* and *Rhophalosiphum maidis* by which *Soybean mosaic virus* spread was reduced under field conditions (Gunasinghe et al. 1988). Studies conducted in Nigeria indicated that cowpea lines (TVU 408-2; 201, 410 and 3273) although resistant to aphid (*A. craccivora*) but could not establish resistance for CpAMV (Atiri et al. 1984). In India, Mali (1986) reported that cowpea genotype P-1476 was resistant to *A. craccivora.* Arias et al. (2004) have reported Nataima-31, a cassava cultivar to be resistant to the whitefly *Aleurotrachelus socialis.* Kooner and Cheema (2007) have screened mungbean germplasm against *Bemisia tabaci* to manage *Mungbean yellow mosaic virus* spread. Intensive efforts were made all over the world for developing crops resistant against whitefly-transmitted virus diseases.

As resistance to a vector may result in an increased level of virus resistance, it is incumbent upon those breeding for resistance to consider the probable effect of vector resistance on virus spread. In addition, since resistance to one arthropod species may be associated with altered levels of susceptibility to other species, the potential impact on virus spread and of cross resistance to a vector species should not be ignored. Despite these dangers, the potential for controlling certain types of virus diseases through vector resistance is worthy of consideration. Research in vector behavior (viz., alighting and probing) as it is influenced by the host genotype, may provide further insight into this aspect of resistance to vectors and its relationship to virus spread.

While concluding the aspect of the scope of resistance breeding in plant virus and virus-like diseases management, it is most relevant to go through some of the constraints mentioned by Thresh (2003). These constraints hold true even for the virus diseases on different continents viz., Asia, Africa, and Latin America.

1. Considerable research efforts are needed to develop effective resistance breeding programmes, as these must take into account the other biotic and abiotic constraints, while meeting the often exacting requirements of farmers, consumers and processors. Even if resistant cultivars are developed, they may not be available because of the lack of an effective seed multiplication and distribution system, or because farmers are unaware of the benefits to be gained from their adoption.

2. There have been instances of resistant cultivars being released without adequate on-farm testing to ensure that these cultivars are suitable for adoption and that they meet the often stringent requirements of farmers and consumers. Some of the recent criticism of this so-called 'top down' approach and of the alleged lack of interaction between farmers and researchers is excessive and unjustified. There is undoubtedly more need for outreach for greater involvement of farmers in evaluation and selection and a participatory approach is now a requirement of many donors funding crop improvement programmes in sub-Saharan Africa.
3. Resistance may be associated with undesirable traits or the resistant cultivars may lack some of the desirable attributes of the susceptible cultivars being grown.
4. Resistance may be overcome due to the emergence or increased prevalence of virus strains that damage previously unaffected cultivars. Moreover, cultivars that are resistant in some areas may be susceptible in others. Thus, it may be difficult to develop and exploit broad-based resistance that is also durable.

2.20 Transgenic Approach

The most effective approach to control viruses relies on the use of resistant cultivars and/or rootstocks. Host resistance genes have been extensively exploited by traditional breeding techniques for the development of virus resistant plants. However, host resistance has been identified for a few viruses only and a limited number of commercial elite crop cultivars and rootstocks exhibit useful resistance. Therefore, engineered resistance is attractive to complement conventional breeding approaches, in particular when resistant materials with desired horticultural characteristics have not been developed successfully or when no host resistance sources are known. Therefore, the transgenic approach would be more appropriate in situations where sufficient levels of resistance to the virus are not available in the related germplasm or the resistance is difficult to transfer by normal crossing techniques because of either reproductive isolation or linkage with other undesirable traits. Actually, the deployment of virus resistant transgenic plants has become an important strategy to implement effective and sustainable control measures against major virus diseases.

The introduction and expression of nucleotide sequences of viral and non-viral origin in plants to confer virus resistance were made possible by gene splicing technologies, development of promoters that ensure constructive expression of genes, and advances in techniques that facilitate transformation of a range of crop plants. Genetic transformation for disease resistance (the creation of genetically modified organisms, GMOs) has also proved be quite effective, as virus-induced diseases have lent themselves to this approach far more often than diseases caused by other pathogen groups, and there are some notable success stories. For example, the papaya industry in Hawaii was saved by selections that were genetically engineered for resistance to *Papaya ringspot virus* (Gonsalves 2004; Fuchs et al.

2004). In general, conventional materials are more accepted in the market place than are GMOs, especially in Europe. As consumers become better educated about the benefits and safety of GMOs, a greater acceptance of these products may occur.

There are a number of different strategies for using transgenic resistance against plant viruses (Culver 1995; Mandahar 1999).

(1) Pathogen-derived resistance, relating to the integration of pathogen components that interfere with the normal life cycle of the virus.
(2) Pathogen-targeted resistance, involving the integration of components that specifically target viral genes and their products to render them non-functional.
(3) Integration of existing host resistance (R) genes into non-resistant hosts.

Pathogen-derived resistance (PDR) technique is one of the approaches focused on for the development of resistance in transgenic plants against viral infection. The advantages and setbacks of genetic engineering for crop improvement are clearly exemplified by virus resistances in transgenic plants. Introducing more than one or two transgenes is usually problematic because of "transgenic silencing". Hence, at the present state of the art of genetic engineering it has fallen short of producing a remedy to all the pathogens of a given crop. It is not surprising that genetic transformation for virus resistance became one of the first approaches of its kind to protect plants against pathogens. Advances in the development of virus resistant transgenic plants (VRTPs) have been reviewed from time to time by Gasser and Fraley (1989), Beachy (1993), Fitchen and Beachy (1993), Scholthof et al. (1993), Hackland et al. (1994), Lomonossoff (1995), Culver (1995), Pappu et al. (1995), Baulcombe (1996), Prins and Goldbach (1998), Reimann-Phillipp (1998), Aaziz and Tepfer (1999), Ranjithkumar and Savithri (1999), Bendahmane and Beachy (1999), Walsh (2000), Arif and Hassan (2000), Jain and Varma (2000), Louis (2000), Callaway et al. (2001), Kundu and Mandal (2001), Tepfer (2002), Varma et al. (2002), Baleswar Singh et al. (2003), Dasgupta et al. (2003), Prins (2003), Sreenivasulu and Subba Reddy (2006), Varma (2006), Sudarshana et al. (2007), Fuchs and Gonsalves (2007), Prins et al. (2008), Reddy et al. (2009), Wani and Sanghera (2010a, b).

2.20.1 Protein mediated resistance

The coat protein of virus normally has a protective function, insulating the virus nucleic acid (RNA or DNA) from environmental assaults, such as inactivation by ultra-violet light or digestion by host cell enzymes. In some cases the protein has a role in host recognition in early stages of infection. One of the approaches is introduction of coat protein genes, either as full length or truncated constructs, which will be expressed in genetically engineered (transgenic) plants in order to produce plants protected against virus infection (Hamilton 1980; Hackland et al. 1994; Murray 1995; Galun and Breiman 1997; Varma et al. 2002). This strategy is based on the concept of pathogen-derived resistance where the introduction of viral sequence into plants could interfere with the viral life cycle leading to resistance against the virus (Sanford and Johnston 1985). It may also help in the transformation of plants to enhance their virus resistance. This approach, which is an attempt to simulate natural cross-protection between virus strains, seems applicable to a wide range of viruses. The coat protein strategy results primarily in symptom retardation and is sensitive to the strength of the viral inoculum and to the amount of coat protein expressed (Powell-Abel et al. 1986), this is known coat-protein mediated resistance (CPMR). This approach has been demonstrated in a number of virus-host systems (Table 2.6).

Genetic modification of plants with viral CP genes is an example of pathogen-derived resistance that has been used successfully to produce virus resistant plants (Lomonossoff 1995; Beachy 1997). Throughout the globe, transgenic plants were developed through coat-protein mediated resistance in different crops and are presented in Table 2.6. In certain crops transgenics were developed against a number of virus diseases, for example transgenic squash resistant to *Cucumber mosaic virus*, *Zucchini yellow mosaic virus* and *Watermelon mosaic virus* was released during 1996 (Tricoli et al. 1995; Clough and Hamm 1995; Fuchs et al. 2004). The level of protection conferred by CP genes in transgenic plants varies from immunity to delay and attenuation of symptoms. The coat protein mediated resistance (CPMR) has been reported for more than 35 viruses representing more than 15 different taxonomic groups including the tobamo-, potex-, cucumo-, tobra-, carla-, poty-, luteo-, and alfamovirus groups. The resistance requires that the CP transgene be transcribed and translated.

2.20.2 Movement-Protein-Mediated Resistance

Movement protein gene mediated resistant phenotypes usually include reduction in virus movement throughout the infected plants. This could be brought above by expressing a defective movement protein conferring resistance against viruses. Unlike other strategies, this approach offers attractive possibility to confer broad-spectrum resistance to related and unrelated viruses.

The efficiency of cell-to-cell virus movement is important in determining pathogenicity, virulence and, in some cases, the host range of a plant virus. This aspect has been reviewed by Atabekov and Taliansky (1990), Maule (1991). When the efficiency of the transport function and the rate of virus movement are reduced, the plant acquires a certain level of resistance to virus infection. Production of dysfunctional or partially active movement proteins (MP) in transgenic plants is assumed to confer resistance to the wild-type (wt) virus by competition between wt virus-coded MP and the preformed modified MP (mMP). Intercellular spread involves plasmodesmata, the channels that traverse plant cell walls and provide symplastic continuity between cells and tissues. Because several MPs were shown to accumulate in plasmodesmata, and interacts with them, virus MPs are also used

Table 2.6 Examples of coat protein-mediated resistance (CPMR) in certain plant species against virus infections

Source of CP gene	Plant species transformed	Virus protection exhibited to	References
AMV	Alfalfa	AMV	Halk et al. (1989)
	Tobacco	AMV	Van Dun et al. (1987), Loesch-Fries et al. (1987), Turner et al. (1987), Van Dun et al. (1988a, b), Halk et al. (1989)
	Tomato	AMV	Turner et al. (1987)
	Pea	AMV	Timmerman et al. (2001)
ArMV	Tobacco	ArMV	Cooper et al. (1994)
BaMMV and BaYMV	Barley	BaMMV and BaYMV	Hagio et al. (1996)
BGMV	Common bean	BGMV	Aragao and Faria (2009)
BNYVV	Sugarbeet	BNYVV	Kallerhoff et al. (1990), Mannerlof et al. (1996)
BPMV	Soybean	BPMV	Di et al. (1996)
CMV	Cucumber	CMV	Slightom et al. (1990), Quemada et al. (1991)
	Muskmelon	CMV	Gonsalves et al. (1994)
	Pepper	CMV	Lee et al. (2009)
	Squash	CMV	Tricoli et al. (1995), Fuchs et al. (2004)
	Tobacco	CMV	Cuozzo et al. (1988)
	Tomato	CMV	Provvidenti and Gonsalves (1995), Gielen et al. (1996), Fuchs et al. (1996)
GFLV	Grapevine	GFLV	Mauro et al. (1995), Ling et al. (1997), Fuchs et al. (2007)
IPCV	Peanut	IPCV	Sharma et al. (2006)
LMV	Lettuce	LMV	Zerbini et al. (1995)
MiLV	Lettuce	MiLV	Kawazu et al. (2009)
PBNV	Groundnut	PBNV	Chander Rao et al. (2006), (2012), Sharma et al. (2006), Sandeep Bag et al. (2007)
	Tomato	PBNV	Jagadish et al. (2006), Raja and Jain (2008)
PEMV	Pea	PEMV	Chowrira et al. (1998)
PLRV	Potato	PLRV	Kawchuk et al. (1990, 1991), Vander Wilk et al. (1991), Presting et al. (1995), Tacke et al. (1996)

(continued)

Table 2.6 (continued)

Source of CP gene	Plant species transformed	Virus protection exhibited to	References
PPV	Tobacco	PPV	Regner et al. (1992)
PRSV	Papaya	PRSV	Fitch et al. (1992), Lius et al. (1997), Yeh et al. (1997, 1998), Silva Rosales et al. (2000), Gonsalves (2002), Bau et al. (2003), Davis and Ying (2004)
PSbMV	Peas	PSbMV	Jones et al. (1998)
PVS	Potato	PVS	MacKenzie and Tremaine (1990)
PVX	Potato	PVX	Hoekema et al. (1989)
	Tobacco	PVX	Hemenway et al. (1988)
PVX + PVY	Potato	PVX, PVY	Lawson et al. (1990), Kaniewski et al. (1990)
PVY	Potato	PVY	Lawson et al. (1990)
RTBV	Rice	RTBV	Uma Ganesan et al. (2009)
RTSV	Rice	RTSV	Sivamani et al. (1999), Tyagi et al. (2006)
RYMV	Rice	RYMV	Kouassi et al. (2006)
SCMV	Sugarcane	SCMV	Smith et al. (1996), Joyce et al. (1998)
SMV	Soybean	SMV	Furutani et al. (2006)
	Tobacco	PVY, TEV	Stark and Beachy (1989)
SPFMV	Sweet potato	SPFMV	Murata (2001), Okada et al. (2001)
SrMV	Sugarcane	SrMV	Ingelbrecht et al. (1999)
TLCV	Tomato	TLCV	Raj et al. (2005)
TYLCV	Tomato	TYLCV	Kunik et al. (1994)
TMV	Tobacco	TMV	Powell-Abel et al. (1986), Nelson et al. (1987)
	Tobacco	ToMV, TMGMV	Stark et al. (1990)
	Tobacco	PVX, CMV, AMV, SHMV	Anderson et al. (1989)
	Tomato	TMV, ToMV	Nelson et al. (1988), Sanders et al. (1992)
TRV	Tobacco	TRV	Van Dun and Bol (1988), Angenent et al. (1990), Ploeg et al. (1993)
	Tobacco	PEBV	Van Dun and Bol (1988)

(continued)

Table 2.6 (continued)

Source of CP gene	Plant species transformed	Virus protection exhibited to	References
TSV	Peanut	TSV	Sandeep Bag et al. (2007)
	Sunflower	TSV	Sujatha et al. (2012)
	Tobacco	TSV	Van Dun et al. (1988b)
TSWV	Peanut	TSWV	Yang et al. (2004a)
	Tobacco	TSWV	Gielen et al. (1991), MacKenzie and Ellis (1992)
TuMV	Rape seed	TuMV	Lehmann et al. (2003)
TVMV	Tobacco	TVMV, TEV	Murphy et al. (1990)
TYLCV	Tomato	TYLCV	Kunik et al. (1994)
WMV	Squash, Cantaloupe	WMV	Clough and Hamm (1995)
WMV-2	Squash	WMV-2	Tricoli et al. (1995), Fuchs et al. (2004)
WSMV	Wheat	WSMV	Hansen et al. (1995), Sivamani et al. (2000, 2002), Sharp et al. (2002)
ZYMV	Squash	ZYMV	Fuchs et al. (2004), Clough and Hamm (1995), Fuchs and Gonsalves (1995), Tricoli et al. (1995)
	Cantaloupe	CMV, ZYMV	Clough and Hamm (1995)

to study protein targeting in plant cells as well as the nature and composition of the plasmodesmata.

Geminivirus DNA is replicated in nuclei and transport of the single-stranded genome from nucleus to cytoplasm requires one type of viral protein, whereas a second protein transports the DNA to adjacent cells (Sanderfoot and Lazarowitz 1996). In contrast, RNA-containing viruses replicate in the cytoplasm. Resistance conferred by the movement gene has a wide spectrum. The transgenic plants are not only resistant to infection by the same virus from which the gene is derived, but also resistant to other, even unrelated viruses. This has been demonstrated for at least six group of viruses (Cooper et al. 1995; Hull 2002).

Indeed, transgenic plants that contain dMP from TMV show resistance to several tobamoviruses and also to taxonomically distant viruses like AlMV, *Cauliflower mosaic virus* (CaMV) and other viruses (Lapidot et al. 1993; Cooper et al. 1995). A mutation that disrupted a putative nucleotide binding site of one of three MPs of *White clover mottle virus* (WClMV) conferred resistance to several different potexviruses (Cooper et al. 1995).

Fluorescence microscopy showed that MP accumulates in several different subcellular locations, including microtubules (Heinlein et al. 1995; McLean et al. 1995) and plasmodesmata (Padgett et al. 1996), and that MP is associated with the endoplasmic reticulum; however, the function of MP in each of these sites remains to be determined. MP produced in transgenic plants can enable MP-mutants of TMV (Deom et al. 1990; Holt and Beachy 1991) to move to adjacent cells, and it was predicted that certain defective mutants of MP (dMP) would restrict infection by TMV and perhaps other viruses (Deom et al. 1992).

The possibility of obtaining engineered resistance to viruses having a single MP or a Triple Gene Block (TGB) simultaneously was pursued by Ares et al. (1998) by engineering both the 24 kDa protein of PVX and the non-structural 30 kDa protein of TMV. Several experimental results indicated that 30 kDa is involved in some way with the movement of TMV through the host. Tomenius et al. (1987) have demonstrated the role of MP in both local and systemic spread of the virus in the host. In spite of the lack of similarity between the two proteins, and of the different genome organization of the potex- and tobamoviruses, systemic resistance was indeed obtained in reciprocal challenges. The postulated mechanism relies on the existence of common functional domains shared by the two proteins, and is supported by the possibility of complementing movement-defective PVX and TMV with heterologous MPs. These domains are thought to compete for common cellular factors required for movement and interfere with the movement of the challenging virus. Alternatively, a non-specific host defence response may be activated. In this case, however, over-expression of MP should lead to a broader range of resistance. It is anticipated that knowledge of MP structure and *in vivo* function(s) will lead to development of other mutant proteins or peptides that act as dominant negative inhibitors to block the local and systemic spread of many different viruses with high efficiency.

2.20.3 Nucleic Acid-Based Protection

The mechanism of RNA mediated resistance involves RNA silencing in which sequence-specific RNA degradation occurs (Kawazu et al. 2009). Even homology dependent selective degradation of RNA, RNA-interference (RNAi) or post transcriptional gene silencing (PTGS) is involved in several biological processes including adaptive defense against viruses in plants (Yu and Kumar 2003; Herr 2005). The first demonstration of RNAi mediated virus resistance was shown by Waterhouse et al. (1998) against PVY in transgenic tobacco plants. The utilization of RNAi technology has resulted in inducing immunity reaction against several plant viruses in different hosts viz., rice, tomato, potato, cassava, common bean and the details are presented in the Table 2.7.

The advantages of RNA-mediated resistance are as follows: The silencing of transgenes can impose great drawbacks in the transgenic expression of proteins in plants, but for pursuing virus resistance, it has proven to be very powerful strategy (Finnegan and McElroy 1994). Resistance based on RNA is expected to be more durable than protein mediated resistance. Prins and Goldbach (1996); de Haan (1998) and Wani and Sanghera (2010) have reviewed the RNA-mediated protection.

Table 2.7 RNAi mediated approach for developing transgenic plants against plant viruses

Target plant	Virus	References
Arabidopsis	*Cabbage leaf curl virus*	Turnage et al. (2002)
Barley, wheat	*Barley stripe mosaic virus*	Holzberg et al. (2002), Scofield et al. (2005)
Barley, rice, maize	*Brome mosaic virus*	Ding et al. (2006)
Cassava	*African cassava mosaic virus*	Fofana et al. (2004)
	Cassava brown streak Uganda virus	Yadav et al. (2011)
	Cassava brown streak disease	Patil et al. (2011)
Citrus	*Citrus psorosis virus*	Zanek et al. (2008)
Common bean	*Bean golden mosaic virus*	Kenny Bonfim et al. (2007)
Lettuce	*Mirafiori lettuce virus*	Kawazu et al. (2009)
Pepper	*Cucumber mosaic virus*	Lee et al. (2009)
Pea	*Pea early browning virus*	Constantin et al. (2004)
Potato	*Potato virus X* and *Potato virus Y*	Bai et al. (2009)
Rice	*Rice yellow mottle virus*	Kouassi et al. (2006)
	Rice tungro bacilliform virus	Tyagi et al. (2008), Upadhyaya et al. (2001)

(continued)

Table 2.7 (continued)

Target plant	Virus	References
Tomato	*Tomato golden mosaic virus*	Day et al. (1991)
	Tomato leaf curl virus	Praveen et al. (2010)
	Tomato yellow leaf curl virus	Bendahmane and Gronenborn (1997), Yang et al. (2004b)
N. benthamiana	*Turnip crinkle virus*	Vasudevan et al. (2008)
	Tomato bushy shunt virus	Hou and Qiu (2003)
	Tomato golden mosaic virus	Peele et al. (2001)
	Grapevine leafroll-associated virus-2	Ling et al. (2009)
	Grape vine A	Brumin et al. (2009)
	Poplar mosaic virus	Naylor et al. (2005)
N. benthamiana, S. tuberosum	*Potato virus X*	Ruiz et al. (1998), Faivre-Rampant et al. (2004)
Nicotiana tabacum	*Satellite tobacco mosaic virus*	Gossele et al. (2002)
N. benthamiana, N. tabacum	*Tobacco mosaic virus*	Kumagai et al. (1995)
N. benthamiana, Arabidopsis, tomato, *Solanum* species, Chilli pepper, opium poppy, *Aquilegia*	*Tobacco rattle virus*	Ratcliff et al. (2001), Brigneti et al. (2004), Chung et al. (2004), Hileman et al. (2005), Gould and Kramer (2007)
N. benthamiana, Tomato, *N. glutinosa, N. tabacum*	*Tomato yellow leaf curl China virus-associated b DNA satellite*	Tao and Zhou (2004)

Source Shabir Hussain Wani and Gulzar S Sanghera (2010)

2.20.4 RNA- and Protein-Mediated Resistance

Some researchers have found yet more complex results in their studies on transgenic resistance. Pang et al. (1994) reported that the mechanisms involved in the *CP* gene-mediated resistance against tospoviruses were variable. When the resistance was against closely related isolates, it was RNA-mediated, but when it was against more distantly related tospoviruses, it was protein-mediated. Wintermantel and Zaitlin (2000) suggested that, in tobacco plants transformed with the *Cucumber mosaic virus* (CMV) family *Bromoviridae*, genus *Cucumovirus* replicase gene, the resistance obtained is likely a result of a complex mechanism involving both transgene mRNA and its expressed protein. During 2000, Goregaoker et al. have demonstrated that in TMV replicase-mediated resistance, both RNA and protein are involved in protecting against the challenge virus

2.20.5 Replicase-Mediated Resistance (Rep-MR)

Generation of transgenic plants with virus resistance has been attempted by using Replicase (Rep) protein-mediated resistance against the plant viruses. Such resistance is quite strong but usually limited to the virus strain from which the transgene obscured. Golemboski et al. (1990) were the first to show the application of rep genes in tobacco against TMV in plants containing the 54 kDa putative Rep gene.

Rep-MR to PVY and AMV has also been engineered and the mechanism of resistance was shown to be protein based since only mutant replicase proteins conferred resistance. In contrast, Rep-MR to PVX, *Cowpea mosaic virus* and *Pepper mild mottle virus*, as well as to others, is dependent on the gene transcript and probably involves PTGS. Resistances have been developed for several other viruses namely *Pea early browning virus* (MacFarlane and Davies 1992) PVY (Audy et al. 1994) and CMV (Hellwald and Palukaitis 1995).

Genes that encode complete or partial replicase proteins can confer near immunity to infection that is generally, but not always, limited to the virus strain from which the gene sequence was obtained. Replicase-mediated resistance (Rep-MR) gives nearly immune type reaction and there is a substantial inhibition of virus replication in initially inoculated cells and some cell to cell movement, but the infection does not spread from the inoculated leave and no systemic disease develops. Although it was suggested that certain examples of Rep-MR are RNA-rather than protein-mediated (Baulcombe 1996), other examples require an open reading frame and, apparently, production of protein (Carr and Zaitlin 1991; Zaitlin et al. 1994). A truncated mutant of replicase derived from a *Cucumber mosaic virus* (CMV) subgroup I virus conferred high levels of resistance in tobacco plants to all subgroup I CMV strains, but not to subgroup II strains or other viruses (Zaitlin et al. 1994). In Rep-MR against PVY (Audy et al. 1994) and AMV (Brederode et al. 1995), mutant but not wild-type replicase conferred resistance to infection; a similar approach provided resistance to *Tomato yellow leaf curl virus* (TYLCV), a geminivirus (single-stranded DNA genome) (Noris et al. 1996; Antignus et al. 2004; Yang et al. 2004b).

Gene constructs of Rep genes that have been used for resistance include full-length, truncated or mutated genes. Many of the above resistance responses have now been shown not to require protein synthesis and to be mediated at the RNA level. This type of resistance remains confined only to a narrow spectrum of viruses, the spectrum being narrower than that of CPMR. To make the resistance broad-based, it may be necessary to pyramid such genes from several dissimilar virus sources into the test plant genome. However, the resistance generated by the use of Rep sequences is very tight; a high dosage of input virus can be resisted easily by the transgenic plant.

2.20.6 Transgenic Approaches Against Viroid Diseases

Antisense RNA molecules have been successfully deployed in controlling viroid replication in plants. Molecules targeting the minus strand of a viroid seems to be more effective than those against the plus strand (Atkins et al. 1995). Further they reported that transgenic tomato plants that made antisense RNA against *Citrus exocortis viroid* exhibited slight reduction in viroid accumulation, whereas antisense RNA against the plus strand actually resulted in the increased viroid accumulation.

Ribozymes are short RNA molecules that can catalytically cleave target RNA molecules with high specificity. Yang et al. (1997) showed that ribozymes could be used effectively to control replication of viroid RNA in the nucleus. They have also showed that transgenic potato plants expressing a ribozyme against PSTVd minus-strand RNA were resistant to the viroid. This is in clear contrast to the antisense RNA approach for the same viroid-host interaction. Atkins et al. (1995) also showed that ribozymes targeting the plus strand were not as effective as the once targeting the minus-strand of the viroid. The difference between the antisense RNA approach of Atkins et al. (1995) and ribozyme approach of Yang et al. (1997) may be due to differences in the efficiency of viroid replication in different host plants are stability between the plus- and minus-trend RNAs of viroids.

2.20.7 Performance of Transgenic Plants

PBNV transgenic peanut lines were tested both in glass house and field conditions (Chander Rao et al. 2006). In the subsequent years Chander Rao et al. (2012) have identified three transgenic lines (GNPBNV-B-1-2-1, GNPBNV-A-K-3-4, GNPBNV-B-11-2-3) that showed considerable reduction in PBNV incidence, only one (GNPBNV-B-1-2-1) showed less than 20 % disease incidence and emerged as somewhat resistant to PBNV. Even Zheng et al. (2008) have reported transgenic clones of sugarcane against SCMV, which have grown rapidly with increased tillering ability and the incidence of SCMV was 0–24 % and the control had 91.7 % virus incidence.

Field trials with transgenic tomato have reduced the incidence of CMV, resulting in higher fruit size and yield (Fig. 2.20).

The use of genetically altered squash and cantaloupe cucurbits resulted in a remarkable reduction in incidence of ZYMV and WMV even under conditions of high vector numbers and extreme disease pressure. All of the transformed squash and cantaloupe lines had much smaller disease progress curves than the controls (Fig. 2.18), and the transgenic squash and cantaloupe lines continued to produce significantly greater amounts of marketable fruit than did the non-transformed control (Fig. 2.19) until the experiment was terminated (Clough and Hamm 1995) (Fig. 2.20).

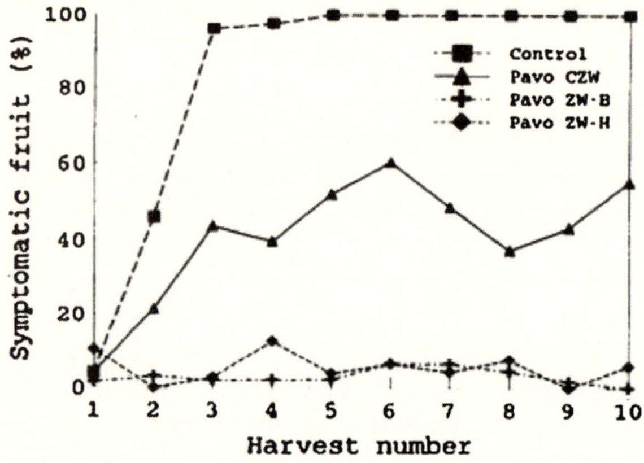

Fig. 2.18 Virus symptom development in transgenic and nontransgenic yellow crookneck squash fruit. *Courtesy* Clough and Hamm (1995)

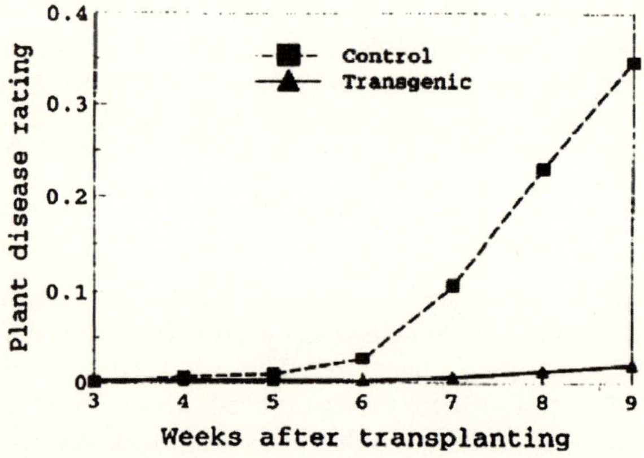

Fig. 2.19 Watermelon mosaic and Zucchini yellows mosaic virus progression in transgenic and nontransgenic cantaloupe plants. Plant rating scale: healthy = 0, possibly infected = 1, and infected = 2. *Courtesy* Clough and Hamm (1995)

At Hawaii between 1992 and 1998 the annual papaya output had declined from 25000 tonnes to under 12000 tonnes, but by 2001 it was back to 20000 tonnes most of which was from transgenic papaya cultivars (Murphy 2011). At Taiwan field evaluation of transgenic papaya lines carrying the coat-protein genes of *Papaya ringspot virus* have indicated positive trend (Bau et al. 2004).

The available evidence indicates that for TMV, the CP confers protection and the mRNA and 3' untranslated sequences do not contribute to the observed virus resistance (Powell et al. 1990). CP plays a major role in vector transmission; CPMR confers additional advantage of resistance to vector inoculation in several

Fig. 2.20 Performance of CMV resistant transgenic tomato plants. *Courtesy* Dennis Gonsalves, Cornell University

cases. e.g., PVX, PVY, and CMV. TSWV nucleocapsid transgene containing tomato plants were resistant to thrips and transgenic rice expressing high level of *Rice stripe virus* CP gene resistant to virus inoculation by plant hopper. A remarkable increase in the yield of the several crops established by CPMR technology has been reported (Dasgupta et al. 2003).

With the aim of broadening resistance to viruses that belong to different groups, CP genes from more than one virus group and genus have been used successfully. Virus resistant transgenic cantaloupe and squash plants with CP genes derived from ZYMV, WMV-2 and CMV, and potato with CP genes of PVX, PVY and PLRV have been produced (Kaniewski et al. 1990; Tacke et al. 1996; Fuchs et al. 1997).

Numerous observations suggest that a viral protein in a transgenic plant does not pose any threat to allergenic safety. Most notable is that virus-infected conventional crops have been consumed by humans with no apparent ill effects since plants and plant products have been consumed as food. Also, cross-protection, the deliberate inoculation of plants with mild virus strains to protect plants from economic damage of severe challenge virus strains has been practiced for many years in some vegetable and fruit crops with no documented adverse effect on human health. In addition, virus-resistant transgenic squash and papaya have been released for a number of years and their products are not known as potential allergens, nor is there evidence of increased allergenicity due to their consumption. Together, allergenicity does not appear to be a significant risk for transgenic crops expressing virus-gene fragments.

The stability and durability of the engineered virus resistance are important from a disease management standpoint. It is desirable to release elite cultivars and rootstocks that perform well over time. This is true whether resistance is achieved through conventional breeding or agricultural biotechnology. Thorough screening prior to the release of resistant plants is important to determine the breadth of resistance to virus infection and predict its durability.

2.20.8 Benefits Derived from Transgenic Crops

Transgenic plants can contributed to:

(a) Increased productivity per unit area and inputs such as water, fertilizer, and energy to meet the demands of the growing and economically ascendant population looking for quality, convenient foods, free from pesticide residues.
(b) Enhanced stability of production by incorporating resistance to various biotic and abiotic stresses.
(c) Development of sustainable production systems minimizing the use of chemical pesticides.
(d) Prevention of post-harvest losses in fruits and vegetables by enhancing shelf life.
(e) Improving the end-use qualities.
(f) Reducing the adverse environmental impact of the production systems.
(g) Reducing production costs.

2.20.9 Risks Associated with Transgenic Crops

Genetically modified crops deal with the growth of plants. Transgenic plants exposed to the open environment and the interaction takes place with other organisms in the field. Many of them contain toxic gene and antibiotic resistance marker with them. Therefore, genetic modification in agriculture has become a more sensitive issue than those experiences with recombinant biotherapeutics. Transgenic plants need special attention because they are exposed to environment. Until now, there is no major risk concern associated with the marketed transgenic crops such as cotton, tomato, corn and soybean. Unfortunately, the public debate over the hazards of transgenic plant or transgenic food suffers from misinformation and misunderstanding of the basis of genetic manipulation in plant system. These GM foods carry a label and have been exposed to extensive field trials for safety and environmental impact before they are approved for commercialization. With the continuing accumulation of evidence of safety and efficiency and no harm to public or the environment, more and more transgenic plants and food are gaining acceptance and are used by people. Nevertheless, thorough assessment of the risk and safety associated with GM crops and food need complete evaluation before they are released to the environment.

The major safety concern associated with GM crops and GM food are as follows:

1. The effect of GM crops on environment and biodiversity
2. Gene pollution (escape of gene through pollen)
3. Toxicity of the GM plant due to altered metabolism
4. Safety, toxicity and allergenicity of the GM food
5. Insect and herbicide resistant cultivars of plants
6. Undesirable effect of transgenic plant on non-targeted or beneficial insect or plant in the environment.

2.21 Bio-Safety Regulations Against GM Crops

Genetically modified plants can spread the transgene to other plants or theoretically even to bacteria. Depending on the transgene, this may pose a threat to the environment by changing the composition of the local ecosystem. Therefore, in most countries environmental studies are required prior to the approval of a GM plant for commercial purposes, and a monitoring plan must be presented to identify potential effects which have not been anticipated prior to the approval.

Application of GM technology for commercial crop production is faced with a number of issues and challenges the most important being safety of environment, and human and animal health (Grumet and Gifford 1998; Martelli 2001; Khetarpal 2002; Tepfer 2002; Varma 2006; Philippe 2007). These concerns are based on the argument that recombinant DNA-based GM technology differs from traditional breeding in that totally new genes using potentially risky technology are transferred between widely unrelated organisms, and the location of these genes on the recipient genome is random, unlike when gene transfer takes place through conventional breeding. These differences demand that adequate laboratory safeguards are used and plants developed by GM technology are rigorously assessed for their performance as also for the likely risks they pose. The present chapter has also provided information regarding GMO's potential risks to the environment and human health and a few other concerns related to the application of biotechnology in agriculture.

Because of the risk associated with genetically modified organisms, there were concerns worldwide to control it. It is also important to protect the biological diversity of the nature while releasing or accidental escape of GMO to the environment. This leads in to the formation of bio-safety protocols dealing with genetically modified organisms. The origin of the Bio-safety Protocol was found in the UN Convention on Biological Diversity, which was signed by over 150 governments at the Rio "Earth Summit" in 1992, and which came into force in December 1993. In the Convention on Biological Diversity (CBD), it was acknowledged that release of GMOs (referred to in the CBD as 'living modified organisms' or LMOs) may have adverse effects on the conservation and sustainable use of biological diversity. All countries that signed up to the CBD were expected to:

(a) "Establish or maintain means to regulate, manage or control the risks associated with the use and release of living modified organisms resulting from biotechnology which are likely to have adverse environmental impacts, taking also into account the risks to human health." and
(b) "Consider the need for and modalities of a protocol setting out appropriate procedures in the field of the safe transfer, handling and use of any living modified organism resulting from biotechnology that may have adverse effect on the conservation and sustainable use of biological diversity."

In accordance with the precautionary approach contained in Principle 15 of the Rio Declaration on Environment and Development, the objective of the protocol is to contribute to ensuring an adequate level of protection in the field of the safe transfer, handling and use of living modified organisms resulting from modern biotechnology that may have adverse effects on the conservation and sustainable use of biological diversity, taking also into account risks to human health, and specifically focusing on transboundary movements.

The Cartagena Protocol on Bio-safety, the first international regulatory framework for safe transfer, handling and use of Living Modified Organisms (LMOs) was negotiated under the aegis of the Convention on Biological Diversity. The Protocol contains reference to a precautionary approach and reaffirms the precaution language in Principle 15 of the Rio Declaration on Environment and Development. The Protocol also establishes a Bio-safety Clearing-House to facilitate the exchange of information on living modified organisms and to assist countries in the implementation of the Protocol.

2.22 Induction of Systemic Resistance

Throughout the world attempts are made to induce resistance/virus inhibition/eliminate the virus from the infected plants by using different natural and chemical compounds. It was also aimed to induce systemic resistance in the healthy plants itself by using different compounds. In this direction some of the positive results are listed below which requires large number of field and laboratory confirmations.

(a) **Effect of Plant growth promoting rhizobacteria (PGPR) on virus diseases**

The use of beneficial soil microorganisms such as Plant growth promoting rhizobacteria (PGPR) for sustainable safe agriculture and management of certain plant diseases has increased globally. For example, studies of Alstrom (1991) and Wei et al. (1991) have indicated that some (PGPR) also act as inducers of systemic resistance in plants due to fungal, bacterial and viral diseases. Bergstrom et al. (1982) showed that resistance in cucumber against *Cucumber mosaic cucumovirus*

(CMV) could be induced by previously inoculating plants with *Colletotrichum orbiculare, Pseudomonas syringae* pv. *lachrymans,* or *Tobacco necrosis virus* (TNV). Raupach et al. (1996) have reported that PGPR strains 89B-27 (*Pseudomonas fluorescens*) and 90-166 (*Serratia marcescens*), have protected against CMV by seed treatment. The treatment has significantly and consistently reduced mean numbers of systematic plants when CMV was inoculated on to cotyledons. Plants treated with PGPR did not develop initial symptoms until 14 days after CMV inoculation and remained symptomless throughout the experimental period. In all experiments, the area under the disease progress curve (AUDPC) was significantly lower with strain 90-166 was also significantly lower than with strain 89B-27. These results suggest that PGPR should be further evaluated for their potential to contribute toward management of viral plant diseases. Even the disease severity in CMV-susceptible tomato cultivar Marion was significantly decreased by treatments with 89B-27 and 90-166 compared to the nonbacterized disease control. It has been demonstrated that in Western Ghats (India) where banana (*Musa* spp.) hill banana, is generally affected by *Banana bunchy top virus* (BBTV), and the studies of Jay Shankar Singh (2013) have indicated that the application of *Pseudomonas fluorescens* strain significantly reduced the BBTV incidence in hill banana under greenhouse and field conditions. Further studies should be conducted to determine whether the PGPR application against number of virus diseases will manage even at the field conditions. From India, Sudhakar et al. (2006) have tried PGPR in *Capsicum annum* against CMV. They have reported that the percentage of chilli plants with CMV symptoms in 2 PGPR treatments ranged from 12–18 % compared with 75.8 % of symptoms in control plants and resulted in significant enhancement of growth and protection against infection by CMV. Current and future progress in understanding of PGPR diversity, colonization ability, mechanisms of action, formulation, and application could facilitate their development as reliable components in plant virus management in sustainable agricultural systems.

In India Srinivasan and Mathivanan (2011) have recorded that plant growth promoting microbes (PGPM) like, *Bacillus licheniformis*, *Pseudomonas aeruginosa* and *Streptomyces fradiae* have effectively prevented the *Sunflower necrosis virus* (SNV) infection in both glasshouse and open field studies. ELISA data indicated lower titre values for localized infection in the PGPM-treated plants than in the control plants.

During 2011 from India, Patil Nisha et al. have studied the control of *Bhendi yellow vein mosaic virus* (BYVMV) in bhendi and also its whitefly vector by using rhizobacteria bioformulations which were applied through seed, soil and foliage. The treatments significantly reduced BYVMV incidence and also whitefly vector population, with a concomitant increase in plant growth and fruit yields under glasshouse condition. Of all the rhizobacterial isolates evaluated, fluorescent Pseudomonas 218 (1) was the most effective strain. It reduced the incidence of BYVMV to the maximum extent (up to 86.6 %) through induced systemic resistance by triggering defense molecules. All the rhizobial isolates resulted in significant reduction in the whitefly population. Pseudomonas 218 (1) showed the highest reduction in the whitefly vector population, to the extent of 83.3 % less than that of the control. Biosynthesis of peroxidase and PALase were improved by 79 and 47 %, respectively over the disease control.

(b) A new biological pesticide ‘Messenger’ that activates natural plant defense and growth systems was developed by Jones (2001a, b). The active ingredient of messenger is ‘harpin’, a protein produced in nature by certain bacterial plant pathogens. It is water soluble and granular formulations degrade rapidly after application. It can repel insect pests and provide effective control of some viral diseases including tomato mosaic and cucumber mosaic in tomato and pepper and *Beet curly top virus* in pepper.
(c) Pappu et al. (2000) have observed the use of plant defense activators such as ‘Actigard’ (acibenzolar-S-methyl and imidacloprid) potentially offers an alternative and effective management tool for TSWV supression in flue-cured tobacco. The effect of this compound (0.2–8 g a.i./7000) plants on pre-transplant application was tried by Csinos et al. (2001b). In tomato this chemical application was proved to be effective in managing TSWV (Awondo et al. 2012; Riley et al. 2012).
(d) Smith et al. (2003) observed resistance to CMV in cantaloupe induced by acibenzolar S methyl (ASM), which is a synthetic analogue of salicylic acid. Disease developed for use in novel strategy crop protection through systemic acquired resistance (SAR). It induces the systemic accumulation of chitinase, a marker protein for SAR in both greenhouse and field grown seedlings. ASM at 50 or 100 μg provides almost complete protection against *Colletotrichum lagenarium* and effectively delayed the spread of CMV.
(e) Anfoka (2000) determined the potential of benzo-(1,2,3)-thiadiazole-7-carbothioic acid S-methyl ester (BTH) to trigger systemic acquired resistance (SAR) in tomato cv. Vollendun plants against a yellow strain of CMV-Y. Application of BTH as a drench 7 days before inoculation of the virus protects the plants from necrosis.
(f) Application of Ozone (O_3) for 2 min has suppressed the wilt symptoms induced by PBNV in cowpea seedlings. However in control (seedlings treated with non-filtered air), the disease severity was higher and it indicated that ozone successfully induces systemic resistance in cowpea against PBNV (Periyar Selvam et al. 2006).
(g) The polysaccharide isolated from the mold, *Trichothecium roseum*, named as *T. poly* has induced resistance against a number of unrelated viruses in systemic and hypersensitive hosts by apparently altering the host-cell susceptibility (Gupta 1985). Chandra et al. (1978) reported that application of *T. poly* either by rubbing or spraying on the basal leaves of *N. glutinosa,* rendered upper untreated leaves resistant to challenge virus (TMV) infection. Finally, local as well as systemic antiviral responses of treated *N. glutinosa* was reversed by timely application of actinomycin D, suggesting that DNA-dependent RNA synthesis is required for the expression of antiviral activity in plants treated with *T. poly*.

(h) Panattoni et al. (2007) have reported the antiviral activity of tiazofurin and mycophenolic acid against Grapevine leafroll associated virus 3 in *Vitis vinifera* explants.
(i) Xia et al. (2006) have identified bitriazolyl compounds as novel antiviral candidates for combating the *Tobacco mosaic virus*. For the majority of the compounds identified as plant virus inhibitors, very limited information is available; more data has to be generated by lab and field trials.

More details on chemotherapy in plant virus research can be obtained from articles of Huffman et al. (1973), De Clercq (2002), Panattoni et al. (2005), D'Anna et al. (2006), Luvisi et al. (2012).

Conclusions

In almost all the crops, resistance sources have been identified and are used in breeding programmes against a majority of the virus and vectors and the approach is likely to be used increasingly in future. However, resistance was broken in certain instances due to introduction of newer virus strains. Hence resistance developed to strains in one location cannot be expected to operate against strains occurring elsewhere, and growing a cultivar in a new area may lead to serious epidemics. Further more evidence is required on the durability of different types of resistance, on possible gene-for-gene relationships, on the relative merits of major and minor genes and on the possible advantages of using mixtures of different genotypes. At least some of this information may soon be forthcoming because of the increased attention being given to breeding for some form of resistance against viruses at many establishments and especially at the International Research Institutions.

The advances made in molecular biology, plant transformation and regeneration technologies have widened the scope of conferring resistance against viruses in plants. It is now possible to engineer resistance to viral pathogens in plants using transgenes from a wide range of organisms. The strategy which involves the introduction of transgenes into a crop in a single step is called horizontal gene transfer (HGT) whereas, conventional breeding involving incorporation of resistance genes from parents to off springs by crossing has been termed as vertical gene transfer (VGT) (Prins and Zadoks 1994).

Even the concept of multi-line cultivars will counter the undesirable and genetic vulnerability of homogenous cultivars of crop species against some of the plant viruses. Attempts to incorporate resistance for more than one pathogen are desirable but are often difficult to accomplish. Besides this, resistance may break down due to evolution of more virulent strains, and thus a flourishing and acceptable cultivar will be discarded in the course of time. Another important factor is that a cultivar having resistance to a given virus may be susceptible to other viruses or pathogens in the field. For example, the bean genotypes possessing monogenic dominant resistance to BCMV revealed severe disease symptoms when infected by comoviruses in South America (Morales and Castano 1992). Another approach is to upgrade the levels of resistance to virus infection by including resistance to a specific vector. Under these circumstances, breeding for general resistance (horizontal type) is of greater significance in long-term breeding programmes for varietal adaptation to broader agro-ecological situations.

Generally, vector resistance may result in a decreased level of virus spread and it would be desirable to breed for vector resistance. In addition, since resistance to one arthropod species may be associated with altered levels of susceptibility to other species, the potential impact of virus spread on crop resistance to a vector species should not be ignored. Despite these dangers, the potential for controlling certain virus diseases through vector resistance is considerable. There exist a sufficient number of examples wherein vector resistance has contributed to disease reduction to justify continued efforts in this area. Research on vector behavior (viz., alighting and probing) influenced by the host genotype may broaden our insight into this aspect of resistance to vectors and their relationship to virus spread. There are obvious advantages in having close collaboration between breeders and virologists to exploit sources of resistance or tolerance and to avoid releasing cultivars that are particularly vulnerable.

To achieve broad based and durable resistance by genetic manipulation, approaches like isolation and transfer of few natural resistance genes (e.g., *N. Rx, Sw5*) and their transfer across the species/breeding lines indicate the potentiality to achieve broad-based resistance in a given crop against different viruses (Frederick et al. 1998). It may be possible to engineer and broaden virus specificity of R genes prior to transfer to susceptible cultivars.

2.23 Quarantines

2.23.1 Introduction

Migrations, military conquests and occupations, voyages of discovery, religious expeditions and germplasm exchange have in many ways contributed to the spread of plant material. A number of diseases that were not present earlier in certain countries have been introduced into new areas either on or in the plant materials which in turn had spread and caused disastrous crop losses. For the increase of the food production by breeding techniques, there is a constant exchange of germplasm throughout the globe and hence a great impact is noticed in agricultural outputs. Most of our major crop plants are now grown in areas where they did not evolve, but have been disseminated by man, are introductions. The phenomenal increase in advanced transport techniques during the recent years has increased the movement of the plant material by man from place to place irrespective of geographic barriers. While traveling by air, a small bud wood stick or cutting can be carried in a viable condition without any protective measures and it has ensured its survival in the new areas. In the case of ship travel involving prolonged duration of time, much propagative material will be lost during the passage and the inspection can be done leisurely at the port of entry. It is also established that nearly 231 viruses are seed transmitted (Sastry 2013a). Further, the vegetative propagules derived from the virus-infected plants are mostly infected. The dissemination of seed-borne viruses is also occurring due to the massive movement of seed, particularly under the 'green revolution'. In the developing countries, import of large quantities of food grains may result in certain inadvertent disease introductions, as at certain times farmers will use the same in smaller amounts as planting material.

All countries are involved in the export and import of agricultural products and are depending heavily on agriculture for their export earnings. The key concept in the new free trade situation is pest risk analysis, an objective assessment of the dangers of invasion by a particular known and unknown pests and diseases including plant virus and virus-like diseases which may damage agriculture production more drastically if they are accidentally introduced.

The scientific community is also introducing the infected genetic stocks for research investigations unaware of national regulations or due to sheer ignorance. On several occasions seeds are sent across the world by post enclosing them in envelopes or are brought by tourists. For scientific purposes, there is no embargo exchange of germplasm when safeguards are adequate and quarantine regulations of every country are playing major role in this regard. Plant quarantine is a government endeavour enforced through legislative measures to regulate the introduction of planting materials, plant products, soil, living organisms, etc. in order to prevent inadvertent introduction of pests (including fungi, bacteria, viruses, viroids, phytoplasmas, nematodes, insects and weeds) harmful to the agriculture of a country/state/region, and if introduced, one has to prevent their establishment and further spread.

It is evident from the Chapter-3 of the Volume-I (Sastry 2013b), that the virus and virus-like pathogens are potentially dangerous in causing a heavy toll on crops. Sometimes the catastrophic consequences suffered by agriculture through the introduction of exotic diseases compelled every nation or a group of governments to impose legal restrictions on international trade of plant materials which sometimes unnoticingly carry pests and diseases. The entry and further spread of these dangerous pests and diseases in each country is controlled by their national regulatory quarantines. The term quarantine is derived from Latin word, *quarantum*, meaning forty. The objective of plant quarantines is to protect the country's agriculture, by preventing the introduction and spread of important pests and diseases by legislative restrictions on the movement of plants and plant products. In other words, quarantines are important disease control measures based on avoidance and exclusion technique. Hewitt and Chiarappa 1977; Khetarpal et al. (2004) and Khetarpal and Gupta (2007) have reviewed various methods for the safe movement of plant germplasm.

Table 2.8 List of plants/planting materials and countries from where import is prohibited along with justifications

S.No	Plant species/variety	Categories of plant material	Prohibited from the countries	Justification for Prohibition Due to pests such as:
1.	Banana, Plantain and Abaca (*Musa spp.*)	Rhizomes/Suckers	Central & South America, Hawaii, Philippines and Cameroon	Moko wilt (*Burkholderia solanacearum*) race 2 and Cameroon marbling (phytoplasmas)
2.	Cassava or tapioca (*Manihot esculenta*)	Seed/Stem cuttings	Africa & South America	Super elongation (*Sphaceloma manihoticola*), Cassava bacterial blight (*Xanthomonas campestris* pv. manihotis) — American strains, Cassava witches' broom (phytoplasma) and several cassava viruses
3.	Cocoa (*Theobroma cacao*) and plants species belong to Sterculiaceae, Bombacaceae and Tiliaceae	Fresh beans)/Pods/Bud wood/Grafts Root stock/Saplings	West Africa, Tropical America and Sri Lanka	Swollen shoot virus and related virus strains of cocoa, Witches' broom (*Crinipellis* (*Marasmius*) *perniciosa* Watery pod rot (*Monilia* (*Moniliopthora*) *roreri*), Mealy pod (*Trachysphaera fructigena*), Mirids (*Sahlbergia singularis* & *Distantiella theobroma*), Cocoa moth (*Acorocercops cramerella*), Cocoa capsid (*Sahlbergiella theobroma*), Cocoa beetle (*Steirastoma brevi*), Seedling damping-off (*Phytophthora cactorum*), Chestnut downy mildew (*Phytophthora katsurae*) and Black pod of cocoa (*Phytophthora megakarya*)
4.	Cocoyam or Dasheen or Taro (Arvi) (*Colocasia esculenta*) and other edible aeroids	Plants/Corms/Cormlets/ Suckers	Cook Islands, Papua New Guinea, Solomon Islands and South Pacific countries	Alomae land Bobone (Rhabdo viruses), Dasheen mosaic virus (South Pacific strains) and Bacterial blight (*Xanthomonas campestric* pv. *dieffenbachiae*)

(continued)

Table 2.8 (continued)

S.No	Plant species/variety	Categories of plant material	Prohibited from the countries	Justification for Prohibition Due to pests such as:
5.	Coconut (*Cocos nucifera*) and related species of Cocoideae	Seed nuts/ Seedlings/ Pollen/Tissue cultures etc.	Africa (Cameroon, Ghana, Nigeria, Togo and Tanzania), North America (Florida in USA, Mexico); Central America and Caribbean (Cayman Islands, Bahmas, Cuba, Dominican Republic, Haiti, Jamaica) Philippines and Gaum Brazil (Atlantic Coast), Trinidad, Tobago, Greneda, St. Vincent, Barbados, Belize, Honduras, Costa Rica, El Salvador, Panama, Columbia, Venezuela and Ecuador, Surinam (Dutch Guyana), Sri Lanka.	Palm lethal yellowing (phytoplasma) and related strains, Cadang cadang & Tinangaja (viroid), Lethal boll rot (*Marasmiellus cocophilus*), Red ring (*Rhadinaphelenchus cocophilus* (*palmarum*), South American Palm weevil (*Rhyncophorus palmarum*), Leaf minor (*Promecotheca cumingi*) and Palm kernel borer (*Pachymerus spp*)
6.	Coffee (*Coffea spp.*) and related species of Rubiaceae	Beans (seeds)/Berries (freshly harvested)/ Grafts/Bud wood/ Seedlings/Rooted cuttings etc	Africa and South America	American leaf spot (*Mycena citricolor*, syn. *Omphalia flavida*), Coffee berry disease (*Colletotrichum coffeanum* var. *virulens*), Tracheomycosis (*Gibberella xylariodes*, syn *Fusarium xylarioids*), Powdery rust (*Hemeleia coffeicola*), Phloem necrosis (*Phytomonas leptovasorum*) and Coffee viruses (coffee ring spot, leaf rugosity, leaf curl, leaf crinkle and mosaic viruses), Coffee berry borer (*Hypothenemus hampei, Sophronica ventralis*) and Coffee thrips (*Diarthrothrips coffeae*)
7.	Date palm (*Phoenix dactylifera*)	Seeds/Off-shoots (suckers)	Algeria and Morocco USA (Florida)	Bayood (*Fusarium oysporum f.sp. albedinis*) and Palm lethal yellowing (*Phytoplasmas*

(continued)

Table 2.8 (continued)

S.No	Plant species/variety	Categories of plant material	Prohibited from the countries	Justification for Prohibition Due to pests such as:
8.	Forest plant species: (i) Chestnut (*Castanea spp.*)	(i) Seeds/Fruits/Grafts and other planting material	North America (USA and Canada)	Chestnut blight or canker (*Cryphonectria* (*Endothia*) *parasitica*)-American strain
	(ii) Elm (*Ulmus spp.*)	(ii) Plants/ planting material	North America (USA and Canada) and Europe and Russia	Dutch elm disease (*Ceratocystis ulmi*)-American and European strains, Elm mottle virus, Elm bark beetles (Scolytidae), Elm phloem necrosis (Phytoplasmas) and White -banded elm leaf hopper (*Scaphoidous luteolus*) - vector of Elm phloem necrosis
	(iii) Oak (*Quercus spp.*)	(iii) Seeds/Root grafts	United States of America	Oak wilt (*Ceratocystis fagacearum*) and Oak bark beetles (*Pseudopityophthorus* spp.)
	(iv) Pine (*Pinus spp.*) and other coniferous species	(iv) (a) Seeds/Saplings	North America (Canada, USA and Mexico)	Pine rusts [Stalactiform blister rust (*Cronartium coleosporioides*), Comandra blister rust (*C. comandrae*), sweet fern blister rust (*C. comptoniae*), Southern fusiform rust (*C. fusiforme*), Western gall rust (*Endocronartium harknessii*), Brown spot needle blight (*Mycosphaerella dearnesii*, syn. *Scirrhia acicola*), Seedling die-back and pitch canker (*Fusarium moniliforme* f.sp. *subglutinans*) and Needle cast (*Lophodermium* spp.)
		(iv) (b) Wood with bark	North America (Canada & USA), Asia (China, Hong Kong, Japan, Korea, Republic of Taiwan)	Pine wood nematode (*Bursaphelenchus xylophilus*)
9.	Oil palm (*Elaeis guineensis*) and related species	Seeds/Pollen/Seed sprouts	Philippines and Guam	Cadang cadang & Tinangaja (viroid)

(continued)

Table 2.8 (continued)

S.No	Plant species/variety	Categories of plant material	Prohibited from the countries	Justification for Prohibition Due to pests such as:
10.	Potato (*Solanum tuberosum*) and other tuber bearing species of Solanaceae	Tubers and other planting material	South America	Potato smut [*Thecaphora (Angiosorus) solani*], Potato viruses *viz.* Andean potato latent, Andean potato mottle, Arracacha B virus, Potato deforming mosaic, Potato T (capillo virus), Potato yellow dwarf, Potato yellow vein, Potato calico strain of Tobacco ring spot virus and Andean potato weevil (*Premnotrypes* spp.)
11.	Rubber (*Hevea spp.*)	Seeds/plants/budwood and any other plant material	Tropical America (Area extending 231/2 degrees North land 231/2 degrees South of the equator (Tropics of Capricorn and Cancer) and includes adjacent islands and longitude 30 degree West land 120 degrees East including part of Mexico, North of the Tropic of Cancer)	South American Leaf Blight of Rubber (*Microcyclus ulei*)
12.	Sugarcane (*Saccharum spp.*)	Cuttings or setts of planting	Fiji, Papua New Guinea, Australia, Philippines and Indonesia	Fiji virus
13.	Sweet potato (*Ipomoea spp.*)	Stem (Vine) cuttings rooted or unrooted/ tubers	South Africa, East Africa, New Zealand, Nigeria, USA, Argentina and Israel	Scab (*Elsinoe batatas*), Scurf (*Moniliochaetes infuscans*), Foot rot (*Plenodomus destruens*), Soil rot (*Streptomyces ipomoeae*), Bacteria wilt (*Pseudomonas batatae*), Sweet potato viruses viz. Russet crack; feathery mottle; internal cork; chlorotic leaf spot; vein mosaic; mild mottle and yellow dwarf, vein clearing; chlorotic stunt; Sheffied's virus A and B etc., Sweet potato witches' broom (phytoplasmas) and seed bruchid (*Mimosestes mimosae*)
14.	Yam (*Dioscorea spp.*)	Tubers for planting or propagation	West Africa and Caribbean region	Yam mosaic virus/green banding virus

2.23.2 Exclusion

Exclusion means the practice of keeping out any materials or objects that are contaminated with pathogens or diseased plants and preventing them from entering the production system for which quarantine rules and regulations are to be implemented.

Throughout the globe, to strenthen the research programmes in any country, generally the germplasm will be exchanged from one country to another based on the indents. In India the legislation of P.Q. Order 2003 prohibiting imports of 14 crops from various regions/countries is presented in the Table 2.8. In every country the entry and further spread of virus and viroid diseases is restricted by the use of virus-free planting material, in a limited quantities received from the importing/exporting countries. The viruses are detected in the planting materials by recently developed molecular techniques by which one can detect the presence of viruses in the planting materials or eliminating the virus from the materials received or the pathogen may be destroyed. The plant species/variety received, category of plant material are to be recorded. The information is also provided regarding prohibiting the infected material from the countries where virus inoculum exists, and is included in the Table 2.8.

At quarantine stations, rapid, accurate and sensitive diagnostic tests have to be used for testing the exporting and importing seed or vegetative plant material for the presence of virus and virus-like diseases besides other pests and diseases. Indexing by the use of molecular tests will also be useful in various contexts viz., (a) indexing to prevent spread of pathogens in international trade, (b) indexing in eradication campaigns, (c) indexing for the maintenance of pest-free areas, (d) Indexing in healthy stock schemes and (e) indexing in advisory work. Exclusion can take many forms, for example, prohibiting the movement of plants from an infested country into one which is free of a particular pest or disease. Indexing is essential for this purpose and implies that some assessment for the presence of a particular pathogen or pest on a given crop or consignment has been made. In most cases, absence of the pathogen or pest must be demonstrated, but in some cases a tolerance may be allowed (Khan 1991). Foster and Hadidi (1998) have reviewed the aspect of exclusion of plant viruses as a means of management of plant viral diseases.

2.23.3 Plant Quarantine Legislation

Measures in plant quarantine in India are enforced under the Central Government's Destructive Insects and Pests Act of 1914 and various notifications are issued under it from time to time. The establishment of directorate of plant protection, quarantine and storage under the Union Ministry of Food and Agriculture in 1946, was an important land mark in the field of plant protection. This was followed by establishment of an ever expending organizations in the constituent states. In India,

"New Policy on Seed Development" was announced liberalizing the import of seed and other planting materials in 1988. Subsequently, plant, fruit and seeds (regulation of imports in to India) Order (PFS Order) first promulgated in 1984 was revised in 1989. Further it was revised due to World Trade Organization (WTO) Agreements and the plant quarantine (Regulation of import in to India) Order 2003. Known as (PQ Order, 2003) came into force with effect from 1st April 2004. This revision was necessary in light of gaps in PFS Order to regulate import of plant germplasm, genetically modified organisms (GMOs) transgenic plant material, live insects and/or to supplement India's legal obligations under various international agreements. This includes incorporations on the basis of Pest Risk Analysis (PRA). The PQ Order, 2003 has been revised 24 times till April, 2011 which now has 22 forms for various plant quarantine regulatory functions and 12 schedules (http://www.plantquarantineindia.org).

In the U.S., the first American law granting plant quarantine authority was passed in California in 1881, and this was followed by the first country plant quarantine ordinance enacted in 1886 (Ryan 1969). Statutory control of plant health is exercised through the Plant Protection and Quarantine programmes (PPQ) planned and executed by the Animal and Plant Health Inspection Service (APHIS) of U.S. Department of Agriculture (USDA). The basic legislation under which USDA operates, authorizes appropriate actions to be taken to prevent the introduction or spread of the plant diseases and pests, as well as measures designed to control or eradicate them. Three primary acts authorize activities dealing with diseases and pests; the plant quarantine act of 1912 (Anon 1912), the federal plant pest act of 1957 (Anon 1957) and the organic act of 1944 (Anon 1944a, b). These acts are for both international and domestic quarantines. There will be a close working relationship between the State and Federal governments and the Federal government has the primary responsibility to prevent the introduction and interstate movement of the plant material which carry pests and diseases. In Australia, the first set of regulations governing plant quarantine came into operation in 1909, when the Federal Quarantine Service was established. The scope of the British Destructive Insects Act of 1877, was broadened in 1907 and also in 1927; this legislation was then consolidated into the 1967 Plant Health Act. Other European countries also took quarantine actions in the early part of the twentieth century. The details of quarantine regulations of foreign countries have been issued by some countries or organizations (USDA 1933–1978; Inter-African Phyto-sanitary Commission 1962; EPPO 1969–1978; Anon 1977–1978).

2.23.4 Plant Quarantine Measures

A number of quarantine measures are available and may be used, either separately or collectively, in an attempt to prevent the establishment of a new plant pest or disease in an area previously free of it. The governments enact regulations and stipulate safeguards in accordance with the risk presented by the importation of the plant materials from areas where hazardous pests and pathogens are known to occur. The quarantines operate between different countries or between adjacement states in one country. The regulations are formulated and implemented by the federal or state government or both. Persons desiring detailed information about the quarantine regulations of foreign countries should request their own country's plant protection and quarantine service. In each country the government regulations issued by ministries or departments of agriculture have been received and abstracted. In general, these regulations, (1) specify requirements of import permits, (2) require phytosanitary certificates, (3) require certificates of origin, (4) stipulate inspection upon arrival, (5) prescribe treatment upon arrival to eliminate a risk, and (6) prescribe quarantine or post-entry quarantine depending on the nature of the risk.

Some quarantine measures comprise the following: embargoes, inspection at the port of entry, inspection at the port of dispatch, field inspection during the growing season (preclearance), controlled entry (i.e. disinfection, disinfectation, treatments) and post-entry quarantine stations. Some details of these quarantine measures are briefly described.

2.23.5 Functions of Plant Quarantine

The prime functions of plant quarantine are : (1) enforcing the measures for the prevention of foreign pests and diseases from overseas and it is defined as international quarantines; (2) quick identification, localization and eradication of exotic pests and diseases which are invaded within a country itself and is called as domestic quarantine; and (3) providing to the country with as many new cultivars/germplasm access as possible without at the same time allowing the entry of new diseases.

2.23.5.1 International Quarantine

This will prohibit or regulate the entry of imports from other countries necessitation international cooperation on phytopathological and entomological problems. For long periods plants and the organisms pathogenic to them in many instances coexisted in balanced equilibrium. Interference by man invariably disturbs this balance and the pathogens of minor significance in one situation may cause epiphytotics when transferred to new environment/places. It may be because of plants developing in the absence of the pathogen, have no opportunity to select resistant factors specific against an introduced pathogen, as a consequence of which when they are grown in a place where they have been imported become extremely vulnerable to disease attack. In certain cases the introduced diseases may also mutate to more destructive forms and cause epidemics.

There are number of examples where the introduced plant viruses in new areas have resulted in to heavy crop losses. For example, Asian greening bacterium and its vectors once established in an area have the potential of totally destroying citrus as an economic crop (Fraser et al. 1966; Salibe and Cortez 1966). There are new strains of tristeza virus which can spread very rapidly (Bar-Joseph and Loebenstein 1973) and can affect many citrus root stocks and scions once considered tolerant to the virus. Even the introduction of exocortis viroid into a country such as Japan or Uruguay, where the predominant root stock is *P. trifoliata*, is a distinct hazard because exocortis can spread by unindexed bud wood or very effectively by cutting tools, and is very damaging to poncirus root stocks. Some of the important virus and virus-like diseases introduced into different countries are discussed under a separate heading.

For the international exchange of germplasm, the involved scientists or commercial nurseries, whether importers or exporters, should be cognizant of the quarantine regulations of the importing country. The quarantine regulations of the importing country will determine the enterability of such germplasm. Even the quarantine service of the exporting country is involved as a source of information about the requirements of the importing country and also in the issuance of phytosanitary certificates, if required by the importing country.

(a) **International standards on phytosanitary measures**

International Standards for phytosanitary measures (ISPMs) are prepared by the Secretariat of the International Plant Protection Convention (IPPC) as part of the United Nations Food and Agriculture Organizations global programme of policy and technical assistance in plant quarantine. The IPPC is an international treaty relating to plant health to which approximately 124 governments currently adhere. The purpose of the IPPC is to secure common and effective action to prevent the spread and introduction of pests of plants and plant products, and to promote appropriate measures for their control.

Because of the global concern that they might be used for trade protection or trade barrier, a specific Agreement on the Application of Sanitary and Phytosanitary (SPS) Measures was negotiated during the Uruguay Round. This Agreement entered into force with the establishment of the World Trade Organization (WTO) on 1 January 1995. It concerns the application of food safety and animal and plant health regulations. The Agreement recognizes that countries have the right to maintain Sanitary and Phytosanitary (SPS) measures for the protection of the population and the agricultural sector. However, it requires them to base their SPS measures on scientific principles and not to use them as disguised restrictions to trade. All countries maintain measures to ensure that food is safe for consumers, and to prevent the spread of pests or diseases among animals and plants. These SPS measures can be in the forms of requiring products to come from a pest-free area, inspection of products, specific treatment or processing of products, setting of allowable maximum levels of pesticide residues or permitted use of only certain additives in food. It therefore, become very important for countries to identify pest-free areas within the country and encourage farmers to produce healthy crops which even can be exported or consumed within the country. Sanitary (human and animal health) and phytosanitary

Table 2.9 List of International Standards on Phytosanitary Measures (ISPMs) published

ISPM number	Title	Year of publication
1	Phytosanitary principles for the protection of plants and the application of phytosanitary measures in international trade	2006
2	Framework for pest risk analysis	2007
3	Guidelines for the export, shipment, import and release of biological control agents and other beneficial organisms	2005
4	Requirements for the establishment of pest free areas	1995
5	Glossary of phytosanitary terms	2010
6	Guidelines for surveillance	1997
7	Phytosanitary certification system	2011
8	Determination of pest status in an area	1998
9	Guidelines for pest eradication programmes	1998
10	Requirements for the establishment of pest free places of production and pest free production sites	1999
11	Pest risk analysis for quarantine pests, including analysis of environmental risks and living modified organisms	2004
12	Phytosanitary certificates	2011
13	Guidelines for the notification of non-compliance and emergency action	2001
14	The use of integrated measures in a systems approach for pest risk management	2002
15	Regulation of wood packaging material in international trade	2009
16	Regulated non-quarantine pests: Concept and application	2002
17	Pest reporting	2002
18	Guidelines for the use of irradiation as a phytosanitary measure	2003
19	Guidelines on lists of regulated pests	2003
20	Guidelines for a phytosanitary import regulatory system	2004
21	Pest risk analysis for regulated non-quarantine pests	2004
22	Requirements for the establishment of areas of low pest prevalence	2005
23	Guidelines for inspection	2005
24	Guidelines for the determination and recognition of equivalence of phytosanitary measures	2005
25	Consignments in transit	2006
26	Establishment of pest free areas for fruit flies (Tephritidae)	2006
27	Diagnostic protocols for regulated pests	2006
28	Phytosanitary treatments for regulated pests	2007
29	Recognition of pest free areas and areas of low pest prevalence	2007
30	Establishment of areas of low pest prevalence for fruit flies (Tephritidae)	2008
31	Methodologies for sampling of consignments	2008
32	Categorization of commodities according to their pest risk	2009
33	Pest free potato (*Solanum* spp.) micropropagative material and minitubers for international trade	2010
34	Design and operation of post-entry quarantine stations for plants	2010
35	Systems approach for pest risk management of fruit flies (Tephritidae)	2012

Source FAO; www.ippc.int

(plant health) measures apply to both domestically produced food or local animal and plant diseases and products coming from other countries.

ISPMs, a global programme makes available to interested countries the standards, guidelines and recommendations to achieve international harmonization of phytosanitary measures with the aim to facilitate trade and avoid the use of unjustified measures as barriers to trade. The standards once developed are periodically harmonized/brought in line with the other more recently developed ISPMs and phytosanitary concepts within the framework of IPPC (Table 2.9).

(b) **Regional plant protection organizations**

In 1929, a first world plant protection convention was held in Rome, but its effect remained small in the face of the then still modest national agency. In 1945, the United Nations and its specialized agency the Food and Agriculture Organization (FAO) made efforts for the development of plant protection at the global level. The real breakthrough in this respect occurred with the establishment of the 1951, FAO International Plant Protection Convention, which at present has 79 signatory nations. With the help of nine regional plant protection organizations, plant protection in different geographical areas is promoting the aims and objectives of the convention within the region. These regional organizations are carrying out useful tasks in providing information for their member countries. The organizations are (1) Asia and Pacific Plant Protection Commission (APPPC) (Region: East Asia, the Indian Subcontinent, Australasia and the Pacific; Secretariat: Thailand); (2) Caribbean Plant Protection Commission (CPPC) (Region: Carribean; Secretariat: Barbados); (3) Comite Regional de Sanidad Vegetal del Cono Sur (COSAVE) (Region: Southern South America; Secretariat: Uruguay); (4) Comunidad Andina (CA) (Region: North-western South America; Secretariat: Peru); (5) European and Mediterranean Plant Protection Organization (EPPO) (Region: Europe and Mediterranean; Secretariat: France (6) Inter-African Phytosanitary Council (IAPSC) (Region: Africa, Madagascar; Secretariat: Cameroon); (7) North American Plant Protection Organization (NAPPO) (Region: North America; Secretariat: Canada); (8) Organismo Internacional Regional de Sanidad Agropecuaria (OIRSA) (Region: Central America; Secretariat: El Salvador); (9) Pacific Plant Protection Organization (PPPO) (Region: Australasia, Pacific; Secretariat: Fiji). Fjelddalen (1974) briefly reviewed the duties of these organizations. They operate as a net work to promote concerted action for combating pests and diseases of international importance and restrict their further spread which are already existing in some parts of the region. In short, the duties of these organizations are to cover the basic aspects like quarantines, technology transfer, surveillance and advice to farmers on the control of harmful organisms and also promote the crop protection research.

(c) **EPPO activities on plant quarantine**

One of the aims of EPPO is to help its member countries to prevent entry or spread of dangerous pests and diseases including plant virus and viroid diseases (plant quarantine). The Organization has therefore been given the task of identifying pests and diseases which may present a risk, and of making proposals on the phytosanitary measures which can be taken for implementation. In recent years, the identification of risk has been formalized, because transparent justifications of phytosanitary measures are required and phytosanitary measures have to be commensurate with the risk. Several EPPO Standards on Pest Risk Analysis (PRA) are now available for major virus diseases of fruit crops and vegetables. To perform these activities, much information on pests presenting a risk to the EPPO region is required and has been collected by the EPPO Organization and most of the data sheets, maps, diagnostic protocols and pictures are available on the virus and viroid diseases of fruit crops, vegetables, ornamentals and other commercially grown crops, for which details on Geographical distribution, Hosts, Transmission, Detection and identification, Economic impact, Control and Phytosanitary risk were described.

For example, TYLCV is listed as an A2 quarantine pest by EPPO. The wide distribution of the vector *Bemisia tabaci*, together with the unreliable means of control available, indicate the likelihood of serious crop damage and economic losses if the virus is introduced into a country. Quarantine requirements to prevent introduction from one country to another within the EPPO region. In some of the tropical countries, it however difficult to apply because of the lack of a cheap, rapid and reliable method of detection and because *B. tabaci* is already widely distributed. If TYLCV is introduced into a country, the only possibilities for avoiding serious damage would be to encourage careful monitoring of the situation within the country and the use of resistant/tolerant cultivars along with prophylactic measures by the growers have to be implemented for effective management of TYLCV (Diez et al. 1999).

2.23.5.2 Domestic Quarantines

Inspite of the best precautions taken and stringent measures of quarantine applied, even the most advanced countries have not been able, on several occasions, to prevent the entry of the diseases. In such cases man has taken recourse to domestic quarantine, which will help in the confinement of the diseases and pests to particular area only due to the restriction in the movement of the plant material between the states. They will primarily deal with the new diseases and pests which have already entered into a country and will be made to eliminate them before they become established. The most effective quarantine action is that which is applied at the source of spread. The problems of domestic quarantine are basically similar to those of international quarantine. In practice, Federal and State governments jointly enforce regulations to prevent intrastate as well as interstate spread.

In each country, even some State Legislations have also passed legislation against certain devastating virus diseases; for example, in California, an ordinance No. 1053 of the county of Monterey was implemented to prevent losses of lettuce crop due to mosaic disease and reads in part as follows:

"Mosaic is a virus disease which infects lettuce crops and is a direct cause of crop failures. It is a disease which is seed-borne and carried by aphids and, if not prevented, spreads from field to field affecting large areas of production and causing great losses to growers. The disease threatens to seriously affect the general economy if prompt and effective measures are not taken at once to eradicate this menace.

It shall be unlawful for any firm or corporation to plant any lettuce seed in the unincorporated area of the county of Monterey which has not been mosaic-indexed.

Any violation of this ordinance constitutes a misdemea nor punishable by a fine of not more than $500.00 or by punishment in the country jail for a period not exceeding six months".

This ordinance mentioned above has been effective in reducing losses in lettuce production for nearly 20 years. Similarly, (*Western*) *celery mosaic virus* is effectively controlled by the establishment of a legally defined period in which no celery can be grown in a given district was described by Milbrath (1948). Due to the implementation of this legislation in the Venice-Sawtelle region of California, celery yields were increased by 50 % in the first 2 years following passage of the law and increased by another 50 % in the subsequent years.

In most of the countries, legislation was passed for nurserymen for the supply of disease-free plants. The nurseries were inspected while growing, by the quarantine staff, so that only certified planting stocks were sold which meets certain prescribed standards. In most of the developed countries virus-free potato seeds are produced in sufficient quantities and CPRI, Simla (India) plays a pivotal role in seed production. In India, the Directorate of Plant Protection, Quarantine and Storage takes necessary steps to regulate the interstate movements of plants and plant materials to prevent the further spread of destructive pests and diseases.

In Sudan, legislation was passed for the pulling of volunteer cotton plants and their regrowth to augment for the control of *Cotton leaf curl virus*. Growing of okra before 15th September is strictly prohibited. Any programme of regulation and free registration can succeed only if it is supported by the growers and nursery men. In California, they have provided outstanding support and cooperation of various regulatory and quarantine programmes. Similarly in South Africa, a cooperative program between growers and the government regulated in legislation passed in 1927 providing for the eradication of all psorosis-affected trees showing bark lesions (Doidge 1939). This programme carried out between 1930 and 1950 effectively reduced psorosis in South Africa from one of the most destructive diseases to one that is now rare. In Philippines during 1962–1974, the area of mandarin, sweet orange and pomello cultivation was greatly reduced in certain parts up to 60 % due to leaf mottling (greening). In 1969, a quarantine measure was issued through an administrative order by the Secretary of the Department of Agriculture to prohibit entry of citrus planting materials into non-infected areas. All citrus for planting would be required to have certification papers from the point of origin for verification in inspection centers and quarantine stations throughout the citrus growing areas of the Philippines (Altamirano et al. 1976). Similarly, in Florida, legislation was passed for certifying the peach nursery stocks. Prior to June 30th, of every year, Phony peach trees should be eradicated which are found one mile within the environs of the nursery. Ultimately the success of quarantine is dependent on a substantial degree of public support, but if regulations can be ignored with impunity by a few, they soon loose the support of many.

2.23.6 Quarantine Status of Plant Importations

A number of precautions are to be undertaken to minimize pest/disease risk. They include the regulations themselves, phytosanitary certificates, with or without added declarations, permits, inspection upon arrival, treatments etc. Based on their quarantine status, when ever a plant or seed material is received at the inspection station or port of entry, they are placed in one of the four categories mentioned here: (1) absolute prohibition, (2) quarantine, (3) post-entry quarantine, and (4) restricted entry. Whenever the risk of introducing pests and pathogens is very high, then the importation of that particular plant material is completely prohibited, even for government services and such material will be included under absolute prohibition category. This is invoked by the importing country when there are no adequate safe guards and when isolation from commercial crop production is not possible. For example, an absolute prohibition against coconut from eight specified geographic areas was recommended by the South West Pacific Commission, because of the diseases like cadang-cadang (South West Pacific Commission 1964). The second category, quarantine, covers the plant materials on which the pests and diseases have already been reported from the country. Admission of plants is usually reserved for government services and not for private agencies. In the third category, post-entry quarantine, the genera are placed on which pests and pathogens have been reported in some, but not all foreign countries, but the importation originates from a country where such risks have not been reported. Post-entry quarantine is usually reserved for government services, in situations and qualified individuals whose facilities meet post-entry quarantine requirements. Safeguards consist of inspection upon arrival and during a special post-entry period, usually for at least two growing seasons, at the premises of the importer. The fourth and last category, restricted entry is usually assigned to only restricted plants, and are received by the general public and they were subjected to inspection and treatment upon arrival at a port of entry or inspection station. In India Schedule IV of PQ Order (2003)specifies the crops and the geographical areas from where imports are prohibited due to presence of pests / diseases which are of quarantine importance to India. There are 14 plant species / variety listed which are prohibited from imports from respective countries with justifications (Table 2.8). Moreover, standard operating procedures (SOPs) for post entry quarantine investigations for various crops have been enumerated for 62 plant species/plant material along with quarantine pest and PEQ requirement for each material (DPPQs, SOP for PEQ instruction, 2007).

2.23.7 *Open Quarantine*

(a) **Role of open quarantine in regional germplasm exchange**

Open quarantine has facilitated safe introduction of large quantities of cassava germplasm, which would not have been possible through other means. The method was cheaper than micro-propagation and plant mortality was also low. This method can be used in germplasm exchange programmes where the climate and pest species are more or less similar. This is the quarantine of plants without using such physical confinement structures as glasshouses or screenhouses. This can be used to reduce or eliminate the risk of spread of pests by adhering to a quarantine protocol. The technique has been used successfully in East and Central Africa to exchange germplasm resistant to *East Africa cassava mosaic virus-Uganda*.

(b) **Location of the open quarantine field**

1. The field should be located in a government-owned institution such as research institute and, 2. The isolation distance should be 200–400 m away from any other cassava field.

(c) **Source of plant materials**

1. Cultivars should be obtained from a reputable source where the history of the cultivars in relation to pests and diseases is well documented.
2. Plant quarantine officers and plant breeders of both importing and exporting countries should be involved in the selection of the materials.
3. Background information (passport data) of the materials to be introduced should be provided to the importing country.
4. The importing country should provide a plant import permit.

(d) **Field establishment**

1. Site selection should be done by the institution where the field will be located and approved by the plant quarantine officers.
2. The field should be fenced and should have only one outlet. A disinfectant should be located at the entry point for use by persons and equipment entering the facility.
3. Prior to introduction the materials, they should be treated with both fungicide and insecticide.

(e) **Field maintenance**

1. Through inspection of the plants should be done daily by authorized staff.
2. Roguing and burning of any plants that sprout with symptoms of diseases such as cassava bacterial blight (CBB), cassava anthracnose, viral diseases and scale insects.
3. No planting materials should be allowed out of the quarantine field until they are certified to be free of quarantine pests and diseases.

4. Quarantine officers bear the responsibility of the field until the materials are certified to be clean for distribution.

(f) **Duration of the quarantine field**

1. Introduced materials should be under strict observation of plant quarantine officer for one crop cycle or a minimum of 1 year.
2. Ratooning of first planting materials can proceed 1 year after introduction.
3. Material may then be distributed for multiplication or for further research.

Open quarantine facilitated germplasm exchange within the East African region with minimum risk of introducing quarantine pests. It was used successfully after the outbreak of the severe form of Cassava mosaic disease (CMD) in the region. Transportation requirements are not as critical as for *in vitro* propagated materials. Unless the materials are present in the laboratory, it takes time from virus indexing until they are ready for exportation and again requires additional time and nurturing before they are ready to be planted in the field.

Open quarantine may, however, present a risk of introducing quarantine pests through germplasm exchange for materials imported from other regions or continents unless strict quarantine measures are adhered to.

2.24 Pest Risk Analysis (PRA)

Pest categorization is the key component of pest risk assessment. This is not just an identification of the pest species, but an analysis of its potential danger as a pest. A pest of quarantine significance refers to a pest of potential economical importance to the area endangered, but not yet present there, or present but not yet widely distributed and being officially controlled. Pest categorization includes the following major elements.

- Identification of the pest.
- Definition of the PRA area.
- Distribution and official control programs within the PRA area.
- Potential of the pest for establishment and spread in the PRA area.
- Potential economic impact potential in the PRA area.
- Endangered areas.

The other major components of PRA include economical impact assessment (basically a thorough examination of the economic risk associated with the process) and the probability of introduction, which looks at the prospects of both entry and establishment of the pest.

Essentially, pest risk management is the process of deciding how to react to a perceived risk, deciding whether action should be taken to minimize the risk, and, if so, what action should be chosen. These steps should provide a clear scientific basis for pest risk assessment, risk management and quarantine decisions, and prevent these from being used in an inconsistent manner and as barriers to international trade.

(a) **Pest and pathogen risk analysis**

Availability of accurate and reliable information on the occurrence of the pests and diseases and the damage they cause is one of the essential requirements for the effective operation of the quarantine services. Unless such information is available, the plant quarantine or plant protection officer has no basis on which to act in regulating the importation of plants. The entry status for the major pests and diseases is to be studied. The entry status refers to the entire range of decisions or policies or regulations that serve as guidelines for rules and regulations of that government whether or not a potential carriers like plants, plant products, cargo, baggage, mail, common carriers etc. are enterable and if enterable, under what safeguards, into one geographic region to another. Pest and pathogen risk is based on an evaluation of biological variables. Examples of such variables are: (1) ecological range of a hazardous organism compared to the ecological range of its hosts in the importing country; (2) hitch hiking activity of the pest; and (3) ease of colonization of the pest. Attitudes towards entry status may range from 'conservative' to 'liberal', but 'liberal' in this context does not imply 'lax'. The most conservative attitude is that the plant material is excluded without any exceptions. Whereas the most liberal attitude is that the plant material may enter freely without agricultural regulatory restrictions. When the points of valid matching's are plotted and connected, a biological curve may be drawn. By illustrating the pathogen and pest risk analysis diagrammatically as a curve on a graph, one can effectively communicate generation philosophy, principles, policies and decisions not only to quarantine officers but to the scientific, commercial or growers. It is also required to explain a quarantine decision or activity that is against the interest of the importer. This curve is also useful in domestic quarantine matters for understanding the interaction of biological, economic and political factors to the trainees. It also helps the quarantine officers in diagnosing priorities or discuss budgetary matters.

(b) **Pest risk analysis for viral diseases of tropical fruits**

Virus and virus-like diseases have been causing serious damage to fruit production in tropical areas. They are not usually transmitted through seeds, but are often carried across national boundaries by infected budwood, vegetatively propagated seedlings and transmission by insect vectors.

The pest and pathogen risk is based on two general precepts, (1) the benefits must exceed the risk, and (2) the benefits must exceed the costs. For international transfer of genetic stocks, the benefits usually consist of the opportunity to introduce new crops or new cultivars of old crops or to introduce new genes to improve the old cultivars and breeding programmes. Such improvements may consist of increased yields through increased pest or pathogen resistance coupled with increased nutritive values. When costs are entered into the pest risk analysis, plant quarantine officers consider the cost of adequate safeguards like virus indexing, heat therapy, meristem tissue culture, shoot tip grafting etc.

With regard to disease risk analysis, the Food and Fertilizer Technology Center (FFTC) training courses on Plant Quarantine, held during 1996 at Guam/Fiji, discussed the major virus diseases of banana, citrus, papaya and passion fruit, in terms of the etiology and epidemiology of the diseases, current diagnosis methods, their geographical distribution and their economic impact. The various strains of important diseases such as banana bunchy-top and banana mosaic were described. Plantlets propagated by tissue culture have been found to be more susceptible to banana mosaic than seedlings of sucker origin. Since tissue culture is widely used to produce disease-free banana seedlings, recommendations were made for intensive vector control in banana plantations where plantlets raised by tissue culture are being grown.

A good example of pathogen risk analysis is the cadang-cadang disease of the coconut palm in the Philippines, whose etiology is proved to be of viroid in nature and the incubation period is long and uncertain. Similarly, there are risks involved in sending carnation and chrysanthemum cuttings to warmer climates to avoid the growing conditions of the northern European winter. At present cuttings are sent from the United Kingdom to overwinter in places such as Malta, Sardinia, the Canary Islands, Kenya and South Africa. The cuttings are grown as mother plants in the warmer climate until large enough to provide cuttings, which are then returned to Britain or other European countries. In these warmer countries, viroid diseases like Chrysanthemum stunt, Chrysanthemum chlorotic mottle, Cucumber pale fruit, Citrus exocortis and Potato spindle tuber are highly hazardous and multiply much more rapidly and reach greater concentrations in plants at temperatures in the range of 30–35 °C. Under these conditions each of them is able to infect the principal host plants of the other viroids. This factor should not be over looked, for example, where chrysanthemums sent to the Canary Islands may be growing near to potato or cucurbit crops, so that infections might be transferred in either direction (Hollings 1957, 1965).

PRA is generally accepted as the principal strategy to make sure that plant quarantine standards are transparent and justified on a sound scientific basis. However, its application to individual pests will require greater discussion and further detailed work. Once this is done, PRA can be expected to remove unnecessary barriers to trade.

There is an urgent need for all countries to reach an equal level in plant quarantine in terms of technology and equipment. An international network on quarantine pest monitoring is also needed to meet the growing danger of exotic pest invasion as a result of growing international tourism and trade, and the long-distance migration of insect pests.

Disease risk analysis has to be followed by control of the diseases. In general, vector-borne systemic diseases can be effectively controlled by integrated control measures, including production and cultivation of virus-free seedlings, elimination of inoculum sources, and prevention of reinfection through IPM of vector insects and cross protection with mild strains. The establishment of pathogen-free nursery systems is the most important way of preventing these diseases from spreading.

2.25 World Trade Organization Regime and Its Implications

The World Trade Organization (WTO), established on January 1, 1995, is the legal and institutional foundation of the multilateral trading system. The main purpose of the WTO is to promote free trade, serve as a forum for trade negotiations and settles disputes based upon the principles of non-discrimination, equivalence and predictability.

The WTO agreement contains more than 60 agreements covered under some 29 individual legal texts encompassing everything from services to government procurement, rules of origin and intellectual property (http://www.wto.org). Of these, the following four agreements have a direct bearing on agriculture and related activities:

- Agreement on agriculture is designed to ensure increased fairness in farm trade.
- Agreement on trade related intellectual property rights is aimed at improving conditions of competition where ideas and inventions are involved.
- Agreement on technical barriers to trade is to ensure that technical regulations and certification procedures do not create obstacles to trade.
- Agreement on the application of sanitary and phytosanitary (SPS) measures is in fact the one which is going to have major implications on seed/plant health in trade (Khetarpal and Ram Nath 1998; Khetarpal and Gupta 2002; Khetarpal et al. 2002, 2003; Khetarpal 2004).

2.26 Plant Biosecurity

Plant pathogens including virus and viroids are emerging and threatening food security. A number of research strategies have been initiated to enhance plant biosecurity capacity at the pre-border, border and post-border frontiers. Plant biosecurity encompasses the policy and regulatory frame works to analyze and manage risks in the sectors of plant life and health, and related environmental concerns. It is a strategic and integrated approach to protect crops from emergency of plant pests (EPPs) at national, regional and individual form levels. Countries are required to comply with international obligations as defined by the World Trade Organization (WTO) agreement on the application of sanitary and phytosanitary measures (WTO 1995). With the emergence of free trade agreements between trading partners, there is a requirement for quality assurance measures for pathogens, including viruses which may occur in both in exporting and importing countries.

The introduction of pests into a new locality is brought through plant hosts which may be carrier viz., vegetative propagule and true seed. The insect vectors and birds may transport the pests (Peters et al. 2012). Inert materials such as packing materials may carry the resting stages of the organisms. Plant/planting material are either imported bulk for sowing or for consumption and as small samples of germplasm for research work. Majority of the vegetative plant materials are virus infected when once mother plant is affected. Similarly, there are nearly 231 plant virus and viroid diseases which are transmitted through seed (Sastry 2013a). There are number of infectious diseases of poultry and livestock are introduced through exporting and importing activities. The list of plant viruses intercepted at NBPGR, New Delhi indicates that any negligence in testing the either imported are exported seed material will lead to entry of new viruses in the country which will establish and causes heavy yield losses. Chalam et al. (2013) have intercepted 34 plant viruses in germplasm including transgenics, which includes 11 viruses not yet reported from India viz; *Barley stripe mosaic virus, Bean pod mottle virus, Broad bean strain virus, Cherry leaf roll virus, Cowpea mottle virus, Cowpea severe mosaic virus, Maize chlorotic mottle virus, Pea enation mosaic virus, Raspberry ringspot virus, Tomato ring spot virus,* and *Wheat streak mosaic virus.* Thus the risk of introduction of 34 seed transmitted viruses or their strains into India was thus eliminated. The plant quarantine stations under DPPQS are under taking quarantine processing and clearance of consignments. NBPGR undertakes the quarantine processing of all plant germ plasm and transgenic planting materials. Khetarpal and Gupta (2007) and Rodoni (2009) have provided exhaustive review on the role of plant biosecurity.

2.27 Role of Bioversity International and NBPGR in Germplasm Maintenance and Exchange

(a) **Role of Bioversity International**

The Bioversity International is an independent international scientific organization that seeks to advance the conservation and use of plant genetic diversity for the well-being of present and future generations. It is one of the 15 Future Harvest Centers supported by the Consultative Group on International Agricultural Research (CGIAR), an association of public and private members who support efforts to mobilize cutting-edge science to reduce hunger and poverty, improve human nutrition and health, and protect the environment. More than 90 % of respondents felt that agricultural biodiversity could help to meet these challenges and that it could make a 'major' contribution to food security and environmental conservation. Bioversity International has its headquarters in Maccarese, near Rome, Italy, with offices in more than 20 other countries worldwide. The institute operates through three programmes: (1) the plant genetic resources programme, (2) the CGIAR genetic resources support programme and (3) the international network for the improvement of banana and plantain (INIBAP). The international status of IPGRI is conferred under an Establishment Agreement which, by January 2003, had been signed by the governments of several countries. Financial support for Bioversity International's research is provided by more than 150 donors, including governments, private foundations and international organizations. For details of donors and research activities please see Bioversity International's Annual Reports, which are available in printed form on request from ipgri-publications@cgiar.org or from IPGRI's web site (www.ipgri.cgiar.org).

(b) **NBPGR's responsibilities**

As the National Bureau of Plant Genetic Resources (NBPGR), New Delhi (India) has been designated as the national nodal agency for India to exchange germplasm material of agri-horticultural and agri-silvicultural crops for research purposes in the country. It has also been entrusted with the quarantine responsibilities in respect to germplasm of these crops. The Director of NBPGR has been empowered to issue 'Permits' for import of seeds/planting materials for research purpose.

(c) **Objectives of NBPGR**

- To plan, organize, conduct and coordinate exploration and collection of indigenous and exotic plant genetic resources.
- To undertake introduction, exchange and quarantine of plant genetic resources.
- To characterize, evaluate, document and conserve crop genetic resources and promote their use, in collaboration with other national organizations.
- To develop information network on plant genetic resources.
- To conduct research, undertake teaching and training, develop guidelines and create public awareness on plant genetic resources.

NBPGR has a separate Division of Plant Quarantine to meet the quarantine requirements in respect of the germplasm materials being exchanged through it. The division has trained scientific and technical staff representing the disciplines of entomology, weed science, nematology and plant pathology, well equipped laboratories, green houses and post-entry isolation growing field facilities to discharge its quarantine responsibilities efficiently. In case of certain crops, after laboratory examination at NBPGR, the exotic material is passed on to the specific crop-based institutes for post-entry isolation growing, before it is released to the indentors. NBPGR has established a regional Plant Quarantine Station at Hyderabad to fulfill the quarantine requirements of the International Crops Research Institute for Semi-Arid Tropics (ICRISAT), Directorate of Rice Research and other research organizations in the region. During the last two decades, a large number of exotic insects and mites, plant parasitic nematodes, plant pathogens and weeds have been intercepted from the imported germplasm materials, many of which are of major quarantine significance and are not yet known to occur in the country. While processing the germplasm for quarantine clearance, all out efforts are made to salvage the infested/infected materials so that valuable exotic germplasm could be made available in a healthy state for exploitation in crop improvement programmes in the country. The data presented in the Table 2.10 includes the seed-borne viruses intercepted in crops like cowpea, mungbean, beans, soybean and broad bean (Khetarpal et al. 2001; Chalam et al. 2004, 2008, 2009a, b; Parakh et al. 2008).

Table 2.10 Viruses intercepted in exotic germplasm at NBPGR and quarantine stations in India

Serial number	Crop	Virus intercepted	Source of country	Reference
1	Beans	TBRV	CIAT, Columbia	Chalam et al. (2004)
2	Broad bean	AMV	ICARDA-Syria, Eritrea, Iraq, Spain	Chalam et al. (2009b, c)
		BYMV, PSbMV	Spain, Syria	Chalam et al. (2009b, c)
		BYMV, BBSV, PSbMV	Bulgaria	Khetarpal et al. (2001)
3	Cowpea	AMV, CMV, TBRV	IITA, Nigeria	Chalam et al. (2008)
		CABMV	Eritrea	Chalam et al. (2008)
		CABMV	USA	Khetarpal et al. (2001)
4	Groundnut/ Peanut	PStV	China, Japan, USA, Myanmar, Philippines	Prasada Rao et al. (2012)
5	Mungbean	CMV	China	Chalam et al. (2008)
		CMV	USA	Chalam et al. (2008)
		CABMV	AVRDC, Taiwan	Chalam et al. (2008)
6	Soybean	AMV	AVRDC-Taiwan, IITA-Nigeria, Brazil, Myanmar, USA	Parakh et al. (2008)
		BCMV	AVRDC-Taiwan, IITA-Nigeria, USA	Parakh et al. (2008)
		CABMV	AVRDC-Taiwan, IITA-Nigeria, Myanmar, USA	Parakh et al. (2008)
		CMV	AVRDC-Taiwan, IITA-Nigeria, Brazil, Myanmar, USA	Parakh et al. (2008)
		PStV	USA	Prasada Rao et al. (2012)
		SBMV	AVRDC-Taiwan, IITA-Nigeria, Australia, Brazil, Hungary, Thailand, USA	Baleshwar Singh et al. (2003), Parakh et al. (2008)
		TRSV	IITA-Nigeria, Myanmar	Parakh et al. (2008)
7	Sunflower	TSV	Spain	Prasada Rao et al. (2012)

The informatics "System for Plant Viruses" system provides genomic information for the identification of intercepted plant viruses, so that proper actions may be taken to prevent the importation of exotic hazardous plant viruses. The information may also be used to trace the infection routes and distribution of the viruses.

Kumar et al. (1994) and Khetarpal et al. (2001) have summarized the viruses intercepted in leguminous crops during 1991–2000. Similar type of information is available on the list of intercepted plant viruses in almost all tropical countries. NBPGR also deals with testing for absence of terminator technology which is

Table 2.11 Projected yield losses due to inadvertent introduction of certain diseases

Crop	Production (million tonnes) (2007–2008)	Losses (1–5 %) (million tonnes)	Producer price-2008 (₹/tonne)	[a]Total predicted losses (₹ in million)
Groundnut	9.18	0.092–0.459	21000	1932–9639
Soybean	10.97	0.110–0.549	13900	1529–7631
Sunflower	1.46	0.015–0.075	22150	332–1661
Sorghum	7.93	0.079–0.395	8600	679–3397
Maize	18.96	0.190–0.948	8400	1596–7963
Tomato	10463 ('000 tonnes)	104.63–523.15 ('000 tonnes) 0.105–0.525 mt	10330	1081–5404 (early) 540–2702 (late)

[a] By considering range of 1–5 % losses in production under each crop (*mt* million tonnes)
₹ Indian rupee. *Source* Prasada Rao et al. (2012)

mandatory as per national legislation. This authorization was vested upon NBPGR for germplasm vide Article 6 of PQ Order 2003 and for transgenic planting material vide Govt. of India Notification No. GSR 1067(E) dated 05.12.1989.

Looking in to the economically important plant virus diseases which were intercepted, Prasada Rao et al. (2012) have predicted yield losses that would have occurred due to introduction of certain diseases were estimated by considering the range of 1–5 % losses in production under each crop against a particular disease (Table 2.11). The estimates were prepared based on the crop production figures of 2007–2008 (GOI, New Delhi 2009) and the producer prices of each crop for the year 2008 (Pricestat 2010) as per the following formula:

Projected crop loss (₹ in million) = Production loss (1–5 %) in mt. × producer price (₹/tonne)

Looking into the number of virus and virus-like diseases intercepted in true seed and also vegetatively propagated material with germplasm exchange, from 1989 onwards, FAO/IPGRI (IBPGR) have published technical guidelines for the safe movement of germplasm of different crops like cocoa, aroids, *Musa* spp., sweet potato, yam, legumes, cassava, citrus, grape vine, vanilla, coconut, sugar-cane, small fruits (*Fragaria, Ribes, Rubus, Vaccinium*), stone fruits, eucaplyptus and potato (Frison and Putter 1988; Frison et al. 1990; Frison 1994; Frison and Diekmann 1998). They have also covered the options for movement of germplasm in relation to the risk of moving pests.

2.27.1 Types of Materials Received

The genetic stocks may be in the form of 'vegetative propagule' or 'seed' type. Vegetative propagative material may be budwood, scion material or unrooted cuttings, all of which although dangerous, carry less risk than does rooted vegetative propagating material and can be transported quickly by air. The vegetative propagating material also includes plant materials like rooted cuttings or bulbs, tubers, corms or rhizomes. The plant material can also be exchanged in the form of 'seed'. Extensive details are provided by Sastry (2013a) in relation to the seed transmitted viruses and the methods to make them virus-free. The introduced plant material should be tested at the quarantine stations which involves some time and labor. The imported plant material should be tested by germination, indexing, serology, by electron microscopy etc. The seeds are less likely to carry viruses than the vegetative propagative material. The propagules derived from virus infected mother plants will be mostly infected. The import of seed is not considered to involve a risk of the same magnitude as that presented by living plant material, except that in some cases viruses are carried through seed. Since no case of seed-borne disease transmitted by whiteflies has been authentically established, the required germplasm can be introduced through seeds. With the discovery of *Potato spindle tuber viroid* and several recently described potato viruses are seed-transmissible, true seed of tuber bearing species of *Solanum* can no longer be regarded as freely interchangeable among breeders and collectors.

FAO-Bioversity International have recommended the following steps for majority of the 'seed' type germplasm movement for majority of the crops and with little modifications, one can plan depending on the size of the consignment. At International Institute of Tropical Agriculture (IITA), Nigeria, the *in vitro* techniques have been applied to the conservation and exchange of cassava germplasm to collaborators outside Nigeria. Plantlets were regenerated from meristem culture were transplanted and indexed for *African cassava mosaic virus* (ACMV). Plantlets regenerated from nodal cuttings are virus tested and used for distribution upon request. National Agricultural Research systems (NARS) has distributed these certified virus-tested plantlets for evaluation in more than 40 countries in Africa (Ng and Ng 1997). The transfer of germplasm should be carefully planned in consultation with quarantine authorities and should be in amounts that allow adequate handling and examination. The material should be accompanied with the necessary phytosanitary and other certificate documentations.

2.28 Role of FAO/Bioversity International in Germplasm Exchange

FAO has a long-standing mandate to assist its member governments to strengthen their Plant Quarantine Services, while Bioversity International's mandate-*inter alia*-is to further the collection, conservation and use of the genetic diversity of useful plants for the benefit of people throughout the world. The aim of the joint FAO/Bioversity International programme is to generate a series of crop specific technical guidelines that provide relevant information on disease indexing and other procedures that will help to ensure phytosanitary safety when germplasm is moved internationally.

The ever increasing volume of germplasm exchanged internationally, coupled with recent, rapid advances in biotechnology, has created a pressing need for crop-specific overviews of the existing knowledge in all disciplines relating to the phytosanitary safety of germplasm transfer. This has prompted FAO and Bioversity International to launch a collaborative programme for the safe and expeditious movement of germplasm reflecting the complementarity of their mandates with regard to the safe movement of germplasm.

Frison et al. (1990) under the instructions of FAO/IBPGR (Bioversity International) have brought out technical guidelines for the safe movement of legume germplasm in which they have furnished the information on virus and virus-like diseases of legume crops and provided the guidelines for international crop improvement programmes, collecting, conservation and utilization of plant genetic resources and their global distribution. In this preparation the experienced crop experts viz., Albrechtsen SE, Johnstone GR, Makkouk KM, Feliu E, McDonald D, Mink GI, Morales FJ, Reddy DVR, Vermeulen H, Rossel HW, and Zhang Zheng have provided the details of virus and virus-like diseases of leguminous crops and participated in the preparation of the guidelines.

The movement of germplasm involves a risk of accidentally introducing plant quarantine pests along with the host plant material; in particular, pathogens that are often symptomless, such as viruses, pose a special risk. To minimize this risk, effective testing (indexing) procedures are required to ensure that distributed material is free of pests that are of quarantine concern. General recommendations on how best to move germplasm of the crop concerned and mentions available intermediate quarantine facilities when relevant are provided.

2.28.1 Conceptual Guidelines for Exchange of Legume Germplasm

(a) **Germplasm**

- All legume germplasm collections should be maintained free of known seed-associated pests (seed-borne or seed-transmitted in the case of fungi and bacteria; seed-transmitted in the case of viruses). Descriptor data should be obtained from pest-free germplasm.
- Only seedlots certified to be free of such pests should be distributed.
- In recipient countries, seedlots should be established and maintained for one generation under conditions of isolation (temporal and/or spatial) or containment, with periodic inspection, testing and roguing.

(b) **Breeding lines**

- Legume seedlots to be exchanged among breeding programmes should be produced under conditions of isolation (with appropriate chemical protection) or containment, with periodic inspection and roguing to eliminate seed-associated pests.
- Seedlots should be tested for seed-associated pests and certified by the appropriate regulatory agency before distribution.

(c) **Commercial seedlots**

- Commercial seedlots should continue to be subject to current regulatory procedures.

2.28.2 The Technical Guidelines for Exchange of Germplasm and Breeding Lines

(a) **General recommendations**

- Vegetative material of legume species should go through intermediate or post entry quarantine and should be tested for absence of viruses.
- Legume seed should not be moved internationally in pods.
- Seed should be harvested at optimal time for the crop and care taken to ensure effective drying.
- Seed samples should be cleaned to eliminate all soil, plant debris, seeds of noxious weeds, and phanerogamic parasites.
- Unless specified otherwise, seeds should be surface-disinfected (with sodium hypochlorite or a similar product) before being given appropriate fungicide and insecticide treatments.
- Seed lots suspected to contain insects should be fumigated with an appropriate fumigant.
- Parcels containing seeds should be unpacked in a closed packing facility and material should be incinerated or autoclaved.

2.28.3 Movement of Germplasm

(a) **Introduction of germplasm**

- Introduction of new germplasm entries should satisfy local regulatory requirements.
- Each new introduction should be grown under containment or isolation.
- Plants should be observed periodically. Plants suspected to be affected with seed-associated pests should be destroyed.

- All symptomless plants should be tested for latent infections by viruses known to occur in the place of origin of the material and in the country of maintenance. Ideally this testing should be carried out at this stage or, if not possible, it should be carried out before the germplasm is distributed (see International distribution of germplasm). Infected plants should be destroyed.
- Seed should be collected from healthy plants only.

(b) **Further multiplication of new introductions or rejuvenation of germplasm accessions**

- Seed should be sown under containment or isolation with appropriate chemical protection.
- Plants should be observed periodically. Plants affected by seed-associated pests should be removed and destroyed.
- Seed should be collected from healthy plants only.

(c) **International distribution of germplasm**

- Germplasm accessions that have been introduced and multiplied according to the procedures described above can be certified and distributed internationally.
- Germplasm accessions which are not yet in a pest-free state should be handled according to the same procedures as described for new introductions.
- Movement of germplasm should comply with regulatory requirements of the importing country.
- In addition to the phytosanitary certificate, a 'germplasm health statement', indicating which tests have been performed to assess the health status of the material should accompany the germplasm accession.

(d) **Movement of breeding material**

- Seeds used for the multiplication of breeding material should be pest-free.
- Breeding material under multiplication should be grown under containment or isolation with appropriate chemical protection.
- Plants should be inspected soon after emergence and periodically thereafter. Plants infected with seed-associated pests should be destroyed. For field grown plants, suitable precautions should be taken to prevent soil spread from infected plants and introduction of possible seed-associated pests from local sources of infection.
- Seeds should be harvested only from symptomless plants.
- Seed samples of appropriate size should be tested for seed-associated pests.
- When non-destructive seed health tests are available, all seeds should be tested accordingly.
- Movement of germplasm should comply with regulatory requirements of the importing country.
- In addition to the phytosanitary certificate, a germplasm health statement, indicating which tests have been performed to assess the health status of the material should accompany the breeding material.

2.28.4 The Steps in Technical Recommendations for "Seed Germplasm"

- Seed production should be carried out in areas which are free from diseases of quarantine significance whenever possible.
- Fruits should be harvested from healthy looking plants.
- Seeds of normal size should be selected from healthy looking fruits.
- Seeds should be treated according to the following recommendations, either in the country of origin or in the country of destination:
  - Immerse the seeds in water and discard any floating seeds.
  - Treat the seeds immersed in water in a microwave oven at full power until the water temperature reaches 73 °C and pour off the water immediately after the treatment.
  - If a microwave oven is not available, treat the seeds with dry heat for 2 weeks at 60 °C.
  - Dry the seeds and treat them with Thiram dust.
  - Pack the seeds in a paper bag.
  - After arrival in the country of destination, the seeds should be inspected for the presence of insect pests. If found to be infested, they should be fumigated or destroyed (if fumigation is not possible).
  - Seeds should be sown under containment or in isolation and kept under observation until the plants are well established and normal healthy leaves are produced.

If bud wood or scion material is to be exported, prior information is required to the stations which enable the officers involved to raise the test plants. Healthy stock seedlings should also be grown in advance, so that they will be suitable for grafting when the importation arrives. Any material received with roots attached; there are chances that the roots may carry nematodes or fungal vectors. Therefore it should be first grown in sterile soil, if possible under glass. Cuttings then can be taken from the shoots of plants whose tops have been found to be healthy and these can be rooted or grafted to healthy root stocks. If it is sap transmitted, it should be tested on susceptible hosts. The original roots must be destroyed.

In the UK, *Carnation necrotic fleck virus* (CNFV) had been detected in the imported carnation cuttings from the Netherlands (Stone and Hollings 1976) During March and April 1977, a series of consignments of Portuguese glasshouse carnation cuttings severely affected by CNFV were intercepted and destroyed at UK. Similarly daphne mosaic plants were found in a lot of 250 *Daphne mezercum* plants at U.S. received from Holland and were promptly destroyed under the supervision of State nursery inspectors. When the unrooted cuttings were received, they should be treated with some insecticide and rooted in an appropriate medium, and transplanted to richer compost when rooted and indexed by all possible methods.

In U.S. between 1957 and 1967, a total of 1277 vegetatively propagated plant material of *Citrus, Ipomea, Prunus, Solanum* and *Vitis* were received from various countries and 62 % of the plant material was infected with different viruses (Kahn et al. 1967). In the next decade, i.e. from 1968 to 1978, the percentage of imported infected plant material was only 50 %, out of the 1913 vegetative propagules of *Saccharam, Theobroma, Vitis, Solanum, Ipomea*, fruit trees, grasses and other plants (Kahn et al. 1967). Some of the virus diseases detected in ornamental plant importations from different countries to U.S. during 1968–1978 have also been reported by Kahn et al. (1967), and Kahn (1988). For breeding programmes, the wild species are exchanged from one country to another and viruses are carried even in these hosts. Kahn and Monroe (1970) recorded 39 % virus incidence in the wild *Solanum* introductions collected as tubers. Similarly, Kahn and Sowell (1970); Kahn (1988, 1989) reported 11 % of the 46 wild *Arachis* species collected from Uruguay, Argentina and Brazil were virus-infected. The above examples clearly indicate the need for effective measures to be taken while introducing any type of plant material.

2.29 Methods of Testing at Quarantine Stations

The testing of the plant material received is done by indexing on indicator hosts either by mechanical/insect transmission, or also by grafting. The symptoms are expressed within a few days or weeks when herbaceous test plants are used, but may require months, or years when woody plants are used in which case the entry of the genetic stocks will be delayed. The various methods useful for detecting the viruses in seeds and seed stocks are described in the book Chapter-5 (Sastry 2013a). Indexing the viruses infecting fruit crops are more difficult than in other crops.

For testing cocoa planting material, Amelonado plants (West African Cocao Research Institute Selection C14) is the best known indicator plant. The surface sterilized bud wood sent from the donor country is budded onto decapitated Amelonado plants under glass house conditions. The health of scion and the indicator shoots from the root stock are checked weekly for virus symptoms. For safety, the scion should be top worked later with an Amelonado bud. If free from symptoms after three or four leaf flushes, bud wood is sent to the recipient country. In certain plant materials, some of the latent viruses cannot be detected during the routine inspection, as they do not exhibit well marked symptoms. Some virus-host combinations have long incubation periods which vary from few weeks to several years. For example, certain of the citrus viruses have 3–8 year incubation period. Inspection and treatment at ports of entry may not be adequate safeguards to prevent the entrance of these lately infected plant materials. Plant experts and plant breeders also introduce the wild/cultivated plants which were not showing the symptoms, considering them as resistant/tolerant source (Khan 1976). Some times more than one virus may become latent or symptomless and this has been demonstrated in certain fruit trees. For example, the grape vine corky bark virus is latent in *Vitis vinifera* CV. 'Emerald Riesling', whereas in 'Pinot Noir' and 'Almeria' it causes severe disease (Hewitt 1975). In many grape cultivars, viruses like fleck (marb rure), vein mosaic and vein necrosis are latent. They can be detected by indexing on *Vitis ruperstris* St. George, *Riparia gloire* and *Berlandieri* x *Rupestris* lloR (Hewitt et al. 1972; Legin and Vuittenez 1973).

Even certain diseases are latent in some ornamentals like chrysanthemum, dahlias, gladiolus, carnations etc. Sometimes the latent virus may not damage the host plant and yet it proves to be very economically important to other crops. For example, Potato virus YN strain which is symptomless or it causes only a very mild symptoms in potatoes, will spread rapidly by several aphid species and can completely destroy the tobacco crop in which it causes severe necrosis. *Tomato spotted wilt virus* in arums, begonias, chrysanthemums, and dahlias often cause no visible symptoms, but if it is introduced it causes a great loss to the crops like tomato, tobacco, lettuce, peas, beans, pineapple and other economic crops. Dodder (*Cusanta californica*) is reported by Bennett (1944) as a symptomless carrier of a virus causing damage to sugarbeets, cantaloupes, tomatoes and several other crops. Even swollen shoot of cocao and sunblotch of avocado are symptomless in some varieties. In such cases, plant materials are sent to special quarantine, where they are grown for one or two seasons under special greenhouses.

Besides the indicator hosts, the other tests comprising histopathological, serological, electron microscopic and molecular techniques are also enabling the quick and authentic identification of the virus and virus-like diseases of the imported plant materials of certain fruit crops. Based on these laboratory tests, the quarantine staff will decide the further measures to be taken against the materials under inspection (James et al. 2001). Only healthy plant material those guaranteed free from infection will be released. The details of the inspection techniques can be obtained from U.S.D.A.Mannual (1971) and also from the hand book of phytosanitary inspectors in Africa (Caresche et al. 1969).

2.30 Important Cases of Introduction

Lack of proper quarantine measures for the introduced plant materials has resulted in the introduction of some very highly devastative plant pathogens including virus and viroid diseases from one country to another, and has proven to be catastrophic. For this human actions are largely responsible for emergence of plant virus diseases. This has resulted into increased cost of food material. Evidence about the entry of disease into new areas is, however, often circumstantial and it is rarely possible to say with certainty as to how a disease was introduced. A good example is tristeza virus of citrus, originated probably in China, and introduced into South Africa by 1900 and probably at times into the U.S., but its presence was marked due to the use of tolerant cultivars and the absence of vector (Broadbent 1964). Prior to 1920s the citrus industry of Argentina and Brazil flourished despite the use of tristeza susceptible root stock, but 20 years after tristeza-infected nursery stock was imported from South Africa and Australia, 20 million trees had perished due to the spread of the virus by an abundant local vector *Aphis citricida* (Costa 1956).

In North America and in the Mediterranean region, no-disease state existed for decades. A few cultivars infected with tristeza virus were introduced into North America at the end of nineteenth century (Olson 1955), but natural spread was encountered only in 1960s in California, and in the 1950s in Florida. In the Mediterranean region, tristeza was introduced in the 1930s (Reichert and Bental 1960), but natural spread was encountered in Spain only in 1960s and in Israel in 1970 (Bar-Joseph et al. 1974). Another example is of Star Ruby isolate of *Citrus ring spot virus* (CRSV-SR) discovered in a 'Star ruby' grape fruit tree (*Citrus paradisi*) and was brought into Florida by a private grower from a commercial sources in Texas, without authorization from the State Department of Agriculture. CRSV-SV has not been found in 'Star Ruby' trees imported officially for testing and released (Garnsey et al. 1976). The other serious diseases of citrus like psorosis, exocortis, xylopsorosis, greening and stubborn are also proved to be quite dangerous as they were introduced through infected bud wood and nursery stock (Klotz et al. 1972; Knorr 1965).

Banana bunchy top which is threatening the banana industry in India has been introduced from Ceylon through the infected suckers during 1940 (Vasudeva 1959). The Fiji disease of sugarcane is quite serious in Fiji Islands and from there this disease was introduced to Madagascar and to other countries. *Potato spindle tuber viroid* has its home in North America and was introduced into Russia and Poland from N. America and barred on to Rhodesia. The infected plants of *Abutilon striatum* var. *thompsonii* with *Abutilon infectious variegation virus*, shows attractive leaf variegation, was brought into Europe from Brazil more than 100 years ago (Hollings and Stone 1979). Circumstantial evidence suggests that *Tomato aspermy virus* in chrysanthemum may have been introduced into the U.S. with chrysanthemum imports from Japan (Brierley 1958) or from Europe in the early 1950s. According to Leppik (1964) *Squash mosaic virus* was introduced into the U.S. by seed from Iran and was disseminated in Iowa by cucumber beetles. After several years of intensive work this disease has been eradicated. From the U.S., the same virus has been introduced into New Zealand through the seed of honey dew rockmelon plants (Thomas 1973). In Great Britain, *Barley stripe mosaic virus* has been isolated only twice in barley seed imported from France (Kassanis and Slykhuis 1959; Watson 1959), and from *Hordeum zeocriton* seed imported from Denmark (Catherall 1972). The introduction of *Tomato yellow leaf curl virus* in to the new world from the old world in 1992 occurred via infected tomato transplants that were probably purchased in Israel by a Dominican Republic grower (Polston et al. 1999). In recent years NBPGR (New Delhi, India) scientists have intercepted 34 plant viruses in the seed of number of vegetable and fruit crops, of which 11 viruses are not reported from India (Table 2.10). Many more examples can be cited where international movement of planting material has helped in distributing pathogens from one part of the world to the other.

Besides the infected plant material, in some cases, the insects which are acting as pathogen vectors were also reported with the early free movement of plant material and also which in due course were established and became potent vectors. For example, *Aphis citriocola* (= spiricola) was recently been introduced to the

Mediterranean region and may also have come from Spanish and Portuguese overseas territories, and now are wide spread (Reichert 1959). *Macrosiphum euphorbiae* was apparently not present in Britain before 1917, and may have been introduced to southern Europe from America sometime before they could have been directly introducted from North America on new potato cultivars. It is now thought that the disease and the vector, the leafhopper *Circulifer tenellus* may have been introduced together to California on live plant material used as animal fodder on ships sailing around Cape Horn during the "Gold Rush" era (Bennett and Tanrisver 1957). During 2004 *Thrips tabaci* which is the most potent vector for tospoviruses, was imported to Vidalia onion production area of Georgia from Peru. This thrips was shown to be the Peruvian biotype (Anon 2004).

The interception of known insect vectors of plant viruses that depend solely on specific insects for spread may prevent the establishment of the viruses, provided no indigenous species are capable of serving in a vector capacity. The best way of restricting the vector entry is by defoliating the plants to be imported and treating with a systemic insecticide, before it is dispatched.

2.31 Important Diseases Restricted to Some Countries

Many diseases have limited distribution and have been reported from only one or two countries. U.S. Plant quarantine regulations (1977) prohibit importation of plants and plant parts from specified countries because of the occurrence in these countries of pests and pathogens of quarantine significance to the U.S. Some of the virus and virus-like diseases infecting citrus are at present confined to one or two countries and some of the diseases mentioned below are infectious and might spread if introduced into new areas. They are bud-union crease in Argentina, abnormal bud-union problems in trees on rough lemon in South Africa, brittle twing yellows in Iran, decline of rangpur lime in Brazil, Dweet mottle and yellow vein in California, gum pocket disease of trees on *Poncirus trifoliata* in Argentina and South Africa, gummy bark in Egypt, gummy pitting of *P. trifoliate* in New South Wales, infectious mottling on navel orange in Japan, leaf curl in Brazil, leaf variegation in Spain and Greece, marchitamiento repentino or sudden wilt in Uruguay, the missiones disease in Argentina, multiple sprouting in South Africa and Rhodesia, narrow leaf in Sardinia, small fruit and stunting in Argentina, and young tree decline in Florida. There should be thorough indexing against these diseases whenever the plant material is dispatched from these countries. The need for strictly enforced quarantines in every citrus producing country is self evident.

Cocao swollen shoot virus and *Cocao necrosis virus* are restricted to West Africa, Ghana and Nigeria. The plants belonging to the families *Sterculiaceae, Bombacaceae,* and *Tiliaceae* are potential carriers of swollen shoot and their export should be strictly controlled. *Abaca mosaic virus* of banana is restricted to Philippines. Yellow dwarf of potato is known only from North America. Cadang-cadang and coconut root wilt diseases are restricted to the Philippines and India, respectively. Another serious disease of coconut, lethal yellowing (Kaincope) is limited to Jamaica, Cuba, Cayman Islands, Bahamas, Haiti, West Africa and Florida. Stunt virus of tobacco is confined to Japan. *Rice yellow mottle virus* in rice is recorded only in certain African countries.

The movement of cassava vegetative material from Africa to American continent should be banned to avoid the introduction of Cassava mosaic in the Western hemisphere. An additional risk would be the introduction of whitefly vector (*B. tabaci*) races better adapted to breed on cassava than those existing in the Americas. In Brazil, *B. tabaci* failed to breed on cassava.

2.32 Effective Methods of Plant Importations

(a) **Phytosanitary certificates**

The risk of introducing the pathogen can be checked by phytosanitary certificates for exporting or importing the plant materials, permits in the form of phytosanitary certificates issued by the quarantines in relation to its health, are highly useful in restricting the entry of the diseased material into new area. Even with these phytosanitary certificates, the imported material may be liable to be destroyed at the port of entry of the importing country as per the international agreement.

"Certified" refers that the plant material is free from virus and virus-like diseases and also from other pests and diseases. In some of the advanced countries, there are approved certified nurseries. For example, Canada and U.S., accept certified *Prunus, Cydonia* and *Malus* nursery stock from the certified nurseries in the U.K., France, Belgium, the Netherlands and West Germany. Similarly, the East African Plant Quarantine Station at Kenya accepts the certification of chrysanthemums by the Nuclear Stock Association in the U.K. and the certification of grapevine by the Foundation Plant materials Service of the University of California. These certificates are issued by the central and state governments after careful examination of the material. Although the variety of phytosanitary certificates varies from country to country, depending on their competence of their regulatory agency, their use has reduced the movement of plant diseases between countries.

For evaluating seed health, there are some international certificates like orange certificate, blue certificate, Rome certificate etc. The Orange certificate has international standard in testifying the quality of the seed for trade transactions which is issued after examining seed lots according to the prescriptions laid down in the international rules of seed testing (ISTA 1966). The Blue certificate is issued for samples which have not been drawn officially. The Rome certificate has the international standard, set up in accordance with FAO model phytosanitary certificate.

Phytosanitary certificates are issued to the plant material which do not show any external symptoms and also tested by mechanical/graft transmission to the indicator hosts. It may also be confirmed by using other techniques like histopathology, serology and electron microscopy. In some of the virus-host combinations, visual observations and indexing do not help in complete restriction of diseased material. For example, during 1950–1960, the citrus budwood from all over the world was introduced into California after through indexing and was cleared with federal, state and county personnel to introduce foreign bud wood into the University of California having quarantine facilities at Riverside. The results of this programme, reported by Roistacher and Nauer (1968) after 9 years, showed that a very high percentage of budwood collected worldwide was infected with one or more viruses. It was difficult to find any old bud lime citrus trees free of all known viruses even though bud wood of many selections had been collected by experienced personnel from apparently healthy trees. The results demonstrated that visual inspections of trees are highly unreliable for detecting viruses. Generally quarantine inspectors depend on symptoms to infer the presence of viruses. However, symptoms alone are totally unreliable if they are not diagnostic and if the plant is infected only recently and the incubation period is incomplete. In some cases the diseases will remain latent.

(b) **Inter-African quarantine and phytosanitary regulatory principles (IAPSC) strategy and entry requirements**

For those viruses which do not occur in Africa, IAPSC legislation prohibits importation of the seeds and lyophilized pollen (where the virus may be pollen transmissible) of plant genera that may be carrying such viruses. Alternatively, importation of high value germplasm is permitted for national plant protection organizations for scientific purposes. An import permit and a phytosanitary certificate with additional declaration are still requirements for importation.

Virus risk can be lowered by specifying the conditions of entry for the importer, i.e., (a) import only by National Plant Protection Organizations (NPPO) for certain types of plant material and (b) entry only through authorized quarantine stations. Because of insufficiently trained manpower, coupled with poor facilities, the Inter-African Phytosanitary Council (IAPSC) as mentioned by Olembo (1999) places responsibility on NPPOs for the importation of some high risk crops including vegetatively propagated plant material such as cassava, potato, *Musa* spp., tea, taro, etc.

With the advent of free trade, many countries are redefining and redrafting their phytosanitary requirements to prevent introduction of pathogens. While this may lead to confusing regulations and standards, unnecessary tests or inspection requirements, and unjustified trade barriers, it has become necessary to standardize seed-health tests and inspection practices (Maddock 1997).

(c) **Closed quarantines**

Because of latent or symptomless behavior of the virus or viroid in a particular consignment, certain serious diseases have escaped inspection in the country of origin. All these imported plant materials should be grown in secured glasshouses in a controlled atmosphere to avoid external contamination and be tested for virus or viroid. This is known as closed quarantines. From the small consignment of plant material accepted for closed quarantine, a few should be grown out and all plants should be inspected daily. If necessary, propagations derived from the indexed mother plants or seeds from the second crop should be released. If suspicious symptoms are observed and the casual virus is determined to be one not yet established within the country, the safest procedure is the destruction of the entire consignments in the station incinerator. If the crop is very important, by means of thermotherapy and tissue culture, the plants can be raised and the healthy plant material can be handed over to the consignee, and the details are discussed in the early part of this chapter. While working in glasshouses, several precautions are taken, such as the treatment of the floor of the entrance cubicle with a disinfectant, using germboots and laboratory coats within the glasshouse unit, using sterilized soil for raising plants, washing hands and instruments with detergents, maintaining distance between plants and use of partition screens to avoid contact transmission etc. The principles of the closed quarantine procedures have been described by Sheffield (1968). This type of plant quarantine stations are at Maguga, Kenya, the Post entry Plant Quarantine Station Ibadan, Nigeria and U.S. Plant Introduction Station, Glenn Dale, Maryland.

(d) **Quantity of plant materials**

The risk of introducing the pathogen and pest can be minimized by reducing the quantity of a given cultivar or clone. For vegetative propagations, small might be construed to be 5–10 roots, tubers or corms; 50 buds, 20 unrooted cuttings or 10 rooted cuttings. In case of seeds, a quantity required for sowing a 10 m row will serve the purpose. Initially these small quantities of planting material, should be maintained free from all destructive pests and diseases. After confirmation as healthy material, it can be tested in the field. The vegetative propagative materials should be defoliated, which will avoid the entry of vectors which may colonize on leaves. The plant material should be examined for the eggs and different stages of a potential vector and should be dusted or sprayed with the insecticide before packing for exporting. The majority of vectors can be controlled by methyl bromide fumigation. It should be sent in unused cloth or paper bags or envelopes. The importers or the exporters of germplasm should ascertain well in advance the requirements of the importing country, for avoiding unnecessary delay at quarantine stations.

(e) **Open quarantine**

This is the quarantine of plants without using such physical confinement structures as glasshouses or screenhouses. This can be used to reduce or eliminate the risk of spread of pests by adhering to a quarantine protocol. The technique has been used successfully in East and Central Africa to exchange germplasm resistant to EACMV-Ug. It has facilitated safe introduction of large quantities of cassava germplasm, which would not have been possible through other means. The method was cheaper than micropropagation and plant mortality was also low. This method can be used in germplasm exchange programmes where the climate and pest species are more or less similar (for more details see 2.23.7 of this chapter).

(f) **Examination of exportable crops during active growth**

The U.S. annually imports large quantities of vegetables from Mexico. Arrangements were made for U.S. and Mexican personnel to conduct field surveys in the vegetable production areas of Mexico. Such co-operative field inspections during growth season at points of origin provide greater protection than checking at ports of entry. The growing season inspection is generally more sensitive than pre-export inspection of planting material or produce. This method will be very effective wherever the vegetatively propagated materials are exported. Stover (1997) described an effective method of exporting bananas for commercial or research purposes. The rhizomes should be taken from a disease-free area, at least for 1 year. They should be peeled free of all root stubs, until only white tissue remains. Later they should be submerged in hot water at 54 °C for 10 min. The cartons having the suckers should be shipped air freight with the required phytosanitary documents at temperature above 14 °C (58 °F). After receiving the rhizomes at quarantine stations the plants should be raised for nearly 1 year and later they can be released.

(g) **The intermediate quarantine**

The genes for resistance to virus and virus-like diseases are likely to be found in regions where the plant in question originated (primary gene centers) or has subsequently been grown (secondary and tertiary gene centers). It is in such places that long association between pathogens and plants has occurred with consequent elimination of susceptible plant genotypes through natural selection. Hence resistance material is generally searched for in what are thought to be as the gene centers of cultivated plants. While collecting the sources of resistance from the gene centers, one should not inadvertently introduce genes for virulence by collecting the new forms of the horizontal and vertical pathotypes. As discussed earlier, many virus diseases can occur in a latent condition, without expressing any visual external symptoms. The risk of introducing new pathogens either from the gene centers or from the exporting country, can be reduced by transferring a genetic stock to a third country instead of sending directly, whereby that crop is not grown and the pathogen would not be establish there. Third-country or intermediate quarantine is an international cooperative effort to lower the risk to country B associated with transferring genetic stocks from country A by passing these stocks through isolation or quarantine in country C. The plants are maintained in country C to test for obscure pathogens, or they are detained to allow incipient infections to surface or to permit treatment at a weak point in the life cycle of hazardous organisms. The salient feature of this type of safeguard is that it advocates the passage through country C only of genera that pose no threat to country C because (1) the crop is not grown in country C, (2) the harmful organisms of that crop have narrow host ranges so they will not attack other crops of country C, (3) the harmful organisms that may gain entry to country C would not become established because susceptible hosts are absent or the climate is unfavourable. The third country quarantine locations include the plant quarantine facility, Glenn Dale, MD., the U.S. Sub-tropical Horticultural Research Unit, Miami, Kew Gardens, U.K., Royal Imperial Institute Waginingen, and Institut National de la Recherche Agronomique (IRAT) at Nogent-Sur-Marne, France. For example, the export of cacoa propagating material primarily from West Africa to the other cacao growing countries is first quarantined at an intermediate quarantine station of third country quarantine. Facilities may exist within the tropics reasonably far from growing cocao as at Mayaguez in Puerto Rico or Salvador in Bahia, but safety usually requires intermediate quarantine in a temperate country, for instance at Kew, Glenn Dale; Miami or Waginengen. Similarly, sugarcane sets entering East Africa are first put into quarantine at the East African Agriculture and Forestry Research Organization (E.A.A., F.R.O.) at Maguga to see if they carry viruses.

(h) **Aseptic plantlet culture**

Another safeguard in the international exchange of germplasm is to import only plantlets established as aseptic cultures. The required plant materials, which are proven to be virus-free after indexing, can be developed by tissue culture technique and can be sent to the other country without any hazards. As the size of the consignment will be small and in aseptic conditions, there won't be any chance of reinfection with other pathogens like fungi, bacteria and also with other pests like insects, mites and nematodes. In this technique, an imported cultivar or clone may be represented by meristem tips or exercised buds or embryos instead of several cuttings, scions, tubers, seeds etc. The tissue culture technique in combination with thermotherapy or chemotherapy or both have been used in the production of virus and virus-like disease-free planting material of number of horticultural crops. During 2009, Wang et al. have developed cryotherapy technique and produced virus-free plants of citrus, grapes, Prunus spp. and certain tuber crops like potato and sweet potato.

Prior to the establishment of tissue cultures the mother plants were indexed and tested serologically for viruses. The use of this technique was first reported by Kahn (1989) for *Asparagus officinalis* L. In 1972, asparagus clones from France were indexed for viruses at Glenn Dale. From the plants that indexed negatively for viruses, aseptic plantlets lets were developed on agar medium in screw cap bottles and were shipped to Kenya. Roca et al. (1979) shipped 340 aseptic cultures of *Solanum* germplasm from the International Potato Center (CIP) in Lima, Peru to the countries like Australia, Boliva, Brazil, Canada, Colombia, Costa Rica, India, Indonesia, Kenya, Mexico, Philippines, Turkey, the UK and the U.S. The clones were successfully established in 12 of the 14 countries. At CIP multiple shoots in tissue culture were produced in shake cultures. Plantlets of potato regenerated from nodal cuttings of multi-meristem shoots were then shipped in culture tubes from CIP to corresponding countries (Roca et al. 1979). While developing the aseptic cultures, heat treatment is also employed for plant materials in which viruses were not easily eliminated. Waterworth and Kahn (1978) developed sugarcane plantlets of three cultivars that indexed negative for *Sugarcane mosaic virus* (SCMV) by hot water treatment and aseptic bud culture. The hot water treatment was three sequential exposures of cuttings at 24 h intervals for 20 min at 52, 57, 57 °C. The SCMV-infected canes were sent from the east African Plant Quarantine Station, Kenya to Glenn Dale where they were subjected to heat therapy followed by bud culture. The aseptic plantlets were returned to Kenya in screw cap bottles for transplanting and indexing (Kahn et al. 1967; Kahn 1980). Aseptic cultures were also developed against crops like ginger, chrysanthemum, sweet potato and banana, and are being used for exchanging virus-free genetic stocks between countries. This technique is also useful to the germplasm explorers, where they can collect the required perishable plant materials right in the field. They have to carry tubes having tissue culture media and after thorough disinfection of the plant material, they can be transferred to the medium, with a higher risk contamination.

(i) **Embryo culture**

As some of the viruses are internally seed-borne, countries which have a zero tolerance against the seed-borne pathogens, have prohibited the entry of the seed from other countries where these specified organisms are known to exist. For commercial purposes the imports are strictly prohibited, however very small quantities are permitted for scientific purposes. Kahn (1979) developed during 1970–1972 an embryo culture technique by using tissue culture methodology. In this technique, instead of complete seed, only embryo axes were exercised and used for culturing and it can be mailed without any risk. This technique is also useful for germplasm exploraters. Guzman and Manuel (1975) used this technique for coconut plant material. Embryos collected in large numbers in diseased area can be maintained in laboratory culture for 16 weeks and grown in screened glass houses for some months before they need to be planted. If no virus, phytoplasma, viroid or other pathogens are found when adequate samples were screened by standard methods, it could be considered safe to plant embryo cultured seedlings of coconut. Braverman (1975) modified this technique so as to include one microbiological test and virus indexing. As a safeguard against the breaking of agar in the glass bottles tubes during transit, it is desirable to add warm sterile agar before shipping, until the container is almost full. The plantlets are submerged in agar, but they ship well and usually arrive in excellent condition with the agar remaining undisturbed.

(j) **Use of shoot tip grafting or micrografting**

Micrografting is another technique which provides a means whereby plants can be transported from one country to another. The culturing of lateral buds *in vitro* to induce multiple shoots may have potentially important applications. This method is proposed by Navarro and Juarez 2007; Navarro et al. (1975). Budwood could be fumigated and shipped in test tubes or in a sealed container and the lateral buds excised in the receiving country and cultured *in vitro* for production of multiple flushes, whose shoot tips then be used for grafting. The method involves, holding the previously trimmed shoot with tweezers under the microscope, the very tip end of the growing bud (0.88 mm) is removed with a razor knife blade and quickly transferred to the cut edge of the inverted "T" on the seedlings. The resulting graft plants after 3–5 weeks could then be indexed for a broad spectrum of pathogens in a special quarantine facility and destroyed if pathogens were found. Through this technique (Nauer et al. 1983; Fifaei et al. 2007) have successfully eliminated *Citrus tristeza virus* from the citrus species. This should pose no serious problems with quarantine regulations to the country of origin or the country processing the bud wood, since all the bud wood, bud or explants would be under continuous cover. This method also helps in exchanging of superior cultivars.

2.33 General Principles for the Overall Effectiveness of Quarantines

The details of quarantine regulations are not included here as they vary from one country to another, but a general set of principles contains the basic guidelines:

(1) Seed rather than vegetative propagation material should be introduced unless clonal propagation is necessary.
(2) For clonal propagations, non-rooted propagative material such as scions or cuttings should take precedence over rooted plants.
(3) Woody plant introductions should not be more than 2 years old.
(4) Consignments of vegetatively propagated material should be small i.e., each cultivar or species should be represented by a few tubers, scions or cuttings.
(5) A stock plant should not be reused for propagation if a foreign bud or scion failed to survive on the stock. Bud or graft union failures may be caused by pathogens such as viruses transmitted from the introduction to the stock.
(6) It should never be assumed that all vegetative propagations of a given species or cultivar were derived from the same mother plant.
(7) Each scion, cutting or tuber of a clonal introduction should be considered as a subclone.
(8) When pest or pathogen detection tests indicate that a particular subclone is eligible for release from quarantine, propagations for release should come only from the subclone that was tested and not from other sub-clones, that were not tested, even though these subclones, constitute part of the original accession.
(9) If introductions are received as roots, such as sweet potato, cuttings derived from the roots should be released rather than the original root itself, which should be destroyed.
(10) Visual observation is not satisfactory for diagnosing virus and virus-like diseases because neither the presence nor absence of virus-like symptoms is necessarily indicative of the presence or absence of virus.

Besides the above ten principles, the following four types of quarantine actions will mostly restrict the entry of the diseases:

(1) The consignment should be followed by embargo and airport permit.
(2) Inspection (field inspection, laboratory tests etc.) in the exporting country before shipment of the consignment and to be on the safe side the material must be given chemotherapy and thermotherapy.
(3) Post-entry growth inspection of the importing country in closed quarantine.
(4) Certification-phytosanitary certificate attested for freedom from diseases and pests.

2.33.1 Quarantine Facilities

The type of quarantine facilities depend on the climate at the introduction station, the crops and its temperature requirements, pests and pathogen risks, and duration of the quarantine period as it affects plant size. Features that can be incorporated into a greenhouse to improve phytosanitation and thus facilitate quarantine include: a series of small glasshouses, air conditioning, filtered air, humidity control, concrete floor with drains, partition screens, soil sterilization, fumigation chambers etc. Other features include a pathogen-free water supply, heat therapy unit, tissue culture rooms, hot water treatment facilities, black-light insect traps, shoe disinfectants and fungicide and insecticide spray programs. Large and well equipped quarantine stations are started when ever a country is both agriculturally and scientifically well advanced. Less advanced countries will have small stations or are nonexistent. Sufficient number of adequately trained and experienced inspectors are required for the effective examination of incoming plant material at airports, railway stations, sea ports and frontier ports.

2.34 Need for Networking for the Developing Countries

Networking within and between the sub-regions, and with institutes in developed countries and international agricultural research centers (IARCs) is critical to strengthen plant virus research in the world. There are numerous advantages of such international networking (Plucknett and Smith 1984). Networking is a rapidly growing mechanism, as funding becomes increasingly scarce, to optimize resource utilization and to facilitate the efficient transfer of technologies to developing countries. It is in developing countries, where the impact of these technologies can be felt almost immediately if the resources are available to utilize them. A good example of such an effort was the collaborative project on identification of cowpea viruses coordinated by the International Institute of Tropical Agriculture (IITA) with monoclonal antibodies provided by, and financial support from, the International Development Research Centre (IDRC), Canada (Thottappilly et al. 1993).

Many existing networks are crop-based and are often divisive from the point of view of the virology community because the networks rarely link together. The networks such as the Cassava Biotechnology Network (CBN), ProMusa, East Africa Root Crops Research Network (EARRNET), and the Southern Africa Root Crops Research Network (SARRNET) have taken a big step in the right direction. However, until capacity and confidence is increased, many of the virologists from developing countries are unable to participate fully and gain maximum benefit from the interactions.

The organization of working group meetings at regular intervals, either at the regional or sub-regional level, is an ideal way to promote the networking concept. These meetings would involve scientists from all over the world who are working on the plant virology problems of their country. It would enable priority constraints to be identified and research needs of national programs to be determined. A framework for collaborative research programs in partnership with IARCs and research institutes in advanced countries must be developed and sustainable funding obtained. Sustainability of funding must be considered a key issue that must be addressed. Working groups on crop-specific virus disease problems have already been developed, for example, the International Working Group on groundnut viruses in Africa and the Virology Working Group of Pro*Musa*. However, a formal network must be developed for the upcoming countries which these collaborative initiatives and other virus-specific subgroups can interact. This will ensure better coordination of efforts and synergy between different partners and stakeholders in addressing a broad range of plant virology needs of the national programs. However, it is vital that any initiative that is taken regarding networks and collaboration is assured of funding for long enough periods for the activities to be self-sustaining. Any plant virology network must be endowed with sufficient funds to have a dynamic coordinator and secretariat with authority (perhaps under the auspices of the steering committee of the network) to initiate appropriate research and donor contacts. Regional and sub-regional organizations should take an active role in encouraging and supporting such initiatives. More information can be also be obtained from different sources on plant quarantines (Chiarappa 1981; Kahn 1989; Ebbels 2003; Khetarpal et al. 2004).

Conclusion

The foregoing discussion clearly stresses the need for effective quarantines. The expenses incurred in establishing and maintaining the quarantines is but a fraction of the economic losses that would be suffered if plant diseases gain free entry into the country. The success of the plant quarantine measures mainly depend on the proper composition of the central and state services of plant quarantine, the scientific background and the qualifications of the inspectors and the specialists, and the availability of necessary equipment at the quarantine stations and laboratories. The post-entry quarantine service for all imported seed and vegetative plant materials must be established, so that it can produce and distribute pathogen-free seed derived from the imported infected material. It will also be useful if world distribution of seed-borne pathogens is mapped. The safest source of the healthy materials should be from a country with efficient quarantine services which has talented staff who can diagnose the indigenous pests and diseases. There is a vital need for public awareness of the importance of quarantine regulations. It is well recognized by quarantine officials that success of a quarantine programme is dependent on co-operation between government agencies and the public. Quarantine action is better received and followed if it is based on cooperation between concerned groups rather than on a unilateral compulsion. In this age of rapid and greatly expanded movement of people and cargoes by air, it is extremely important to alert the public by all means to the danger of bringing in infected or infested plant materials from abroad through advertising leaflets, posters, broadcasting on radio and television. Unless quarantine regulations are scientifically sound and administratively feasible, they cannot be successful and may cause serious political and economic problems. On the other hand, there must be adequate legal authority and appropriate enforcement in prosecuting willful violators. The enforcement of quarantines requires considerable expenditure of money, much interference in trade travel and other normal activities of man. The international cooperation and strengthening of regional organizations are of paramount importance in attaining the objectives of plant quarantine.

2.35 Integrated Approach

The viruses have been recognized over several decades earlier and are world wide in distribution through seed and vectors. Subsequent studies revealed that percentages of seed transmission varied depending upon virus-host combinations. Majority of the virus and virus-like pathogens are transmitted in vegetative planting material at a high percentage and furthermore, new viruses are continuously being identified in vegetatively propagated plants. It would not be surprising that with further research this number may increase more than one third of the recognized plant viruses.

Since definite viricides are not yet available, an effective management of seed-borne and vegetative plant material-borne virus and virus-like diseases is possible by development and implementation of certification schemes by following certain measures like indexing for virus and virus-like pathogens, destroying the sources of infections like weed and volunteer plants, roguing the infected plants wherever disease incidence is low, and controlling the virus spread through oils and or insecticides in reducing vector population etc. In the case of infection among fruit crops, sometimes the virus may be latent and propagation from the infected plant is done inadvertently. Since certain fruit tree viruses are also seed-transmitted, seedlings used as root stocks serve as sources of inoculum. Hence, in the production of nursery stocks, it is essential that both scion and root stocks be derived from trees that are indexed and known to be virus-free.

The available experimental data on the management of virus and virus-like diseases of different crops in tropics indicate that substantial virus management can be achieved by the application of a combination of the measures like the use of cultural practices, utilization of pathogen-free planting material or application of insecticides and or oils for vector control. Since the pathologists/virologists/entomologists felt that no single strategy will help in achieving maximum virus and virus-like diseases management, research has also been conducted by embracing multidisciplinary strategy including virus and vector identification, ecology and epidemiology of virus and vector, chemical methods, genetic methods, and implementation of exclusion and eradication techniques along with phytosanitary measures. In short, this approach is a multidisciplinary approach and is called integrated disease management (IDM).

In the entomology discipline the term 'Integrated Pest management (IPM) concept has been used which combines and integrates biological and chemical control measures for most effective management of different pests and diseases. Similarly, the pathologists have coined the technology as "Integrated Disease Management (IDM). Even for plant virus disease management, various workers have used IDM techniques for virus disease management in different virus-host combinations. Some of the available information on plant virus and virus-like diseases was presented here which have followed the aspects like the use of resistant/tolerant cultivars, biological control, crop rotations, selected planting and harvesting dates and cultural or environmental control. The governments of every country should provide grants to the Agricultural departments and research organizations for developing innovative IDM techniques and subsidies to the farming community to meet the exorbitant production expenditure and training on recent IDM technology to the farming community through extension education programmes.

Most definitions of IDM further stress that IDM should first rely as far as possible on cultural and biological methods. The real challenge of IDM is how the various methods discussed above can be best combined to give the farmers an acceptable and socially and environmentally desirable form of prevention of crop losses (Van Emden 1982; Maelzer 1986; Jones 2004, 2009; Makkouk and Kumari 2009).

In an IDM programme against virus diseases, the first requisite is identification of crop damaging virus or virus-like agents. At present there is no universal strategy for the detection of plant viruses and the main methods currently in use for the detection and diagnosis of viruses can be grouped as biological, physical, serological and molecular. Bioassays based on knowledge of symptoms produced and the diversity of experimentally infectable hosts provides an understanding for detection or diagnosis of virus/viroid diseases. The biological methods are labor intensive and require a lot of space. Physical methods like electron microscopy of planting material are more rapid and frequently allows for tentative identification which can be confirmed by using other criteria. However, these methods require costly equipment which is often difficult to maintain in good working conditions. The serological methods overcome the disadvantages of biological and physical methods since they use only a small proportion of the information encoded by the virus. In serology, a diverse array of tests exploiting antigen-antibody recognition are available to detect viruses even at lower concentrations and in few cases, it also helps in detection of localized infections. Enzyme-linked immunosorbent assays using chromogenic, fluorogenic, chemiluminescent or radioactive substrates are some of the choices. Molecular methods involving nucleic acid hybridization, which can use whole or selected parts of the viral genome, have been developed. Some of the basic principles of IDM and successful stories of IDM against plant virus diseases are discussed.

2.35.1 The Tropical Whitefly IPM Project (TWFP)

There are more than 200 different plant viruses which are transmitted by *Bemisia tabaci* alone and are responsible for the loss of millions of hectares of valuable food and industrial crops in tropical, subtropical and temperate regions (Morales 2007). Glasshouse whitefly *Trialeurodes vaporariorum* also transmits virus diseases like: *Potato yellow vein virus* in potato; *Tomato infectious chlorosis* and *Tomato torrado virus* and is responsible for heavy yield losses.

As the whitefly-transmitted viruses of cassava, sweet potato and other vegetable crops are limiting factors for successful cultivation, the Tropical Whitefly IPM Project (TWFP) was launched in 1996 by the system wide IPM programme of Consultative Group on International Agricultural Research (CGIAR) to manage whitefly pests and whitefly-transmitted viruses in the tropics. The international center for tropical agriculture (CIAT) was the co-ordinating center; the International Institute of Tropical Agriculture (IITA) in Nigeria; the International Potato Center (CIP) in Peru; the Asian Vegetable Research and Development Center (AVRDC) in Taiwan; and the International Center of Insect Physiology and Ecology (ICIPE) in Kenya were the participating members.

The goal of the project was to promote sustainable agriculture and socio-economic growth in resource-poor farming communities possessing mixed cropping systems affected by whitefly pests and whitefly-transmitted viruses in Sub-Saharan Africa (SSA), Southern Asia and Tropical Latin America.

The project was planned in 3 phases in order to:

(i) Survey the tropics and determine the regions and the crops affected by whiteflies and whitefly transmitted viruses;
(ii) Evaluation of the most viable IPM practices available to control these pests; and
(iii) Dissemination of knowledge acquired during phases 1 and 2 in the affected regions.

The project was divided into 6 sub projects:

(1) Whiteflies as the vectors of cassava and sweet potato viruses in sub-Saharan Africa,
(2) Whiteflies as pests and vectors of plant viruses in Eastern Africa,
(3) Whiteflies as pests and vectors of plant viruses in Middle America,
(4) Whiteflies as pests in the Andean highlands of South America,

(5) Whiteflies as virus vectors in mixed cropping systems of South East Asia, and
(6) Whiteflies as cassava pests in South America.

The main crops investigated in this project were: cassava, sweet potato, common bean, tomato and peppers. The outcome of this project was: Genetic resistance was identified as the most important component of this IPM programme followed by phytosanitary, cultural and legal measures. The objectives of this project were well attempted in terms of etiological agent identification, diagnostic techniques, virus disease management including the identification of cultivars/tolerant varieties in cassava and sweet potato, and training of the farmers. The most outstanding results from this project were (1) Development of world's first whitefly-resistant cassava; (2) TWP research has provided regionally specific IPM measures in 31 countries; (3) In El Salvador, IPM strategies increased tomato yields by 40 % and sweet pepper yields by 236 %. Improved bean varieties are yielding 800 kg/ha compared to around 60 kg/ha from susceptible landraces. Morales (2007) furnished the full details of this project in his review article.

IPM against thrips transmitted virus diseases:
Around the world, tospo- and ilarviruses transmitted by thrips infect large number of host plants and cause heavy losses in a majority of the crops infected, and has been discussed in Chapter-3 of Volume-I. The type member of *Tospovirus* genus, *Tomato spotted wilt virus* (TSWV) has a very wide host range, infecting vegetable, fruit crops and ornamentals. Given the seriousness of TSWV, the role of integrated virus disease management were worked out in different solanaceous crops by Diez et al. 1999; Adkins 2000; Jones 2004; Brown et al. 2005; Pappu et al. 2009; Awondo et al. 2012; and Riley et al. 2012. Various effective recommended integrated virus disease management strategies based on epidemiological data, proposed by Jones (2004), Pappu et al. (2009) is provided in Table 2.12.

Perusal of the Table 2.12 provides a listing of all the aspects of plant virus management techniques that were suggested against TSWV and some suggestions are quite effective at nurseries, some against protected crops and some against field crops. With the vast experience in plant virology, Prof. R.A.C. Jones of Australia, has suggested approaches that can be planned for other virus-vector-host combinations in different countries. As no single control measure in likely to be effective on its own in field situations, steps must be taken to ensure that a IDM approach is followed.

Even the trials conducted at Spain by Diez et al. (1999) by using resistant cultivars along with raising in tunnels 10 × 5 m, protected with a mesh of 14 × 10 threads/cm has proved to be one of the effective integrated approach for the management of TSWV in tomato. The environment created under mesh stimulated growth, neutralizing the effect of the infection. The results were positive in reducing the disease incidence when the clean TSWV resistant transplants were raised under thrips proof mesh houses. On the other hand when susceptible tomato cultivars were grown under mesh, the TSWV infection was enhanced and observed reduced yields. From U.S., Awondo et al. (2012) have proved that by

Table 2.12 Integrated disease management (IDM) approaches to minimize infection with TSWV in vegetable crops within seedling nurseries, protected cropping situations and field crops

Control measures	Nurseries	Protected crops	Field crops
Phytosanitary			
Avoid spread from finished crops	No	Yes	Yes
Avoid spread from ornamental plants	Yes	Yes	Yes
Minimize spread from weeds or 'volunteer' crops plants	Yes	Yes	Yes
Use roguing within crops	No	Yes	Yes
Introduce healthy transplants	No	Yes	Yes
Avoid spread within seedlings trays	Yes	No	No
Certification of seedlings nurseries	Yes	No	No
Cultural			
Isolate from other susceptible crops	No	(Yes)	Yes
Promote early canopy cover and high plant density	No	Yes	Yes
Manipulate planting date	No	No	Yes
Use mulches or minimum tillage	No	No	Yes
Employ windbreaks, and barrier or cover crops	No	No	Yes
Diminish vector population growth	(Yes)	Yes	No
Install fine nets	(Yes)	Yes	No
Obtain advance warning of outbreaks	Yes	Yes	Yes
Institute susceptible crop and seed-free period' (Ultimate measure when all else fails)	No	Yes	Yes
Resistance			
Deploy virus resistant cultivars	No	Yes	Yes
Chemical			
Apply insecticides	Yes	Yes	Yes
Apply oils or film-forming products	(Yes)	(Yes)	No
Biological			
Introduce thrips predators	Yes	Yes	Yes

(YES) implies a likely method of control
Source Jones (2004), Pappu et al. (2009)

using the combination of Actigard and reflective mulch was shown to be effective in achieving reduced TSWV incidence with increased yields in tomato. Since TSWV is a problem in majority of the vegetable crops, the efficacy of similar type of treatment in other crops has to be tried.

There is a need to adopt a diverse array of control measures to be applied before, during and after planting and to provide effective education and information transfer to growers.

Another member of *Tospovirus* genus, *Watermelon bud necrosis virus* (WBNV) is responsible for severe out breaks wherever the watermelon crop is grown and Rajasekharam (2010) recommended an integrated approach consisting of (1) seed treatment with imidacloprid plus three rows of maize as border crop, and use of silver colour UV reflective mulch plus alternate sprays of imidacloprid and (3) + (2) thiomethoxam plus spraying of imidacloprid on bunds plus keeping weed-free, which have proven to be effective. In these treatments, the initial PDI was 1.25 at 30 DAS and have delayed the WBNV incidence beyond flowering initiation stage, only 10 % at 60 DAS and 34.3 % disease incidence was recorded at 90 DAS while the WBNV incidence in control treatment was 95 %. From IIHR, Bangalore (India), Krishnakumar et al. (2006b) have also successfully reduced the incidence of WBNV in watermelon, by using the insecticides and black polythene mulch viz., covering the soil with black polythene, seed treatment with imidacloprid (Gaucho 75 WS) and acephete spray at 10 day interval, were the effective treatments.

In central India, even *Stem necrosis disease* in potato caused by PBNV was successfully reduced by integrated approach comprising of (1) delaying the planting of potato crop from 4th October to 3rd November, (2) two foliar sprays of imidacloprid (0.07 %) at 21 and 35 days after planting and (3) use of polyethylene sheet or paddy straw. For a number of years intensive studies carried out by Green (1991) from AVRDC (Taiwan) resulted in the development of integrated approach comprising of use of resistant lines along with reflective mulches in crops like chillies, potato, tomato, onion and other vegetables against major virus diseases and the extension activities have made for the successful implementation under the field conditions by a large number of farmers.

IDM and virus diseases:

Rice tungro virus is one of the major destructive viruses affecting rice crop whereever it is grown. Among the many possible methods of controlling tungro, the most practical at present are (1) to grow resistant cultivars (host plant resistance), and (2) cultural practices (cultural control). Among the latter, it has been well demonstrated that synchronous crop establishment is very important (Chancellor and Holt 2008). The use of insecticides for the management of tungro through the control of green leafhoppers (GLH) (*Nephotettix virescence*) is often not effective. This is because: (1) tungro spreads rapidly (GLH is a very efficient vector of tungro viruses), (2) leafhoppers can transmit tungro viruses in very short feeding time; and (3) there is continuous movement of leafhoppers from surrounding fields.

Some of the cultural control measures are proven to be effective in reducing the tungro virus incidence and they are:

1. Timing of planting: with prior knowledge of seasonal patterns of GLH abundance and tungro incidence, crop establishment can be done when the vector population is low.
2. Synchrony of planting: this refers to farmers in a given area establishing their rice crops at about the same time. This reduces the spread of tungro: inoculum sources are minimized and leafhoppers cannot transmit the disease from one crop to another. Late crop establishment, relative to the average date in a given area, carries a very high diseases risk. Such late fields then pose a risk to early plantings in the next season. Farmers who cannot avoid late crop establishment should plant resistant cultivars.
3. Plowing under infected stubbles: this reduces inoculum sources, and destroys the eggs and breeding sites of GLH. This is recommended if the previous crop

was diseased and should be done immediately after harvest. The practice may be difficult for some farmers, because it involves extra expenses and depends on the availability of water.

4. Roguing: this may help to reduce tungro incidence, but at low incidence only. Removing diseased plants is generally ineffective because there may be other plants that appear healthy but are already infected. Actually, roguing infected plants may disturb leafhoppers and accelerate disease spread.
5. Direct seeding: that tungro incidence is often lower in direct-seeded than in transplanted rice. This is because dense plant populations reduce the chances for GLH to find and feed on diseased plants.

With optimum timing and synchrony of planting, early maturing rice cultivars may escape tungro infection. This is because by the time GLH population reaches its peak in the season, rice plants start to mature and become less suitable for GLH. However, the practice of optimum timing and synchrony of planting depends on the availability of irrigation water in the area and the access of farmers to inputs, credit and labor. To be effective, a concerted effort among farmers in a village or within large contiguous farm units is required (Chancellor et al. 1997; Tiongco et al. 2008). In West Africa, Integrated management of *Rice yellow mottle virus* (RYMV) in low land and irrigated ecosystems were followed (WARDA 1999).

IPMCRSP for virus disease management:
Similar to the above discussed projects, in the U.S., a Global Theme project on insect-transmitted viruses with integrated pest management collaborative research support program (IPMCRSP), initiated during 2010 to provide science-based knowledge for developing sustainable and eco-friendly integrated disease management strategies to reduce crop losses due to virus diseases in South and Southeast Asia and interacts closely with a companion Global Theme project on insect-transmitted viruses in Central and South America and Eastern and Western Africa and other Global Theme Projects. The project had strategic partnerships with AVRDC - the World Vegetable Center, public and philanthropic agencies seeking international public good, national research organizations and universities, private sector organizations, and NGOs. USAID was instrumental in facilitating such a unified approach in moving from boutique projects to partnerships for progress in economic growth and poverty reduction in developing countries.

(a) **Specific objectives of IPMCRSP were**:

1. Strategic research on tospoviruses and thrip vectors, and develop host plant resistance.
2. Applied and adaptive research to deploy eco-friendly integrated disease management strategies to control tospovirus diseases.
3. Strategies for strengthening institutional capacities within host countries to conduct problem-oriented research on virus diseases.

This project has been carried out by Dr. Naidu Rayapati as project director from Washington State University. The research outcomes of this project has been practiced by the farming community. Similar types of funded projects which help in framing suitable virus management measures with extension moto are to be initiated at different parts of the globe.

With the rich experiences while researching in sub-Saharan Africa, Prof. JM Thresh during 2003 recommended that integrated control measures have evident benefits and have to be fostered and promoted as a means of enhancing crop productivity to meet the increasing demands of a burgeoning human population (Thresh 2003).

2.35.2 Integrated Management of Insect-Transmitted Plant Virus and Viroid Diseases

Research conducted in different countries on virus and viroid disease management have attempted to utilize some of the established procedures in their field trials to include cultural, chemical, and cultivar resistance factors. Depending on the vector involved and the time of conducting the field trial and the crop age at which the treatment was implemented, are factors to be considered in determining if control will be minimum, medium or maximum depending on virus-host combinations. Since single treatment are not providing expected results in terms of virus disease management, so attempts were made to combine two or three treatments. Implementation of certain cultural practices like use of nylon net cover in nursery beds, soil application of systemic insecticides/pyrethroids or oils/alluminium or plastic mulches/use of resistant or tolerant cultivars has been tried in almost all continents.

For example at Coimbatore (India), soil application of carbofuran at 1 kg a.i./ha coupled with covering the nursery bed with nylon net of 400 mesh followed by 2 sprays of 2 % neem oil at 15 and 35 DAP recorded lower virus incidence and highest yields, with a cost: benefit ratio of 1:33 in tomato crop against virus diseases. At a number of places the field operations like covering the nursery beds, soil application of carbofuran at 1 kg a.i./ha in combination with foliar application of 0.1 % imidacloprid at 10 days interval effectively reduced the vector population as well as the disease incidence in different virus-host combinations (IPMCRSP, Naidu Rayapati unpublished data).

At Sabour (India) use of nylon netting and soil application of carbofuran at 1 kg a.i./ha in the seed bed and application of carbofuran at 1.5 kg a.i./ha in the main field a week after transplanting the seedlings of vegetable crops like tomato, chillies, etc. plus 3 sprays of Nuvacron at 1 ml/l was most effective in managing virus diseases with cost: benefit ratio of 1: 5.6.

Similarly, Parvatha Reddy (2009) while discussing in his text book on various management measures against *Cucumber mosaic virus*, which occurs on number of vegetable and fruit crops, he has listed the following integrated measures.

- Select seeds from absolutely virus-free plants.
- Remove weed hosts in around the field.
- Roguing of infected plants from time to time.
- Use of barrier crops like sunflower, sorghum, pearl millet or maize around the cucumber crop to prevent the vectors entering the field.
- Mulch the soil with saw dust or wood chips or paddy straw or with aluminium foil which repels the aphid vectors and reduce the CMV infection up to 87.9 %.
- Spray with either dimethoate or monocrotophos each at 0.1 % at weekly interval to control aphid vectors. Spraying of mineral oils at 0.75–1.0 % at weekly intervals was also found promising.

At present new methods for the analysis of plant viral genomes are exemplified by the analysis of avirulence genes in some of the vegetatively transmitted plant viruses. The information gained from the application of these new methods is likely to generate new types of resistance mechanism that may be effective against a wide range of viruses. More and more novel developments are likely to emerge through the use of biotechnological techniques. This new area of research is already well enough developed for single genes to be transferred at least in dicotyledons, but do pose some difficulties in handling with monocotyledonous plants. However, it is likely that suitable methods will soon be devised to overcome these problems, which opens more options for virus management.

Intensive researches have been carried out involving chemicals, sometimes with the combination of irradiation or heat, to eliminate viruses from the infected seed without loss of viability but in many cases the control obtained has not been absolute.

After going through the examples cited regarding plant virus management aspects in this chapter, the reader will realize that generally no single procedure offers maximum control of virus and virus-like diseases. Hence a number of experiments were carried out throughout world by using different combinations having aspects like seed dressing and foliar spray with insecticides, use of nylon nets and aluminum mulches along with tolerant cultivar if available, were tried and based on the exponential data best combination of treatments, which has resulted in effective vector control, with low disease incidence along with higher yields were recommended to the farmers through extension workers and other methods. A rational combination of procedures weighing the efficacy of one procedure against others and utilizing their supplementary effect is the strategy for an integrated control programme and encouraging results have been obtained to combat virus diseases. In this direction, virologists of different countries have combined two or three effective treatments for effective plant virus disease management. Some of the recommendations which proved to be very effective in plant virus disease management in vegetable, cereals and fruit crops are mentioned below.

(1) In India the incidence of *Papaya ring spot virus* in papaya was delayed by a combination of reflective row cover, mineral oil and imidacloprid and deltamethrin sprays (Kalleshwaraswamy et al. 2009).

(2) Management of sunflower necrosis disease (caused by TSV) transmitted by thrips, was effectively minimized by bordering the sunflower crop with sorghum and sunflower seeds treated with imidacloprid (Gaucho 70 W.S., 5 g/kg) along with spraying the sunflower crop with imidacloprid (Confidor 200 S.L., 0.05 %, three sprays at 15, 30 and 45 days after sowing (Shirshikar 2008). The IDM also involved the sowing of sorghum of six rows, sown 15 days prior to the sunflower crop. Both thrips and the incidence of TSV were reduced with a concomitant increase in yields.

(3) The rice tungro disease has been managed through the use of insecticide and resistant cultivars which have reduced virus incidence. Resistance to vector provides good protection against rice tungro causal viruses under field conditions. Green leafhopper resistant rice cultivar IR-28 treated with insecticides had a very low incidence of tungro disease, whereas susceptible cultivar IR-22 under similar treatment had an extremely high incidence. In contrast, only insecticide treated plants of the moderately resistant cultivar IR-36 showed lower infection of causal viruses than untreated plots in which tungro incidence was high (Heinrichs et al. 1986).

(4) Sastry et al. (1977) have successfully reduced the TLCV incidence in tomato by having 5 rows of maize as border crop followed by 3–4 sprays with systemic insecticide. Even, Nilakshi Kakati and Nath (2006) have reduced the TLCV incidence by taking care both at the nursery and field level. The seedlings raised in the nursery bed covered with nylon net showed no incidence of TLCV disease as against 8.0 % incidence in the uncovered nursery bed. In the main field where the seedlings raised under nylon net covering when 2 sprays of Nimbicidine @ of 2.0 % at 20 and 35 days after transplantation showed no TLCV incidence and a high yield of tomato (318.89 q/ha). Whereas in untreated control (nursery raised in open field without spraying in the main field) recorded highest TLCV disease incidence 89.33 % with a corresponding low yield of 98.89 q/ha.

(5) In Israel, *Tomato yellow leaf curl virus* is managed primarily through the use of resistant cultivars, pesticides, cultural practices and exclusion through the use of 50 mesh screens, and regular or UV absorbing plastics in the case of protected production. In the open fields the virus is managed through cultural practices and heavy reliance on insecticides (Cohen and Antignus 1994; Lapidot and Friedmann 2002; Polston and Lapidot 2007). In Dominican Republic, *Tomato yellow leaf curl virus*, transmitted by *Bemisia tabaci* was a major problem and Gilbertson et al. (2007) developed successful management strategy based on IPM principles.

(6) *Tomato spotted wilt virus* (TSWV) in tomato was successfully managed by the using resistant tomato cultivar 'RDD' and raising the crop in tunnels protected with a mesh of 14 × 10 threads/cm. The thirps vector population was less than the control and yield of tomato of 'RDD' was significantly higher under tunnel cultivation than in open air (Diez et al. (1999). The TSWV incidence in pepper and tomato in southeastern U.S. was

successfully managed by Awondo et al. (2012), and the treatments were, reflective mulch, resistant cultivar, imidacloprid, and Actigard. A majority of the farmers were satisfied in using the combination of Actigard and reflective mulch as this treatment was effective in decreasing thrips vector population and disease incidence.

Another recommendation provided by Momol et al. (2001) for effective TSWV management in tomato, is the combined use of a reduced-risk insecticide Spinosad, a systemic acquired resistance inducer (Actigard), and UV-reflective metalized mulch. The metalized mulch was most effective in reducing disease incidence. Actigard reduced incidence of TSWV on the standard black mulch but not on metallized mulch. Spinosad was as effective as methamidophos in reducing the spread of the disease during mid- and late-season. The regimen of metalized mulch, Actigard, and insecticides reduced TSWV by as much as 76 %.

(7) At Malawi, Subrahmanyam et al. (2002) have successfully reduced the groundnut rosette disease by following cultural practices and use of resistant genotypes. Culbreath et al. (2010) have reported that planting date and use of moderately resistant peanut cultivars will reduce the incidence of *Tomato spotted wilt virus* (TSWV).

(8) In Africa, the Sweet potato virus disease (SPVD) caused by dual infection of sweet potato with the whitefly-borne *Sweet potato chlorotic stunt virus* (SPCSV) and aphid-borne *Sweet potato feathery mottle virus* (SPFMV) causes heavy yield losses. The intensive research conducted by Gibson et al. (2004) have developed a recommendation of a dual approach of deploying both resistant cultivars like NOSPOT-1 and phytosanitation which has helped the farmers of Uganda to doubling the yield of sweet potato. The phytosanitation measures included isolation and roguing of infected plants which are nearby and following other cultural practices. Aritua et al., (2003) have followed similar IPM measures to manage the sweet potato virus disease. In China, Feng et al. (2000) have recommended early roguing of infected sweet potato plants and use of virus resistant cultivars to manage virus diseases.

(9) Latham and Jones (2004) have effectively managed Olpidium brassicae transmitted lettuce big vein disease (LBVD) spread by using partially resistant genotypes and plastic mulch on the soil surface.

(10) The use of integrated pest management scouting programs have provided encouraging results in plant virus disease management particularly in open field production, scouts can be used very effectively to minimize incidence of virus-infected plants. For example, although action thresholds for whitefly control have been developed, these are useful for limiting direct damage caused by whiteflies but are insufficient for the purposes of reducing incidences of plants infected with whitefly-vectored viruses. In Florida, many scouts take a zero tolerance approach to whiteflies especially early in the tomato production cycle. Scouts monitor whitefly populations weekly or twice weekly for number of adults and for whitefly development on the tomato

Table 2.13 IDM elements for management of virus and viroid diseases

Traditional system			
No IDM	Low IDM	Medium IDM	High IDM (bio-intensive)
–	Crop rotation, use of resistant cultivars, virus-free seed, attractant bait/crops, scouting, selective insecticidal sprays, barrier/border crops	Weather-based disease prediction, use of composts and green manures, use of induced resistance activators, vector biotype monitoring	Release of bio control agents, use of pheromones, trap cards, soil solarization, use of pest/weather/crop models

plants. They can recommend the optimal time to begin applying insecticides as well as alternative chemical controls. Similar procedures have to be developed against leafhopper, thrips and aphid-transmitted viruses. The above cited IDM technology was successfully applied against some economically important virus diseases in different parts of the world. Similar IDM lines may also aid in managing economically important viroid diseases of plants. The management of a wide range of virus and viroid diseases based on IDM elements is given in the Table 2.13.

2.35.3 Learning IDM

IDM success depends on farmer's knowledge and understanding of plant virus diseases, their vectors and ecology and epidemiology of the vectors, and on how well they use that knowledge in the choice of IDM options. Knowledge and understanding of which options to use and at which crop growth stage can be critical to IDM success. Participatory learning approaches promote effective communication between extension workers and farmers, encourage learning by pursuing IDM, and promote informed decision-making by farmers on how to solve plant virus disease problems.

Farmer participatory learning processes enable participants to:

- Increase their knowledge and understanding of biodiversity associated with crop plants and also of pathogens.
- Adopt regular field monitoring for timely detection of changes in pest status.
- Increase their understanding of ecological and socio-economic aspects of virus disease problems.
- Critically assess and adapt traditional, new knowledge and IDM options.
- Promote hands-on learning of crop and pest management techniques and skills.
- Select, adopt, combine and apply IDM options as integral components of crop production.
- Facilitate IDM learning and informed decision-making by colleagues.
- Promote wider adoption of IDM.

Farmer participatory learning processes make IDM easy to understand, apply and use at farm level. The farmer field school (FFS) is a common experiential learning model for IDM in almost all countries where in small groups of farmers and extension workers learn to integrate scientific information and technologies, with the primary aim of disease management to achieve the higher crop yields.

The readers may also get more clarity about the subject of IDM strategies in plant virus disease management by additional readings of the research and review articles: Nono-Womdim et al. (1999), Thresh (2003), Jones (2004), Varma (1993, 2006), Pedersen et al. (2007), Ambang et al. (2009), Pappu et al. (2009), El-Khoury and Makkouk (2010), Fereres and Moreno (2011).

2.36 Challenges for the Future

Viruses having no physiology of their own and cannot be controlled directly by chemical applications, though a number of attempts have been made. Therefore, until an effective viricide is developed against virus diseases, indirect methods of managing virus diseases has to involve cultural practices like crop rotation, plant population, date of planting, etc. These management tools have been shown to be most promising in minimizing virus spread under field conditions. If these cultural practices are well planned, they are for the most part low cost management tactics aimed at minimizing vector populations and subsequent virus spread. These practices should be factored into management decisions that are made prior to the crop production season. Success in virus disease management so far gained with a few crops augurs well, but it cannot be assumed that methods successful at one place will work at another; this necessitates the study of the epidemiology of virus disease wherever they are prevalent at higher incidence.

Plant resistance in crop plants having transmission through vegetative plant material has the great advantage for a component that it usually enhances the effectiveness of other virus management measures. Plant resistance can reduce virus infection and disease development and a large number of disease resistance genes were identified throughout world in number of economically important crops against major virus diseases (Varma et al. 1991; Khetarpal et al. 1998; Byoung-Cheorl et al. 2005; Pitrat 2012). Although the scientific and technical advances have been rapid, additional studies must be carried out for the better understanding of the molecular basis of resistance of producing transgenic plants.

It is well documented that host plant and vector resistance are the most effective control measures against certain seed-borne diseases. Usefulness and success of the resistance strategy depends on our knowledge of the mechanism(s) of resistance and its effects on the virus-vector-host interactions. More complex and probably more durable resistance can be more difficult to establish and certainly more difficult to achieve by conventional plant breeding. This does not preclude their existence or possible future utilization. Paradoxically, application of knowledge of the genetics of major gene resistance in breeding programs may have militated against breeding for polygenic resistance by more empirical approaches.

It is clear that before horizontal or polygenic resistances can be exploited in crop protection, there again should be an attempt at a cost-benefit assessment. The evidence for horizontal resistance against plant viruses is still sparse. It may be more effective to combine the known major genes which are genetically well understood and therefore can be handled on a rational rather than empirical basis, to construct robust oligogenic systems. Novel methods for revealing or creating variation and for transferring it between genotypes by nontraditional methods are already available and are increasingly being applied to resistance genes.

Biotechnologies should be seen not only as a means of solving problems when traditional methods have failed, but also as a way of generating a better understanding of crop plants and the genes of plant pathogens through the cooperation of scientists from different disciplines, who for the first time are basing the model for their studies on plants. Molecular biologists, biochemists, physiologists and virologists, working with plant geneticists, breeders, pathologists, and entomologists, can form research teams capable of bringing about a revolution in the agricultural sciences. However, the need for a combination of several disciplines may constitute a barrier to the worldwide application of biotechnology in developing countries. For this reason, biotechnology research is being conducted mainly in developed countries.

Biotechnology represents the latest front in the ongoing scientific progress of this century. However, its increasing importance, at least in plant improvement, should not obscure the fact that traditional plant breeding, based on hybridization followed by selection and evaluation of a large population in the field, accounts for over 50 % of the global increase in agricultural productivity. Not only have particularly important new genotypes been bred in Asia through the so-called "Green Revolution" but, worldwide, new cultivars have been bred in response to the changing needs of agriculture.

Public responsibilities, that is, responsibilities of society at large, concern (1) national legislation and international agreement as through the international Plant Protection Convention, (2) implementation of collective measures such as inspection at import and export, quarantine, sometimes eradication campaigns, supervision of quality certification of seed material, (3) teaching and extension and (4) research, both basic and applied. The extent to which public institutions, including governments should be involved remains a matter of political debate. During a period of socialization, governments were thought to have and to be able, to take care of everything with innate risks of bureaucracy. The balance now seems to have moved to the other extreme of utmost privatization for the sake of efficiency with the ensuing risks of dominance of interests in short-term benefits. When looking for a balance when dealing with evil in nature and human society, the approach appears to be one of trial and error, which can imply that this is the only way to get something done.

Since there are number of catastrophic virus diseases in vegetables, cereals, fruits and other commercial crops, the phytosanitation/integrated approach should be implemented at community level for controlling the diseases. The three key elements of phytosanitation that can be used to control virus/viroid diseases are: (1) establishing crops using healthy planting material; (2) maintaining the health of crops during the active growth period; and (3) avoiding growing/planting healthy plant materials of seed/vegetative propagules near to diseased crops. In Tanzania, Mtunda et al. (2013) have implemented the community-wide phytosanitation programme in controlling cassava brown streak disease in cassava with the help of Cooperatives, DRD, IITA, other local government agencies in respective districts of Tanzania. Attempts should be made for the implementation of community action for plant virus and viroid management in almost all countries.

There are countries where breeding for resistance to viruses and the production of commercial seed with tolerated levels of virus infection are taken care of by government agencies, and there are other countries where such activities are in the hands of private breeders and seed producers or companies under final supervision and quality inspection and certification by government agencies. Both extremes require complex socio-infrastructures and none ever are perfect because of human involvement.

Governments and international organizations must remain committed to the management of virus diseases, and in the support of long-term research to sustain such control or further improve it. IARCs and development-aid agencies intensively involved in changing crop genetic make-up and farming systems in the developing parts of the world cannot get away from their responsibilities for ensuing phytopathological consequences of the agricultural modernization they advocate and sometimes impose. Governments and international organizations must also remain involved in spite of the present tendency towards commercial privatization and concentration on projects that promise rapid return of investment.

Large number of international organizations and private institutions have funded for number of research projects to bring out solutions to some of the catostrophic plant diseases and insects pests. The out come from such projects has been benefited by farming community after implementing the research recommendations. Very recently a new recombinant geminivirus *Cassava brown streak disease* (CBSD) presently is confined to East Africa and possibility exists that this virus may spread and become threat to rest of Africa. CBSD is caused by two species of *Ipomoviruses* and transmitted by whiteflies and cuttings and reduces a minimum 30 % of the total cassava harvest. The Global Cassava Partnership for the twenty-first century (GCP 21), a recognized global organizations within the cassava community, is declaring war to combat the cassava viruses in Africa. GCP21, with the number of experts, have been involved in drawing up a comprehensive and co-ordinated plan to decrease these constants and to prevent CBSD from reaching West Africa, the largest cassava producing region in Africa and in the world (Fauquet 2013).

The organizations with resources from the developed world that have a concern for food security in developing countries, and must continue their mission by supporting and funding national programs in developing countries. The Consultative Group for International Agricultural Research (CGIAR) is of primary importance, with its mission to bring the benefits of modern science to poor farmers "living on the fringes of the global economy" (Anon 2004). Eleven of its globally distributed International Agricultural Research Centers have programs in the management of plant disease, each Center being mandated to handle certain crops: Africa Rice Center (WARDA), Centro International de Agricultura Tropical (CIAT), Center for International Forestry Research (CIFOR), Centro International de Mejoramiento de Maiz y Trigo (CIMMYT), Centro International de la Papa (CIP), International Center for Agricultural Research in the Dry Areas (ICARDA), International Crops Research Institute for the Semi-Arid Tropics (ICRISAT), International Institute of Tropical Agriculture (IITA), International Plant Genetic Resources Institute (IPGRI), International Rice Research Institute (IRRI), and World Agro forestry Centre (ICRAF).

Internationally, the Food and Agriculture Organization (FAO) of the United Nations plays a key role in addressing the challenge of food security, including the depravations of plant disease (FAO 2000). Throughout the developed world, national governments have development-assistance agencies that support agricultural development in developing countries, sometimes with specific attention to plant disease issues.

References

Aaziz R, Tepfer M (1999) Recombination in RNA viruses and in virus-resistant transgenic plants. J Gen Virol 80:1339–1346

Abbott EV (1961) A new strain of sugarcane mosaic virus. Phytopathology 51:642 (Abstract)

Abdalla N, Lear B (1975) Lethal dosages of methyl bromide for four plant-parasitic nematodes and the effect of soil temperature on its nematicidal activity. Plant Dis Rep 59:224–228

Abdalla NA, Raski DJ, Lear B, Schmitt RV (1974) Movement, persistence and nematocidal activity of a pesticide containing 1,3-dichloropropene in soils treated for nematode control in replant vineyards. Plant Dis Rep 58:562–566

Abe N (1987) Studies of the ecology and control of *Polymyxa betae* Keskin as a fungal vector of the causal virus (beet necrotic yellow vein virus) of rhizomania disease of sugar beet. Bull Hokkaido Prefectural Agric Exp Stn 60:81

Abo ME, Sy AA, Alegbejo MD, Afolabi AS, Onasanya A, Nwilene FE, Sere Y (2013) The mode of transmission of *Rice yellow mottle virus*. In: 12th International Symposium on plant epidemiology, 28th Jan–1st Feb 2013, Arusha, Tanzania (Abstract, pp 041, p 142)

A'Brook J (1964) The effect of planting date and spacing on the incidence of groundnut rosette disease and of the vector Aphis craccivora, Koch, at Mokawa, Northern Nigeria. Ann Appl Biol 54:199–208

Abu Salih HS, Ishag HM, Siddig SA (1973) Effect of sowing date on incidence of *Sudanese broad bean mosaic virus* and yield of Vicia faba. Ann Appl Biol 74:371–378

Acedo VZ, Labana CU (2008) Rapid propagation of released Philippine cassava varieties through tissue culture. J Root Crops 34:108–114

Adam AV (1962) An effective program for the control of banana mosaic. Plant Dis Rep 46:366–370

Addy ND (1988) Opportunities and challenges for private industry. Am Potato J 65:221–227

Adejare GO, Coutts RHA (1981) Eradication of cassava mosaic disease from Nigerian cassava clones by meristern-tip culture. Plant Cell Tissue Organ Cult 1:25–32

Adipala E, Warren HL, Epieru G, Takan JP, Khyamanywa S, Wilson H (1998) Comparative performance of cv *Igola-1* and other local groundnut cultivars for the control of rosette disease. In: Proceedings IPM/CRSP workshop, 15–18 May 1998, Blacksburg, Virginia Technology, USA, 290 pp

Adkins S (2000) Tomato spotted wilt virus-positive steps towards negative success. Mol Plant Pathol 1:151–158

Adlerz WC (1972a) Momordica charantia as a source of Watermelon mosaic virus 1 for cucurbit crops in Palm Beach Country, Florida. Plant Dis Rep 56:563–564

Adlerz WC (1972b) *Melothria pendula* plants infected with *Watermelon mosaic virus 1* as a source of inoculum for cucurbits in Collier Country, Florida. J Econ Entomol 65:1303–1306

Adlerz WC, Everett PH (1968) Aluminum foil and white polyethylene mulches to repel aphids and control water melon mosaic. J Econ Entomol 61:1276–1279

Adsuar J (1961) Deleterious effect of Simazine on mosaic-infected sugarcane. J Agric Univ PR 45:191

Agarwal VK, Sinclair JB (1996) Principles of seed pathology, 2nd edn. CRC Press Inc., Boca Raton

Agarwal VK, Nene YL, Beniwal SPS, Verma HS (1979) Transmission of *Bean common mosaic virus* through urdbean (*Phaseolus mungo* L.) seeds. Seed Sci Technol 7:103–108

Aghora TS, Mohan N, Krishna Reddy M (2010) Evaluation of French bean germplasm for resistance to *Mungbean yellow mosaic virus*. Indian Phytopathol 63:238–239

Ahlawat YS (2010) Diagnosis of plant viruses and allied pathogens. Studium Press (India) Pvt. Ltd., New Delhi, 224 pp

Ahmed NE, Kanan HO, Sugimoto Y, Ma YQ, Inanaga S (2001) Effect of imidacloprid on incidence of *Tomato yellow leaf curl virus*. Plant Dis 85:84–87

Ahohuendo BC, Sarkar S (1995) Partial control of the spread of *African cassava mosaic virus* in Benin by intercropping. Z Pfl Krankh Pfl Schutz 102:249–256

Ahoonmanesh A, Shalla T (1981) Feasibility of cross-protection for control of tomato mosaic virus in fresh market field-grown tomatoes. Plant Dis 65:56–58

Akhtar KP, Aslam M, Haq MA, Jamil FF, Khan AI, Elahi MT (2005) Resistance to *Cotton leaf curl virus* (CLCuV) in a mutant cotton line. J Cotton Sci 9:175–181

Akhtar KP, Kitsanachandee R, Srinives P, Abbas G, Asghar MJ, Shah TM, Atta BM, Chatchawankanphanich O, Sarwar G, Ahmad M, Sarwar N (2009) Field evaluation of mungbean recombinant inbred lines against mungbean yellow mosaic disease using new disease scale in Thailand. Plant Pathol J 25:422–428

Akram M, Jain RK, Chaudhary Vikas, Ahlawat YS, Khurana SMP (2004) Comparison of *Groundnut bud necrosis virus* isolates based on movement protein (NSm) gene sequences. Ann Appl Biol 145:285–289

Alam SN, Cohen MB (1998) Detection and analysis of QTL's for resistance to brown plant hopper Nilaparvata lugens, in a double haploid population. Theor Appl Genet 97:1370–1379

Albanese G, La Rosa R, Tessitori M, Fuggetta E, Catara A (1996) Long-term effect of CVd-III of plants on citrange, trifoliate and sour orange. In: da Graca JV, Moreno P, Yokomi RK (eds) Proceedings of the 13th conference of the International Organization of Citrus Virologists, Riverside, CA, pp 367–369

Albrechtsen SE (2006) Testing methods for seed transmitted viruses: principles and protocols. CAB1 publishing, Wallingford, pp 1–268

Aliyu TH, Balogun OS (2011) Effect of variety and planting density on the incidence of common viral diseases of cowpea (*Vigna unguiculata*) in a Southern Guinea Savannah Agro ecology. Asian J Plant Pathol 5:126–133

Allavena A (1989) Modification of the seed coat color associated to the gene conferring resistance to BCMV. Annu Rept Bean Improv Co-op 32:90–91

Allen TC Jr (1965) Field spread of *Potato virus A* inhibited by oil. Plant Dis Rep 49:557

Allen DJ, Thottappilly G, Rossel HW (1982) *Cowpea mottle virus*: field resistance and seed transmission in virus tolerant cowpea. Ann Appl Biol 100:331–336

Allen WR, Tehrani B, Luft R (1993) Effect of horticultural oil, insecticidal soap, and film-forming products on the western flower thrips and the Tomato spotted wilt virus. Plant Dis 77:915–918

Almekinders C, Louwaars N (1999) Farmers 'Seed Production': new approaches and practices. Intermediate Technology Publications, London, pp 291

Al-Musa A (1982) Incidence, economic importance and control of tomato leaf curl virus in Jordan. Plant Dis 66:561–563

Alper M, Levy S, Loebenstein G (1973) Mulches to repel aphids from peppers. Hassadeh 53:529–531 (In Hebrew)

Alshami AA, Nasher MMA, Almezgagi M, Alyaremi A (2008) Evaluation of some non-chemical methods for reducing infection with *Tomato leaf curl virus* transmitted by *Bemisia tabaci* (Gennadius). Arab J Plant Prot 26:45–49

Alstrom S (1991) Induction of disease resistance in common bean susceptible to halo blight bacterial pathogen after seed bacterization with rhozosphere pseudomonas. J Gen Appl Microbiol 37:495–501

Al-Taleb MM, Hassawi DS, Abu-Rommau SM (2011) Production of virus-free potato plants using meristem culture from cultivars grown under Jordanian environment. J Agric Environ Sci 11:467–472

Altamirano DM, Gonzales CI, Vinas RC (1976) Analysis of the devastation of leaf mottling (Greening) disease of citrus and its control program in the Philippines. In: Proceedings of the 7th conference of the IOCV, Riverside, pp 22–26

Ambang Z, Ndongo B, Amayana D, Djile B, Ngoh JP, Chewachong GM (2009) Combined effect of host plant resistance and insecticide application on the development of cowpea viral diseases. Aust J Crop Sci 3:167–172

Amin PW (1985) Apparent resistance of groundnut cultivar Robut 33–1 to bud necrosis disease. Plant Dis 69:718–719

Anandam RJ, Doraiswamy S (2002) Role of barrier crops in reducing the incidence of mosaic disease in Chilli. J Plant Dis Prot 109:109–112

Anderson EJ, Stark DM, Nelson RS, Powell PA, Tumer NE, Beachy RN (1989) Transgenic plants that express the coat protein genes of tobacco mosaic virus or alfalfa mosaic virus interfere with disease development of some non related viruses. Phytopathology 79:1284–1290

Anfoka GH (2000) Benzo-(1,2,3)-thiodiazole-7 carbothioic acid 5-methyl ester induces systemic resistance in tomato (*Lycopersicon esculentum* Mill. cv. Vollendung) to Cucumber mosaic virus. Crop Prot 19:401–405

Angenent GC, Van der Ouweland JMW, Bol JF (1990) Susceptibility to virus infection of transgenic tobacco plants expressing structural and non-structural genes of *Tobacco rattle virus*. Virology 175:191–198

Anithacherian K (1998) Management of *Tomato leaf curl virus* in tomato through the control of its whitefly vector, *Bemisia tabaci* Genn. by entamopathogenic fungi. Ph.D. Thesis, UAS, Bangalore, pp 226

Anjaneyulu A (1977) Donors of resistance to rice tungro. IRRN 2:3

Anjaneyulu A, Singh SK, Shenoi MM (1982) Evaluation of rice varieties to tungro resistance by field screening technique. Trop Pest Manag 28:147–155

Anjaneyulu A, Bhaktavatsalam G, Mohanty SK (1994) Prevention and control of rice tungro disease through new synthetic pyrethroids. In: Rishi N, Ahuja KL, Singh BP (eds) Virology in the tropics. Malhotra Publishing House, New Delhi, pp 660–671

Anon (1912) The Plant quarantine act, August 20, 1912, as amended March 4, 1913, and March 4, 1917. http://www.archive.org/details/plantquarantinea00unit

Anon (1944a) Entomological investigations. 17th Report Council for Scientific and Industrial Research, Australia, 1942–1943, pp 15–20

Anon (1944b) Entomological investigations. 18th Report Council for Scientific and Industrial Research, Australia, 18–22, pp 14–20

Anon (1957) Common names of virus diseases used in the Review of Applied Mycology. Rev Appl Mycol 35(Suppl):43

Anon (1971) Diseases of cultivated plants observed in Greece from 1962 to 1967. Ann Inst Phytopathol Benaki 10:70–118

Anon (1977–1978) List of intercepted plant pests. United States. Plant Quarantine and Control Administration. Canada Agriculture

Anon (1992) Quarantine implications: Cassava Program 1987–1991. Working Document No. 116, CIAT, Colombia

Anon (1994) Sky's reflection repels aphids. Calif Agric 48:4–5. doi:10.3733/ca.v048n04p4 July–Aug 1994

Anon (2004) Entomological society of America—Annual Meeting, Nov 2004

Antignus Y, Ben-Yakir D (2004) Ultraviolet-absorbing barriers, an efficient integrated pest management tool to protect greenhouses from insects and virus diseases. In: Horowitz AR, Ishaaya I (eds) Insect pest management field and protected crops. Springer, Berlin, pp 319–337

Antignus Y, Cohen S, Mor N, Messika Y, Lapidot M (1996a) The effects of UV-blocking greenhouse covers on insects and insect-borne virus diseases. Plasticulture 112:15–20

Antignus Y, Mor N, Ben-Joseph R, Lapidot M, Cohen S (1996b) UV-absorbing plastic sheets protect crops from insect pests and from virus diseases vectored by insects. Environ Entomol 25:919–924

Antignus Y, Lapidot M, Hadar D, Messika Y, Cohen S (1997) The use of UV-absorbing plastic sheets to protect crops against insects and spread of virus diseases. In: Ben-Yehosua S (ed) CIPA, Proceedings of international congress for plastics in agriculture. Jerusalem, Laser Pages, pp 23–33

Antignus Y, Lachman O, Leshem Y, Matan E, Yehezkel H, Messika Y (1999) Protection efficiency of UV-absorbing films in greenhouse with vertical walls. In: Zeidan O (ed) Summary of research projects and field experiments in tomato crops for 1999. Israeli Ministry of Agriculture: Bulletin of the Israeli Extension Service, Bet Dagan, pp 29–39

Antignus Y, Lapidot M, Cohen S (2001a) Interference with ultraviolet vision of insects to impede insects pests and insect-borne plant viruses. In: Kerry FH, Oney PS, Duffus JE (eds) Virus-insect-plant interactions. Academic Press, San Diego, pp 331–334

Antignus Y, Nestel D, Cohen S, Lapidot M (2001b) Ultraviolet deficient greenhouse environment affects whitefly attraction and flight behavior. Environ Entomol 30:394–439

Antignus Y, Lachman O, Pearlsman M (2005) Light manipulation by soil mulches protect crops from the spread of Begomoviruses. Abstracts of the IX International Plant Virus Epidemiology Symposium, Lima Peru

Antignus Y, Lachman O, Pearlsman M (2007) The spread of *Tomato apical stunt viroid* (TAS Vd) in green house tomato crops is associated with seed transmission and bumble bee activity. Plant Dis 91:47–50

Antignus Y, Vunsh R, Lachman O, Pearlsman M, Maslenin L, Hananya U, Rosner A (2004) Truncated Rep gene originated from Tomato yellow leaf curl virus Israel [Mild] confers strain-specific resistance in transgenic tomato. Ann Appl Biol 144:39–44

Appel O (1934) Vitality and vitality determination in potatoes. Phytopathology 24:482–494

Aragao FJL, Faria JC (2009) First transgenic Gemini virus-resistant plant in the field. Nat Biotechnol 27:1086–1089

Ares X, Calamante G, Cabral S, Lodge J, Hemenway P, Beachy RN, Mentaberry A (1998) Transgenic plants expressing potato virus X ORF2 protein (p24) are resistant to tobacco mosaic virus and Ob tobamoviruses. J Virol 72:731–738

Arias BV, Bellotti AC, Vargas HLB (2004) Nataima—31, a cassava (Manihot esculenta) variety resistant to the whitefly Aleurotrachelus socialis. In: 6th International science meeting of the Cassava Biotechnology Network, CIAT, Cali, Columbia, pp 66

Arif M, Hassan S (2000) Pathogen-derived resistance against plant viruses: postscript and prospects. Pak J Biol Sci 3:1–9

Arif M, Hassan S (2002) Evaluation of resistance in soybean germplasm to *Soybean mosaic poty virus* under field conditions. J Biolog Sci 2:601–604 (Online). ISSN 1608–4127

Arif M, Ibrahim M, Ahmad A, Hassan S (2005) Elimination of citrus tristeza closterovirus from citrus bud-wood through thermotherapy. Pak J Bot 37:423–430

Aritua V, Adipala E, Carey EE, Gibson RW (1998a) The incidence of *Sweet potato virus disease* and virus resistance of sweet potato grown in Uganda. Ann Appl Biol 132:399–411

Aritua V, Alicai T, Adipala E, Carey EE, Gibson RW (1998b) Aspects of resistance to *Sweet potato virus disease* in sweet potato. Ann Appl Biol 132:387–398

Aritua V, Bua B, Gibson RW, Mwanga ROM, Adipala E (2003) Toward integrated management of *Sweet potato virus disease*: lessons from Uganda. African Crop Sci Conf Proc 6:307–314

Arora SK, Dhanju KC, Sharma BR (1992) Resistance in okra (*Abelmoschus esculentus* L. Moench) genotypes to yellow vein mosaic virus. Plant Dis Res 7:221–225

Arumugam R, Chelliah S, Muthukrishnan CR (1975) Abelmoschus manihot: a source of resistance to bhindi yellow vein mosaic. Madras Agric J 62:310–312

Ashfaq M, Aslam Khan M, Mughal SM, Javed N, Mukhtar T, Bashir M (2007) Evaluation of urdbean germplasm for resistance against *Urdbean leaf crinkle virus*. Pak J Bot 39:2103–2111

Asjes CJ (1972) Virus disease in narcissus. Gard Chron, pp 28–33

Asjes CJ (1974) Control of the spread of brown ring formation virus disease in the lily. Mid century hybrid 'Enchantment' by mineral oil sprays. Acta Hort 36:85–92

Asjes CJ (1975) Control of the spread of tulip breaking virus in tulips with mineral-oil sprays. Neth J Plant Path 81:64–70

Atabekov JG, Taliansky ME (1990) Expression of a plant virus coded transport function by different viral genomes. Adv Virus Res 38:201–247

Atiri GI, Ekpo EJA, Thottappilly G (1984) The effect of aphid-resistance in cowpea on infestation and development of *Aphis craccivora* and the transmission of *Cowpea aphid-borne mosaic virus*. Ann Appl Biol 104:339–346

Atkins D, Young M, Uzzell S, Kelly L, Fillatti J, Gerlach WL (1995) The expression of antisense ribozyme genes targeting *Citrus exocortis viroid* in transgenic plants. J Gen Virol 76:781–790

Attafuah A (1965) Embryo abortion in virus infected sweet cherry. FAO Plant Prot Bull 13:79–82

Audy P, Palukaitis P, Slack SA, Zaitlin M (1994) Replicase-mediated resistance to potato virus Y in transgenic plants. Mol Plant Microbe Interact 7:15–22

Avilla C, Collar JL, Duque M, Hernaiz P, Martın B, Fereres A (1996) Cultivos barrera como metodo de control de virus no persistentes en pimiento. Bol San Veg Plagas 22:301–307

AVRDC (2000) Report 2000, Asian vegetable research and development center, Shanhua, Tainan, Taiwan, 2001

Awan AR, Mughal SM, Iftikhar Y, Khan HZ (2007) *In vitro* elimination of Potato leaf roll polerovirus from potato varieties. Eur J Sci Res 18:155–164

Awondo SN, Fonsah EG, Riley D, Abney M (2012) Effectiveness of tomato-spotted wilt virus management tactics. J Econ Entomol 105(3):943–948

Ayabe M, Sumi S (2001) A novel and efficient tissue culture method-"Stem-disc dome culture"-for producing virus-free garlic (*Allium sativum* L.). Plant Cell Rep 20:503–507

Bacon PE (1980) Effects of dwarfing inoculations on the growth and productivity of Valencia oranges. J Hortic Sci 55:49–55

Badak KS, Nirmal DD, Mahatale PV, Shinde VB (2006) Management of *Papaya ring spot virus* by cross protection. Ann Plant Physiol 20:116–118

Badara Gueve, Lava Kumar P, Peter Kulakow, Michael Abberton (2013) Virus checks for safe exchange of cassava germplasm for crop improvement and food security. In: 12th International Symposium on plant virus epidemiology, 28th Jan–1st Feb 2013, Arusha, Tanzania (Abstract, pp 035, p 136)

Bag S, Singh RS, Jain RK (2007) Agrobacterium-mediated transformation of groundnut with coat protein gene of *Tobacco streak virus*. Indian J Virol 18:65–69

Baggett JR, Kean D (1988) Seven pea seed-borne mosaic virus resistant pea breeding lines. Hortic Sci 23:630–631

Bagnall RH (1953) The spread of *Potato virus Y* in seed potatoes in relation to the date of harvesting and the prevalence of aphids. Can J Agric Sci 33:509–519

Bai Y, Guo Z, Wang X, Bai D, Zhang W (2009) Generation of double-virus-resistant marker-free transgenic potato plants. Prog Nat Sci 19:543–548

Balamuralikrishnan M, Doraisamy S, Ganapathy T, Viswanathan R (2002) Combined effect of chemotherapy and meristem culture on sugarcane mosaic virus elimination in sugarcane. Sugar Tech 4:19–25

Balaraman K (1981) Pre-immunization for control of *Citrus tristeza virus* on citron. Z Pfl Krankh Pfl Schtz 88:218–222

Balaraman K (1987) Cross protection performance trials on acid lime against tristeza virus in South India. Mysore J Agric Sci 21:193–195

Balaraman K, Ramakrishnan K (1977) Studies on strains and strain interaction in citrus trizteza virus. Univ Agric Sci Bangalore. Tech Series No 19:1–62

Balaraman K, Ramakrishnan K (1979a) Transmission studies with strains of *Citrus tristeza virus* on acid lime. Z Pfl Krankh Pfl Schutz 86:653–661

Balaraman K, Ramakrishnan K (1979b) Two new additional hosts of *Citrus tristeza virus*. Curr Sci 48:453–454

Balaraman K, Ramakrishnan K (1980) Strain variation and cross protection in *Citrus tristeza virus* on acid lime. In: Calavan EC et al (eds) Proceedings of the 8th conference of the International Organization of Citrus Virologists, FL, pp 60–68

Baleshwar Singh R, Bhalla S, Challam VC, Pandey BM, Sateesh Singh SK, Kumar N, Khetarpal RK (2003) Quarantine processing of imported transgenic planting material. Indian J Agric Sci 73:97–100

Bankina TF (1976) The importance of early harvest in the preservation of potato from virus infection. Znachenie rannel uborki v. Sokhranenii Kartofelya at porazheniya virusami. Tr Tulsk Gos Skh Opyt St 5:125–134

Banttari EE, Khurana SMP (1998) Serological procedures in plant virology. In: Khurana SMP (ed) Pathological problems of economic crop plants and their management. Scientific Publishers, Jodhpur, pp 603–624

Baranwal VK, Verma HN (2000) Antiviral phytoproteins as biocontrol agents for efficient management of plant virus diseases. pp 71–79. In: Upadhyay RK (ed) Biocontrol potential and its exploitation in sustainable agriculture. Kluwer Academic/Plenum Publishers, New York

Barba M (1998) Virus certification of fruit tree propagative material in Western Europe. In: Hadidi A, Khaterpal RK, Koganezawa H (eds) Plant virus disease control. APS Press, St. Paul, pp 288–293

Bar-Joseph M, Loebenstein G (1973) Effects of strain, source of plant and temperature on the transmissibility of Citrus tristeza virus by melon aphid. Phytopathology 63:716–720

Bar-Joseph M, Loebenstein G, Gren Y (1974) Use of electron-microscopy in eradication of tristeza sources recently found in Israel. In: Weathers LG, Cohen M (eds) Proceedings of the 6th Conference of the IOCV, Berkeley, pp 83–85

Barker H, Solomon RM (1990) Evidence of simple genetic control in potato of ability to restrict *Potato leaf roll virus* concentration in leaves. Theor Appl Genet 80:188–192

Barker H, Waterhouse PM (1999) The development of resistance to luteoviruses mediated by host genes and pathogen-derived transgenes. In: Smith HG, Barker H (eds) The Luteoviridae. CAB International, Wallingford, pp 169–210

Barnes GL (1959) Herbicidal agents as possible aids for roguing diseased seed potato plants. Am Potato J 36:212–218

Barone A, Frusciante L (2007) Molecular marker-assisted selection for resistance to pathogens in tomato. In: Guimaraes E, Ruane J, Scherf BD, Sonnino A, Dargie JD (eds) Marker-assisted selection: current status and future perspectives in crops, livestock, forestry and fish/. FAO, Rome (Italy). Agric and Consum Prot Dept., pp 151–164

Bartels W (1955) Untersuchungen uber die inaktivierung des Tabak-mosaik virus durch Extrakte und Sekrete Non hoheren Pflanzen und einigen Microorganismen. Ein Beitrag Zur Frage der Kompostierung tabak mosaic virusshaltigen. Pflanzen Materials Phytopathol Z 25:72–152

Bartha L (1963) French bean seed certification. J Agric Vic 419:408–411

Bashir M, Ahmad Z, Ghafoor A (2005) Sources of genetic resistance in mungbean and blackgram against *Urdbean leaf crinkle virus* (ULCV). Pak J Bot 37:47–51

Basky Z (1984) Effect of reflective mulches and oil sprays on mosaic virus incidence in seed cucumbers. Prot Ecol 7:243–248

Bau HJ, Cheng YH, Yu TA, Yang JS, Yeh SD (2003) Broad spectrum resistance to different geographic strains of Papaya ringspot virus in coat protein gene transgenic papaya. Phytopathology 93:112–120

Bau HJ, Cheng YH, Yu TA, Yang JS, Liou PC, Hsiao CH, Lin CY, Yeh SD (2004) Field evaluation of transgenic papaya lines carrying the coat protein gene of *Papaya ringspot virus* in Taiwan. Plant Dis 88:594–599

Baulcombe DC (1996) Mechanisms of pathogen-derived resistance to viruses in transgenic plants. Plant Cell 8:1833–1844

Baulcombe D (2005) RNA silencing. Trends Biochem Sci 30:290–293

Beachy RN (1993) Transgenic resistance to plant viruses. Semin Virol 4:327–416

Beachy RN (1997) Mechanisms and applications of pathogen-derived resistance. Curr Opin Biotechnol 8:215–220

Beck BDA (1982) Historical prospectives of cassava breeding in Africa. In: Hahn SK (ed) Proceedings of a workshop on root crops in Eastern Africa, Kigali, Rwanda. Ker ADR IDRC, Canada, pp 13–18

Beier H, Siler DJ, Russell ML, Bruening G (1977) Survey of susceptibility to *Cowpea mosaic virus* among protoplasts and intact plants from *Vigna sinensis* lines. Phytopathology 67:917–921

Bendahmane M, Beachy RN (1999) Control of tobamovirus infections via pathogen derived resistance. Adv Virus Res 53:369–386

Bendahmane M, Gronenborn B (1997) Engineering resistance against Tomato yellow leaf curl virus (TYLCV) using antisense RNA. Plant Mol Biol 33:351–357

Bennett CW (1944) Latent virus of dodder and its effect on sugarbeet and other plants. Phytopathology 34:77–91

Bennett CW (1960) Sugar beet yellows disease in the United States. U S Dep Agric Tech Bull 1218:1–63

Bennett CW (1967) Epidemiology of leaf hopper transmitted viruses. Annu Rev Phytopathol 5:87–108

Bennett CW (1971) The curly top disease of sugarbeet and other plants. Monograph. No. 7. American Phytopathological Society, St. Paul, p 81

Bennett CW, Tanrisever A (1957) Sugarbeet curly top disease in Turkey. Plant Dis Rep 41:721–725

Berg LA, Bustamante M (1974) Heat treatment and meristem culture for the production of virus-free bananas. Phytopathology 64:320–322

Bergstrom GC, Johnson MC, Kuc J (1982) Effects of local infection of cucumber by *Colletotrichum lagenarium, Pseudomonas lachrymans,* or Tobacco necrosis virus on systemic resistance to Cucumber mosaic virus. Phytopathology 72:922–926

Berlinger MJ, Dahn R (1987) Breeding for resistance to virus transmission by whiteflies in tomatoes. Insect Sci Appl 8:783–784

Berlinger MJ, Mordechi SL (1996) Physical methods for the control of Bemisia. In: Gerling D (ed) Bemisia 1995: taxonomy, biology, damage control and management. Andover, Intercept, pp 617–634

Berlinger MJ, Taylor RAJ, Lebiush Mordechi S, Shalhevet S, Spharim I (2002) Efficiency of insect exclusion screens for preventing whitefly transmission of *Tomato yellow leaf curl virus* of tomatoes in Israel. Bull Entomol Res 92:367–373

Bernardo R, Charcosset A (2006) Usefulness of gene information in marker-assisted recurrent selection: a simulation appraisal. Crop Sci 46:614–621

Bertaccini A, Botti S, Tabanelli D, Dradi G, Fogher C, Previati A, Re F (2004) Micropropagation and establishment of mite-borne virus free garlic (Allium sativum). Acta Hortic 631:201–206

Besedina VA (1985) Sources of resistance in tomatoes to *Tobacco mosaic virus*. Sholnik Nauchnykh Tendov Po Prikloduoi Botanike Genetike i Selektsii 92:81–84 (RPP 66, 25037)

Bethke JA, Parrella MP (1990) Control of the sweet potato whitefly on poinsettia under glasshouse conditions, 1989. Insectic Acaricide Tests 15:330

Bevington KB, Bacon PE (1977) Effect of rootstocks on the response of navel orange trees to dwarfing inoculations. In: Proceedings of the International Society of Citriculture, Orlando, FL, pp 567–570

Bevington KB, Bacon PE (1978) Effect of rootstocks on the response of navel orange trees to dwarfing inoculations, 1977. Proc Int Soc Citricult 2:570–571

Bhaktavatsalam G, Anjaneyulu A (1984) Green house evaluation of synthetic pyrethroids for tungro prevention. Int Rice Res Newsl 9:11

Bharatan N, Reddy DVR, Rajeswari R, Murthy VK, Rao VR (1984) Screening of peanut germplasm lines by enzyme linked immunosorbent assay for seed transmission of Peanut mottle virus. Plant Dis 68:757–758

Bhardwaj SV, Thakur PD (1993) Virus diseases of temperate fruits, pp 1447–1452

Bhardwaj SV, Rai SJ, Thakur PD, Handa A (1998) Meristem tip culture and heat therapy for production of Apple mosaic virus free plants in India. Acta Hortic 472:65–68

Bhargava KS, Khurana SMP (1969) Papaya mosaic control by oil sprays. Phytapathol Z 64:338–343

Bhat ZA, Dhillon WS, Rashid R, Bhat JA, Alidar W, Ganaie MY (2010) The role of molecular markers in improvement of fruit crops. Not Sci Biol 2:22–30

Bhat Bharati N, Raja Ram Reddy D, Chander Rao S, Singh TVK (2012) Integrated disease management of sunflower necrosis disease. Indian J Plant Prot 40:99–104

Bhatoa GS, Sekhon PS, Pathak D, Monga D (2013) Molecular screening of American cotton (*Gossypium hirsutum* L.) genotypes for resistance to cotton leaf curl disease (CLCuD). J Cotton Res Dev 27:104–110

Bhojwani SS, Razdan MK (1983) Plant tissue culture: theory and practice. Elsevier Science Publishers, Amsterdam

Birkett MA, Agelopulos N, Jensen KMV, Jespersen JB, Pickett JA, Prijs HJ, Thomas G, Trapman JJ, Wadhams LJ, Woodcock CM (2004) The role of volatile semiochemicals in mediating host location and selection by nuisance and disease-transmitting cattle flies. Med Vet Entomol 18:313–322

Biswas MK, Hossain M, Islam R (2007) Virus-free plantlets production of strawberry through meristem culture. World J Agric Sci 3:757–763

Biswas KK, Tarafdar A, Jayakumar BK, Pun KB (2009) Strategy for the production of virus-free mandarin (Citrus reticulata) planting materials in the Darjeeling hills of India. Indian Phytopathol 62:376–380

Bitters WP, Murashige T, Ranga TS, Nauer E (1972) Investigation on establishing virus-free citrus plants through tissue culture. In: Proceedings of the 5th International Organization of Citrus Virologists, pp 267–271

Black LL (1980) Aluminium' mulch less virus disease, higher vegetable yields. Lousiana Agric 23:16–18

Black LL, Rolston LH (1972a) Aluminium foil mulch reduces virus infection of peppers. Lousiana Agric 15:6–7

Black LL, Rolston LH (1972b) Aphids repelled and virus diseases reduced in peppers planted on aluminium foil mulch. Phytopathology 62:747 (Abstr)

Black LL, Hobbs HA, Gatti JM (1991) Tomato spotted wilt virus resistance in *Capsicum chinense* PI152225 and 159236. Plant Dis 75:863

Black LL, Hobbs HA, Kammer-lohr DS (1996) Resistance of Capsicum chinense lines to *Tomato spotted wilt virus* isolates from Louisiana, USA and inheritance of resistance. Acta Hortic 431:393–401

Blanchard RO (1974) Electrotherapy: a new approach to wound healing. Proc Am Phytopathol Soc 1:133–134

Blattny C, Pozdena J, Prochazkova Z (1965) Virus-induced rough dwarf and stripe disease in Z. mays. Phytopathol Z 52:105–130

Bleeker PM, Diergaarde PJ, Ament K, Guerra J, Weidner M, Schutz S, de Both MTJ, Haring MA, Schuurink RC (2009) The role of specific tomato volatiles in tomato-whitefly interaction. Plant Physiol 151:925–935

Blencowe JW, Tinsley TW (1951) The influence of density of plant population on the incidence of yellows in sugar-beet crops. Ann Appl Biol 38:395–401

Bock KR (1973) *Peanut mottle virus* in East Africa. Ann Appl Biol 74:171–179

Bock KR (1983) Epidemiology of cassava mosaic disease in Kenya. In: Plumb RT, Thresh JM (eds) Plant virus epidemiology. Blackwell Scientific Publications, Oxford, pp 337–347

Boiteux LS, Nagata T, Dutra WP, Fonseca MEN (1993a) Sources of resistance to tomato spotted wilt virus (TSWV) in cultivated and wild species of Capsicum. Euphytica 67:89–94

Boiteux LS, Nagata T, De Giordano B (1993b) Field resistance of tomato *Lycopersicon esculentum* lines to tomato spotted wilt disease. Tomato Genet Coop Rep 43:7–9

Boize L, Gudin C, Purdue G (1976) The influence of leaf surface roughness on the spreading of oil spray drops. Ann Appl Biol 84:205–211

Bond DA, Jellis GJ, Rowland GG, Le Guen J, Robertson ID, Khalil SA, Li Juan I (1994) Present status and future strategy in breeding faba bean (Vicia faba L.) for resistance to biotic and abiotic stresses. In: Muelbauer FJ, Kaiser WJ (eds) Expanding the production and use of cool season legumes. Kluwer Academic Publishers, Dordrecht, pp 592–616

Bonfim K, Faria JC, Elsa OPL, Nogueira EO, Mendes EA, Aragao FJL (2007) RNAi-mediated resistance to *Bean golden mosaic virus* in genetically engineered common bean (*Phaseolus vulgaris*). Microbe Plant Mol Interact 20(6):717–726

Bonnemaison L (1961) Protection des betteravesracines et des porte graines contre la jannisse basee sur la lutte contre les vecteurs. Ann Epiphyties 12:153–217

Bora GC, Saikia AK, Shadeque A (1992) Screening of okra genotypes for resistance to yellow vein mosaic virus disease. Indian J Virol 8:55–57

Borah RK (1995) Effect of synthetic pyrethroids and organophosphorus insecticides on the incidence of whitefly, Bemisia tabaci (Genn) and yellow mosaic virus in greengram, Vigna radiate (L.) Wilczek. Indian J Virol 11:75–76

Boulcombe D (1994) Novel strategies for engineering virus resistance in plants. Curr Opin Biotechnol 5:117–124

Bowers GR Jr, Goodman RM (1982) New sources of resistance to seed transmission of Soybean mosaic virus in soybeans. Crop Sci 22:155–156

Bowers GR, Goodman RM (1991) Strain specificity of *Soybean mosaic virus* seed transmission in soybean. Crop Sci 31:1171–1174

Bradley RHE (1956) Effects of depth of stylet penetration on aphid transmission of *Potato virus Y*. Can J Microbiol 2:539–547

Bradley RHE (1963) Some ways in which a paraffin oil impedes aphid transmission of *Potato virus Y*. Canad J Microbiol 9:369–380

Bradley RHE, Wade CV, Wood FA (1962) Aphid transmission of *Potato virus Y* inhibited by oils. Virology 18:327–328

Bradley RHE, Moore CA, Pond DD (1966) Spread of *Potato virus Y* curtailed by oil. Nature 209:1370–1371

Braverman SW (1975) Aseptic culture of soybean and peanut embryonic axes to improve phytosanitation of plant introductions. Seed Sci Technol 3:725–729

Brederode FT, Taschner PEM, Posthumus E, Bol JF (1995) Replicase mediated resistance to alfalfa mosaic virus. Virology 207:467–474

Brierley P (1958) *Tomato aspermy virus* in chrysanthemums from Asia and Europe. Plant Dis Rep 42:61–62

Brigneti G, Martin-Hernandez AM, Jin H, Chen J, Baulcombe DC, Baker B, Jones JD (2004) Virus-induced gene silencing in *Solanum* species. Plant J 39:264–272

Broadbent L (1948) Aphids migration and the efficiency of the trapping method. Ann Appl Biol 35:379–394

Broadbent L (1957) Investigations of virus diseases of brassica crops. A.R.C. Report Series No. 14. Cambridge University Press, London
Broadbent L (1963) The epidemiology of tomato mosaic. III cleaning virus from hands and tools. Ann Appl Biol 52:225–232
Broadbent L (1964) Control of plant virus diseases. In: Corbett MK, Sisler HD (eds) Plant virology. University of Florida Press, Gainsville, pp 330–364
Broadbent L, Fletcher JT (1963) The epidemiology of tomato mosaic. IV. Persistance of virus on clothing and glasshouse structures. Ann Appl Biol 52:233–241
Broadbent L, Tinsley TW, Buddin W, Roberts ET (1951) The spread of lettuce mosaic in the field. Ann Appl Biol 38:689–706
Broadbent L, Heathcote GD, Mc Dermott N, Taylor CE (1957) The effect of date of planting and harvesting potatoes on virus infection and on yield. Ann Appl Biol 45:603–622
Broadbent L, Heathcote GD, Burt PE (1960) Field trials on the retention of potato stocks in England. Eur Potato J 3:251–262
Broadbent L, Read WH, Last FT (1965) The epidemiology of tomato mosaic. 12. Sources of TMV in commercial tomato crops under glass. Ann Appl Biol 57:113–120
Broadbent DE, Cooper PF, FitzGerald P, Parkes KR (1982) The cognitive failures questionnaire (cfq) and its correlates. Br J Clin Psychol 21:1–16
Broadbent P, Forsyth JB, Hutton RJ, Bevington KB (1992) Guidelines for the commercial use of graft-transmissible dwarfing in Australia—potential benefits and risks. In: Proceedings of the International Society of Citriculture. Acireale, Italy, pp 697–701
Brown RH (1978) The distribution of Tylenchulus semipenetrans and other parasitic nematodes associated with citrus in northern Victoria. Aust J Exper Agric Anim Husb 18:148–151
Brown JK (2008) Plant resistance to viruses. Encycl Virol 4:164–170
Brown JE, Dangler JM, Woods FM, Henshaw MC, Griffy WA, West MW (1993) Delay in mosaic virus onset and aphid vector reduction in summer squash grown on reflective mulches. Hortic Sci 28:895–896
Brown SL, Culbreath AK, Todd JW, Gorbet DW, Baldwin JA, Beasley JP (2005) Development of a method of risk assessment to facilitate integrated management of spotted wilt of peanut. Plant Dis 89:348–356
Bruce TJA, Wadhams LJ, Woodcock CM (2005) Insect host location: a volatile situation. Trends Plant Sci 10:269–274
Bruckbauer H (1969a) Observations on the reaction of grapevine after the application of the soil fumigant Shell D-D. Weinberg u Keller 16:65–70
Bruckbauer H (1969b) Economic importance and transmission of grapevine virus diseases. Dt Weinbau 24:774–777
Brumin M, Stukalov S, Haviv S, Muruganantham M, Moskovitz Y, Batuman O, Fenigstein A, Mawassi M (2009) Post-transcriptional gene silencing and virus resistance in *Nicotiana benthamiana* expressing a Grapevine virus A minireplicon. Transgenic Res 18:331–345
Bruna A (1997) Effect of thermotherapy and meristem-tip culture on production of virus-free garlic in Chile. Acta Hortic 433:631–634
Brunt AA, Jackson GVH, Frison EA (eds) (1989) FAO/IBPGR Technical Guidelines for the safe movement of yam germplasm. FAO, Rome/International Board for Plant Genetic Resources, Rome
Brust GE (2000) Reflective and black mulch increase yield in pumpkins under virus disease pressure. J Econ Entomol 93:828–833
Bryan JE (1988) Implementation of rapid multiplication and tissue culture methods in the third world countries. Am Potato J 65:199–207
Bua B, Adipala E, Gibson RW (2009) Quantification of the effects of roguing on growth and tuber yields of sweetpotato from Uganda. Afr Crop Sci Conf Proc 9:283–287
Buitelaar K (1974) Choice of tomato variety for heated and cold glasshouse culture. Groenten en Fruit 30:845
Burt PE, Heathcote GD, Broadbent L (1964) The use of insecticides to find where leaf roll and Y viruses spread within potato crops. Ann Appl Biol 54:13–22

Butter NS, Rataul HS (1973) Control of *Tomato leaf curl virus* (TLCV) in tomatoes by controlling the vector whitefly Bemisia tabaci Gen. by mineral oil sprays. Curr Sci 42:864–865

Buzzell RI, Tu JC (1984) Inheritance of soybean resistance to *Soybean mosaic virus*. J Hered 75:82

Byoung-Cheorl K, Inhwa Y, Molly MJ (2005) Genetics of plant virus resistance. Annu Rev Phytopathol 43:581–621

Cabunagan RC, Castilla N, Coloquio EL, Tiongco ER, Truong XH, Fernandez J, Du MJ, Zaragosa B, Hozak RR, Savary S, Azzam O (2001) Synchrony of planting and proportions of susceptible varieties affect rice tungro disease epidemics in the Philippines. Crop Prot 20:499–519

Cadman CH, Chambers J (1960) Factors affecting the spread of aphid-borne viruses in potato in Eastern Scotland. III. Effects of planting date, rogueing and age of crop on the spread of potato leaf roll and Y viruses. Ann Appl Biol 48:729–738

Calavan EC, Weathers JM (1959) Cachexia disease of citrus in California. Phytopathology 49:113

Calavan EC, Goheen AC, Nyland G (1970) Technique for rapid multiplication of clean foundation stocks. In: Proceedings of the 18th international horticultural congress, vol III, pp 74–79

Calavan EC, Roistacher CN, Nauer EM (1972) Thermotherapy of citrus for inactivation of certain viruses. Plant Dis Rep 56:976–980

Callaway A, Giesman-Cookmeyer D, Gillock ET, Sit TL, Lommel SA (2001) The multifunctional capsid proteins of plant RNA viruses. Annu Rev Phytopathol 39:419–460

Calpouzos L (1966) Action of oil in the control of plant disease. Annu Rev Phytopathol 4:369–390

Calvert EL, Harrison BD (1963) Out breaks of tomato black ring virus in onion and leek crops in Northern Ireland. Hortic Res 2:115–120

Campbell RN (1980) Effect of benomyl and ribavirin on the lettuce big vein agent and its transmission. Phytopathology 70:1190–1192

Campbell RN (1996) Fungal transmission of plant viruses. Annu Rev Phytopathol 34:87–108

Campbell RN, Greathead AS, Westerlund FV (1980) Big-vein of lettuce: infection and methods of control. Phytopathology 70:741–746

Cann HJ (1952) Bunchy top virus of bananas' at all time low' in New South Wales. Agr Gaz (N S W) 63:73–76

Capoor SP, Rao DG (1967) Tristeza virus infection of citrus in India. In: Proceedings of the international symposium on sub tropical and horticulture, Bangalore, India, pp 723–736

Caranta CA, Palloix A, Gebre – Salassie K, Lefebvre V, Moury B, Daubeze AM (1996) A complementation of two genes originating from susceptible Capsicum annuum lines confers a new and complete resistance to *Pepper veinal mottle virus*. Phytopathology 86:739–743

Caresche L, Cotterell GS, Peachy JE, Fayner RW, Jacques-Felix H (1969) Hand book for phytosanitary inspectors in Africa. Organization of African Unity/STRC, Lagos

Carey EE, Gichuki ST, Ndolo G, Turyamureeba G, Kapinga R, Lutaladio G, Teri JM (1997) Collaborative sweet potato breeding in eastern central and southern Africa. In: CIP program report 1995–1996. CIP, Lima, Peru, pp 49–57

Carr JP, Zaitlin M (1991) Resistance in transgenic tobacco plants expressing a nonstructural gene sequence of tobacco mosaic virus is a consequence of markedly reduced virus replication. Mol Plant Microbe Interact 4:579–585

Carroll TW (1980) *Barley stripe mosaic virus*: its economic importance and control in Montana. Plant Dis 64:135–140

Carroll TW (1983) Certification schemes against barley stripe mosaic. Seed Sci Technol 11:1033–1042

Carroll TW, Hockett EA, Zaske SK (1983) Registration of mobet barley germplasm. Crop Sci 23:599–600

Cartier JJ (1966) Aphid responses to colors in artificial rearings. Bull Entomol Soc Am 12:378–380

Castle WS (1978) Controlling citrus tree size with rootstocks and viruses for higher density plantings. Proc Fla State Hortic Soc 91:46–50

Castle WS (1987) Citrus rootstocks. In: Rom RC, Carlson RF (eds) Rootstocks for fruit crops. Wiley, New York, pp 361–399

Castle S, Palumbo J, Prabhakar N (2009) Newer insecticides for plant virus disease management. Virus Res 141:131–139

Catherall PL (1972) *Barley stripe mosaic virus*. Report, Welsh Plant Breeding Station, 1971, p 62

Catherall PL (1984) Resistance to sap and seed transmission of *Barley stripe mosaic virus* (BSMV) in some spring barley cultivars. Report, Welsh Plant Breeding Station, 1883, pp 167–169

Cebolla-Cornejo J, Soler S, Gomer B, Soria DM, Nuez F (2003) Screening *Capsicum* germplasm for resistance to *Tomato spotted wilt virus* (TSWV). Ann Appl Biol 143:143–152

Cembali T, Folwell R, Wandschneider P, Eastwell K, Howell W (2003) Economic implications of a virus prevention program in deciduous tree fruits in the US. Crop Prot 22:1149–1156

Chalam VC, Khetarpal RK, Mishra A, Jain A, Gupta GK (2004) Seed transmission of *Soybean mosaic potyvirus* in soybean cultivars. Indian J Mycol Plant Pathol 34:86–87

Chalam VC, Parakh DB, Khetarpal RK, Maurya AK, Jain A, Singh S (2008) Interception of seed-transmitted viruses in cowpea and mungbean germplasm imported during 2003. Indian J Virol 19:12–16

Chalam VC, Bhalla S, Gupta K, Singh B, Rajan, Goswami S, Chhabra R, Maurya AK, Randhawa GJ, Kapur ML, Khetarpal RK (2009a) Methodology for processing transgenic germplasm in quarantine. In: Tyagi RK, Bhag Mal, Sharma SK, Khetarpal RK, Brahmi P, Singh N, Tyagi V, Archak S, Gupta K, Agarwal A (eds) National symposium on recent developments in the management of plant genetic resources, 17–18 Dec 2009. Souvenir and Abstracts, Indian Society of Plant Genetic Resources, New Delhi, pp 320–321

Chalam VC, Khetarpal RK, Parakh DB, Deepti S, Kapoor P, Maurya AK (2009b) Detection and management of risk of introducing seed-transmitted viruses associated with legume germplasm imported into India. Plant Dis Res 24:98–99

Chalam VC, Parakh DB, Khetarpal RK, Maurya AK, Pal D (2009c) Interception of seed transmitted viruses in broad bean germplasm imported into India during 1996–2006. Indian J Virol 20:83–87

Chalam VC, Parakh DB, Maurya AK, Singh S, Khetarpal RK, Agarwal PC (2013) Diagnostics in biosecuring agriculture from transboundary plant viruses: a case study of India. In: 12th International symposium on plant virus epidemiology, 28th Jan–1st Feb 2013. Arusha, Tanzania (Abstract, pp 123, 224)

Chalfant RB, Jaworski CA, Johhson AW, Summer DR (1977) Reflective film mulches, millet barriers and pesticides: Effects on watermelon mosaic virus, insect, nematodes, soil-borne fungi and yield of yellow summer squash. J Am Soc Hortic Sci 102:11–15

Chamberlain EE, Baylis GJS (1948) Onion yellow dwarf. Successful eradication. N Z J Sci Technol A.29:300–301

Chamberlain RB, Atkinson JD, Hunter JA (1964) Cross protection between strains of Apple mosaic virus. N Z J Agric Res 7:480–490

Chan MS, Jeger MJ (1994) An analytical model of plant-virus disease dynamics with roguing and replanting. J Appl Ecol 31:413–427

Chancellor TCB, Holt J (2008) Tungro disease dynamics. In: Tiongco ER, Angeles ER, Sebastian LS (eds) *Rice tungro virus disease*: a paradigm in disease management. Science City of Munoz, Nueva Ecija: Philippine Rice Research Institute and Chiba, Japan: Honda Research Institute, 2008, pp 92–115

Chancellor TCB, Teng PS, Heong KL (eds) (1997) Rice tungro disease epidemiology and vector ecology: the development of sustainable and cost effective pest management practices to reduce yield losses in intensive rice cropping systems. IRRI Discussion Paper Series no. 19, International Rice Research Institute, Philippines

Chancellor TCB, Holt J, Villarreal S, Tiongco ER, Venn J (2006) Spread of plant virus diseases to new plantings: a case study of rice tungro disease. Adv Virus Res 66:1–30

Chander Rao S, Bhatnagar-Mathur P, Lavakumar P, Sudarshana Reddy A, Sharma KK (2012) Pathogen-derived resistance using a viral nucleocapsid gene confers only partial non-durable protection in peanut against *Peanut bud necrosis virus*. Arch Virol 158:133–143

Chander Rao S, Lavakumar P, Reddy AS, Swamykrishna T, Waliyar F, Nigam S, Laxminarasu M, Sharma KK (2006) Development and evaluation of transgenic peanut plants against *Peanut bud necrosis disease* (PBND) under green house and field conditions. Indian J Virol 17:135

Chandra K, Gupta BM, Varma HN (1978) Influence of age and host plant on the expression of acquired local and systemic antiviral resistance induced by treatment with *Trichothecium* polysachride in *N. glutinosa*. Curr Sci 47:168–170

Chandra R, Birhman RK (1994) In vitro micropropagation in relation to pedigree in potato. J Indian Potato Assoc 21:87

Chandra R, Dhingra MK (1990) Innovations in potato seed production. In: Grewal JS et al (eds) Current facets in potato research. Indian Potato Association and CPRI, Shimla, pp 66–76

Chandra R, Dodds JH, Tover P (1988) In vitro tuberization in potato (Solanum tuberosum L.). Int Assoc plant Tissue Cult Newsl 55:10–20

Chandra R, Chaudhuri DR, Birhman RK (1992a) Genotypic variability for microtubers production in potato. J Indian Potato Assoc 19:50–54

Chandra R, Randhawa GJ, Chaudhuri DR (1992b) Use of ordinary sugar in vitro production of microtubers. J Indian Potato Assoc 19:87–89

Chandra R, Randhawa GS, Chaudhuri DR, Upadhya MD (1992c) Efficacy of triazoles for in vitro microtuber production in potato. Potato Res 35:339–341

Chandra R, Chaudhuri DR, Upadhya MD (1994) A simplified low cost medium for potato micropropagation. J Indian Potato Assoc 20:270–272

Channon AG, Cheffins NJ, Hitchon GM, Barker J (1978) The effect of inoculation with an attenuated mutant strain of *Tobacco mosaic virus* on the growth and yield of early glasshouse tomato crops. Ann Appl Biol 88:121–129

Chatenet M, Delage C, Ripolles M, Irey M, Lockhart BEL, Rott P (2001) Detection of *Sugarcane yellow leaf virus* in quarantine and production of virus-free sugarcane by apical meristem culture. Plant Dis 85:1177–1180

Chattopadhyay SB, Bhowmik TP (1965) Control of 'foorkey' disease of larger cardamom in west Bengal. Indian J Agric Sci 35:272–275

Chatzivassiliou EK (2008) Management of the spread of *Tomato spotted wilt virus* in tobacco crops with insecticides based on estimates of thrips infestation and virus incidence. Plant Dis 92:1012–1020

Cha-um S, Thi-Thanh Hien N, Kirdmanee C (2006) Disease-free production of sugarcane varieties (*Saccharum officinarum* L.) using in vitro meristem culture. Biotechnology 5:443–448

Cheema SS, Lore JS, Gosal SS (1999) Management of citrus ring spot disease through thermotherapy. Ind Phytopath 52:354–356

Chen CC (1980) Rice wilted stunt in Taiwan: transmission by the brown planthopper Nilaparvata lugens and effect on the yield. Paper present on the 2nd Southeast Asian Sym. on Plant Disease in the Tropics, 20–26 Oct 1980, Bangkok, Thailand

Chen MJ, Chen CC (1980) Plastic mulches for the control of viruses in peppers. In: Abstracts of experimental report on agriculture. Department of Agriculture and Forestry, Taiwan Provincial Government, Taiwan, Republic of China, p 72

Chen CM, Chen CC (1983) Plastic cloth mulching for the control of viruses in peppers. In: Abstracts of experimental report on agriculture. Department of Agriculture and Forestry, Taiwan Provincial Government, Republic of China, p 106

Chen WQ, Sherwood JL (1991) Evaluation of tip culture, thermotherapy and chemotherapy for elimination of *Peanut mottle virus* from Arachis hypogea. J Phytopathol 132:230–236

Chen TH, Lu YT (2000) Application of *Ribavirin* in tissue culture of green bamboo (*Bamboosa oldhami* Muro) for elimination of Bamboo mosaic virus. Plant Prot Bull 42:159–168

Chen ZC, White RF, Antoniw JF, Lin Q (1991) Effect of pokeweed antiviral protein (PAP) on the protection of plant viruses. Plant Pathol 40:612–620

Chen LZ, Du ZS, Terao H, Tsuzuki E (2004) Establishment of propagation method in large quantities to produce virus-free plants of sweet potato, *Ipomoea batatas* (L.) Lam. by means of cultures of shoot apex and sucker. Bull Fac Agri Univ Miyazaki 50:1–9 (In Japanese)

Chen Z, Wang X, Song B, Wang H, Bhadury PS, Yan K, Zhang H, Yang S, Jin L, Hu D, Xue W, Zeng S, Wang J (2008) Synthesis and antiviral activities of novel chiral cyanoacrylate derivatives with (E) configuration. Bioorg Med Chem. 16(6):3076–3083

Cheng MF, Tung BK (1987) Studies on the management of sweet pepper virus diseases in the field. J Chin Soc Hortic Sci (Republic of China) 33:125–131

Chiarappa L (1981) Manmade epidemiological hazards in major crops of developing countries. In: Maramorosch K, Harris KF (eds) Plant diseases and vectors ecology and epidemiology. Academic Press, New York, pp 319–339

Chiari A, Bridgen MP (2002) Meristem culture and virus eradication in Alstromeria. Plant Cell Tissue Organ Cult 68:49–55

Chitturi A, Riley DG, Joost PH (2007) Effect of pine pollen on settling behavior of *Frankliniella occidentalis* and *Frankliniella fusca* (Thysanoptera: Thripidae) on tomato and peanut. Environ Entomol 35:1396–1403

Cho EK, Goodman RM (1979) Strains of soybean mosaic virus: classification based on virulence in resistant soybean cultivars. Phytopathology 69:467–469

Cho EK, Goodman RM (1982) Evaluation of resistance in soybean mosaic virus strains. Crop Sci 22:1133–1136

Cho JJ, Ullman DE, Wheatley E, Holly J, Gonsalves D (1992) Commercialization of ZYMV cross protection for zucchini production in Hawaii. Phytopathology 82:1073

Choudhary N, Kapoor HC, Lodha ML (2008) Cloning and expression of antiviral/ribosome-inactivating protein from Bougainvillea xbuttiana. J Biosci 33:91–101

Chowfla SC, Parmar GS (1995) Management of bell pepper mosaic complex by cultural practices. In: Gupta VK, Sharma RC (eds) Integrated disease management and plant health. Scientific Publishers, Jodhpur, pp 187–192

Chowrira GM, Cavileer TD, Gupta SK, Surquin PF, Berger PH (1998) Coat protein-mediated resistance to *Pea enation mosaic virus* in transgenic *Pisum sativum* L. Transgenic Res 7:265–271

Chu PWG, Waterhouse PM, Martin RR, Gerlach WI (1989) New approaches to the detection of microbial plant-pathogens. Biotechnol Genet Eng Rev 7:45–111

Chung E, Seong E, Kim YC, Chung EJ, Oh SK, Lee S, Park JM, Joung YH, Choi D (2004) A method of high frequency virus induced gene silencing in chili pepper *Capsicum annuum* L. cv. *Bukang*). Mol Cell 17:377–380

CIAT (2006) Recent products and developments at the International Center for Tropical Agriculture. A Newsletter from http://www.ciat.cgiar.org

Cieslinska M (2002) Elimination of Apple chlorotic leaf spot virus (ACLSV) from pear by in vitro thermotherapy and chemotherapy. Acta Hortic 596:481–484

Cieslinska M (2003) Elimination of Strawberry mottle virus (SMoV) from Fragaria virginiana UC-11 indicator plants by thermotherapy and chemotherapy. Phytopathol Pol 30:51–59

Clarke WC (1937) The close planting of ground nuts. E Abr Agri J 2:380

Clark MF (1981) Immunosorbent assays in plant pathology. Annu Rev Phytopathol 19:83–106

Clement SL, El. Din Sharaf N, Weigand S, Lateef SS (1994) Research achievements in plant resistance to insect pest of cool season legumes. Euphytica 73:41–50

Clough GH, Hamm PB (1995) Coat protein transgenic resistance to *Watermelon mosaic* and *Zucchini yellow mosaic virus* in squash and cantaloupe. Plant Dis 79:1107–1109

Cohen M (1968) Exocortis virus as a possible factor in producing dwarf citrus trees. Proc Fla State Hortic Soc 81:115–119

Cohen S (1982) Control of whitefly vectors of viruses by color mulches. In: Harris KF, Maramorosch K (eds) Pathogens, vectors and plant diseases: approaches to control. Academic Press, New York, pp 45–56

Cohen S, Antignus Y (1994) *Tomato yellow leaf curl virus*—a whitefly-borne geminivirus of tomatoes. Adv Dis Vector Res 10:259–288

Cohen S, Berlinger MJ (1986) Transmission and cultural control of whitefly-borne viruses. Agric Ecosyst Environ 17:89–97

Cohen S, Marco S (1973) Reducing the spread of aphid-transmitted viruses in peppers by trapping the aphids on sticky yellow polyethylene sheets. Phytopathology 63:1207–1209

Cohen S, Marco S (1979) Reducing virus spread in vegetables and potato by net cover. Phytoparasitica 7:40–41

Cohen S, Gertman E, Kedar N (1971) Inheritance of resistance to *Melon mosaic virus* in cucumbers. Phytopathology 61:253–255

Colvin J, Muniyappa V (1999) Sustainable management of the whitefly, Bemisia tabaci, and tomato leaf curl virus (ToLCV) on tomato in India. DFID Crop Protection Programme, Final Technical Report, Project R6627. Natural Resources Institute, University of Greenwich, Chatham, Kent, UK, and Department of Plant Pathology, University of Agricultural Sciences, GKVK, Bangalore, Karnataka, India, 79 pp

Colvin J, Nagaraju N, Moreno-Leguizamon C, Govindappa MR, Manjunatha R, Joshi N, Seal SE, Muniyappa V (2012) Socio-economic and scientific impact created by white-fly transmitted, plant virus disease resistance tomato varieties in southern India. J Integr Agric 11:337–345

Conover RA, Litz RE, Malo SE (1986) 'Cariflora'- a *Papaya ringspot virus* - tolerant papaya for south Florida and the Caribbean. Hort Science 21:1072

Constantin GD, Krath BN, MacFarlane SA, Nicolaisen M, Johansen IE, Lund OS (2004) Virus-induced gene silencing as a tool for functional genomics in a legume species. Plant J 40:622–631

Converse RH (1970) Effect of oil sprays on spread of strawberry viruses. Plant Dis Rep 54:479–482

Cooper JL, Jones RAC, Harrison BD (1976) Field and glasshouse experiments on the control of potato mop top virus. Ann Appl Biol 83:215–230

Cooper JI, Edwards ML, Rosenwasser O, Scott NW (1994) Transgenic resistance genes from nepoviruses: efficacy and other properties. N Z J Crop Hortic Sci 22:129–137

Cooper B, Lapidot M, Heick JA, Dodds JA, Beachy RN (1995) Multivirus resistance in transgenic tobacco plants expressing a dysfunctional movement protein of tobacco mosaic virus. Virology 206:307–313

Cornwell PB (1958) Movement of vectors of virus disease of cacao in Ghana. I. Canopy movement in and between trees. Bull Entomol Res 49:613–630

Costa AS (1956) Anthocyanosis, a virus of cotton in Brazil. Phytopathol Z 28:167–186

Costa CL (1972) Controle de molestias de virus de plantas com superficies reflectivas repelentes as vector. Rev Per Entomol 15:135–139

Costa CL, Costa AS (1971) Reducao da disseminacao de mosaico em abobora de moita com superficies reflectivas replentes aos afidios vectors. XIth Reuniao Annal da Soc. Oler. do Brasil, Piracicaba, 12–17 Julho

Costa AS, Grant TJ, Moreira S (1954) Behavior of various citrus rootstock-scion combinations following inoculation with mild and severe strans of tristeza. Proc Fla State Hortic Soc 67:26–30

Costa CL, Muller GW, Costa AS, Teofilo Sobrinho J (1974) Reducing tristeza infection of citrus seedlings by repelling Toxoptera citricidus with rice husk mulch. In: Proceedings of the 6th conference of the IOCV, pp 97–100

Costa HS, Robb KL, Wilen CA (2002) Field trials measuring the effects of ultraviolet-absorbing green house plastic films on insect populations. J Econ Entomol 95:113–120

Coundriet DL, Prabhaker N, Meyerdirk DE (1985) Sweetpotato whitefly (Homoptera: Aleyrodidae): effects of neem-seed extract on oviposition and immature stages. Environ Entomol 14:776–779

Courshee RJ (1960) Some aspects of the application of insecticides. Annu Rev Entomol 5:237–252

Cousin MT, Grison C (1969) Note concernant I'action dúne huile minerale sur la transmission de la mosaique commune du pois (Pea virus Z) par Acyrtosiphum pisi. Ann Phytopathol 1:315–318

Coutts BA, Jones RAC (2005) Suppressing spread of *Tomato spotted wilt virus* by drenching infected source or healthy recipient plants with neonicotinoid insecticides to control thrips vectors. Ann Appl Biol 146:95–103

Coutts BA, Thomas-Carroll ML, Jones RAC (2004) Patterns of spread of *Tomato spotted wilt virus* in field crops of lettuce and pepper: spatial dynamics and validation of control measures. Ann Appl Biol 145:231–245

Coutts BA, Prince RT, Jones RAC (2008) Further studies on *Pea seed-borne mosaic virus* in cool-season crop legumes: responses to infection and seed quality defects. Aust J Agric Res 59:1130–1145

Cradock KR, da Graça JV, Laing MD (2001) Control of aphid virus-vectors in *Cucurbita pepo* L. in KwaZulu-Natal, South Africa. Subtrop Plant Sci 53:49–54

Craig GW, Stephen JW, Michael GKJ (2004) Diagnosis of plant viral pathogens. Curr Sci 86:12–25

Crane GL (1968) Mineral oil as a foliar spray for the potential control of cercospora leaf spot and virus yellows diseases by Beta vulgaris L. Ph.D. Thesis. University of Minnesota, 100 pp

Crane GL, Calpouzos L (1967) Mineral oil suppressed symptoms of virus yellows on sugar beets. Phytopathology 57:807–808

Crane GL, Calpouzos L (1969) Suppression of symptoms of sugar beet virus yellows by mineral oil. Phytopathology 59:697–698

Crnko J (1969) Obtainment of mosaic free seed in lettuce cultivars of Argentina. Revta Investnes Agrop Ser 2(6):103–129

Cropley R (1979) The production and practical benefits of virus-free propagating material of fruit crops. In: Hebbels DL, King JE (eds) Plant Health. Blackwell Scientific Publishers, Oxford, pp 121–127

Crosse JE (1976) Plant path. Section—Virus and virus-like diseases. East Malling Research Station Report for 1976

Crowdy SH, Posnette AF (1947) Virus diseases of cacao in West Africa. II. Cross immunity experiments with viruses 1A, 1B and 1C. Ann Appl Biol 34:403–411

Csinos AS, Pappu HR, McPherson RM, Stephenson MG (2001a) Management of Tomato spotted wilt virus in flue-cured tobacco with acibenzolar-S-methyl and imidacloprid. Plant Dis 85:292–296

Csinos AS, Pappu HR, McPherson RM, Stephenson MG (2001b) Acibenzolar S methyl (0.2 to 8 g a.i./7000 plants) free transplant application for suppression of TSWV. Plant Dis 85:290–296

Culbreath AK, Todd DW, Gorbet FM, Shokes FM, Pappu HR (1997) Field response of new peanut cultivar UF 91108 to *Tomato spotted wilt virus*. Plant Dis 81:1410–1415

Culbreath AK, Todd JW, Brown SL (2003) Epidemiology and management of tomato spotted wilt in peanut. Annu Rev Phytopathol 41:53–75

Culbreath AK, Tillman BL, Tubbs RS, Beasley JP Jr, Kamerait RC Jr, Brenneman TB (2010) Ineractive effects of planting date and cultivar on Tomato spotted wilt of peanut. Plant Dis 94:898–904

Culver JN (1995) Molecular strategies to develop virus resistant plants. In: Singh RP, Singh US (eds) Molecular methods in plant pathology. CRC/Lewis Publishers, Boca Raton, pp 235–248

Culver JN, Sherwood JL, Melouk HA (1987) Resistance to *Peanut stripe virus* in Arachis germplasm. Plant Dis 71:1080–1082

Cuozzo M, O'Connell KM, Kaniewski W, Fang R, Chua NH, Tumer NE (1988) Viral protection in transgenic tobacco plants expressing the *Cucumber mosaic virus* coat protein or its antisense RNA. Bio Technology 6:649–557

Dader B, Moreno A, Vinuela E, Fereres A (2012) Spatio-temporal dynamics of viruses are differentially affected by parasitoids depending on the mode of transmission. Viruses 4(11):3069–3089

Dahal G, Neupane FP, Baral DR (1992) Effect of planting and insecticides on the incidence and spread of yellow vein mosaic of okra in Nepal. Int J Plant Dis 10:109–124

Daiber CC, Donaldson JNI (1976) *Watermelon mosaic virus* in Vegetable marrows: the effect of aluminum foil on the vector. Phytoparasitica 8:85–86

Dale JL (1987) Banana bunchy top: an economically important tropical plant virus disease. Adv Virus Res 33:301–325

Dale JL, Harding RM (1998) Banana bunchy top disease: current and future strategies for control. In: Hadidi A, Khetarpal RK, Koganezawa H (eds) Plant virus disease control. APS Press, St. Paul, pp 659–669

Daly P, Tomkins B (1995) Production and post harvest handling of Chinese cabbage (*Brassica rapa* var. *pekinensis*). RIRDC Publication No. 97/1, Barton, pp 1–39

Dalmasso A (1971) La desinfection nematicide du sol dans la prevention de la degenerescence infectieuse de la vigne (Soil disinfection for preventing infectious degeneration). Union Agric 315:55–66

Dalmasso A, Boubals D, Bongiovanni M, Pistre R (1972a) Distribution des nematodes propagateurs du court-noue dans les vignobles mediterraneens français (Distribution of nematode vectors of fanleaf in vineyards of southern France). Progr Agric Vitic 89:456–462

Dalmasso A, Cuany A, Joubert J (1972b) Utilisation de l'aldicarbe contre les nematodes vecteurs de maladies a virus (Use of aldicarb against nematodes that transmit viruses). Compte Rendus des Journees Aldicarbe, Versailles, Decembre 1972:221–230

D'Anna F, Panattoni A, Triolo E (2006) Emerging antiviral drugs in plant chemotherapy. J Plant Pathol 88:39

Damsteegt V (1976) A naturally occurring corn virus epiphytotic. Plant Dis Rep 60:858–861

Dasgupta I, Malathi VG, Mukherjee SK (2003) Genetic engineering for virus resistance. Curr Sci 84:341–354

Davies JC (1976) The incidence of rosette disease in groundnut in relation to plant density and its effect on yield. Ann Appl Biol 82:489–501

Davis RF, Shifriss O (1983) Natural virus infection in silvery and non silvery lines of *Cucurbita pepo*. Plant Dis 67:379–380

Davis MJ, Ying Z (2004) Development of papaya breeding lines with transgenic resistance to papaya ring spot virus. Plant Dis 88:352–358

Dawson WO, Bubrick P, Crantham GL (1988) Modifications of *Tobacco mosaic virus* coat protein gene affecting replication, movement and symptomalogy. Phytopathology 78:783–789

Day AG, Bejarano ER, Buck KW, Burrell M, Lichtenstein CP (1991) Expression of an antisense viral gene in transgenic tobacco confers resistance to the DNA virus tomato golden mosaic virus. Proc Natl Acad Sci U S A 88:6721–6725

De Clercq E (2002) Highlights in the development of new antiviral agents. Mini-Rev Med Che 2:163–175

De Haan P (1998) Mechanisms of RNA-mediated resistance in plant viruses. In: Methods in Molecular Biology, Plant Virology Protocols, Humana Press, Totowa, New Jersey 81:533–546

Delanoy M, Salmon M, Kummert J, Frison E, Lepoivre P (2003) Development of real-time PCR for the rapid detection of episomal *Banana streak virus* (BSV). Plant Dis 87:33–38

Delfosse P, Reddy AS, Devi KT, Legreve A, Risopoulos J, Doucet D, Devi PS, Maraite H, Reddy DVR (2002) Dynamics of Polymyxa graminis and *Indian peanut clump virus* (IPCV) infection on various monocotyledonous crops and groundnut during the rainy season. Plant Pathol 51:546–560

Demski JW (1975) Source and spread of *Peanut mottle virus* in soybean and peanut. Phytopathology 65:917–920

Demski JW, Kuhn CW (1975) Resistant and susceptible reaction of soybeans to *Peanut mottle virus*. Phytopathology 65:95–99

Demski JW, Lovell GR (1985) *Peanut stripe virus* and the distribution of peanut seed. Plant Dis 69:734–738

Deogratias JM, Lutz A, Dosba F (1986) In vitro micrografting of shoot tips from juvenile and adult Prunus avium (L.) and Prunus persica (L.) Batsch to produce virus-free plants. Acta Hortic 193:139–145

Deol GS, Rataul HS (1978) Role of various barrier crops in reducing the incidence of *Cucumber mosaic virus* in chilli, *Capsicum annuum* Linn. Indian J Ent 40:261–264

Deom CM (2004) Plant resistance to viruses: engineered resistance. In: Encyclopedia of virology, pp 1307–1314

Deom CM, Schubert KR, Wolf S, Holt CA, Lucas WJ, Beachy RN (1990) Molecular characterization and biological function of the movement protein of tobacco mosaic virus in transgenic plants. Proc Natl Acad Sci U S A 87:3284–3288

Deom CM, Lapidot M, Beachy RN (1992) Plant virus movement proteins. Cell 69:221–224

Dewar AM, Smith HG (1999) Forty years of forecasting virus yellows incidence in sugar beet. In: Smith HG, Barker H (eds) The Luteoviridae, Chapter 9, Epidemiology and control. CABI Publishing, Wallingford, pp 228–261

de Wijs JJ (1980) The characteristics of mineral oils in relation to their inhibitory activity on the aphid transmission of potato virus Y. Neth J Plant Pathol 86:291–300

de Wijs JJ, Saturm E, Schurinn FJ (1979) The Viscosity of mineral oils in relation to their ability to inhibit the transmission of stylet-borne viruses. Neth J Plant Pathol 85:19–22

Dhanju KC, Lal T, Dhaliwal MS (1995) Screening muskmelon hybrids against mosaic viral complex. Indian J Virol 11:71–73

Dhankar BS, Saharan BS, Hooda VS, Singh P (1991) Exploitation of alien genes for yellow vein mosaic resistance in okra. In: Prakash J, Pierik RLM (eds) Horticulture new technologies and applications. Kluwar Academic Publishers, Dordrecht, pp 45–48

Dhillon NPS, Jellis GJ, Boulton RE, Jackson EA, Jack PL, Lacey CND (1995) Isozyme and RFLP mapping of Sbm-4, a gene in pea (Pisum sativum) conferring resistance to P4 pathotype of Pea seed-borne mosaic virus. Adv Hortic Sci 9:159–161

Dhital SP, Lim HT, Sharma BP (2008) Electrotherapy and chemotherapy for eliminating double-infected potato virus (PLRV and PVY) from in vitro plantlets of potato (*Solanum tuberosum* L.). Hortic Environ Biotechnol 49:52–57

Di R, Purcell V, Collins GB, Ghabrial SA (1996) Production of transgenic soybean lines expressing the bean pod mottle virus coat protein precursor gene. Plant Cell Rep 15:746–750

Diaz-Pendon JA, Fernandez-Munoz R, Gomez-Guillamon ML, Moriones E (2005) Inheritance of resistance to *Watermelon mosaic virus* in *Cucumis melo* that impairs virus accumulation, symptom expression and aphid transmission. Phytopathology 95:840–846

Diaz-Perez JC, Batal KD, Cranberry D, Bertrand D, Giddings D, Pappu H (2003) Growth and yield of tomato on plastic film mulches as affected by *Tomato spotted wilt virus*. Hort Sci 38:395–399

Diaz JA, Mallor C, Soria C, Camero R, Garzo E, Fereres A, Alvarez JM, Gomez-Guillamon ML, Luis-Arteaga M, Moriones E (2003) Potential sources of resistance for melon to non-persistently aphid-borne viruses. Plant Dis 87:960–964

Dickson RC, Laird EF Jr (1966) Aluminum foil to protect melons from *Watermelon mosaic virus*. Plant Dis Rep 50:305

Dickson RC, Johnson MMcD, Flock RA, Laird EF (1956) Flying aphid populations in Southern California citrus groves and their relation to the transmission of the tristeza virus. Phytopathology 46:204–210

Diez MJ, Rosello S, Nuez F, Costa J, Lacasa A, Catala MS (1999) Tomato production under mesh reduces crop loss to *Tomato spotted wilt virus* in some cultivars. Hortic Sci 34:634–637

Dimov A, Stefanov I (1975) Interdependence between sowing times and infection of wheat by virus diseases. Zavisimost mezhdn srokovete na seitba I napadenieto na pshenitsatot virusni bolesti. Rastitelna Zashchita 23:31–33

Ding XS, Schneider WL, Chaluvadi SR, Rouf Mian RM, Nelson RS (2006) Characterization of a Brome mosaic virus strain and its use as a vector for gene silencing in monocotyledonous hosts. Mol Plant Microbe Interact 19:1229–1239

Dobie ND, Vaughan EK, Miler PW, Waldo GF (1958) New facilities and procedures used for growing disease-free strawberry plants in Oregon. Plant Dis Rep 42:1048–1050

Doidge EM (1939) Scaly bark or psorosis of citrus trees in South Africa. Dept Agr For Sci Bull 208:1–16

Dogimont C, Slama S, Martin J, Lecoq H, Pitrat M (1996) Sources of resistance to *Cucurbit aphid borne yellows luteo virus* in melon germplasm collection. Plant Dis 80:1379–1382

Domfeh O, Dzahini-Obiatey H, Ameyaw Akumfi G, Opoku IY (2009) The effect of mild strains on growth and yield of cocoa. 2008/2009 Progress Report and Work plan, Cocoa Research Institute of Ghana, July 2009, Akim Tafo pp 151–153

Domier LL, Steinlage TA, Hobbs HA, Wang Y, Herrera Rodriguez G, Haudenshield J, McCoppin NK, Hartman GL (2007) Similarities in seed and aphid transmission among soybean virus isolates. Plant Dis 91:546–550

Donald M, Bonde R (1954) Dissemination of spindle tuber by contaminated tractor wheels and by foliage contact with diseased potato plants. Phytopathology 44:111

Doncaster JP, Gregory PH (1948) The spread of virus diseases in the potato crop. Agricultural Research council Report Series No. 7, London, H.M.S.O

Drinkwater TW, Walters MC, Van Rensburg JBJ (1979) The application of systemic insecticides to the soil for the control of maize stalk borer, *Busseola fusca* (fuller) (Lep.: Noctuidae), and of *Cicadulina mbila* (Naude) (Hem.: Cicadellidae), the vector of maize streak virus. Phytophylactica 11(1):5–11

Dubey GS, Nene YL (1974) Aphid transmission of *Cowpea mosaic virus* as influenced by oil sprays. Indian Phytopathol 27:325–330

Duffus JE (1964) Host relationships of beet western yellows virus strains. Phytopathology 54:736–738

Duffus JE (1971) Role of weeds in the incidence of virus diseases. Annu Rev Phytopathol 9:319–340

Duncan JM, Torrance L (eds) (1992) Techniques for the rapid detection of plant pathogens. Blackwell Scientific, Oxford

Duncan JH, Sproule RS, Bevington KB (1978) Commercial application of virus induced dwarfing. In: Carey PR (ed) Proceedings of the International Society of Citriculture, pp 317–319

Dunn JA, Kirkley J (1966) Studies on the aphid, Cavariella aegopodii scop. II. On secondary hosts other than Carrot. Ann Appl Biol 58:213–217

Duriat AA, Karyadi AK, Miura M, Sukarna E (1990) Influence of border crops on virus incidence of potato tuber. Bul Penel Hortic 19:94–108

Dutrecq A, Vanderveken J (1969) Remanence de l'effect inhibiteur d'une huile minerale a l'egard de la transmission aphidienne du virus de la jannisse de la betterave. Bull Rech Agron Gembloux 4:66–75

Dutta OP (1978) I.I.H.R. Annual Report

Dzanini-Obiatey H, Domfeh O, Amoah FM (2010) Over seventy years of a viral disease of cocoa in Ghana: from researchers' perspective. Afr J Agric Res 5:476–485

Dziedzic E (2008) Elimination of Prunus necrotic ring spot (PNRSV) from plum 'Earliblue' shoot through thermotherapy in vitro. J Fruit Ornam Plant Res 16:101–109

Ebbels DL (2003) Principles of plant health and quarantine. CAB International, Wallingford

Edwards MC, Steffenson BJ (1996) Genetics and mapping of *Barley stripe mosaic virus* resistance in barley. Phytopathology 86:184–187

El-Dougdoug KhA, Dawood RA, El-Shamy MMM, Korkar HM (2009) Eradication of banana viruses from naturally infected plants. 3 Disease management of banana through the use of virus free plant materials. J Appl Sci Res 5:1872–1880

El-Khoury W, Makkouk K (2010) Integrated plant disease management in developing countries. J Plant Pathol 92(4):S4.35–S4.42

El Far Mervat MM, Ashoub A (2009) Utility of thermotherapy and meristem tip for freeing sweetpotato from viral infection. Aust J Basic Appl Sci 3:153–159

EPPO (1969–1978) Phytosanitary regulations of EPPO member countries. EPPO, Paris

EPPO (1991) Certification scheme. Virus-free or virus-tested fruit trees and rootstocks. Part I: basic scheme and its elaboration. EPPO Bull 21:267–277

EPPO (1992) Certification scheme. Virus-free or virus-tested fruit trees and rootstocks. Part II, III, IV: tables of viruses and vectors. EPPO Bull 22:255–283

EPPO (1994) Certification scheme. Pathogen-tested material of grapevine varieties and rootstocks, Ribes, Rubus, Strawberry. EPPO Bull 24:347–367

EPPO/CABI (1996) Quarantine pests for Europe, 2nd edn. CAB Int, Wallongford

Esparza-Duque J, Verhoyen M (1975a) Lutte contre le virus de la mosaique du celeri par des pulverizations d'huile et des surfaces reflechissantes repulsives pour vecteurs. Meded Fakulteit Landbonw wetenschappen Genet 40:779–789

Esparza-Duque J, Verhoyen M (1975b) Reduction des infections de virus de la mosaique, du celeri par des pulverizations d'huile et des surfaces reflechissantes repulsives pour vecteurs. Meded Fakulteit Landbonwwetenschappen Genet 40:779–789

Eulitz EG (1977) Aluminium foil for the control of watermelon mosaic in vegetable marrow. Phytoparasitica 9:23

Evans SG (1972) Trials with the Dutch mild strain of *Tomato mosaic virus*. Forward 56:13–14

Evans TA, Miller LC, Vasilas BL, Taylor RW, Mulrooney RP (2007) Management of *Xiphinema americanum* and soybean severe stunt in soybean using crop rotation. Plant Dis 91:216–219

Ewusie EA, Parajulee MN, Aba Adabie-Gomez D, Wester D (2010) Strip cropping: a potential IPM tool for reducing whitefly, *Bemisia tabaci* Gennadius (Homoptera: Aleyrodidae) infestations in cassava. West Afr J Appl Ecol 17:109–119

Faccioli G (2001) Control of Potato viruses using meristem and stem-cutting cultures, thermotherapy and chemotherapy. In: Loebenstein G et al. (ed) Virus and virus-like disease of potatoes and production of seed-potatoes. Kluwer Academic Publishers, Dordrecht, pp 365–390

Faccioli G, Colombarini A (1991) PVS and PVM elimination by meristem culture, chemo- and thermotherapy. Phytoparasitica 19:3

Faccioli G, Marani F (1998) Virus elimination by meristem tip culture and tip micrografting. In: Hadidi A, Khetarpal RK, Koganezawa H (eds) Plant virus disease control. APS Press, St. Paul, pp 346–380

Fagbola O, Boulton RE, Jellis GJ, Dhillon NPS (1996) The reaction of some varieties of fababean (Vicia faba) to *Pea seed-borne mosaic potyvirus*. Plant Var Seeds 9:43–51

Faivre-Rampant O, Gilroy EM, Hrubikova K, Hein I, Millam S, Loake GJ, Birch P, Taylor M, Lacomme C (2004) Potato virus X-induced gene silencing in leaves and tubers of potato. Plant Physiol 134:1308–1316

Fajardo TG (1930) Studies on the mosaic disease of the bean (*Phaseolus vulgaris* L.). Phytopathology 20:469–494

Fajardo J, Lutaladio NB, Larinde M, Rosell C, Barker I, Roca W, Chujoy E (2010) Quality declared planting material. Protocols and standards for vegetatively propagated crops. FAO, Italy Rome 195:33–35

Fallah M, Mozafari M, Sokhandan-Bashir N, Hashemi M (2007) Elimination of a DNA virus associated with yellow leaf curl disease in tomato using an electrotherapy technique. 2nd International symposium on tomato diseases, Kusadasy, Turkey, 8–12 Oct 2007

Fang Y, Dhital SP, Li K, Khu D, Kim H, Song Y, Lim H (2005) Utilization of single nodal cuttings and therapies for eradicating double-infected potato virus (PLRV, PVY) from in vitro plantlets of potato (Solanum tuberosum). J K Soc Hort Sci 48:119–125

FAO (2000) The stage of food insecurity in the world (SOFI). FAO, Rome. www.fao.org/focus/e/of/100/s0f1001.htm

FAO (2010) Quality declared planting material. Protocols and standards for vegetatively propagated crops. FAO Plant Production and Protection Paper 195. FAO, Rome

Fargette D, Fauquet CM (1988) A preliminary study on the influence of intercropping maize and cassava on the spread of African cassava mosaic by whiteflies. Ann Appl Biol 17:195–202

Fargette D, Jeger M, Fauquet C, Fishpool L (1994) Analysis and modelling of the temporal spread of African cassava mosaic virus. Phytopathology 84:91–98

Fargette D, Leslie M, Harrison BD (1996) Serological studies on the accumulation and localization of three tomato leaf curl Gemini viruses in resistant and susceptible Lycopersicon species and tomato cultivars. Ann Appl Biol 128:317–328

Farrell JAK (1976) Effects of groundnut sowing date and plant spacing on rosette virus disease in Malawi. Bull Entomol Res 66:159–171

Fauquet CM (2013) War on African cassava viruses: a novel strategy against mighty foes of cassava. Paper presented at 12th International on plant virus epidemiology, 28 Jan–1st Feb 2013, Arusha, Tanzania (Abstract, OP-71), p 96

Fauquet CM, Fargette D, Thouvenel JC (1987) Selection of healthy cassava plants obtained by reversion in cassava fields. In: Proceedings of the International Seminar on *African Cassava Mosaic Disease* and its control Yamoussoukro, Cote d'Ivoire 4–8 May 1987. CTA/FAO/ORSTOM/IITA/IAPC, pp 146–149

Felix Nweke (2009) Controlling cassava mosaic disease. IFPRI discussion paper 00912 (Nov, 2009)

Feng G, Vifu G, Pinto Z (2000) Production and development of virus-free sweet potato in China. Crop Prot 19:105–111

Fenigstein A, Eliyo M, Kletter E, Ganmor S, Veierov D (2001) Relative activities and field evaluation of light vegetable oils for the control of sweet potato whitefly, *Bemisia tabaci* on cotton. In: Proceedings of the European Whitefly Symposium, Ragosa, Italy, 27th Feb–3rd March 2001, p 84

Fereres A (2000) Barrier crops as a cultural control measure of non-persistently transmitted aphid-borne viruses. Virus Res 71:221–231

Fereres A, Moreno A (2011) Integrated control measures against viruses and their vectors. In: Caranta C, Aranda MA, Tepfer M, Lopez-Moya JJ (eds) Recent advances in plant virology. Academic press, Caister, 412 pp

Ferreira SA, Mau RFL, Manshardt R, Pitz KY, Gonsalves D (1992) Field evaluation of *Papaya ringspot virus* cross protection. In: Proceedings of the 28th annual Hawaii Papaya Industry Association conference, Honolulu, pp 14–19, 25–26 Sept

Fifaei R, Golein B, Taheri H, Tadjvar Y (2007) Elimination of Citrus tristeza virus of Washington Navel orange (*Citrus sinensis* L. Osbeck) through shoot-tip grafting. Int J Agric Biol 9:27–30

Finnegan J, McElroy D (1994) Transgene inactivation: plants fight back! Biotechnology 12:883–884

Fitch MM, Manshardt RM, Gonsalves D, Slightom JL, Sanford JC (1992) Virus resistant papaya derived from tissues bombarded with the coat protein gene of Papaya ringspot virus. Bio/Technology 10:1466–1472

Fitch MMM, Lehrer AT, Komor E, Moore PH (2001) Elimination of Sugarcane yellow leaf virus from infected sugarcane plants by meristem tip culture visualized by tissue blot immunoassay. Plant Pathol 50:676–680

Fitchen JH, Beachy RN (1993) Genetically engineered protection against viruses in transgenic plants. Annu Rev Microbiol 47:739–763

Fjelddalen J (1974) Organization of plant protection in Norway and research orientation. Bull Org Eur Med Plant Prot 4:241–249

Fletcher JT, Rowe JM (1975) Observations and experiments on the use of an avirulent mutant strain of *Tobacco mosaic virus* as means of controlling tomato mosaic. Ann Appl Biol 81:171–179

Fletcher PJ, Fletcher JD, Lewthwaite SL (1998) In vitro elimination of onion yellow dwarf and shallot latent viruses in shallot (Allium cepa var. ascalonicum L.). N Z J Crop Hort 26:23–26

Fofana IB, Sangare A, Collier R, Taylor C, Fauquet CM (2004) A geminivirus-induced gene silencing system for gene function validation in cassava. Plant Mol Biol 56:613–624

Fondong VN, Thresh JM, Zok S (2002) Spatial and temporal spread of *Cassava mosaic virus* disease in cassava grown alone and when intercropped with maize and/or cowpea. J Phytopathol 150:1–10

Foster JA, Hadidi A (1998) Exclusion of plant viruses. In: Plant Virus Disease Control (ed. Hadidi A, Khetarpal RK, Koganezawa H). The American Phytopathological Society Press, St. Paul, Minnesota, USA. pp. 208–229

Foxe MJ (1992) Breeding for viral resistance: Conventional methods. Netherlands J Plant Pathol 98:13–20

Franklin RM (1966) Purification and properties of replicative intermediate of the RNA bacteriophage R17. Proc Natl Acad Sci U S A 55:1504–1511

Fraselle J (1968) L'utilisation des produits huileux dans les traitements antiparasitaires. Annls Gembloux 74:37–46

Fraser RSS (1982) Biochemical aspects of plant resistance to virus disease: a review. Acta Hortic 127:101–116

Fraser RSS (1990) The genetics of resistance to plant viruses. Annu Rev Phytopathol 28:179–200

Fraser RSS (1998) Biochemistry of resistance to plant viruses. Breeding for resistance to plant viruses. In: Hadidi A, Khetarpal RK, Koganezawa H (eds) Plant virus disease control. AP Press, St. Paul, pp 56–64

Fraser LR, Broadbent P (1979) Virus and related diseases of citrus in New South Wales. Beatty and Sons Pty Ltd, Chipping Norton

Fraser LR, Levitt EC (1959) Recent advances in the study of exocortis (scally butt) in Australia. In: Proceedings of the conference on citrus virus diseases, Riverside, CA, pp 129–133

Fraser RSS, Vanloon LC (1986) Genes for resistance to plant viruses. CRC Crit Rev Plant Sci 3:257–294

Fraser LR, Levitt EC, Cox J (1961) Relationship between exocortis and stunting of citrus varieties on Poncirus trifoliata rootstocks. In: Proceedings of the 2nd conference of the IOCV. University of Florida Press, Gainesville, pp 34–39

Fraser LR, Singh D, Capoor SP, Nariani TK (1966) Greening virus-the likely cause of citrus die back in India. FAO Plant Prot Bull 14:127–130

Frederick RD, Thilmony RL, Sessa G, Martin GB (1998) Recognition specificity for the bacterial avirulence protein AvrPto is determined by Thr-204 in the activation loop of the tomato Pto kinase. Mol Cell 2:241–245

Freitas AJ, Rezende JAM (1998) Odontoglossum ringspot virus-free Cymbidium obtained through meristem tip chemotherapy. Fitopatol Bras 23:158–160

French N, Wilson WR (1976) Influence of crop rotation, weed control and nematicides on sparing in potatoes. Plant Pathol 25:167–172

Freytag AH (1965) A 'hot knife' to prevent virus spread in orchid. Bull Am Orchid Soc 34:501–502

Fridley R (1977) High density orchards facilitate harvest. Calif Agric 31:12–13

Fridlund PR (1968) IR-2, a germplasm bank of virus-free fruit tree clones. Hortic Sci 3:227–229

Fridlund PR (1980) The IR-2 program for obtaining virus-free fruit trees. Plant Dis 64:826–830

Frison EA (1994) IBPGR concerns on seed-borne disease. In: Proceedings of the Conference on the potential of biotechnology to minimize seed-borne diseases, 28 Sept–1 Oct 1989. Research and Extension Center, Washington State University, Puyallup, WA. pp 32–40

Frison EA, Diekmann M (1998) IPGRI's in controlling virus disease in plant germplasm. In: Hadidi A, Khetarpal RK, Koganezawa H (eds) Plant virus disease control. APS Press, St. Paul, pp 230–236

Frison EA, Putter CAJ (1988) FAO/IBPGR technical guidelines for the safe movement of germplasm, p 279. In: Abstract for the 5th International congress of plant pathology, 20–27 Aug, 1988, Kyoto

Frison EA, Bos L, Hamilton RI, Mathur SB, Taylor JW (1990) FAO/IBPGR Technical guide lines for the safe movement of legume germplasm. FAO/IBPGR, Rome 88 pp

Fuchs M, Gonsalves D (1995) Resistance of transgenic hybrid squash ZW-20 expressing the coat protein genes of zucchini yellow mosaic virus and watermelon mosaic virus 2 to mixed infections by both potyviruses. Bio/Technology 13:1466–1473

Fuchs M, Gonsalves D (2007) Safety of virus-resistant transgenic plants two decades after their introduction: lessons from realistic field risk assessment studies. Annu Rev Phytopathol 45:173–202

Fuchs M, Provvidenti R, Slighton JL, Gonsalves D (1996) Evaluation of transgenic tomato plants expressing the coat protein gene of *Cucumber mosaic virus* strain WL under field conditions. Plant Dis 80:270–275

Fuchs M, Ferreira S, Gonsalves D (1997) Management of virus diseases by classical and engineered protection. Molecular plant pathology on line. http://www.Bspp.org.uk.mppol/1997/01161fuchs

Fuchs M, Chirco EM, McFerson J, Gonsalves D (2004) Comparative fitness of a free-living squash species and free-living x virus-resistant transgenic squash hybrids. Environ Biosaf Res 3:17–28

Fuchs M, Cambra M, Capote N, Jelkmann W, Kundu J, Laval V, Martelli GP, Minafra A, Petrovic N, Pfeiffer P, Pompe-Novak M, Ravelonandro M, Saldarelli P, Stussi-Garaud C, Vigne E, Zagrai I (2007) Safety assessment of transgenic plum and grapevines expressing viral coat protein genes: new insight into real environmental impact of perennial plants engineered for virus resistance. J Plant Pathol 89:2–12

Fuji S, Iida T, Nakamae H (2000) Selection of an attenuated of *Japaneese yam mosaic virus* and its use for protecting yam plants against severe strains. Jpn J Phytopathol 66:35–39

Fulton RW (1986) Practices and precautions in the use of cross protection for plant virus disease control. Annu Rev Phytopathol 24:67–81

Fulton RA, Romney VE (1940) The chloroform-soluble components of beet leafhoppers as an indication of the distance they move in the spring. J Agr Res 61:737–743

Fulton JP, Seymour C (1957) The Arkansas strawberry certification program. Plant Dis Rep 41:749–754

Funderburk J (2009) Management of the western flower thrips (Thysanoptera: Thripidae) in fruiting vegetables. Fla Entomologist 92:1–6

Furmidge CGL (1968) Use of hydrocarbon oils in agricultural sprays. J Proc Inst Agric Eng 23:173–182

Furutani N, Hidaka S, Kosaka Y, Shizukawa Y, Kanematsu S (2006) Coat protein gene-mediated resistance to soybean mosaic virus in transgenic soybean. Breed Sci 56:119–124

Gabova RN (1989) Virus free pome fruits through meristem tip culture. Acta Hortic 235:69–76

Gal-On A, Shiboleth YM (2006) Cross-protection. In: Loebenstein G, Carr JP (eds) Natural resistance mechanisms of plants to viruses. Springer, Dordrecht, pp 261–288

Galun E, Breiman A (1997) Transgenic plants. Imperial College Press, London, pp 124–211

Ganesan U, Suri SS, Shanmugam R, Rajam MV, Indranil I (2009) Transgenic expression of coat protein gene of *Rice rungro bacilliform virus* in rice reduces the accumulation of viral DNA in inoculated plants. J Virus Genes 39:113–119

Garcia-Cano E, Castillo JN, Moriones E, Fernandez-Munoz R (2010) Resistance to *Tomato chlorosis virus* in wild tomato species that impair virus accumulation and disease symptom expression. Phytopathology 100:582–592

Gardner RG, Panthee DR (2012) *Tomato spotted wilt virus*-resistant fresh-market tomato breeding lines: NC 585, NC 1235, NC 1275 and NC 1325. Hortic Sci 47:531–532

Garnsey SM, Jones WJ (1968) Relationship of symptoms to the presence of tatter leaf virus in several citrus hosts. In: Childs JFL (ed) Proceedings of the 4th conference of the International Organization of Citrus Virologists. University of Florida Press, Gainesville, pp 207–212

Garnsey SM, Randles JW (1987) Biological interactions and agricultural implications of viroids. In: Semancik JS (ed) Viroids and viroid-like pathogens. CRC Press, Boca Raton, pp 127–160

Garnsey SM, Weathers LG (1972) Factors affecting mechanical spread of exocortis virus. In: Price WC (ed) Proceedings of the 5th conference of the International Organization of Citrus Virologists. University of Florida Press, Gainesville, pp 105–111

Garnsey SM, Whidden R (1971) Decontamination treatments to reduce the spread of *Citrus exocortis virus* (CEV) by contaminated tools. Proc Fla State Hortic Soc 84:63–67

Garnsey SM, Youtsey CO, Bridges GD, Burnett HC (1976) A necrotic ring spot-like virus found in a 'Star Ruby' grapefruit tree imported without authorization into Florida. Proc Fla State Hortic Soc 89:63–67

Garnsey SM, Gottwald TR, Yokomi RK (1998) Control strategies for citrus tristeza virus. In: Hadidi A, Khetarpal RK, Koganezawa H (eds) Plant virus disease control. APS Press, St. Paul, pp 639–658

Garrett RG, McLean GD (1983) The epidemiology of some aphid-borne viruses in Australia. In: Plumb RT, Thresh JM (eds) Plant virus epidemiology. Blackwell, Oxford, pp 199–209

Gasser CG, Fraley TT (1989) Genetically engineering plants for crop improvements. Science 244:1293–1299

Gay JD, Johnson AW, Chalfant RB (1973) Effects of a trap-crop on the introduction and distribution of cowpea virus by soil and insect vectors. Plant Dis Rep 57:684–688

Geering ADW, Thomas JE (1996) A comparison of four serological tests for the detection of *Banana bunchy top virus* in banana. Aus J Agric Res 47:403–412

Genereux H (1973) L'emploi diinsecticides et de l'huile "Control" comme mogen repressif des viruses de la pomme de terre. Phytoprotection 54:91

George EF (1996) Commercial micropropagation In: George EF (ed) Plant propagation by tissue culture. Exegeties Ltd., London, pp 761–833

George WL Jr, Kring JB (1971) Virus protection of late-season summer squash with aluminium mulch. Circular of the Connecticut Agricultural Experiment Station No. 239

Ghanekar AM (1980) Groundnut virus research at ICRISAT. In: Gibbons RW, Mertin JV (eds) Proceedings of the international workshop on groundnuts. ICRISAT, Patancheru, pp 211–216

Ghanekar AM, Sheila VK, Beniwal SPS, Reddy MV, Nene YL (1992) Sterility mosaic of pigeonpea. In: Singh US, Mukuopadhyay AN, Kumar J, Chaubey HS (eds) Plant diseases of international importance: diseases of cereals and pulses, vol-1. Prentice Hall, Eaglewood Cliffs, pp 415–428

Ghosh DK, Ahlawat YS, Gupta MD (1997) Production of virus free garlic (Allium sativum) plants by thermotherapy and meristem tip culture. Ind J Agr Sci 67:391–393

Ghosh D, Latha SK, Biswas NK (2009) Effect of different pesticides on incidence of *Mungbean yellow mosaic* incidence. Int J Plant Prot 2:67–70

Giacometti DC, Araujo CM (1965) Cross protection from tristeza in different species of citrus. In: Rice WC (ed) Proceedings of the 3rd conference of the International Organization of Citrus Virologists. University of Florida Press, Gainesville, pp 14–17

Gibson RW, Plumb RT (1976) The transmission and effect on yield of *Ryegrass mosaic virus* in a filtered air environment. Ann Appl Biol 82:79–84

Gibson RW, Pickett JA, Dawson GW, Rice AD, Stribley MF (1984) Effects of aphid alarm pheromone derivatives and related compounds on non- and semi-persistent plant virus transmission by Myzus persicae. Ann Appl Biol 104:203–209

Gibson RW, Aritua V, Byamukama E, Mpembe I, Kayongo J (2004) Control strategies for sweet potato virus disease in Africa. Virus Res 100:115–122

Gichuki S, La Bonte D, Burg K, Kapinga R, Simon JC (2005) Assessment and genetic diversity, farmer participatory breeding and sustainable conservation of Eastern Africa sweet potato germ-plasm. Annual report April 2004–March 2005

Gielen J, de Haan JLP, Kool AJ, Peters D, van Grinsven MQJM, Goldbach (1991) Engineered resistance to *Tomato spotted wilt virus*, a negative-strand RNA virus. Bio/Technology 9:1363–1367

Gielen J, Ultzen T, Bontems S, Loots W, Schepen VA, van Schepen A (1996) Coat protein mediated protection to *Cucumber mosaic virus* infection in cultivated tomato. Euphytica 88:139–149

Gilbert RZ, Kyle MM, Munger HM, Gray SM (1994) Inheritance of resistance to *Watermelon mosaic virus* in Cucumis melo L. Hortic Sci 29:107–110

Gilbert-Albertini F, Lecoq H, Pitrat M, Nicolet JL (1993) Resistance of Cucurbita moschatato *Watermelon mosaic virus* type 2 and its genetic relation to resistance to *Zucchini yellow mosaic virus*. Euphytica 69:231–237

Gilbertson RL (2011) Implementation and success of host-free periods for managing tomato-infecting *Begomoviruses* in developing countries. Phytopathology 101:S210

Gilbertson RL, Rojas MR, Kon T, Jaquez J (2007) Introduction of *Tomato yellow leaf curl virus* into the Dominican Republic: the development of a successful integrated pest management strategy. In: Czosnek H (ed) Tomato yellow leaf curl virus disease. Springer, Dordrecht, pp 279–303

Gill SS, Singh H, Sandhu KS (1982) Effect of sowing dates and plant density on the incidence of yellow vein mosaic virus on seed crop of okra (*Abelmoschus esculentus* L. Moench). Indian J Ecol 9:160–162

Gillaspie AG Jr (2001) Resistance to *Cucumber mosaic virus* in Cowpea and implications for control of cowpea stunt disease. Plant Dis 85:1004–1005

Ginarte A (2010) Evaluacion de variedades frente al insecto Tagosodes orizicolus (Muir) y el virus de la hoja blanca. I Curso de Capacitacion en mejoramiento genetico del arroz. 2006 [cited 2010]. http://agr.unne.edu.ar/fao/Cuba-ppt

Gippert R, Schmelzer K (1973) Experience with meristem tip cultures of pelargonium (*Pelargonium zonale* hybrids). Archiv fur Phytopathologie und Pflanzenechutz 9:353–362

Goheen AC (1955) Methods for determining the success of graft unions between plants in strawberry virus studies. Plant Dis Rep 39:31–32

Goheen AC (1980) Grape pathogens and prospects for controlling grape diseases. In: Proceedings of the Grape and Wine Centenary symposium, Davis, CA, pp 24–27

Goheen AC, Luhn CF, Hewitt WB (1966) Inactivation of grape viruses *in vivo*. In: Proceedings of the International conference on virus and vector on Perennial Hosts. University of California, Davis, pp 255–265

Goldbach RW, Bucher EC, Prins AH (2003) Resistance mechanisms to plant viruses: an overview. Virus Research 92:207–212

Golemboski DB, Lomonossoff GP, Zaitlin M (1990) Plants transformed with a tobacco mosaic virus nonstructural gene sequence are resistant to the virus. Proc Natl Acad Sci U S A 87:6311–6315

Golino DA, Savino V (2005) Certification and international regulation of planting material. In: Wilcox WF, Gubler WG, Uyemoto JK (eds) Compendium of grape diseases. APS Press, St. Paul

Gonsalves D (1989) Cross protection techniques for control of plant virus diseases in the tropics. Plant Dis 73:592–597

Gonsalves D (1998) Control of *Papaya ringspot virus* in papaya: a case study. Annu Rev Phytopathol 36:415–437

Gonsalves D (2002) Coat protein transgenic papaya "acquired" immunity for controlling papaya ringspot virus. Curr Top Microbiol Immunol 266:73–83

Gonsalves D (2004) Transgenic papaya in Hawaii and beyond. Agric Bio Forum 7:36–40

Gonsalves D, Garnsey SM (1989) Cross-protection techniques for control of plant virus diseases in the tropics. Plant Dis 73:592–597

Gonsalves C, Xue B, Yepes M, Fuchs M, Ling K, Namba S, Chee P, Slightom JL, Gonsalves D (1994) Transferring cucumber mosaic virus-white leaf strain coat protein gene into *Cucumis melo* L. and evaluating transgenic plants for protection against infections. J Am Soc Hortic Sci 119:345–355

Goodman RM, Nene YL (1976) Identification of virus-resistant tropically adopted soybean germplasm. Annual Report on contract AID/TA-C-1294, 1 April 1977–21 March 1978, pp 73–103

Goodman RM, Oard JH (1980) Seed transmission and yield losses in tropical soybeans infected by soybean mosaic virus. Plant Dis 64:913–914

Goregaoker SP, Eckhardt LG, Culver JN (2000) *Tobacco mosaic virus* replicase-mediated cross-protection: Contributions of RNA and protein-derived mechanisms. Virology 273:267–275

Gossele VV, Fache II, Meulewaeter F, Cornelissen M, Metzlaff M (2002) SVISS-a novel transient gene silencing system for gene function discovery and validation in tobacco. Plant J 32:859–866

Goto T, Komochi S, Oshima N (1966) Study on control of plant virus diseases by vaccination with attenuated virus. 2. Effects of concentration and time elapsed after inoculation of tomato with attenuated TMV against infection with virulent parent strain. Ann Phytopathol Soc Jpn 32:221–226

Gould B, Kramer EM (2007) Virus-induced gene silencing as a tool for functional analyses in the emerging model plant *Aquilegia* (columbine, Ranunculaceae). BMC Plant Methods 12:6

Gracia O, Boninsegna JA (1977) Use of mineral oils to control viruses on pepper. J IDIA 349/354:34–40

Grancini P (1958) I, sintomi del 'Namismo ruvido' del Mais. Maydica 3:67–79

Grant TJ (1959) Tristeza virus strains in relation to different citrus species used as test plants. Phytopathology 49:823–827

Grant TJ, Costa AS (1951) A mild strain of the tristeza virus of citrus. Phytopathology 41:114–122

Gray SM, Bergstrom GC, Vanghan R, Smith DM, Kalb DW (1996) Insecticidal control of cereal aphids and its impact on the epidemiology of the barley and yellow dwarf luteoviruses. Crop Prot 15:687–697

Gray SM, Moyar JW, Kennedy GG, Campbell CL (1986) Virus suppression and aphid resistance effects on spatial and temporal spread of watermelon mosaic virus-2. Phytopathology 76:1254–1259

Greathead A (1966) An effective program for controlling lettuce mosaic. Mimeo. Agricultural experiment series 4. California University

Greber RS (1966) *Passionfruit woodiness virus* as cause of passion vine tip blight disease. Qld J Agric Animal Sci 23:533–538

Green SK (1991) Integrated control of virus diseases of vegetables in Taiwan. In: Proceedings of the 1990 International Workshop Implementation Integrated Control of Virus Disease Important Crops, Taichung, Taiwan, pp 35–68

Green SK, Lee DR (1989) Occurrence of *Peanut stripe virus* (PStV) on soybean in Taiwan-effect on yield and screening for resistance. Trop Pest Manag 35:123–126

Green SK, Luo CY, Wu SF (1992) Elimination of leafcurl virus of sweet potato by meristem tip culture, heat and ribavirin. Plant Prot Bull Taipei 34:1–7

Greenough DR, Black LL, Bond WP (1990) Aluminum-surfaced mulch: an approach to control of tomato spotted wilt virus in solanaceous crops. Plant Dis 74:805–808

Greer L, Dole JM (2003) Aluminum foil, aluminum-painted, plastic, and degradable mulches increase yields and decrease insect-vectored viral diseases of vegetables. Hortic Tech 13:176–184

Gregorimi G, Lorenzi R (1974) Meristem-tip culture of potato plants as a method of improving productivity. Potato Res 17:24–33

Gregory PH (1948) The multiple-infection transformation. Ann Appl Biol 35:412–417

Griffiths HM, Slack SA, Dodds JH (1990) Effect of chemical and heat therapy in virus concentrations *in vitro* potato plantlets. Can J Bot 68:1515–1521

Grogan RG (1980) Control of lettuce mosaic virus with virus-free seed. Plant Dis 64:446–449

Grogan RG (1983) Control of *Lettuce mosaic virus* by use of virus indexed seed. Seed Sci Technol 11:1043–1049

Grogan RG, Welch JE, Bardin R (1952) Common lettuce mosaic and its control by the use of mosaic free seed. Phytopathology 42:573–578

Grout BWW (1990) Meristem-tip culture for propagation and virus elimination. In: Hall RD (ed) Plant cell culture protocols. Humana Press Inc, Totowa, pp 115–125

Grumet R (1994) Development of virus resistant plants via genetic engineering. Plant Breed Rev 12:47–79

Grumet R, Gifford F (1998) Plant biotechnology in the United States: issues and challenges en route to commercial production. Hortic Sci 33:187–192

Gudin C, Syratt WJ, Boize L (1976) The mechanisms of photosynthetic inhibition and the development of scorch in tomato plants treated with spray oils. Ann Appl Biol 84:213–219

Gunasinghe UB, Irwin ME, Kampmeier GE (1988) Soybean leaf pubescence affects aphid vector transmission and field spread of soybean mosaic virus. Ann Appl Biol 112:259–272

Gupta BM (1985) Development in antiviral chemotherapeutic research. Indian Phytopath 38:401–412

Gupta Y, Chowfla SC (1990) Screening of French bean germplasm for resistance to *Bean common mosaic virus*. Indian Phytopathol 43:434–436

Gururaj S, Kenchanagoudar PV, Naragund VB (2002) Identification of sources for field resistance to *Peanut bud necrosis disease* in groundnut. Karnataka J Agric Sci 15:646–648

Guta IC, Buciumeanu EC, Gheorghe RN, Teodorescu A (2010) Solutions to eliminate grapevine leafroll associated virus serotype 1 + 3 from *V. vinifera* L. cv. Ranai Magaraci, Romanian Biotechnol Lett 15:72–78

Guzman EV, de Manuel GC (1975) Improved root growth in embryo and seedling cultures of coconut "Makapuno" by the incorporation of charcoal in the growth medium. Paper Fourth Session, FAO Technical Working Party on Coconut Production, Protection and Processing, Kingston (Jamaica)

Hackland AF, Rybicki EP, Thomson JA (1994) Coat protein mediated resistance in transgenic plants. Arch Virol 139:1–22

Haddad NI, Muehlbauer FJ, Hampton RO (1978) Inheritance of resistance to *Pea seed-borne mosaic virus* in lentils. Crop Sci 18:613–615

Hadidi A (1995) Elimination of apple scar skin viroid from pears by in vitro thermotherapy and apical meristem culture. Acta Hortic 386:536–543

Hadidi A, Hansen AJ, Parish CL, Yang X (1991) Scar skin and dapple apple viroids are seed borne and persistant in infected apple trees. Res Virol 142:289–296

Hagio T, Kasiawazaki S, Hirabayashi T (1996) Genetic transformation of barley (Hordium vulgare L.) using coat protein genes of *Barley mild mosaic virus* (BaMMV) and *Barley yellow mosaic virus* (BaYMV) *in vitro*. Plant J Sivb 32:309

Hakkaart FA (1967) Effects of aluminum stripes on the spread of two aphid borne chrysanthemum viruses. Neth J Plant Pathol 73:181–185

Hakkaart FA (1968) Het tegengaan van de versprelding van non-persistente plante virussen door toepassing van oliepreparaten. Meded Dir Tuinb 31:262–268

Hakkaart FA, Quak F (1964) Effect of heat treatment of young plants on freeing chrysanthemums from virus B by means of meristem culture. Neth J Plant Pathol 70:154–157

Halk EL, Merlo DJ, Liao LW, Jarvis NP, Nelson SE, Krahn KJ, Hill KK, Rashka KE, Loesch-Fries LS (1989) Resistance to *Alfalfa mosaic virus* in transgenic tobacco and alfalfa. In: Staskawicz BJ, Ahlquist P, Yoder O (eds) Molecular biology of plant pathogen interactions. Alan R. Liss Inc, New York, pp 283–296

Hamilton RI (1980) Defenses triggered by previous invaders: viruses. In: Horsfall JG, Cowling EB (eds.) Plant disease: an advanced treatise, vol 5. Academic Press, New York, pp 279–303

Hamilton RI (1983) Certification schemes against seed-borne viruses in leguminous hosts, present status and further areas for research and development. Seed Sci Technol 11:1051–1062

Hamilton RI, Nichols C (1978) Serological methods for detection of *Pea seed-borne mosaic virus* in leaves and seeds of *Pisum sativum*. Phytopathology 68:539–543

Hampton RO (1967) Seed transmission of viruses in red clover. Phytopathology 57:98

Hampton RO (1980) Pea seedborne mosaic symptom variation among Pisum plant introduction accessions: expressions and variation in symptom expression among different genotypes of Pisum and pathological implications. Pisum News Lett 12:29–30

Hampton RO, Braverman SW (1979a) Occurrence of *Pea seed-borne mosaic virus* and new virus-immune germplasm in the plant introduction collection of *Pisum sativum*. Plant Dis Rep 63:95–99

Hampton RO, Braverman SW (1979b) Occurrence of *Pea seed-borne mosaic virus* in North American pea breeding lines and new virus-immune germplasm in the plant introduction collection of *Pisum sativum*. Plant Dis Rep 79:631–633

Hampton RE, Sill WH Jr, Hansing ED (1957) Barley stripe mosaic virus in Kansas and its control by greenhouse seed-lot testing technique. Plant Dis Rep 41:735–740

Hampton R, Ball E, De Boer S (1990) Serological methods for detection and identification of viral and bacterial plant pathogens. APS Press, St Paul 389 pp

Hanafi A, Bouharoud R, Miftah S, Amouat S (2003) Performance of two insect screens as a physical barrier against *Bemisia tabaci* and their impact on TyLCV incidence in greenhouse tomato in Souss valley of Morocco. IOBC Wprs Bull 26:39–42

Handa A, Thakur PD (1999) Virus and virus-like diseases of Apple. In: Verma LR, Sharma RC (eds) Diseases of horticultural crops fruits. Indus Publishing Company, New Delhi, pp 120–139

Handizi J, Legorburu FT (2002) Escaping from *Potato virus Y*: Aphid repellants and planting dates. In: Proceedings of the 8th international plant viruses epidemiology symposium: Aschersleben, 12–17 May, Germany

Hansen HP (1950) Investigations on virus yellows of beets in Denmark. Trans Danish Acad Technol Sci 1:1–68

Hansen AJ (1984) Effect of *Ribavirin* on green ring mottle causal agent and necrotic ringspot virus in prunus species. Plant Dis 68:216–218

Hansen AJ, Lane WD (1985) Elimination of apple chlorotic leafspot virus from apple shoot cultures by ribavirin. Plant Dis 69:134–135

Hansen JL, Shiel PS, Zenetra RS, McCarthy PL, Wyatt SD, Berger PH (1995) Expression of barley yellow dwarf or wheat streak mosaic virus coat proteins in transgenic wheat. Phytopathology 85:1146

Hanson PM, Dernacchi D, Green SD, Tanskley SD, Muniyappa V, Padmaja AS, Chen HM, Kuo George, Fang Denise, Tzu Chen Jen (2000) Mapping wild tomato introgression associated with *Tomato yellow leaf curl virus* resistance in cultivated tomato line. J Am Hortic Sci 125:15–20

Hansing ED (1943) A study of the control of the yellow-dwarf disease of potatoes. Cornell Univ Agric Exp Stn Bull 792:1–28

Hare WW, Lucas GB (1959) Control of contact transmission of *Tobacco mosaic virus* with milk. Plant Dis Rep 43:152–154

Harpaz I (1961) *Calligypona marginata*, the vector of maize rough dwarf virus. FAO Plant Prot Bull 9:144–147

Harpaz I (1982) Nonpesticidal control of vector-borne viruses. In: Harris KF, Maramorosch K (eds) Pathogens, vectors and plant diseases: approaches to control. Academic Press, New York, pp 1–21

Harris D (2004) On-farm seed priming reduces risk and increases yield in tropical crops. In: New directions for a diverse planet. Proceedings of the 4th International crop science congress, 26 Sep–1 Oct 2004, Brisbane, Australia

Harrison BD (1966) Further studies on a British form of pea early browning virus. Ann Appl Biol 57:121–129

Harrison BD (1977) Ecology and control of viruses with soil-inhabiting vectors. Annu Rev Phytopathol 15:331–360

Harrison BD, Hopper DJ (1963) Longevity of *Longidorus elongates* (de Man) and other nematodes in soil kept in polythene bags. Nematologica 9:158–160

Hartmann HT, Kester DE, Davies FT, Geneve RL (2002) Plant propagation, principles and practices, 7th edn. Prentice Hall, Upper Saddle River

Harvey TL, Seifers DL, Kofoid KD (1996) Effect of sorghum hybrid and imidacloprid seed treatment on infestations by corn leaf aphid and greenbug (Homoptera: Aphididae) and the spread of *Sugarcane mosaic virus* strain MDMV-Bl. J Agric Entomol 13:9–15

Hassall KA (1969) World crop protection. Vol. 2. Pesticides, lliffe, London, 249 pp

Hawkes R, Niday E, Gordon J (1982) A dot-immunobinding assay for monoclonal and other antibodies. Anal Biochem 119:142–147
Heathcote GD (1968) Protection of sugarbeet stecklings against aphids and viruses by cover crops and aluminum foil. Plant Pathol 17:158–161
Heathcote GD (1970) Weeds, herbicides and plant virus diseases. In: Proceedings of the 10th British weed control conference, pp 934–941
Heathcote GD (1972) Influence of cultural factors on incidence of aphids and yellows in beet. J Int Inst Sugar Beet Res 6:6–14
Heathcote GD (1973) Control of viruses spread by invertebrates to plants. In: Gibbs AJ (ed) Viruses and invertebrates'. North-Holland Publishing Company, London, pp 587–609
Heathcote GD (1978) Review of losses caused by virus yellows in English sugar beet crops and the cost of partial control with insecticide. Plant Pathol 27:12–17
Hegde V, Makeskumar T, Ganga G, Edison S (2005) Meristem culture and virus indexing for management of cassava mosaic disease. J Mycol Plant Pathol 35:482
Hein A (1964) Die wirkung eines Milchfilms auf die. Z. Pfl. Krankh. 71:267–270 (RPP 44:309)
Hein A (1965) Die Wirkung emul gierter Fette auf die Ubertragung nicht-persistenter viren durch Myzus persicae Sulz. Phytopath Z 52:29–36
Hein A (1971) Zur wirkung von oel auf die Virusuebertragung durch Blattlaeuse. Phytopath Z 49:313–324
Hein A (1972) Untersuchungen zur Wirkung von Olen bei der Virusubertragung durch Blattlause 1. Wirkung von Olen auf mechanische virusubertragung. Phytopath Z 74:126–130
Hein A (1975) Spritzungen in Nelken-Mutter-Pflanzenbestanden zum schutz for virus infektionen. Z Pflanzenkrankh 82:77–83
Heinlein M, Epel BL, Padgett HS, Beachy RN (1995) Interaction of tobamovirus movement proteins with the plant cytoskeleton. Science 270:1983–1985
Heinrichs EA, Aequino GB, Palis F (1986) Integration of host plant resistance and insecticides in the control of *Nephotettix virescence* (Homoptera: Cicadellidae), a vector of rice tungro virus. J Econ Entomol 79:437–443
Heinze K (1966) Die Vergilbungs – krankheit der Konl – und wasserrube als Krankheitsursache auf Zierpflanzen. Mitt boil Bund Aust Ld-u Forstw 121:132–139
Helliot B, Panis B, Hernandez R, Swennen R, Lepoivre P, Frison E (2004) Development of in vitro techniques for the elimination of Cucumber mosaic virus from banana (Musa spp.). In: Mohan JS, Swennen R (eds) Banana improvement. Cellular, molecular biology, and induced mutations. Sci Publishers, Inc, Enfield, pp 183–191
Hellwald KH, Palukaitis P (1995) Viral RNA as a potential target for two independent mechanisms of replicase-mediated resistance against cucumber mosaic virus. Cell 83:937–946
Hemenway C, Fan RX, Kaniewski WK, Chua NH, Tumer NE (1988) Analysis of the mechanism of protection in transgenic plants expressing the *Potato virus x* coat protein or its antisense RNA. EBO J 7:1273–1280
Henson JM, French R (1993) The polymerase chain reaction and plant disease diagnosis. Annu Rev Phytopathol 31:81–109
Henzell RG, Persley DM, Greber RS, Fletcher DS, Van Slobbe I (1982) Development of grain sorghum lines with resistance to sugarcane mosaic and other sorghum diseases. Plant Dis 66:900–901
Hermence C, Mtunda K, Muhanna M, Boniface S, Jeremiah S, Kanju E, Legg JP (2013) Managing the spread of cassava viruses through clean 'seed' systems. In: 12th International symposium on plant virus epidemiology, 28th Jan–1st Feb 2013, Arusha, Tanzania (Abstract, pp 076, p 177)
Hernandez R, Fontanella J, Noa JC, Pichardo T, Manzo R, Cardenas H (1997) Electrotherapy a novel method for eliminating viruses from garlic (*Allium sativum* L.). Hortic Argent 16:68–71
Hernandez R, Bertrand H, Lepoivre P, Gonzalez JE, Rojas X, Pairol A, Gonzalez Y, Gonzalez GY, Cortes C (2002) Diagnostico y saneaniento de *Banana Streak Virus* (BSV) en *Musa spp.* Centro Agricola 2:42–47

Herr AJ (2005) Pathways through the small RNA world of plants. Fed Eur Biol Soc Lett 579:5879–5888

Hewitt WB (1975) Graft transmission of a grapevine wood pitting and a flat trunk disease. Plant Dis Rep 59:845–848

Hewitt WB, Chiarappa L (1977) Plant Health and Quarantine in International Transfer of Genetic Resources. CRC, Cleveland, Ohio

Hewitt WB, Goheen AC, Raski DJ, Gooding GV (1962) Studies on virus diseases in the grapevine in California. Vitis 3:57–83

Hewitt WB, Goheen AC, Cory L, Luhn CF (1972) Grapevine fleck disease, latent in many varieties, is transmitted by graft inoculation. Ann Phytopathol Hors Ser:43–47

Hidaka Z (1960) Proceedings of the symposium of the soil-borne viruses, July 1960

Hileman LC, Drea S, Martino G, Litt A, Irish VF (2005) Virus induced gene silencing is an effective tool for assaying gene function in the basal eudicot species *Papaver somniferum* (opium poppy). Plant J 44:334–341

Hilje L, Stansly PA (2008) Living mulch ground covers for management of *Bemisia tabaci* (Gennadius) (Homoptera: Aleyrodidae) and *Tomato yellow mottle virus* (ToYMoV) in Costa Rica. Crop Prot 27:10–16

Hilje L, Costa HS, Stansly PA (2001) Cultural practices for managing *Bemisia tabaci* and associated viral diseases. Crop Prot 20:801–802

Hill SA (1977) Virus testing of home-saved seed potatoes. Agricultural Development and Advisory Service Agricultural Science Service Annual Report for 1976, pp 195–200

Himel CM (1969) The optimum size for insecticide spray droplets. J Econ Entomol 62:919–925

Hochmuth GJ, Kostewicz SR, William S (2000) Row covers for commercial vegetable culture in Florida, Florida Cooperative Extension Service, University of Florida, Circular 728

Hoekema A, Huisman MJ, Molendijk L, van der Elzen PJM, Cornelissen BJC (1989) The genetic engineering of two commercial potato cultivars for resistance to *Potato virus X*. Bio/ Technology 7:273–278

Hollings M (1957) Investigations of chrysanthemum viruses. 11. Virus B (mild mosaic) and chrysanthemum latent virus. Ann Appl Biol 45:589–602

Hollings M (1965) Disease control through virus free stock. Annu Rev Phytopathol 3:367–396

Hollings M, Stone OM (1979) Viruses of forage crops. Annual report, The Glasshouse Crops Research Institute, Littlehampton, UK, pp 160–16 l

Holt CA, Beachy RN (1991) In vivo complementation of infectious transcripts from mutant tobacco mosaic virus cDNAs in transgenic plants. Virology 181:109–117

Holzberg S, Brosio P, Gross C, Pogue GP (2002) Barley stripe mosaic virus-induced gene silencing in a monocot plant. Plant J 30:315–327

Honma S, Muraskishi HH, Wetter SH (1968) A *Tobacco mosaic virus* resistant greenhouse tomato. Q Bull Mich St Univ Agric Exp Stn 50:285–287

Hooks CR, Fereres A (2006) Protecting crops from non-persistently aphid-transmitted viruses: a review on the use of barrier plants as a management tool. Virus Res 120:1–16

Horak I, Schlosser E (1978) Wirkung von Prothiocarb auf *Polymyxa betae* und *Olpidium brassicae*. Med Fac Landbouww Rijksuniv Gent 43:979–987

Hormozi-Nejad MH, Mozafari J, Rakhshandehroo F (2010) Elimination of *Bean common mosaic virus* using an electrotherapy technique. J Plant Dis Prot 117:201–205

Horowitz AR, Isshaaya I (2004) Biorational insecticides-mechanisms, selectivity and importance in pest management. In: Horiwitz AR, Ishaaya I (eds) Insect pest management-field and protected crops. Springer, Berlin, pp 1–28

Hosokawa M, Ohishi K, Sugawara Y, Hayashi T, Yazawa S (2004a) Rescue of shoot apical meristems of chrysanthemum by culturing on root tips. Plant Cell Rep 22:443–448

Hosokawa M, Otake A, Ohishi K, Ueda E, Hayashi T, Yazawa S (2004b) Elimination of *Chrysanthemum stunt viroid* from an infected chrysanthemum cultivar by shoot regeneration from a leaf primordium-free shoot apical meristem dome attached to a root tip. Plant Cell Rep 22:859–863

Hou HW, Qiu W (2003) A novel co-delivery system consisting of a tomato bushy stunt virus and a defective interfering RNA for studying gene silencing. J Virol Methods 111:37–42

Howell WE, Mink GI (1971) The relationship between volunteer sugarbeets and occurrence of *Beet mosaic* and *Beet western yellow viruses* in Washington beet fields. Plant Dis Rep 55:676–678

Howell WE, Mink GI (1977) The role of weed hosts, volunteer carrots, and overlapping growing seasons in the epidemiology of *Carrot thin leaf* and *Carrot motley dwarf viruses* in central Washington. Plant Dis Rep 61:217–222

Huang SC, Millikan DF (1980) In vitro micrografting of apple shoot tips. Hortic Sci 15:741–743

Huffman JH, Sidweell RW, Khare GP, Witkowski JT, Allen LB, Robins RK (1973) In vitro effect of 1-ß-D-ribofuranosyl-1,2,4-triazole-3-carboxamide (Virazole, ICN 1229) on deoxyribonucleic acid and ribonucleic acid viruses. Antimicrob Agents Ch 3:235–241

Hughes Jd'A, Ollennu AA (1994) Mild strain protection of cocoa in Ghana against *Cocoa swollen shoot virus*—a review. Plant Pathol 43:442–457

Hull R (1952) Control of virus yellows in sugarbeet seed crops. J R Agric Soc 113:86–102

Hull R (1964) Spread of groundnut rosette virus by Aphis craccivora (Koch). Nature 202:213–214

Hull R (1965) Virus diseases of sweet peas in England. Plant Pathol 14:150–153

Hull R (1969) The spray warning scheme for the control of virus yellows. Br Sug Beet Rev 37:169–172

Hull R (1990) Non-conventional resistance to viruses in plants—concepts and risks. In: Gustafson JP (ed) Gene manipulation in plant improvement II. Plenum Press, New York, pp 289–303

Hull R (2002) Matthews' plant virology. Academic Press, London, p 1001

Hull R, Davies JW (1992) Approaches to non-conventional control of plant virus diseases. Crit Rev Plant Sci 11:17–33

Hussein MY, Samad NA (1993) Intercropping chilli with maize or brinjal to suppress populations of *Aphis gossypii* Glov., and transmission of chilli viruses. Int J Pest Manag 39:216–222

Hutton RJ, Broadbent P, Bevington K (2000) Viroid dwarfing for high density citrus planting. Hortic Rev 24:277–317

Idris AM (1990) Cotton leaf curl virus disease in the Sudan. Mededelingen van de Faculteit Landbouwwetenschappen, Rijksuniversiteit Gent, 55(2a):263–267

Igarza-Castro J, Hernandez PR, Cruz Castellanos B (2001) Electrotherapy as an alternative for elimination of the *Dasheen Mosaic Virus* in Tania. Span Manejo Integrado de plagas 60:57–60

IIPR (2006) Annual report of 2005-2006. Indian Institute Pulses Research (IIPR), Kanpur

Ingelbrecht IL, Irvine JE, Mirkov TE (1999) Post transcriptional gene silencing in transgenic sugarcane. Dissection of homology-dependent virus resistance in a monocot that has a complex polyploid genome. Plant Physiol 119:1187–1197

Inouye T (1962) Studies on barley stripe mosaic in Japan. Ber Ohara Inst Landw Biol 11:413–496

Inter-African Phyto-sanitary Commission (1962) A memorandum for phytosanitary procedures in Africa. Scientific Council for African South of the Sahara. Publication no. 82. London, CCTA

Ioannou N (1987) Cultural management of yellow leaf curl disease in Cyprus. Plant Pathol 36:367–373

Irwin ME, Goodman RM (1981) Ecology and control of *Soybean mosaic virus* in soybeans. In: Maramorosch K, Harris KF (eds) Plant diseases and vectors: ecology and epidemiology. Academic Press, New York, pp 182–215

Irwin ME, Nault LR (1996) Virus/vector control. In: GJ Persley (ed) Biotechnology and integrated pest management. CAB International, London, pp 304–322

ISTA (1966) International rules for seed testing. Proc Int Seed Test Assoc 32:1–152

Iwo GA, Olorunju PE (2009) Yield stability and resistance to leaf spot diseases and rosette in Groundnut. Czech J Genet Plant Breed 45:18–25

Jackson AO, Lane LC (1981) Hordeiviruses. In: Kurstakk E (ed) Hand book of plant viurs infections and comparative diagnosis. Elsevier/North-Holland, Amsterdam, pp 565–625

Jadot R, Vanderveken J (1973) Action de pulverizations huilenses sur la dissemination de virus de la betterave. Medad Fakulteit Landbonwwetenschappen Gent 38:1677–1683

Jaeger S (1966) Milch-und olspritzungen Zur Einsch rankung der Blattlau Suber tragung des Slatmosaik virus bei Freiland Salat. Nachr Bl Dt PflSchuz dienstes (Braunschweig) 18:82–84

Jagadish K, Krishnamurthy KB, Chandrasekara KN, Rashmi S, Akella Vani (2006) Generation and evaluation of transgenic tomato cv. Arka Vikas resistant to Peanut bud necrosis virus. In: XVI annual convention of IVS and international symposium on management of vector-borne viruses, held at ICRISAT, Patancheru, Hyderabad (India). Abstract OP 10/11

Jain RK, Varma A (2000) Biotechnological management of viral diseases of plants. In: Trivedi PC (ed) Plant diseases. Pointer Publisher, Jaipur, pp 1–20

Jain RK, Sharma Jyoti, Sivakumar AS, Sharma PK, Byadgi AS, Varma A (2004a) Variability in the coat protein gene of *Papaya ringspot virus* isolates from multiple locations in India. Arch Virol 149:2435–2442

Jain RK, Khurana SMP, Bhat AI, Chaudary Vikas (2004b) Nucleocapsid protein gene sequence studies confirm that *Potato stem necrosis disease* is caused by a strain of *Groundnut bud necrosis virus*. Indian Phytopathol 57:169–173

Jambhale ND, Nerkar YS (1986) 'Parbhani Kranti', a yellow vein mosaic resistant okra. Hortic Sci 21:1470–1471

James AP, Fowler AR, Hekmeijer S, Braithwaite KS, Whittle PJL, Smith GR (2001) Implementation of molecular assays for the routine screening of quarantined germplasm. In: Hogarth DM (ed) Proceedings of the 24th International Society of Sugarcane Technology, vol II. ASSCT, Brisbane, pp 604–606

Jameson JD, Thomas DG (1954) Groundnut rosette disease record of investigations of the Department of Agriculture, Uganda, No. 3, pp 17–20

Janeckova M (1993) Elimination of a virus complex (PPV, PNRSV, PDV) from plum varieties using a combination of in vivo and in vitro cultures. Vedecke Prace Ovocnrske 13:51–64

Jarret RL, Rodriguez W, Fernandez R (1985) Evaluation, tissue culture propagation and dissemination of 'Saba' and 'Pelipita' plantains in Costa-Rica. Sci Hortic 25:137–147

Jayasinghe U, Salazar LF (1997) Virus eradication: tissue culture of meristem, thermotherapy and chemotherapy. In: Jayasinghe U, Salazar LF (eds) Techniques in plant virology. International Potato Center, La Molina, pp 1–8

Jay Shankar Singh (2013) Plant Growth Promoting Rhizobacteria. Resonance 18:275–281

Jeeva ML, Balakrishnan S, Edison S, Rajmohan K (2004) Meristem culture and thermotherapy in the management of Sweet Potato Feathery Mottle Virus (SPFMV). J Root Crops 30:135–142

Jeger MJ, Chan MS (1995) Theoretical aspects of epidemics: uses of analytical models to make strategic management decisions. Can J Plant Pathol 17:109–114

Jeger MJ, Eden-Green S, Thresh JM, Johannson A, Waller JM, Brown AE (1995) Banana diseases. In: Gowen S (ed) Bananas and plantains. Chapman and Hall, London, pp 3117–3381

Jenkinson JG (1955) The incidence and control of cauliflower moaic in broccoli in South-west England. Ann Appl Biol 43:409–422

Johnson CG (1969) Migration and dispersal of insects by light. Methuen, London 763 pp

Johnson BV, Bing A, Smith FF (1967) Reflective surfaces used to repel dispersing aphids and reduce spread of aphid-borne cucumber mosaic virus in gladiolus plantings. J Econ Entomol 60:16–18

Johnson WC, Guzman P, Mandala D, Mkandawire ABC, Temple S, Gilbertson RL, Gepts P (1997) Molecular tagging of the bc-3 gene for introgression into Andean common bean. Crop Sci 37:248–254

Johnstone GR, Rapley PEL (1979) The effect of time of sowing on the incidence of *Subterranean clover red leaf virus* infection in broad bean (*Vicia faba*). Ann Appl Biol 91:345–351

Johnstone GR, Rapley PEL (1981) Control of *Subterranean clover red leaf virus* in broad bean crops with aphicides. Ann Appl Biol 99:135–141

Jonah PM, Bello LL, Lucky O, Midau A, Moruppa SM (2011) Review: the importance of molecular markers in plant breeding programmes. Glob J Sci Frontier Res 11(5):5–12

Jones DR (1987) Seed-borne diseases and the international transfer of plant genetic resources; an Australian perspective. Seed Sci Technol 15:765–776

Jones DR (1991) Status of banana disease in Australia. In: Valmayor RV et al (eds) Banana diseases in Asia and the Pacific. INIBAP/ASPNET, Los Banos, pp 21–37

Jones RAC (1993) Effects of cereal borders, admixture with cereals and plant density on the spread of bean yellow mosaic potyvirus into narrow-leafed lupins (Lupinus angustifolius). Ann Appl Biol 122:501–518

Jones J (2001a) Harpin. Pestic Outlook 12:134–135

Jones RAC (2001b) Developing integrated disease management strategies against nonpersistently aphid borne viruses. Integr Pest Manag Rev 6:15–46

Jones RAC (2004) Using epidemiological information to develop effective integrated virus disease management strategies. Virus Res 100:5–30

Jones RAC (2009) Plant virus emergence and evolution: origins, new encounter scenarios, factors driving emergence, effects of changing world conditions, and prospects for control. Virus Res 141:113–130

Jones LK, Burnett G (1935) Virus diseases of green house grown tomatoes. Wash State Univ Agric Exp Stn Bull 308:1–36

Jones FR, Chapman RK (1968) Aluminum foil and other reflective surfaces to manipulate the movement of aphid vectors of plant viruses. Proc North Central Branch Entom Soc Am 23:146–148

Jones RAC, Cowling WA (1995) Resistance to seed transmission cucumber mosaic in narrow-leafed lupins (*Lupinus angustifolius*). Aust J Agric Res 46:1339–1352

Jones AL, Johansen IE, Bean SJ, Bach I, Maule AJ (1998) Specificity of resistance to *Pea seed-borne mosaic potyvirus* in transgenic peas expressing the viral replicase (Nlb) gene. J Gen Virol 79:3129–3137

Jones RAC, Pearce RM, Prince RT, Coutts BA (2008) Natural resistance to Alfalfa mosaic virus in different lupin species. Aust Plant Pathol 37:112–116

Joshi RD, Gupta UP (1974) Prevention of natural spread of *Soybean mosaic virus* with some oils. Proc Natl Acad Sci (India) B 44:1–4

Joyce PA, McQaualter RB, Handley JA, Dale JL, Harding RM, Smith GR (1998) Transgenic sugarcane resistant to *Sugarcane mosaic virus*. Proc Aust Soc Sugar Cane Technol 20:204–210

Jutras PJ, Coppock GE (1958) Mechanization of citrus fruit picking. Proc Fla State Hortic Soc 71:201–204

Kabierson W (1962) Seed potato production in Western Germany. Outlook Agric 3:268–273

Kadian MS, Iiangantileke S, Arif M, Hossain M, Roder W, Sakha BM, Singh SV, Farooq K, Mazeen ACM (2007) Status of potato seed systems in South West Asia (SWA). Potato J 34:25–30

Kahn RP (1979) A concept of pest risk analysis. EPPO Bull 9:119–130

Kahn RP (1980) Ornamental plants as virus vectors: regulatory considerations. Acta Hortic 110:153–160

Kahn RP (1988) The importance of seed health in international exchange seed exchange. In: Rice seed health. IRRI, Manila, pp 7–20

Kahn RP (1989) Plant protection and quarantine: selected pests and pathogens of quarantine significance, vol II. CRC Press Inc, Boca Raton, 265 pp

Kahn RP, Monroe RL (1970) Virus infection in plant introductions collected as vegetative propagations: 1. wild vs. cultivated solanum species. FAO Plant Prot Bull 18:97–101

Kahn RP, Sowell G Jr (1970) Virus infection in plant introductions collected as vegetative propagations, 2, wild vs cultivated (Arachis) sp. FAO Plant Prot Bull 18:142–144

Kahn RP, Monroe RL, Hewitt WB, Goheen AC, Wallace JM, Roistacher CN, Nauer EM, Ackerman WL, Winters HF, Seaton CA, Pifer W (1967) Incidence of virus detection in vegetatively propagated plant introduction under quarantine in the United States, 1957–1967. Plant Dis Rep 51:715–741

Kairo MTK, Kiduyu PK, Mutinda CJM, Empig LT (1995) *Maize streak virus*: evidence for resistance against Cicadulina mbila Naude, the main vector species. Euphytica 84:109–114

Kaiser WJ, Teemba LR (1979) Use of tissue culture and thermotherapy to free East-African cassava cultivars of African cassava mosaic and cassava brown streak diseases. Plant Dis Rep 63:780–784

Kalita MK (2003) Epidemiology and management of yellow vein mosaic and leaf curl begomovirus diseases of okra. Ph.D. thesis, CCS Haryana Agricultural University, Hisar, 129 pp

Kallerhoff J, Perez P, Bouzoubaa S, Ben Tahar S, Perret J (1990) *Beet necrotic yellow vein virus* coat protein mediated protection in sugarbeet (*Beta vulgaris* L.) protoplasts. Plant Cell Rep 9:224–228

Kalleshwaraswamy CM, Krishna Kumar NK, Dinesh MR, Chandrashekar KN, Manjunatha M (2009) Evaluation of insecticides and oils on aphid vectors for the management of papaya ringspot virus (PRSV). Karnataka J Agric Sci 22(3-Spl Issue):552–553

Kalyani G, Sonali S, Reddy AS, Reddy AGS, Waliyar F, Nigam SN (2005) Resistance to *Tobacco streak virus* in groundnut, *Arachis hypogaea* L. J Oilseeds Res 22:105–107

Kalyani G, Reddy AS, Kumar PL, Rao RDVJP, Aruna R, Waliyar F, Nigam SN (2007) Sources of resistance to *Tobacco streak virus* in wild Arachis (Fabaceae: Papilionoidae) germplasm. Plant Dis 91:1585–1590

Kamala V, Prasad Rao RDVJ, Varaprasad KS (2007) Resistance to *Black eye cowpea mosaic virus* in cowpea germplasm collections from Andhra Pradesh. Indian J Plant Prot 35:144–145

Kamanna BC, Shankarappa TH, Arunkumar GS (2010) Evaluation of fungicides for the management of chrysanthemum leaf blight caused by *Aternaria alternata* (Fr.) Keissler. Plant Arch 10:595–597

Kang BC, Yeam I, Jahn MM (2005) Genetics of plant virus resistance. Annu Rev Phytopathol 43:581–621

Kaniewski WK, Lawson EC, Sammons B, Haley L, Hart J, Delannay X, Tumer NE (1990) Field resistance of transgenic Russet Burbank potato to effect of infection by *Potato virus X* and *Potato virus Y*. Bio/Technology 8:750–754

Karchi Z, Cohen S, Govers A (1975) Inheritance of resistance to *Cucumber mosaic virus* in melons. Phytopathology 65:479–481

Kartha KK (1981) Tissue culture technique for virus elimination and germplasm preservation. Genetic Engineering Crop Improvement, Rock feller Foundation Conference, 1980, pp 123–141

Kartha KK, Gamborg OL, Constabel F, Shyluk JP (1974) Regeneration of Cassava plants from apical meristems. Plant Sci Lett 2:107–113

Kassanis B (1957) The use of tissue cultures to produce virus-free clones from infected potato varieties. Ann Appl Biol 45:422–427

Kassanis B, Govier DA (1971) New evidence on the mechanism of aphid transmission of *Potato C* and *Potato aucuba mosaic viruses*. J Gen Virol 10:99–101

Kassanis B, Posnette AF (1961) Thermotherapy of virus infected plants. Recent Adv Bot 1:557–563

Kassanis B, Slykhuis JT (1959) Some properties of *Barley stripe mosaic virus*. Ann Appl Biol 47:254–263

Kassanis B, Tinsley TW (1958) The freeing of tobacco tissue cultures from potato virus Y by 2-thiouracil. In: Proceedings of the 3rd conference on potato virus disease, Lisse-Wageningen, 1957, pp 153–155

Kawazu Y, Ryoi F, Yuji N (2009) Transgenic resistance to Mirafiori lettuce virus in lettuce carrying inverted repeats of the viral coat protein gene. Transgenic Res 18:113–120

Kawchuk LM, Martin RR, McPherson J (1990) Resistance in transgenic potato expressing the *Potato leafroll virus* coat protein gene. Mol Plant-Microbe Interact 3:301–307

Kawchuk LM, Martin RR, McPherson J (1991) Sense and antisense RNA-mediated resistance to *Potato leafroll virus* in Russet Burbank potato plants. Mol Plant-Microbe Interact 4:247–253

Keener PD (1956) Virus diseases of plants in Arizona. II. Field and experimental observations on curly top affecting vegetable crops. Bull Ariz Agric Exp Stn 271:1–28

Kelly JD (1997) A review of varietal response to bean common mosaic potyvirus in *Phaseolus vulgaris*. CAB Abstracts AN. 971605689

Kemper A (1962) Zum anfttreten Von Salat mosaik virus an salt (*Lactuca sativa* L.) und an Krenz krant (*Senecio vulgaris* L.). Z pfl kranckh pfl schutz 69:653–663

Kenganal MY, Amudai JA, Patil FS, Kulikarni UG (2008) Feasibility of meristem culture for management of Banana streak virus (BSV) through micropropagation. Res Crop 9:605–609

Kennedy JS (1976) Host plant resistance and the spread of plant viruses. Environ Entomol 5:827–832

Kennedy GG (2003) Tomato, pests, parasitoids, and predators: tritrophic interactions involving the genus *Lycopersicon*. Annu Rev Entomol 48:51–72

Kennedy JS, Booth CO, Kershaw WJS (1961) Host finding by aphids in the field. III. Visual attraction. Ann Appl Biol 49:1–21

Khan RP (1991) Exclusion as a disease control strategy. Annu Rev Phytopathol 29:219–246

Khan MA, Mukhopadhyay S (1985) Effect of different pesticide combinations on the incidence of yellow vein mosaic virus disease of okra (*Abelmoschus esculentus*) and its whitefly vector Bemisia tabaci Genn. Indian J Virol 1:147–151

Khanam D, Khatun MM, Faisal SM, Hoque MA (1998) *In vitro* propagation of strawberry through meristem tip culture. Plant Tissue Cult 8:35–39

Khatri HL, Sekhon IS (1973) Effect of oil spray on aphid transmission of *Chilli-mosaic virus*. Indian J Agric Sci 43:667–669

Khetarpal RK (2002) Biosafety of genetically modified crops and its implications. In: Chadha KL, Choudhary ML, Prasad KV (eds) Hitech horticulture. Horticulture Society of India, New Delhi, pp 444–453

Khetarpal RK (2004) A critical appraisal of seed health certification and transboundary movement of seeds under WTO regime. Indian Phytopathol 57:408–427

Khetarpal RK, Gupta K (2002) Implications of sanitary and phytosanitary agreement of WTO on plant protection in India. Annu Rev Plant Pathol I:1–26

Khetarpal RK, Gupta K (2007) Plant biosecurity in India-Preparedness. Indian J Plant Prot 35:168–178

Khetarpal RK, Nath R (1998) Evolving national plant quarantine system: harmonization of current quarantine regulations with sanitary and phytosanitary (SPS) agreement. In: National dialogue (ISPGR dialogue 1998). Losses on plant Genetic Resources, 1–2 Dec 1998. National Bureau of plant Genetic Resources, New Delhi, 47 pp

Khetarpal RK, Maury Y, Cousin R, Burghoufer A, Varma A (1990) Studies on resistance of pea to *Pea seed-borne mosaic virus* and new pathotypes. Ann Appl Biol 116:297–304

Khetarpal RK, Maisonneuve B, Maury Y, Chaloub B, Dinant S, Lecoq H, Varma A (1998) Breeding for resistance to plant viruses. In: Hadidi A, Khetarpal RK, Koganezawa H (eds) Plant virus disease control. APS Press, St. Paul, pp 14–32

Khetarpal RK, Singh Shamsher, Parakh DB, Maurya AK, Chalam VC (2001) Viruses intercepted in exotic germplasm during 1991–2000 in quarantine. Indian J Pl Genet Resour 14:127–129

Khetarpal RK, Chalam VC, Dhillon BS (2002) Diagnostics and development of seed certification protocols for management of seed-transmitted viral diseases of grain legumes Leaflet on activities and achievements of the project. NBPGR, New Delhi, 4 pp

Khetarpal RK, Chalam VC, Prakash HS, Shetty HS, Mishra A (2003) Diagnostics and development of seed certification protocols for management of seed-transmitted viral diseases of grain legumes. In: Singh A, Trivedi TP, Sharma HR, Sharma OP, Sabir N (eds) Recent advances in integrated pest management (Proceedings and recommendations of the NATP interactive workshop on Integrated Pest Management, 26–28 Feb 2003). National Centre for Integrated Pest Management, New Delhi, pp 220–223

Khetarpal RK, Nath R, Raghunathan V (2004) Safe movement of plant germplasm. In: Dhillon BS, Tyagi RK, Lal A, Saxena S (eds) Plant genetic management. Narosa Publishing House, New Delhi, pp 172–178

Khurana SMP (2004) Potato viruses and their management. In: Naqvi SAMH (ed) Diseases of fruits and vegetables, vol II. Kluwer Academic Publishers, Dordrecht, pp 389–440

Khurana SMP, Garg ID (1993) New techniques for detection of viruses and viroids. In: Chadha KL, Grewal JS (eds) Advances in horticulture, vol 7-Potato. Malhotra Publishing House, New Delhi. pp 529–566

Khurana SMP, Pandey SK, Srivastava KK (2000) Tuber treatment with imidacloprid is effective for control of *Potato stem necrosis disease*. Indian Phytopathol 53:142–145

Kimble KA, Grogan RG, Greathead AS, Paulus AO, House JK (1975) Development, application and comparison of methods for indexing lettuce seed for mosaic virus in California. Plant Dis Rep 59:461–464

Kiranmai G, Sreenivasulu P, Nayudu MV (1996) Comparison of three different tests for detection of *Cucumber mosaic cucumovirus* (CMV) in banana (*Musa paradisiaca*). Curr Sci 71:764–767

Kiritani K, Su HJ (1999) Papaya ringspot, banana bunchy top and citrus greening in the Asia and Pacific region: occurrence and control strategy. JARQ 33:23–30

Kitaba GN (1979) Preliminary attempts to control the insect transmitted *Watermelon mosaic virus* (WMV-K) on baby marrows. Kenya Entomologists Newsl 9:8–9

Klesser PJ, Le Roux PM (1957) Groundnut rosette prevention the best method of control. Fmg S Afr 33:40–41

Klotz LJ, Calavan EC, Weathers LG (1972) Virus and virus-like diseases of citrus. Circular No. 559, Division of Agricultural Science, University of California, 42 p

Knapp E, Hanzer V, Weiss HH, da Camara Machado A, Weiss B, Wang Q, Katinger H, da Camara Machado ML (1995) New aspects of virus elimination in fruit trees. Acta Hortic 386:409–418

Knorr LC (1965) Serious diseases of citrus foreign to Florida. Bull Fla Dep Agric 5:1–59

Knutson KW, Bishop GW (1964) Potato leaf roll virus-effect of date of inoculation on percentage of infection and symptoms expression. Am Potato J 41:227–238

Koehler B, Bever WM, Bennett OT (1952) Soil-borne wheat mosaic. Bull Ill Agric Exp Stn 556:567–599

Koike H, Tippett RL (1972) Roguing mosaic diseased ratoon seed plots with herbicides in Louisiana. Sugarcane Path Newsl 8:20

Kolbanova EV, Kukharchik NV, Barai VN, Zinchenko AI (2004) Production of black currant free from raspberry ringspot virus using virazole in in vitro culture. Vestsi Natsyyanal nai Akademii Navuk Belarusi Seryya Biyalagichnykh Navuk 2:90–93

Komar V, Vigne V, Demangeat G, Lemaire O, Fuchs M (2008) Cross protection as control strategy against *Grapevine fan leaf virus* in naturally infected vineyards. Plant Dis 92:1689–1694

Konate G, Fargette D (2001) *Rice yellow mottle virus*: an overview. Plant virology in sub-Saharan Africa, 4–8 June, Ibadan, Nigeria, p 2

Konate G, Barro N, Fargette D, Swanson MM, Harrison BD (1995) Occurrence of whitefly transmitted geminiviruses in crops in Burkina Faso and their serological detection and differentiation. Ann Appl Biol 126:121–129

Kooner BS, Cheema HK (2007) Screening of mungbean germplasm against whitefly (Bemisia tabaci Genn.) and *Mungbean yellow mosaic virus*. Acta Hortic 752:307–309

Kosaka Y, Fukunishi T (1993) Attenuated isolates of *Soybean mosaic virus* derived at a low temperature. Plant Dis 77:882–886

Kosaka Y, Fukunishi T (1994) Application of cross-protection to the control of black soybean mosaic disease. Plant Dis 78:339–341

Kosaka Y, Ryang B, Kobori T, Shiomi H, Yasuhara H, Kataoka M (2006) Effectiveness of an attenuated *Zucchini yellow mosaic virus* isolate for cross-protecting cucumber. Plant Dis 90:67–72

Koshimizu Y, Iizuka N (1963) Studies on soybean virus diseases in Japan. Bull Tohoku Natl Agric Expt Stn 27:1–103

Kostiw M, Iskrzycka T (1976) Mozliwosc organiczenia szerzenia sie nietrwalych wirusow Zimniaka przy pomocy opryskiwania substancjami obejowymi. Biueletyn Instytutu Ziemniaka 18:59–64

Kouassi NK, Chen L, Sire C, Bangratz-Reyser M, Beachy RN, Fauquet CM, Brugidou C (2006) Expression of *Rice yellow mottle virus* coat protein enhances virus infection in transgenic plants. Arch Virol 151:2111–2122

Koubouris GC, Maliogka VI, Efthimiou K, Katis NI, Vasilakakis MD (2007) Elimination of Plum pox virus through in vitro thermotherapy and shoot tip culture compared to conventional heat treatment in apricot cultivar Bebecou. J Gen Plant Pathol 73:370–373

Kousalya G, Ayyavoo R, Krishnamurthy CS, Bhaskaran S (1971) Effect of spacing, roguing and weeding on the incidence of rosette disease of groundnut with observation on the aphid vector, Aphis craccivora Koch. Madras Agric J 58:495–505

Kraft JM, Coffman VA (2000) Registration of 97–261 and 97–2154 pea germplasm lines. Crop Sci 40:302

Kramer W, Sass O, Fritzsche R, Proesler G (1978) Mineralol-insektizid-Kombinationen Zur Eioschrankung der virus ausbreitung. bei der production Von Beta Ruben-Saatgut. Nachirichtenblatt fur den Pflanzen schutz in der DDR 32:29–30

Krczal G (1998) Virus certification of ornamental plants-the European strategy. In: Plant virus disease control. APS Press—The American Phytopathological Society, St Paul, pp 277–287

Kring JB (1964) New ways to repel aphids. Frontiers Sci Conn Agric Exp Stn 17:6–7

Krishnakumar NK, Eswara Reddy SG (2006) Incidence and extent of yield loss due to Tomato leafcurl gemivirus (ToLCV) in tomato grown under protected and open field cultivation. Abstract PP 9/16. In: XVI Annual convention of IVS and International symposium on management of vector-borne viruses, held at ICRISAT, Patancheru, Hyderabad (India)

Krishnakumar NK, Venkatesh N, Kalleshwaraswamy CM, Ranganath HR (2006a) Seasonal incidence of thrips and bud necrosis virus on watermelon. Pest Manag Hortic Eco-Syst 12:85–92

Krishnakumar NK, Ranganath HR, Venkatesh N, Kalleshwaraswamy CM (2006b) Thrips vectors of watermelon bud necrosis virus (tospovirus) and their management. Abstract, OP 9/11. In: XVI Annual convention of IVS and International symposium on management of vector-borne viruses, held at ICRISAT, Patancheru, Hyderabad (India)

Krishnareddy M (1989) Studies on yellow mosaic and leaf crinkle diseases of blackgram. Ph.D. Thesis submitted to the P.G. School, IARI, New Delhi, India, 263 pp

Kuhn CW, Demski JW (1975) The relationship of *Peanut mottle virus* to peanut production. Res Rep Georgia Agric Exp Stn 213:1–18

Kuhn CW, Sowell G Jr, Chakley JH, Stubbs HF (1968) Screening for immunity. Plant Dis Rep 56:467–468

Kulps G (1968) Untersuchungen zur Schutzwirkung von Olen bei der Virusubertragung durch Blattlause. Z Pflkrankh Pfl Schutz 75:213–217

Kulps G (1969) Untersuchungen zum Verhalten von Mineralolen im Hinblick aufihre Wirking bei der Virussu – bertragung durch Blattlanse. Universitat Hohenneim, Dissertation, 85 pp

Kulps G (1971) Untersuchungen zum Verhalten von Mineralolen im Hinblick auf ihre Wirkung bei der Virusubertragung durch Blattlanse. I. Verhalten von Mineralol auf. Blattoberflachen Z Pflkrankh Pfl Schutz 78:126–129

Kulps G (1972) Untersuchungen zum verhalten von Mineralolen im Hinblick auf ihre Wirkung bei der Virusubertragung durch Blattlanse III. Verteilung und Wirkungs daner von Mineralol in der Pflanze. Phytopathol Z 73:263–276

Kulps G, Hein A (1972) Untersuchungen zum Verhalten von Mineralolen im Hinblick auf ihre Wirking bei der Virus ubertragung durch Blattlanse II. VErhalten von Mineralol in der Epidermis. Phytopathol Z 73:149–162

Kumagai MH, Donson J, Della-Cioppa G, Harvey D, Hanley K, Grill LK (1995) Cytoplasmic inhibition of carotenoid biosynthesis with virus-derived RNA. Proc Natl Acad Sci U S A 92:1679–1683

Kumar P, Singh DK (2013) Evaluation of okra germplasm against yellow vein mosaic virus during rainy season under Tarai region. Pantnagar Journal Res 11:112–117

Kumar NKK, Ullman DE, Cho JJ (1993) Evaluation of *Lycopersicon* germ plasm for *Tomato spotted wilt tospovirus* resistance by mechanical and thrips transmission. Plant Dis 77:938–941

Kumar S, Khan MS, Raj SK, Sharma AK (2009) Elimination of mixed infection of *Cucumber mosaic* and *Tomato aspermy virus* from Chrysanthemum morifolium Ramaty. cv. Pooja by shoot meristem culture. Scientia Horticulturae 119:108–112

Kumar CA, Khetarpal RK, Parakh DB, Singh S, Nath R (1994) Check-list on seed-transmitted viruses: Leguminous hosts. Technical Bulletin, NBPGR, New Delhi

Kundu P, Mandal RK (2001) Transgenic approaches for producing virus resistant plants. Proc Indian Natl Sci Acad 67:53–80

Kunik T, Salmon R, Zamir D, Navot N, Zeidan M, Michelson I, Gafni Y, Czosnek H (1994) Transgenic tomato plants expressing the *Tomato yellow leaf curl virus* capsid protein are resistant to the virus. Biotechnology (N Y) 12:500–504

Kunkel LO (1934) Tobacco and aucuba-mosaic infections by single units of virus. Phytopathology 24:13

Kuo FY, Wu SW (1989) The problem of crop succession and management of disease and pest on the vegetable facility culture. In: Sheen TF, Hsu MM (eds.) Proceedings of second congress on horticultural production under structure, 23–24 Feb, TARI, Taichung. FTHES, pp 172–191, 252 pp (in Chinese)

Kvist K, Morner J (1977) Oil sprays in seed potato fields—a method to control spread of virus? Växtskyddsnotiser 41:123–124

Kwon SH, Oh JH (1980) Resistance to a nenotic strain of Soybean mosaic virus in soybean. Crop Sci 20:403–404

Ladipo JL, Allen DJ (1979) Identification of resistance to *Cowpea aphid-borne mosaic virus*. Trop Agric (Trinidad) 56:353–360

Laimer M, Hanzer V, Kriston V, Toth EK, Mendonca D, Kirilla Z, Balla I (2006) Elimination and detection of pathogens from tissue cultures of Prunus sp. Acta Hort 725:319–323

Laird EFJ, Dickson RC (1972) Turnip mosaic virus vector relationships in field grown statice *Limonium-Perezii*. Plant Dis Rep 56:722–725

Lamberti F (1981a) Combating nematode vectors of plant viruses. Plant Dis 65:113–117

Lamberti F (1981b) Plant nematode problems in the Mediterranean region. Helminthol Abstr Ser B Plant Nematol 50:145–166

Lana AF (1976) Mosaic virus and leaf curl diseases of okra in Nigeria. PANS 22:474–478

Lange E, Hammi M (1977) Oil spray trials against virus diseases of Capsicum transmitted by aphids in Tunisia. Phytopath Medit 16:18–21

Lapidot M (2002) Screening common bean (*Phaseolus vulgaris*) for resistance to *Tomato yellow leafcurl virus*. Plant Dis 86:429–432

Lapidot M, Friedmann M (2002) Breeding for resistance to whitefly transmitted geminiviruses. Ann Appl Biol 140:109–127

Lapidot M, Gafny R, Ding B, Wolf S, Lucas WJ, Beachy RN (1993) A dysfunctional movement protein of tobacco mosaic virus that partially modifies the plasmodesmata and limits virus spread in transgenic plants. Plant J 2:959–970

Lapidot M, Friedmann M, Lachman O, Yehezkel A, Nahon S, Cohen S, Pilowsky M (1997) Comparison of resistance level to *Tomato yellow leafcurl virus* among commercial cultivars and breeding lines. Plant Dis 81:1425–1428

Lapidot M, Ben-Joseph R, Cohen L, Machbash Z, Levy D (2006) Development of a scale for evaluation of *Tomato yellow leafcurl virus* resistance level in tomato plants. Phytopathology 96:1404–1408

Larentzaki E, Plate J, Nault BA, Shelton AM (2008) Impact of straw mulch on populations of onion thrips (Thysanoptera: Thripidae) in onion. J Econ Entomol 101: 1317–1324

Larew HG (1988) Limited occurrence of foliar-, root-, and seed-applied neem seed extract toxin in untreated plant parts. J Econ Entomol 81:593–598

Latham LJ, Jones RAC (1998) Selection of resistance breaking strains of *Tomato spotted wilt tospovirus*. Ann Appl Biol 133:385–402

Latham LJ, Jones RAC (2001a) Incidence of virus infection in experimental plots, commercial crops and seed stocks of cool season crop legumes. Aust J Agric Res 52:397–413

Latham LJ, Jones RAC (2001b) Alfalfa mosaic and pea seed-borne mosaic viruses in cool season crop, annual pasture and forage legumes: Susceptibility, sensitivity and seed transmission. Aust J Agric Res 52:771–790

Latham LJ, Jones RAC (2004) Deploying partially resistant genotypes and plastic mulch on the soil surface to suppress spread of lettuce big-vein disease in lettuce. Australian Journal of Agricultural Research 55:131–138

Lawson C, Kaniewski W, Haley L, Rozman R, Newell C, Sanders P, Tumer NE (1990) Engineering resistance to mixed virus infection in a commercial potato cultivar. Resistance to *Potato virus X* and *Potato virus Y* in transgenic Russet Burbank. Bio/Technology 8:127–134

Lazar J, Mikulas J, Farkas G, Kolber M (2002) Certification programme for production of virus-free propagation material of grapevine and its results in Hungary. Int J Hortic Sci 8:39–43

Lecoq H (1998) Control of plant virus diseases by cross protection. In: Hadidi A, Khetarpal RK, Koganezawa H (eds) Plant virus disease control. APS Press, St. Paul, pp 33–41

Lecoq H, Pitrat M (1983) Field experiments on the integrated control of aphid-borne viruses in muskmelon. In: Plumb RJ, Thresh JM (eds) Plant virus epidemiology—the spread and control of insect-borne viruses. Blackwell Scientific publication, London, pp 169–183

Lecoq H, Cohen S, Pitrat M, Labonne G (1979) Resistance to *Cucumber mosaic virus* transmission by aphids in *Cucumis melo*. Phytopathology 69:1223–1225

Lecoq H, Lemaire JM, Wipf-Scheibel C (1991) Control of zucchini yellow mosaic virus in squash by cross protection. Plant Dis 75:208–211

Lee RF (2004) Certification programs for citrus. In: Diseases of fruits and vegetables: diagnosis and management, vol I. Kluwer Academic Publishers, Dordrecht, pp 247–290

Lee HY, Lee V (1973) Plant protection. Agric Hong Kong 1:161–164

Lee RF, Lehman PS, Navarro L (1999) Nursery practices and certification program for budwood and rootstocks. In: Citrus health management. APS Press, St. Paul, pp 35–45

Lee YH, Jung M, Shin SH, Lee JH, Choi SH, Her NH, Lee JH, Ryu KH, Pack KY, Ham CH (2009) Transgenic peppers that are highly tolerant to a new CMV pathotype. Springer. Genetic transformation and Hybridization. Plant Cell Rep 28:223–232

Legg JT (1979) The campaign to control the spread of *Cocoa swollen shoot virus* in Ghana. In: Ebbels DL, King JE (eds) Plant health. Black well Scientific Publications, London, pp 285–293

Legin R, Vuittenez A (1973) Comparison des symptomes et transmission par greffage d'une mosaique nervaire de Vitis vinifera, de la marbrure de V. rupestris et d'une affection necrotique des nervures de l'hybride Rupestris Berlandieri 110R. Rivista di Patologia Vegetale 9(Suppl):57–63

Lehmann P, Jenner CE, Kozubek E, Greenland AJ, Walsh JA (2003) Coat protein-mediated resistance to *Turnip mosaic virus* in oilseed rape (*Brassica napus*). Mol Breeding 11:83–94

Lenardon SL, Bazzalo ME, Abratti G, Cimino C, Galella MT, Grondona M, Giolitti F, Leon AJ (2005) Screening sunflower for resistance to *Sunflower chlorotic mottle virus* and mapping the *Rcmo-1* resistance gene. Crop Sci 45:735–739

Leppik EE (1964) Some epiphytotic aspects of squash mosaic. Plant Dis Rep 48:41–42

Leu LS (1978) Apical meristem culture and re-differentiation of callus masses to free some sugarcane systemic diseases. Plant Prot Bull Taiwan 20:77–82

Lewandowski DJ, Hayes AJ, Adkins S (2010) Surprising results from a search for effective disinfectants for Tobacco mosaic virus-contaminated tools. Plant Dis 94:542–550

Li YW, Xu C, Chen JS (2002) Establishment of virus-free taro (*Colocasia esculenta* cv Fenghuayunaitou) by meristemtip culture combined with thermotherapy. Pak J Plant Pathol 1:40–43

Lim M (1985) Resistance to *Soybean mosaic virus* in soybeans. Phytopathology 75:199–201

Linford MB (1943) Influence of plant populations upon incidence of pineapple yellow spot. Phytopathology 33:408–410

Ling KC (1972) Rice virus diseases. International Rice Research Institute, Los Banos

Ling KS, Zhu HY, Drong RF, Slighton JL, Gonsalves D (1997) Nucleotide sequencing and genome organisation of grapevine leafroll associated closterovirus 3 and development of transgenic plants expressing its coat protein and other genes. In: Extended abstracts 12th meeting ICVG, Lisbon, Portugal, 28 Sept–2 Oct, p 18

Ling KS, Zhu HY, Gonsalves D (2009) Resistance to Grapevine leafroll associated virus-2 is conferred by post-transcriptional gene silencing in transgenic *Nicotiana benthamiana*. Transgenic Res 17:733–740

Linn MB (1940) The yellows disease of lettuce and endive. Cornell Univ Agric Exp Stn Bull 742:1–33

Lius S, Manshardt M, Fitch MMM, Slightom JL, Sanford JC, Gonsalves D (1997) Pathogen-derived resistance provides papaya with effective protection against *Papaya ringspot virus*. Mol Breeding 3:161–168

Loebenstein G, Alper M, Dentsch M (1964) Preventing aphid spread *Cucumber mosaic virus* with oils. Phytopathology 54:960–962

Loebenstein G, Alper M, Levy S (1970) Field tests with oil sprays for the prevention of aphid-spread viruses in peppers. Phytopathology 60:212–215

Loebenstein G, Deutsch M, Frankel H, Sabar Z (1966) Field tests with oil sprays for the prevention of *Cucumber mosaic virus* in cucumbers. Phytopathology 56:512–516

Loebenstein G, Alper M, Levy S, Palevitch D, Menagem E (1975) Protecting peppers from aphid-borne viruses with aluminium foil or plastic mulch. Phytoparasitica 3:43–53

Loebenstein G, Berger PH, Brunt AA, Lawson RH (2001) Virus and virus-like diseases of potatoes and production of seed potatoes. Kluwer Academic Publisher, Dortrecht

Loesch-Fries LS, Merlo D, Zinnen T, Burhop L, Hill K, Krahn K, Jarvis N, Nelson S, Halk E (1987) Expression of *Alfalfa mosaic virus* RNA 4 in transgenic plants confers virus resistance. EMBO J 7:1845–1851

Loevinsohn ME (1984) The ecology and control of rice pests in relation to the intensity and synchrony of cultivation. Ph.D. thesis, University of London, London, UK

Lokko Y, Danquah EY, Offei SK, Dixon AGO, Gedil MA (2005) Molecular markers associated with a new source of resistance to the *Cassava mosaic disease*. Afr J Biotechnol 4:873–881

Lomonossoff GP (1995) Pathogen-derived resistance to plant viruses. Annu Rev Phytopathol 33:323–343

Long RD, Cassells AC (1986) Elimination of viruses from tissue cultures in the presence of antiviral chemicals. In: Withers LA, Alderson PG (eds) Plant tissue culture and its agricultural applications. Butterworths, London, pp 239–248

Long JK, Fraser LR, Cox JE (1972) Possible value of close-planted, virus dwarfed orange trees. In: Price WC (ed) Proceedings of the 5th conference on international organization of citrus virologist. University of Florida Press, Gainesville, pp 262–267

Louie R (1968) Epidemiology of onion yellow dwarf. Ph.D. thesis, Cronell University, Ithaca, NY, 112 pp

Louie R (1980) *Sugarcane mosaic virus* in Kenya. Plant Dis 64:944–947

Louis HE (2000) Genetically modified plants in developing countries. Plant Physiol 124:923–926

Louws FJ, Rademaker JLW, Brujin FJ (1999) The three Ds of PCR-based genomic analysis of phytobacteria: diversity, detection, and disease diagnosis. Annu Rev Phytopathol 37:81–125

Love SL, Rhodes BB, Moyer JW (1989) Meristem-tip culture and virus indexing of sweet potatoes, 2nd edn. International Board for Plant Genetic Resources, Rome

Lowery DT, Eastwell KC, Smirle MJ (1997) Neem seed oil inhibits aphid transmission of *Potato virus Y* to pepper. Ann Appl Biol 130:217–225

Lozoya-Saldana HF, Abello J, Garcia G (1996) Electrotherapy and shoot-tip culture eliminate *Potato Virus X* in potatoes. Am J Potato Res 73:149–154

Lu J (1970) Essai in vivo de repression de la mosaique due soja par l'huile. Phytoprotection 51:149

Lutman PJW (1979) The control of volunteer potatoes in cereal stubbles. I. Factors affecting potato regrowth. Ann Appl Biol 93:41–47

Luvisi A, Panattoni A, Triolo E (2012) Eradication trials of Tobacco mosaic virus using chemical drugs. Acta Virol 56:155–158

Ma SS, Shii CT (1974) Growing banana plantlets from adventitious buds. J Chin Soc Hortic Sci 20:6–12

Ma SS, Shii CT, Wang SO (1978) Regeneration of banana plants from shoot meristem tips and inflorescence sections in vitro. In: Abstract of the 20th international horticulture congress, Sydney, Australia, 15–23 Aug 1978. Abstract No. 1639

MacFarlane SA, Davies JW (1992) Plants transformed with a region of the 201-kilodalton replicase gene from pea early browning virus RNA 1 are resistant to virus infection. Proc Natl Acad Sci U S A 89:5829–5833

MacKenzie DJ, Ellis PJ (1992) Resistance to *Tomato spotted wilt virus* infection in transgenic tobacco expressing the viral nucleocapsid gene. Mol Plant-Microbe Interact 5:34–40

MacKenzie DJ, Tremaine JH (1990) Transgenic *Nicotiana debneyi* expressing viral coat protein are resistant to *Potato virus S* infection. J Gen Virol 71:2167–2170

Mackie AE, McKirdy SJ, Rodoni B, Kumar S (2002) *Potato spindle tuber viroid* eradicated in Western Australia. Australasian Plant Pathology 31:311–312

Maddock DA (1997) Implications of new technologies for seed health testing and the worldwide movement of seed. In: Symposium on seed biology and technology: applications and advances, 13–16 August 1997, Fort Collins, Colorado. Seed Sci Res (1998) 8(2):227–284

Madhavi Reddy K, Krishna Reddy M, Kumar R, Rai AB, Mathura R, Singh HB (2010) Breeding for virus resistance. In: Kumar R, Rai AB, Mathura R, Singh AB (eds) Advances in chilli research. Studium Press, New Delhi, pp 119–132

Maelzer DA (1986) Integrated control of vectors of plant virus diseases. In: Plant virus epidemics: monitoring, modelling and predicting out breaks. Academic Press, Sydney, p 550

Magalhaes LC, Hunt TE, Siegfried BD (2009) Efficacy of neonicotinoid seed treatments to reduce soybean aphid populations under field and controlled conditions in Nebraska. J Econ Entomol 102:187–195

Magee CJ (1948) Transmission of bunchy top to banana cultivars. J Aust Inst Agric Sci 14:18–24

Magee CJP (1967) Report to the Government of Western Samoa. The control of banana bunchy top. Soth Pacific Commission, Technical paper no 150, 13 pp

Mahbub A, Zahoor A, Talib H, Muhammad T (1992) *Cotton leaf curl virus* in the Punjab 1991–1992. Hoechst Pakistan Limited

Mahmoud YM, Hosseny MH, Abdel-Ghaf MH (2009) Evaluation of some therapies to eliminate *Potato Y potyvirus* from potato plants. Int J Virol 5:64–76

Mahungu NM, Dixon AGO, Kumbira JM (1994) Breeding for multiple pest resistance to Africa. Afr Crop Sci J 2:539–552

Makkouk KM, Kumari SG (2009) Epidemiology and integrated management of persistently transmitted aphid borne virus of legume and cereal crops in West Asia and North America. Virus Res 141:209–218

Malathi VG, Thankappan M, Nair NG, Nambisan B, Ghosh SP (1987) Cassava mosaic disease in India. In: Proceedings of the international seminar on *African Cassava Mosaic Disease* and its control. Yamoussoukro, Cote d'Ivoire 4–8 May. CTA/FAO/ORSTOM/IITA/IAPC, pp 189–198

Malaurie B (1998) *In vitro* storage and safe international exchange of yam (*Dioscorea* spp.) germplasm. http://ejb.org/content/voll.issue3:full.2.bip.index.asp

Mali VR (1986) Identification of source of combined resistance to some viruses and seed transmission in cowpea. 2nd annual convention of Indian Virological Society, New Delhi, 3–5 Oct 1986

Mali VR, Kulthe KS, Patil FS, Mundhe GE, Dhond VM, Deshmukh RV (1987) Identification of source of resistance to some viruses and seed transmission in cowpea. Indian J Virol 3:99–109

Mandahar CL (1999) Molecular biology of transgenic plants. In: Mandahar CL (ed) Molecular biology of plant viruses. Kluwer Academic Publishers, Boston, pp 241–269

Mannan MA (2003) Some aspects of integrated management of potato aphid, *Myzus persicae* (Sulz.) (Homoptera: Aphididae). Thai J Agric Sci 36:97–103

Mannerlof M, Lennerfors BL, Tenning P (1996) Reduced titre of BNYVV in transgenic sugar beets expressing the BNYVV coat protein. Euphytica 90:293–299

Mansour A, Akkawi M, Al-Musa A (2000) A modification of aluminum foil technique for controlling *Aphid borne mosaic diseases* of Squash. Dirasat Agric Sci 27:1–9

Mantell SH (1980) Apical meristem-tip culture for eradication of flexuous rod viruses in yams (*Dioscorea alata*). Trop Pest Manag 26:170–179

Mantell SH, Haque SQ, Whitehall AP (1978) Clonal multiplication of *Dioscorea alata* L. and *Dioscorea rotundata* Poir. yams by tissue culture. J Hortic Sci 53:95–98

Mantell SH, Haque SQ, Whitehall AP (1979) A rapid propagation system for yams. Yam Virus Project Bull. No. 1. CARDI. 19 pp

Marco S (1993) Incidence of nonpersistently transmitted viruses in pepper sprayed with whitewash, oil, and insecticide, alone or combined. Plant Dis 77:1119–1122

Marenaud G, Dunez J, Bernhard R (1976) Identification and comparison of different strains of *Apple chlorotic leaf spot virus* and possibilities of cross protection. Acta Hortic 67:219–226

Mariappan V, Saxena RC (1983) Effect of custard apple oil on survival of *Nephotettix virescens* (Homoptera: cicadellidac) and in rice tungro virus transmission. J Econ Entomol 76:573–576

Maris PC, Joosten NN, Goldbach RW, Peters D (2003) Restricted spread of *Tomato spotted wilt virus* in Thrips-Resistant Pepper. Phytopathology 93:1223–1227

Martelli GP (1992) Classification and nomenclature of plant viruses: state of the art. Plant Dis 76:436–442

Martelli GP (2001) Virus diseases of grapevine. Wiley Online Library. doi:10.1038

Martelli GP, Walter B (1998) Virus certification of grape vines. In: Khetarpal RK, Koganezawa H (eds) Plant virus disease control Hadidi A. APS Press, St Paul, pp 261–276

Martin WJ, Kantack EJ (1960) Control of internal cork of sweet potato by isolation. Phytopathology 50:150–152

Maruthi MN, Muniyappa V, Green SK, Colvin J, Hanson P (2003) Resistance of tomato and sweet-pepper genotypes *Tomato leaf curl Bangalore virus* and its vector *Bemisia tabaci*. Int J Pest Management 49:297–303

Maruthi MN, Alam SN, Kader KA, Rekha AR, Cork A, Colvin J (2005) Nucleotide sequencing, whitefly transmission and screening of tomato for resistance against two newly described begomoviruses in Bangladesh. Phytopathology 95:1472–1482

Maruthi MN, Mohammed IU, Hillocks R (2013) Exploiting reversion and tissue culture techniques for eliminating cassava brown streak viruses from infected cassava plants. In: 12th International symposium on plant virus epidemiology, 28th Jan–1st Feb 2013, Arusha, Tanzania (Abstract, pp 051, p 152)

Matthiew JL, Verhoyen M (1979) Lutte contre le virus de la mosaique du Celeri par l'application d'une haile emulsifiable. Parasitica 35:107–122

Mau RFL, Gonsalves D, Bautista R (1989) Use of cross protection to control *Papaya ringspot virus* at Waianae. In: Proceedings of the 25th annual Papaya Industry Association conference, Hilo, pp 77–84

Maude RB (1996) Seedborne diseases and their control: principles and practice. CAB International, Tucson

Maule A (1991) Virus movement in infected plants. Crit Rev Plant Sci 9:457–473

Maule AJ, Caranta C, Boulton M (2007) Sources of natural resistance to plant viruses: status and prospects. Mol Plant Pathology 8:223–231

Mauro MC, Toutain S, Walter B, Pinck L, Otten L, Coutos-Thevenot P, Deloire A, Barbier P (1995) High efficiency regeneration of grapevine plants transformed with the GFLV coat protein gene. Plant Sci 112:97–106

Maury Y, Dubey C, Khetarpal RK (1998a) Seed certification for viruses. In: Hadidi A, Khetarpal RK, Koganezawa H (eds) Plant virus disease control. APS Press, St Paul, pp 237–248

Maury Y, Khetarpal RK, Albrechtsen SE (1998b) Challenge in standardization of diagnostic method for quality control of seed for viruses in relation to world seed trade. ISAT News Bull 116:42–43

Maury Y, Khetarpal RK, Chaube HS (1992) Pea seed borne mosaic virus. In: Kumar J, Mukhopadhyay AN, Singh US (eds) Plant disease of international importance, vol II. Diseases of vegetables and oil seeds crops. Prentice Hall, Inc., Englewood Cliffs, pp 74–92

Mba REC, Dixon AGO (1997) Heritability estimates for ACMD resistance for some newly developed cassava clones in Nigeria. Afr J Root Tuber Crops 2:49–52

McClean APD (1977) Tristeza disease of citrus trees, and sources of tristeza virus that cause the disease. Citrus Subtrop Fruit J 523:7–19

McKenry MV (1978) Selection of preplant fumigation. Calif Agric 32:15–16

McKenry MV, Thomason IJ (1974) Dosage values obtained following pre-plant fumigation for perennials. I 1,3-dichloropropene nematicides in eleven field situations. Pestic Sci 7:521–534

McKinney HH (1929) Mosaic disease in the Canary Islands, West Africa, and Gibraltar. J Agric Res 39:557–578

McKinney HH, Paden WR, Koehler B (1957) Studies on chemical control and over seasoning of, and natural inoculation with the soil borne viruses of wheat and oat's. Plant Dis Rep 41:256–266

Mckirdy SJ, Jones RAC (1995a) Bean yellow mosaic poty virus infection of alternative hosts associated with subterranean and narrow leafed lupins (*Lupinus angustifolius*); Field Screening procedure, relative, subceptibility/resistance rankings, seed transmission and persistence between growing season. Aust J Agri Res 46:135–152

Mckirdy SJ, Jones RAC (1995b) Occurance of *Alfa-alfa mosaic* and *Subterranean clover red leaf viruses* in legume pastures in western Australia. Aust J Agri Res 46:763–774

McKirdy SJ, Jones RAC (1996) Use of imidacloprid and newer generation synthetic pyrethroids to control the spread of *Barley yellow dwarf luteovirus* in cereals. Plant Dis 80:895–901

McKirdy SJ, Jones RAC (1997) Effect of sowing time on Barley yellow dwarf virus infection in wheat: virus incidence and grain yield losses. Australian Journal of Agricultural Research 48(2):199–206

McKnight T (1953) The woodiness virus of the passion vine (*Passiflora edulis* Sims). Ql J Agric Sci I0:4–35

McLean GD, Burt JR, Thomas DW, Sproul AN (1982) The use of reflective mulch to reduce the incidence of *Watermelon mosaic virus* in Western Australia. Crop Prot 1:491–496

McLean BG, Zupan J, Zambryski PC (1995) Tobacco mosaic virus movement protein associates with the cytoskeleton in tobacco cells. Plant Cell 7:2101–2114

Meijneke CAR (1982) Ideal schemes and associated problems in the production, maintenance, multiplication, distribution and certification of fruit crops. Acta Hortic 130:29–31

Meister G, Tsuchl T (2004) Mechanisms of gene silencing by double-stranded RNA. Nature 431:343–349

Melouk HA, Sanborn MR, Banks DJ (1984) Sources of resistance to *Peanut mottle virus* in Arachis germplasm. Plant Dis 68:563–564

Mendel K (1968) Interrelations between tree performance and some virus diseases. In: Childs JFL (ed) Proceedings of the 4th international conference on Organization of Citrus Virologists. University of Florida Press, Gainesville, pp 310–313

Miano DW, La Bonte DR, Clark CA (2008) Identification of molecular markers associated with sweet potato resistance to *Sweet potato virus disease* in Kenya. Euphytica 160:15–24

Miao H, Di D, Wu H (1998) A preliminary study on the effects of seed-borne virus rate and resistance level of maize by inoculating MDMV in different growth periods. J Hebrew Agric Univ 21:27–30

Migliori A, Ginous G, Peyriere J, Marrou J, Musard M, Fauvel M (1972) The use of immunisation against tobacco mosaic virus in glass house tomato crops. Application de la premunition centre le virus de la mosaique du tabac dans les cultuies de tomate sous serre. Pepinieristes Horticulteurs Maraichers 132:15–19

Mignouna HD, Dixon AGO (1997) Genetic relationships among cassava clones with varying levels of resistance to *African mosaic disease* using RARD markers. Afr J Root Tuber Crops 2:28–32

Milbrath DG (1948) Control of western celery mosaic. State Calif Dep Agric Bull 37:3–7

Milkus BN, Avery JD, Pinska VN (2000) Elimination of grapevine viruses by heat treatment and meristem shoot tip culture. In: Proceedings of the 13th international council of virus and virus-like of Grapevine, Adelaide (Australia), 12–17 March, 174 pp

Miller SA, Martin RR (1988) Molecular diagnosis of plant diseases. Annu Rev Phytopathol 26:409–432

Milosevic S, Subotic A, Bulajic A, Dekic I, Jevremovic S, Vucurovic A, Krstic B (2011) Elimination of TSWV from Impatiens hawkerii Bull and regeneration of virus-free plant. Electron J Biotechnol 14 (1). doi:10.2225, ISSN: 0717-3458

Mink GI, Wample R, Howell WE (1998) Heat treatment of perennial plants to eliminate phytoplasmas, viruses and viroids while maintaining plant survival. In: Hadidi A, Khetarpal RK, Koganezawa H (eds) Plant virus disease control. APS Press, St Paul, pp 332–345

Minnini F (2007) Hot water treatment and field coverage of mother plant vineyards to prevent propagation material from phytoplasma infections. Bull of Insectlo 60:311–312

Mishra MD, Ghosh A, Niazi FR, Basu AN, Raychudhuri SP (1973) The role of graminaceous weeds in the perpetuation of rice Tungro virus. J Indian Bot Soc 52:176–183

Mishra S, Singh D, Tiwari AK, Lal M, Rao GP (2010) Elimination of *Sugarcane mosaic virus* and *Sugarcane streak mosaic virus* by tissue culture. Sugar Cane Intern 28:119–122

Moericke V (1954) New investigations on color in relation to Homoptera. In: Proceedings of the 2nd conference on potato virus diseases, Lisse Wageningen, June 25–29, pp 55–69

Mohammad Roff MN, Ho BL (1991) Maize as a barrier crop in reducing aphids, the virus vector of chilli. MARDI Res J 19(1):251–258

Mohanty SK, Bhaktavatsalam G, Anjaneyulu A (1989) Identification of field resistant rice cultivars for tungro disease. Trop Pest Manag 35:48–50

Momol MT, Funderburk JE, Olson S, Stavisky J (2001) Management of TSWV on tomatoes with UV-reflective mulch and acibenzolar-S-methyl. In: Thrips and Tospoviruses: Proceedings of the 7th international symposium of Thysanoptera, Reggio Calabria, Italy, 2–7th July 2001, pp 105–110

Momol T, Olson SM, Funderburk JE, Stavisky J, Marois JJ (2004) Integrated management of tomato spotted wilt in field-grown tomato. Plant Dis 88:882–890

Moniau M, Ionita C, Coman T, Gheorghiu E, Suta V, Parnia P, Modoran I, Casavela S, Cocia V, Ivan I, Pattatynus K (1977) The system of production of virus-free planting material in fruit growing. Probleme de Protectia Plantelor 5:1–15

Moore WD, Smith FF, Johnson GV, Wolfenbarger DO (1965) Reduction of aphid populations and delayed incidence of virus infection on yellow straight neck squash by use of aluminum foil. Proc Fla State Hortic Soc 78:187–191

Morales FJ (2007) Tropical whitefly IPM Project. Adv Virus Res 69:249–306

Morales FJ, Castano M (1987) Seed transmission characteristics of selected *Bean common mosaic virus* strains in differential bean cultivars. Plant Dis 71:51–53

Morales FJ, Castano M (1992) Increased disease severity induced by some comoviruses in bean genotypes possessing monogenic dominant resistance to *Bean common mosaic poty virus*. Plant Dis 76:570–573

Morales FJ, Niessen AI (1988) Comparative responses of selected *Phaseolus vulgaris* germplasm inoculated artificially and naturally with *Bean golden mosaic virus*. Plant Disease 72: 1020-1023

Mori R (1971) Production of virus-free plants by means of meristem culture. Jpn Agri Res Q 6:1–7

Mossop DW, Procter CH (1975) Cross protectionof glasshouse tomatoes against *Tobacco mosaic virus*. N Z J Exp Agric 3:343–348

Mound LA (1962) Studies on the olfaction and colour sensitivity of *Bemisia tabaci* (Genn.) (Homoptera: Aleyrodidae). Entomol Exp et Applic 5:99–104

Mowat WP, Woodford JAT (1976) Virology section. Scottish Horticulture Research Institute, 22nd Annual report for the year 1975, pp 65–80

Mtunda KJ, Legg JP, Ndyetabura I, Jeremiah S, Mvanda K, Mutasingwa G, Shumbusho E (2013) Community action in cassava brown streak disease control through clean seed. In: 12th International Symposium on plant virus epidemiology, 28th Jan–1st Feb 2013, Arusha, Tanzania (Abstract pp 077, p 178)

Muehlbauer FJ (1983) Eight germplasm lines of pea resistant to *Pea seed-borne mosaic virus*. Crop Sci 23:1019

Mukhopadhyay S (1980) Ecology of *Nephotettix* spp. and its relation with *Rice tungro virus*. Final report Ecology of Indian Council of Agricultural Research, New Delhi

Mukhopadhyay S (1984) Ecology of *Rice tungro virus* and its vectors. In: Mishra A, Polasa H (eds) Virus ecology. South Asian Publishers, New Delhi, pp 139–164

Mukhtar SK, Hassim A, Petersehmitt M, Abdrahman MK (2009) Field screening and molecular identification of *Tomato leaf curl virus* in Sudan. Arab J Plant Prot 27:95–98

Mulholland RI (1962) Control of the spread of mechanically transmitted plant viruses. Commonw Phytopathol News 8:60–61

Muller H (1964) The relation of recombination to mutational advance. Mutat Res 1:2–9

Muller GM, Costa AS (1968) Further evidence on protective interference in citrus tristeza. In: Childs JF (ed) Proceedings of the 4th International conference of the International Organization of Citrus Virologists. University of Florida Press, Gainesville, pp 71–82

Muller GW, Costa AS (1972) Reduction in yield of Galego lime avoided by preimmunization with mild strains of tristeza virus. In: Price WC (ed) Proceedings of the 5th International conference of the International Organization of Citrus Virologists. University of Florida Press, Gainesville, pp 171–75

Muller GW, Costa AS (1977) Tristeza control in Brazil by preimmunization with mild strains. Proc Int Soc Citric 3:868–872

Muller GW, Costa AS (1979) Utilizacao da laranja Azeda como cavalo em presenca da tristeza. Summa Phytopatologica 5:31–32

Muller GW, Rezende JAM (2004) Preimmunization: application and perspectives in viral disease control. In: Diseases of fruits and vegetables, diagnosis and management (SAMH Naqvi, ed.), pp 361–396. Kluwer Academic Publishers, Dordrecht

Muniyappa V, Saikia AK (1983) Prevention of the spread of *Tomato leaf curl disease*. Indian Phytopathol 36:183–186

Muniyappa V, Jalikop SH, Saikia AK, Chennarayappa, Shivashankar G, Bhat AI, Ramappa HK (1991) Reaction of Lycopersicon cultivars and wild accessions to *Tomato leaf curl virus*. Euphytica 56:37–41

Muniyappa V, Padmaja AS, Venkatesh HM, Sharma A, Chandrasekhar S, Kulkarni RS, Hanson PM, Chen JT, Green SK, Calvin J (2002) *Tomato leafcurl virus* resistant tomato lines TLB 111, TLB 130 and TLB 182. Hortic Sci 37:603–606

Munster J (1958) Methods zur Beobachtung der Entwicklung der virus bertregenden Blattlause zwecks Austzung des Fruherntetermins und dessen Ruckwirkungen auf den Ertrag an Saatkartoffeln. Eur Potato J 1:31–41

Murashige T (1974) Plant propagation through tissue culture. Annu Rev Plant Physiol 25:135–166

Murata T (2001) Virus resistance in transgenic sweetpotato [*Ipomoea batatas* L. (Lam)] expressing the coat protein gene of *Sweet potato feathery mottle virus*. Theor Appl Genet 103:743–751

Murphy M (2011) Plants, biotechnology and agriculture. CAB International, Wallingford 310 pp

Murphy JF, Hunt AG, Rhoads RE, Shaw JG (1990) Expression of potyvirus genes in transgenic tobacco plants. Abstracts of VIIIth International congress of virology, Berlin, 24–26 Aug, 471 pp

Murray LE (1995) Use of plant virus genes to produce disease resistant crops. In: Genetically modified foods. ACS Symposium Series, vol 605, Chapter 10. American Chemical Society, Washington DC, pp 113–123

Murugan M, Cardona PS, Duraimurugan P, Whitefield AE, Schneweis D, Starkey S, Smith CM (2011) Wheat curl mite resistance: interactions of mite feeding with *Wheat streak mosaic virus* infection. J Econ Entomol 104:1406–1414

Musa AM, Harris D, Johansen C, Kumar J (2001) Short duration chickpea to replace fallow after aman rice: the role of on-farm seed priming in the High Barind Tract of Bangladesh. Exp Agric 37:509–521

Musil M, Jurik M (1990) Resistance of Ceskoslovenskeho allotment of pea and field pea to four important viruses of legumes in Czechoslovakia. Ochrana-Rostlin 26:167–173

Mwanga ROM, Odongo B, Smit N, Ocitti P' Obwoya C, Gibson RW (1995) Submission to the variety release committee for release of sweet potato varieties. Application for inclusion of a crop/variety in the National cultivar list No. F 00004 – 9. The Ugandan Ministry of Agriculture, Animal Industries and Fisheries, Uganda

Murata T (2001) Virus resistance in transgenic sweetpotato [*Ipomoea batatas* L. (Lam)] expressing the coat protein gene of sweet potato feathery mottle virus. Theor Appl Genet 103:743–751

Mwanga ROM, Odongo B, Ocitti p'Obwoya C, Gibson RW, Smit NEJM, Carey EE (2001) Release of five sweet potato cultivars in Uganda. Hortic Sci 36:385–386

Mzira CN (1984) Assessment of effects of maize streak virus on yield of maize. Zimbabwe J Agric Res 22:141–149

Nagai Y (1977) Control of the mosaic vidsease of tomato by seedling inoculation with the attenuated strain of TMV. Trop Agric Res Ser 10:179–183

Nagaich BB, Pushkarnath Bharadwaj VP, Giri BK, Anand SR, Upreti GC (1969) Production of disease free seed potatoes in the Indo-Gangetic plains. Indian J Agric Sci 39:238–243

Nagaich BB, Chaubey IP, Singh SJ, Upreti GC, Kaley DM (1972) Effect of herbicides on infection, concentration and spread of potato virus X. Indian J Hortic 29:105–108

Nagaraja R, Venugopal R, Murthy KVK, Jagadish KS, Nagaraju (2005) Evaluation of groundnut genotypes against *Peanut bud necrosis virus* (PBNV) and its thrips vector at Bangalore. Environ Ecol 23:118–120

Nagata R (1991) The utility of meristem culture plants as controlling the russet-crack like symptom of sweet potatoes and the prevention of re-infection of them. Agric Technol 46:71–74 (in Japanese)

Nagib A, Hossain SA, Alam MF, Hossain MM, Islam R, Sultana RS (2003) Virus-free potato tuber seed production through meristem culture in tropical Asia. Asian J Plant Sci 2:616–622

Naik PS, Chandra R (1994) Response of Indian potato cultivars to micro-propagation. Potato: present and future. In: Proceedings of the national symposium, Modipuram, 1–3 March 1993, pp 308–310

Nair NG, Kartha KK, Gamborg OC (1979) Effect of growth regulators and plant generation from shoot epical meristem of cassava and on culture of internodes *in vitro*. Zeitschrift Pflanzen Physiologie 95:51–56

Najjar KF (1993) Commercialization of the patented microtuber multiplication system of potato. The second Arab conference on perspectives of modern biotechnology, 24–28 April 1993, Amman, Jordan

Nakazono–Nagaoka E, Takahashi T, Shimizu T, Kosaka Y, Natsuaki T, Omura T, Sasaya T (2009) Cross protection against *Bean yellow mosaic virus* (BYMV) and *Clover yellow vein virus* by attenuated BYMV isolate M 11. Phytopathology 99:251–257

Nameth ST, Dodds JA, Paulus AO (1986) Cucurbit viroses of California: an ever-changing problem. Plant Dis 70:8–11

Napometh B (1988) Status of biological control in non-rice crops in Southeast Asia. In: Teng PS, Heong KL (eds) Pesticide Management and Integrated Pest Management in Southeast Asia. Proceedings of the Southeast Asia Pesticide Management and Integrated Pest Management work shop, 23–27 Feb 1987, Pattaya, Thailand. Consortium for International Crop Protection, College Park, pp 123–129

Narasimha Prasad BC, Madhavi Reddy K, Sadasiva AT (2001) Development of F1 hybrids with resistance to *Cucumber mosaic virus* (CMV) in chilli (*Capsicum annuum* L.). Capsicum Eggplant Newsl 20:78–81

Nariani TK, Seth ML (1958) Reaction of *Abelmoschus* and *Hybiscus* species to yellow vein mosaic virus. Indian Phytopathol 11:137–143

Nascimento LC, Pio Ribeiro G, Willadino L, Andrade GP (2003) Stock indexing and Potato virus Y elimination from potato plants cultivated in vitro. Sci Agric 60:525–530

Natarajan K (1990) Natural enemies of *Bemisia tabaci* (Gennadius) and effect of insecticides on their activity. J Biol Control 4:86–88

Nauer EM, Roistacher CN, Reuther W, Calavan EC, Godale JH, Blue RL, Atkin DR, Wagner RL (1975) Virus-free bud wood sources: II. The Valencia orange. Citrograph 61:53–60

Nauer EM, Roistacher CN, Carson TL, Murashigue T (1983) In vitro shoot tip grafting to eliminate citrus viruses and virus-like pathogens produces uniform budlines. Hortic Sci 18:308–309

Navarro L (1993) Citrus sanitation, quarantine and certification programs. In: Proceedings of the 12th conference of the IOCV. IOCV, Riverside, pp 383–391

Navarro L, Juarez J (2007) Shoot-tip grafting in vitro: impact in the citrus industry and research applications. In: Khan IA (ed) Citrus genetics, breeding and biotechnology. CAB International, Wallingford, pp 353–364

Navarro L, Roistacher CN, Murashige T (1975) Improvement of shoot-tip grafting in vitro for virus-free citrus. J Am Soc Hortic Sci 100:471–479

Nawrocka BZ, Eckenrode CJ, Uyemoto JK, Young DH (1975) Reflective mulches and foliar sprays for suppression of aphid borne viruses in lettuce. J Econ Entomol 68:694–698

Naylor M, Reeves J, Cooper JI, Edwards ML, Wang H (2005) Construction and properties of a gene-silencing vector based on Poplar mosaic virus (genus *Carlavirus*). J Virol Methods 124:27–36

Ndiaye M, Bashir M, Keller KE, Hampton RO (1993) Cowpea viruses in Senegal, West Africa: identification, distribution, seed transmission and sources of genetic resistance. Plant Dis 77:999–1003

Ndunguru J, Kapooria RG (1999) Identification and incidence of virus diseases of *Capsicum annuum* in the Lusaka Province of Zambia. EPPO Bull 29:183–189

Neergaard P (1977a) Quarantine policy for seed in transfer of genetic resources. In: Hewitt WB, Chiarappa L (eds) Plant health and quarantine in international transfer of genetic resources. CRC Press Inc, Cleveland, pp 309–314

Neergaard P (1977b) Seed-borne viruses, chapter 3. In: Seed pathology, vol I. MacMillan Press, London, 839 pp

Nelson MR, Tuttle DM (1969) The epidemiology of cucumber mosaic and watermelon mosaic 2 of cantaloupes in an arid climate. Phytopathology 59:849–856

Nelson RS, Powell-Abel P, Beachy RN (1987) Lesion and virus accumulation in inoculated transgenic tobacco plants expressing the coat protein gene of *Tobacco mosaic virus*. Virology 158:126–132

Nelson R, McCormick SM, Delannay X, Dube P, Layton J (1988) Virus tolerance, plant growth and field performance of transgenic tomato plants expressing the coat protein from *Tobacco mosaic virus*. Bio/Technology 6:403–409

Nene YL (1972) A survey of viral diseases of pulse crops in Uttar Pradesh. G.B. Pant Univ Agric Technol Pantnagar Res Bull 4:1–191

Nene YL (1973) Control of *Bemisia tabaci* Genn: a vector of several plant viruses. Indian J Agric Sci 43:433–436

Nene YL (1988) Multiple disease resistance in grain legume. Annu Rev Phytopathol 26:203–217

Nene YL (1995) Sterility mosaic of pigeon pea: the challenge continuous. J Mycol Plant Pathol 23:1–22

Nesi B, Trinchello D, Lazzereschi S, Grassopti (2009) Production of Lily symptomless virus-free plants by fruit meristem tip culture and *in vitro*. Hortic Sci 44:217–219

Ng SYC, Ng NQ (1997) Cassava in vitro germplasm management at the International Institute of Tropical Agriculture. Afri J Root Tuber Crops 2:232–234

Ng SYC, Thottappilly G, Rossel HW (1992) Tissue Culture in disease elimination and micropropagation. In: Thottappilly G, Monti LM, Mohan-Raj D, Moore AW (eds) Biotechnology: enhancing research on tropical crops in Africa. CTA/IITA copublication, IITA, Ibadan, pp 171–182

Nigam SN, Prasada Rao RDVJ, Bhatnagar-Mathur P, Sharma KK (2012) Genetic management of virus diseases in peanut, In: Janick J (ed) Plant breeding reviews, vol 36. Wiley, New York, pp 293–356

Nilakshi Kakati, Nath PD (2006) Integration of physical barrier with bio pesticides against whitefly for management of *Tomato leafcurl virus*. Abstract OP 9/26. In: XVI Annual convention of IVS and international symposium on management of vector-borne viruses, held at ICRISAT, Patancheru, Hyderabad (India)

Nitzany FE (1960) Tests for *Tobacco mosaic virus* inactivation on tomato trellis wires. Ktavim 10:59

Nitzany FE (1966) Tests for the control of field spread of pepper viruses by oil sprays. Plant Dis Rep 50:158–160

Nono-Womdim R, Palloix A, Gebre-Selassic K, Marchoux G (1994) Sources of resistance to *Tomato spotted wilt virus* in Capsicum species. In: Proceedings of the congress of Mediterranean Phytopathological Union, 18–24 Sept, Turkey, pp 237–239

Nono-Womdim R, Swai IS, Green SK, Chadha ML (1999) *Tomato yellow leaf curl virus* and *Tomato leaf curl-like virus* in Eastern and Southern Africa. Paper presented at final IPM workshop held at the International Center of Insect Physiology and Ecology (ICIPE), Nairobi, KE, 2 pp

Noris E, Accotto GP, Tavazza R, Brunetti A, Crespi S, Tavazza M (1996) Resistance to tomato yellow leaf curl Geminivirus in Nicotiana benthamiana plants transformed with a truncated viral C1 gene. Virology 224:130–138

NRSP-5 (1997) Develop and distribute deciduous fruit tree clones that are free of known graft-transmissible pathogens. Prepared for the approval of the NRSP-5 project for the period 1997–2002. Irrigated Research and Extension center of Washington State University, Prosser WA

Nutman FJ, Roberts FM, Williamson JG (1964) Studies on varietal resistance in the groundnut (*Arachis hypogaea* L.) to rosette disease. Rhodesian J Agric Res 2:63–77

Nyland G, Goheen AC (1969) Heat therapy of virus diseases of perennial plants. Annu Rev Phytopathol 7:331–354

Obura MS, Legg JP, Isutsa DK, Kamau AW (2003) Use of cultural methods for the management of *Cassava mosaic disease* in Western Kenya. Afr Crop Sci Conf Proc 6:367–370

Ogero KO, Gitonga NM, Mwangi M, Ombori O, Ngugi M (2012a) Cost-effective nutrient sources for tissue culture of cassava (*Manihot esculenta* Crantz). Afr J Biotechnol 11:12964–12973

Ogero KO, Mburugu GN, Mwangi M, Ngugi MM, Ombor O (2012b) Low cost tissue culture technology in the regeneration of sweet potato (*Ipomoea batata* (L.) Lam). Res J Biol 2:51–52

Ogliari JB, Castarro M (1992) Identification of resistant germplasm to the *Bean common mosaic virus*—BCMV. Pesquisa Agropecuaria Brasileira 27:1043–1047

O'Herlihy EA, Croke JT, Cassells AC (2003) Influence of in vitro factors on titre and elimination of model fruit tree viruses. Plant Cell Tissue Org 72:33–42

Okada Y, Saito A, Nishiguchi M, Kimura T, Mori M, Hanada K, Sakai J, Miyazaki C, Matsuda Y, Murada T (2001) Virus resistance in transgenic sweetpotato [*Ipomoea batatas* L. (Lam)] expressing the coat protein gene of sweet potato feathery mottle virus. Theor Appl Genet 103:743–751

Olembo SAH (1999) Inter-African Plant Quarantine and Phytosanitary Guidelines and Regulations, vol 2. The revised phytosanitary regulations and schedules. Laurier Linguistic services, Nairobi

Ollennu LA, Owusu GK (2003) Field evaluation of the protective capability of CSSV mild strain N1 against severe strain New Juaben (1A) isolate. Ghana J Agric Sci 36:3–12

Ollennu LA, Owusu GK, Thresh JM (1989) The control of *Cocoa swollen shoot disease* in Ghana. Cocoa Grow Bull 42:25–35

Ollennu LAA, Hughes Jd'A, Owusu GK (1996) Mild strain cross-protection of cocoa against *Cocoa swollen shoot badnavirus*. Trop Sci 36:116–128

Ollennu LA, Owusu GK, Dzahini-Obiatey H (1999) Recent studies of mild strain cross-protection with *Cocoa swollen shoot virus*. J Ghana Sci Assoc (Special Edition) 2:5–11

Ollennu LA, Osei Bonsu K, Acheampong K, Aneani F, Ackonor JB (2003) The use of non-host crops as barrier between cocoa plantings. Progress Report 2002–2003, Cocoa Research Institute of Ghana, pp 180–181

Olson EO (1955) A survey for tristeza virus in Texas. In: Proceedings of the 9th annual Rio Grande Valley Horticulture Institute. Rio Grande Valley Horticulture Club, Weslaco, pp 51–60

Olson EO (1958) Response of lime and sour orange seedlings and four scion root stock combinations to infection by strains of the tristeza virus. Phytopathology 48:454–459

Olubayo F, Kibaru A, Nderitu J, Njeru R, Kasina M (2010) Management of aphids and their vectored diseases on seed potatoes in kenya using synthetic insecticides, mineral oil and plant extracts. J Innov Dev Strateg 4:1–5

O'Malley PJ, Hartmann RW (1989) Resistance to *Tomato spotted wilt virus* in lettuce. Hortic Sci 24:360–362

Omori H, Matsumoto H (1972) The cause of stem pitting and small fruit in Nasudaidai trees. In: Proceedings of the 5th Conference of the IOCV. University of Florida Press, Gainesville, pp 143–146

Ong CA (1975) Studies on a mosaic disease of chilli (*Capsicum annuum* L.) in Peninsular Malaysia. M. Agricultural Science Thesis, Faculty of Agriculture, University of Malaya, 100 pp

Ong CA (1984) Reducing the spread of *Chilli veinal mottle virus* using reflective surfaces. MARDI Res Bull 12:200–204

Onwughalu JT, Abo ME, Okoro JK, Onasanya A, Sere Y (2011) *Rice yellow mottle virus* infection and reproductive losses in rice (*Oryza sativa* Linn.). Trends Applied Sci Res 6:182–189

Oostenbrink M (1964) Harmonious control of nematode infection. Nematologia 10:49–56

Oshima N (1981) Control of tomato mosaic disease by attenuated virus. JARQ 14:222–228

Oshima N, Komochi S, Goto T (1965) Study on control of plant virus diseases by vaccination of attenuated virus (1) control of tomato mosaic disease Hokkaido. Natl Agric Exp Stn Bull 85:23–33

Otim-Nape GW, Bua A, Baguma Y (1994) Accelerating the transfer of improved production technologies: controlling African cassava mosaic virus disease in Uganda. Afr Crop Sci J 2:479–496

Over de Linden AJ, Chamberlain EE (1970) Production of virus-free grapevines in New Zealand. N Z J Agric Res 13:991–1000

Ozturk G, Azeri FN, Yildrim Z (2012) Field performance of in vitro sweet potato [*Ipomoea batatas* L. (Lam)] plantlets derived from seed stocks. Turkish J Field Crops 17:1–4

Padgett HS, Epel BL, Heinlein MH, Watanabe Y, Beachy RN (1996) Distribution of tobamovirus movement protein in infected cells and implications for cell-to-cell spread of infection. Plant J 10:1079–1099

Page WW, Smith MC, Holt J, Kyetere D (1999) Intercrops, Cicadulina spp., and maize streak virus disease. Ann Appl Biol 135:385–393

Painter RH (1951) Insect resistance in crop plants. Macmillan, New York 520 p

Palaniswami MS, Makesh Kumar T, Hegde V (2005) Management of vector borne diseases in horticultural crops. In: Chadha KL et al (eds) Crop improvement and production technology of horticultural crops, vol 1. Horticultural Society of India, New Delhi, pp 698–726

Palaniswamy P, Thirumurthi S, Subramaniam TR (1973) Effects of systemic granular insecticides on the incidence of the yellow mosaic disease of okra Abelmoschus esculentus (L). South Ind Hortic 21:104–106

Palloix A, Ordon F (2011) Advanced breeding for virus resistance in plants. In: Caranta C, Aranda MA, Tepfer M, Moya JJL (eds) Recent advances in plant virology. Caister Academic Press, Norfolk, pp 198–214

Paludan N (1975) Tobak mosaik virus (TMV). Knlturfor 50 g, Smittetidspunkt og udbytte med Svaekkede formatlinier of TMV hos tomato. Tobacco mosaic virus (TMV): Experiments with growing condition, time of infection and yield in tomato as influenced by attenuated tomato strains of TMV. Tdsskr Plant Reavol 79:273–287

Panattoni A, D'Anna F, Triolo E (2005) Differential antiviral activity of some IMPDH inhibitors. J Plant Pathol 87:270–280

Panattoni A, D'Anna F, Triolo E (2007) Antiviral activity of tiazofurin and mycophenolic acid against *Grapevine leafroll associated virus 3* in Vitis vinifera explants. Antivir Res 73:206–211

Panattoni A, Luvisi A, Triolo E (2013) Review. Elimination of viruses in plants: twenty years of progress. Span J Agric Res 11:173–188

Pandey PK, Pandey KK (2001) A note on source of resistance on *Watermelon bud necrosis virus* caused by tospovirus. Veg Sci 28:199–200

Pang S-Z, Bock JH, Gonsalves C, Slightom JL, Gonsalves D (1994) Resistance of transgenic *Nicotiana benthamiana* plants to tomato spotted wilt and impatiens necrotic spot tospoviruses: evidence of involvement of the N protein and *N* gene RNA in resistance. Phytopathology 84:243–249

Pappu HR (1977) Managing tospoviruses through biotechnology progress and prospects. Biotechnol Dev Monit 32:14–17

Pappu HR, Csinos AS, McPherson RM, Jones DC, Stephenson MG (2000) Effect of acibenzolar-S-methyl and imidacloprid on suppression of tomato spotted wilt tospovirus in flue-cured tobacco. Crop Prot 19:349–354

Pappu HR, Jones RAC, Jain RK (2009) Global status of tospovirus epidemics in diverse cropping systems: success achieved and challenges ahead. Virus Res 141:219–236

Pappu HR, Niblett CL, Lee RF (1995) Application of recombinant DNA technology to plant protection: molecular approaches to engineering virus resistance in crop plants. World J Microbiol Biotechnol 1:426–437

Parakh DB, Khetarpal RK, Chalam VC (2008) Risk of seed transmitted viruses associated with exchange of soybean germplasm and the South Asian scenario. Indian J Virol 19:47–49

Paris HS, Edlstein M, Baumkoler F, Porat A, Saar U, Hanan A Burger Y, Antignus Y (2005) Plastic mulches for increasing yields and delaying symptoms of whitefly-borne viruses in cucurbits. 2nd Joint Conference of the International working group on legume (IWGLV) and vegetable viruses (IWGVV). April 10–14

Parmessur Y, Aljanabi S, Saumtally S, Dookun-Saumtally A (2002) Sugarcane yellow leaf virus and Sugarcane yellow phytoplasma: elimination by tissue culture. Plant Pathol 51:561–566

Parvatha Reddy P (2009) Advances in integrated pest and disease management in horticultural crops, vol I: Fruit crops, vol 2: vegetable crops. Studium Press (India) Pvt ltd., New Delhi

Parveen S, Tripathi S, Varma A (2001) Isolation and characterization of an inducer protein (Crip-31) from *Clerodendrum inerme* leaves responsible for induction of systemic resistance against viruses. Plant Sci 161:453–459

Patezas KG, Schuerger AC, Wetter C (1989) Management of *Tomato mosaic virus* in hydroponically grown pepper (*Capsicum annum*). Plant Dis 73:570–573

Patil BL, Ogwok E, Wagaba H, Mohammed IU, Yadav JS, Bagewadi B, Taylor NJ, Alicai T, Kreuze JF, Gowda MN, Fauquet CM (2011) RNAi mediated resistance to diverse isolates belonging to two virus species involved in Cassava brown streak disease. Mol Plant Pathol 12:31–41

Patil Nisha M, Jagadeesh KS, Krishnaraj PU, Patil MS, Vastrad AS (2011) Plant Growth Promoting Rhizobacteria (PGPR) Mediated Protection in Bhendi against Bhendi Yellow Vein Mosaic Virus. Indian J Plant Protect 39:48–53

Pazhouhande M, Mozafari J (2001) *In vitro* development of potato virus-free germplasm. Dissertation, Tarbiat Modarres University

Pedersen P, Grau C, Cullen E, Koval N, Hill JH (2007) Potential for integrated management of soybean virus disease. Plant Dis 91:1255–1259

Peele C, Jordan CV, Muangsan N, Turnage M, Egelkrout E, Eagle P, Hanley-Bowdoin L, Robertson D (2001) Silencing of a meristematic gene using Geminivirus-derived vectors. Plant J 27:357–366

Pennazio S, Redolfi P (1973) Factors affecting the cultures *in vitro* of potato meristem tips. Potato Res 16:20–29

Pensuk V, Jogloy S, Wongkaew S, Patanothai A (2004) Generation means analysis of resistance to Peanut bud necrosis caused by *Peanut bud necrosis tospovirus* in peanut. Plant Breeding 123:90–92

Perez R, Rodriguez R, Gonzales A, Del Valle N, Duran-Vila N (1992) Dwarf citrus trees for high density plantings. In: Proceedings of the International Society of Citriculture, Acireale, Italy, pp 712–713

Periyar Selvam S, Nagendra Prasad D, Sudhakar N, Mohan N, Murugesan K (2006) Developing induced resistance against Peanut bud necrosis virus by using Ozone. Abstract PP 10/05. In: XVI Annual convention of IVS and International symposium on management of vector-borne viruses, held at ICRISAT, Patancheru, Hyderabad, India

Perrin RM, Gibson RW (1985) Control of some insect-borne plant viruses with the pyrethroid rate. Int Pest Control 27:142–145

Perring TM, Farrar CA, Blua MJ, Wang HL, Gonsalves D (1995) Cross protection of cantaloupe with a mild strain of zucchini yellow mosaic virus: effectiveness and application. Crop Prot 14:601–606

Perring TM, Gruenhagen NM, Farrar CA (1999) Management of plant viral diseases through chemical control of insect vectors. Annu Rev Entomol 44:457–481

Peters D (1977) The use of oil to prevent the spread of virus diseases and its mode of action. In: Proceedings of the British Insecticide Fungicide conference, Brighton, England, 21–24 Nov, vol 3, pp 823–830

Peters D, Lebbink G (1973) The effect of oil on the transmission of *Pea enation mosaic virus* during short inoculation probes. Entomol Exp Appl 16:185–190

Peters D, Lebbink G (1975) The inhibitory action of mineral oil on the number of local lesions on *Nicotiana glutinosa* L. leaves inoculated with *Tobacco mosaic virus*. Virology 65:574–578

Peters D, Engels C, Sarra S (2012) Natural Spread of Plant Viruses by Birds. Journal of Phytopathology 160:591–594

Peterson CJ, Coats JR (2001) Insect repellents—past, present and future. Pestic Outlook 12:154–158

Philippe V (2007) Trends in GM crop, food and feed safety literature. Nat Biotechnol 25:6–624

Picard D, Kao CC, Hudak KA (2005) Pokeweed antiviral protein inhibits Brome mosaic virus replication in plant cells. J Biol Chem 280:20069–20075

Pichersky E, Noel JP, Dudareva N (2006) Biosynthesis of plant volatiles: nature's diversity and ingenuity. Science 311:808–811

Pico B, Diez M, Nuez F (1998) Evaluation of whitefly-mediated inoculation techniques to screen *Lycopersicon esculentum* and wild relatives to tomato yellow leaf curl virus. Euphytica 101:259–271

Pink DAC, Walkey DGA (1984) Resistance in marrow (*Cucurbita pepo*) to different strains of *Cucumber mosaic virus*. J Agric Sci 103:519–521

Pirone TP, Raccah B, Madden LV (1988) Suppression of aphid colonization by insecticides: effect on the incidence of potyvirus in tobacco. Plant Dis 72:350–353

Pitrat M (2012) Vegetable crops in the Mediterranean Basin with an overview of virus resistance. Adv Virus Res 84:1–29

Pitrat M, Lecoq H (1984) Inheritance of *Zucchini yellow mosaic virus* resistance in *Cucumis melo* L. Euphytica 33:57–61

Pitz K, Ferreira S, Mau R, Gonsalves D (1994) Papaya cross protection: the near commercialization experience on Oahu. In: Proceedings of the annual Hawaii Papaya Industry Association conference, Maui, pp 4–6

Plobner L, Leiser RM (1990) Induction of virus resistance by carnation proteins. In: Proceedings of the international congress on virology, Berlin 26-21

Ploeg AT, Mathis A, Bol JF, Brown DJF, Robinson DJ (1993) Susceptibility of transgenic tobacco plants expressing tobacco rattle virus coat protein to nematode-transmitted and mechanically inoculated *Tobacco rattle virus*. J Gen Virol 74:2709–2715

Plucknett DL, Smith NJH (1984) Networking in International Agricultural Research. Science 225:989–993

Polak J, Hauptmanova A (2009) Preliminary results of in vivo thermotherapy of plum, apricot and peach cultivars artificially infected with PPV-M and PPV-D strains of Plum pox virus. Hortic Sci 36:92–96

Polizzi G, Albanese G, Azzaro A, Davino M, Catara A (1991) Field evaluation of dwarfing effect of two combinations of citrus viroids on different citrus species. In: Brlansky RH, Lee RF, Timmer LW (eds) Proceedings of the 11th Conference of the International Organization of Citrus Virologists. IOCV, Riverside, pp 230–233

Polston JE, Lapidot M (2007) Management of *Tomato yellow leaf curl virus*: US and Israel perspectives. In: Czosnek H (ed) Tomato yellow leaf curl virus disease. Springer, New York, pp 251–262

Polston JE, McGovern RJ, Brown LG (1999) Introduction of *Tomato yellow leaf curl virus* in Florida and implications for the spread of this and other geminiviruses of tomato. Plant Dis 83:984–988

Ponz F, Russell ML, Rowhani A, Bruening G (1988) A cowpea line has distinct genes for resistance to *Tobacco ring spot virus* and *Cowpea mosaic virus*. Phytopathology 78:1124–1128

Pop IV (1995) Scheme and methods for production and certification of virus-free grapevine planting material. EPPO Bull 25:403–409

Posnette AF (1945) Cacao virus research in West Africa. Report and Proceedings of the Cocoa research conference colonial Office, No. 192, H.M. Stationery Office

Posnette AF (1947) Virus diseases of cacao in West Africa. I. Cacao viruses. 1A, 1B, 1C, and 1D. Ann Appl Boil 34:388–402

Posnette AF, Cropley R (1952) A preliminary report on strains of the *Apple mosaic virus*. Rep East Malling Res Stn 1951:128–130

Posnette AF, Cropley R (1956) *Apple mosaic virus*: host reactions and strain interference. J Hortic Sci 31:119–133

Posnette AF, Todd JMcA (1955) Virus diseases of cacao in West Africa IX. Strain variation and interference in virus 1A. Ann Appl Biol 43:433–453

Postman JD (1997) Blueberry scorch carlavirus eliminated from infected blueberry (Vaccinium corymbosum) by heat therapy and apical meristem culture. Plant Dis 81:111

Powell PA, Sanders PR, Tumer NE, Beachy RN (1990) Protection against *Tobacco mosaic virus* infection in transgenic plants requires accumulation of capsid protein rather than coat protein RNA sequences. Virology 175:124–130

Powell-Abel P, Nelson RS, De B, Hoffman N, Rogers SG, Fraley RT, Beachy RN (1986) Delay of disease development in transgenic plants that express the *Tobacco mosaic virus* coat protein gene. Science 232:738–743

Pozzer L, Silva JBC, Dusi AN, Kitajima EW (1995) Performance of micropropagated sweetpotato plants after two field propagations and rate of reinfection by *Sweetpotato feathery mottle virus*. Fitopatol Bras 20(3):464–468

Prasad V, Srivastava S, Verma, Varsha HN (1995) Two basic proteins isolated from *Clerodendrum inerme* Gaertn. are inducers of systemic antiviral resistance in susceptible plants. Plant Sci 110:73–82

Prasad SM, Sharma RB, Kudada N (2004) Effect of density of plant population on incidence of leaf crinkle disease in urd bean. RAU J Res 14:161–162

Prasada Rao RDVJ, Reddy AS, Chakravarthy SK, Sastry KS, Reddy DVR, Ramanath, Moses JP (1989) *Peanut stripe virus* research in India. II. Peanut stripe coordinator meeting held at ICRISAT, Hyderabad, India

Prasada Rao RDVJ, Reddy AS, Chakravarthy SK, Reddy DVR, Rao VR, Moss JP (1991) Identification of *Peanut stripe virus* resistance in wild Arachis germplasm. Peanut Sci 18:1–2
Prasada Rao RDVJ, Pio-Ribeiro G, Pittman R, Reddy DVR, Demski JW (1995) *In vitro* culture to eliminate *Peanut stripe virus* from peanut seeds. Peanut Sci 22:54–56
Prasada Rao RDVJ, Reddy AS, Reddy SV, Tirumala Devi K, Chander Rao S, Manoj Kumar V, Subramanyam K, Yellamanda Reddy T, Nigam SN, Reddy DVR (2003a) The host range of *Tobacco streak virus* in India and transmission by thrips. Ann Appl Biol 142:365–368
Prasada Rao RDVJ, Reddy DVR, Nigam SN, Reddy AS, Waliyar F, Yellamanda Reddy T, Subramanyam K, John Sudheer M, Naik KSS, Bandyopadhyay A, Desai S, Ghewande MP, Basu MS, Somasekhar (2003b) Peanut stem necrosis: a new disease of groundnut in India. Information Bulletin No. 67, CRIDA, Hyderabad and ICRISAT, Patancheru (India), 16 pp
Prasada Rao RDVJ, Anitha K, Chakrabarty SK, Girish AG, Sarath Babu B, Ramesh K, Abraham B, Varaprasad KS (2012) Quarantine pathogen interceptions on crop germplasm in India during 1986-2010 and their possible economic impact. Indian J Agric Sci 82:436–441
Prasanna BM, Firoz Hossain (2010) Molecular marker-assisted breeding for improving disease resistance in crop plants. In: Dasharantha Reddy V, Nagaraja Rao P, Venkateswara Rao K (eds) Pests and pathogens management strategies. BS Publications, Hyderabad, pp 353–378
Praveen S, Ramesh SV, Mishra AK, Koundal V, Palukaitis P (2010) Silencing potential of viral derived RNAi constructs in *Tomato leaf curl virus*-AC4 gene suppression in tomato. Transgenic Res 19:45–55
Pressman E, Moshkovitz A, Rosenfeld K, Shaked R (1996) The effects of U.V. blocking films on quality parameters of tomato and pepper fruits. Final report submitted to Ginegar Plastic Industries, Israel
Presting GG, Smith OP, Brown CR (1995) Resistance to potato leafroll virus in potato plants transformed with the coat protein gene or with vector control constructs. Phytopathology 85:436–442
Pricestat (2010) http://faostat.fao.org/
Prins MW (2003) Broad virus resistance in transgenic plants. Trends in Biotechnology 21(9):373–375
Prins M, Goldbach R (1996) RNA-mediated virus resistance in transgenic plants. Arch Virol 141:2259–2276
Prins M, Goldbach R (1998) The emerging problem of tospovirus infection and nonconventional methods of control. Trends Microbiol 6:31–35
Prins TW, Zadoks JC (1994) Horizontal gene transfer in plants, a biohazard out come of literature review. Euphytica 76:133–138
Prins M, Laimer M, Noris E, Schubert J, Wassenegger M, Tepfer M (2008) Strategies for antiviral resistance in transgenic plants. Mol Plant Pathol 9:73–83
Proeseler G, Karl E, Zschiegner HJ (1975) Cutting down field bean infection by aphid transmissible viruses in a plot experiment. Arch Phytopathol u Pflanzenschutz 11:405–412
Proeseler G, Fritzsche R, Karl E, Geissler K, Zschiegner HJ (1976) Systemic insecticides and citol oil for controlling aphids as virus transmitters in plot and large-scale experiments with Beta beet. Arch Phytopathol U Pflanzenschutz Berlin 12:127–134
Provvidenti R (1975) Resistance in *Glycine max* to isolates of *Bean yellow mosaic virus* in New York State. Plant Dis Rep 59:917–919
Provvidenti R (1977) Evaluation of vegetable introductions from the Peoples Republic of China for resistance to viral diseases. Plant Dis Rep 61:851–855
Provvidenti R (1991) Inheritance of resistance to the NL-8 strain of bean common virus in *Pisum sativum*. J Heridity 82:352–355
Provvidenti R, Alconero R (1988) Inheritance of resistance to a third pathotype of *Pea seed-borne mosaic virus* in *Pisum sativum*. J Hered 79:76–77
Provvidenti R, Braverman SW (1976) Seed transmission of *Bean common mosaic virus* in phasemy bean. Phytopathology 66:1274–1275
Provvidenti R, Gonsalves D (1995) Inheritance of resistance to cucumber mosaic virus in a transgenic tomato line with the coat protein gene of the white leaf strain. J Hered 86:85–88

Provvidenti R, Schroeder WT (1973) Resistance in *Phaseolus vulgaris* to the severe strain of *Bean yellow mosaic virus*. Phytopathology 63:196–197

Provvidenti R, Robinson RW, Munger HM (1978) Multiple virus resistance in Cucurbita species. Cucurbit Genetic Coop 1:26–27

Pun KB, Doraiswamy S, Jeyarajan R (2005) Management of *Okra yellow vein mosaic virus* disease and its whitefly vector. Indian J Virol 16:32–35

Purcifull DE, Zitter TA (1971) Virus diseases affecting lettuce and endive in Florida. Proc Fla State Hortic Soc 84:165–168

Pushkarnath VP (1959) Producing healthy seed potatoes in the plains: a new approach. Indian Potato J 1:63–72

Pushkarnath VP, Nagaich BB, Singh Jagpal, Srivastava SNS, Giri BK (1967) The basis of the disease-free seed potato production in India. In: Proceedings of the 1st international symposium on plant pathology, New Delhi, pp 289–295

Quacquerelli A, Gallitelli O, Savinow V, Piazzolla P (1980) The use of electrical corrient (RACE) for obtaining mosaic free almonds. Acta Phytopathologica. Academic Scientianrum Hungaricae 15:251–255

Quak F (1977) Meristem culture and virus-free plants. In: Reinert J, Bajaj YPS (eds) Applied and fundamental aspects of plant cells tissue and organ culture. Springer, Berlin, pp 598–615

Quemada HD, Gonsalves D, Slightom JL (1991) Expression of coat protein gene from *Cucumber mosaic virus* strain C in tobacco: protection against infection by CMV strains transmitted mechanically or by aphids. Phytopathology 81:794–802

Quemener J (1976) Essais de protection centre le virus Y par des traitments a base d'huile minerale. Le pomme de terre francaise 371:5–16

Quint M, Lubberstedt T (2008) Application of resistance gene analogs in breeding for virus resistance. In: Rao GP, Valverde RA, Dovas CI (eds) Techniques in diagnosis of plant viruses. Studium Press LLC, Houston, pp 265–285

Raccah B (1985) Use of combination of mineral oils and pyrethroids for control of non-persistent viruses. Phytoparasitica 13:280

Raccah B (1986) Nonpersistent viruses: epidemiology and control. Adv Virus Res 31:387–429

Raccah B, Fereres A (2009) Plant virus transmission by insects. In: Encyclopedia of life sciences. Wiley, Chichester. doi:10.1002/9780470015902.A0021525.a0000760.pub2

Ragetli HWJ, Weintraub M (1962) Purification and characteristics of a virus inhibitor from *Dianthus caryophyllus* I. Purification and activity. Virology 18:232–240

Raid RN, Nagata RT (1996) A recent out break of lettuce mosaic potyvirus in commercial lettuce production in Florida. Plant Dis 80:343

Raj SK, Singh R, Pandey SK, Singh BP (2005) *Agrobacterium*-mediated tomato transformation and regeneration of transgenic lines expressing *Tomato leafcurl virus* coat-protein gene for resistance against TLCV infection. Curr Sci 88:1675–1679

Raja P, Jain RK (2008) Development of nucleocapsid gene-mediated resistance in tomato against *Groundnut bud necrosis virus*. J Plant Pathol 90:S2442

Rajamony L, More TA, Seshadri VS (1990a) Inheritance of resistance to *Cucumber green mottle mosaic virus* in muskmelon (*Cucumis melo* L.). Euphytica 47:93–97

Rajamony L, More TA, Seshadri VS, Varma A (1990b) Reaction of muskmelon collections to *Cucumber green mottle mosaic virus*. J Phytopathol 129:237–244

Rajapakse RHS, Janaki CR (2006) Insecticidal properties of Chenopodium-based botanical against virus vectors, *Myzus persicae* and *Frankliniella schultezi*. Abstract OP 9/14. In: XVI Annual convention of IVS and International symposium on management of vector-borne viruses, held at ICRISAT, Patancheru, Hyderabad (India)

Rajasekharam T (2010) Biological and molecular characterization and management of *Watermelon bud necrosis virus*. PhD thesis, University of Agricultural Sciences, Dharwad, 142 pp

Ramgareeb S, Snyman SJ, Van Antwerpen T, Rutherford RS (2010) Elimination of virus and rapid propagation of disease free sugarcane (*Saccharum* spp. cultivar NCO 376) using apical meristem culture. Plant Cell Tissue Org 100:175–181

Ramakrishnan K (1963) Control of virus disease in plants. Bull Natl Inst Sci India 24:78–91

Ramallo JC, Garci AE, Went S, Roncedo LR (1975) Applicaciones de aceites minerals para retardar las infecciones del virus Y de la para (PVY) en cultivos de pimiento. Revista Agronomica del Noroeste Argentino 12:181–182

Ramkat RC, Wangai AW, Ouma JP, Rapando PN, Lelgut DK (2008) Cropping system influences *Tomato spotted wilt virus* disease development, thrips population and yield of tomato (*Lycopersicon esculentum*). Ann Appl Biol 153:373–380

Rampersad SN, Umaharan P (2003) Identification of resistance to *Potato yellow mosaic virus*-Trinidad isolate (PYMV-TT) among lycopersicon species. Plant Dis 87:686–691

Randhawa GJ, Chandra R (1990) In vitro tuberization in six Indian potato cultivars. J Indian Potato Assoc 17:199–201

Ranjithkumar CT, Savithri HS (1999) Advances in the development of virus resistant transgenic plants. J Plant Biol 26:97–110

Rao GM, Anjaneyulu A (1978) Host range of *Rice tungro virus*. Plant Dis Rep 62:955–957

Rao GP, Maneesha Singh (2008) Techniques in diagnosis of plant viruses. In: Rao GP, Varverde RA, Dovas CI (eds) Techniques in diagnosis of plant viruses. Studium Press LLC, Houston, pp 1–47

Rao GP, Chateret M, Gigard JG, Rott P (2006) Distribution of *Sugarcane mosaic* and *Sugarcane streaked mosaic virus* in India. Sugar Tech 8:79–81

Rashid MH, Yasmin L, Kibria MG, Mollik AKMSR, Monowar Hossain SM (2002) Screening of okra germplasm for resistance to yellow vein mosaic virus under field conditions. Pak J Plant Pathol 1:61–62

Rashid A, Harris D, Hollington PA, Ali S (2004a) On-farm seed priming reduces yield losses of mungbean (*Vigna radiata*) associated with *Mungbean yellow mosaic virus* in the North West Frontier Province of Pakistan. Crop Prot 23:1119–1124

Rashid A, Harris D, Hollington PA, Rafiq M (2004b) Improving the yield of mungbean (*Vigna radiata*) in the North West Frontier Province of Pakistan using on-farm seed priming. Exp Agric 40:233–244

Rashid MH, Hossain I, Hannan A, Uddin SA, Hossain MA (2008) Effect of different dates of planting time on prevalence of *Tomato yellow leaf curl virus* and whitefly of tomato. J Soil Nat 2:1–6

Raski DJ, Lear B (1962) Influence of rotation and fumigation on root-knot nematode population on grape replants. Nematologica 8:143–151

Raski DJ, Hewitt WB, Gohenn AC, Tayor CE, Taylor RH (1965) Survival of Xiphinema index and reservoirs of fan leaf virus in followed vine yard soil. Nematologia 11:349–352

Rast ATB (1972) M II-16, an artificial symptomless mutant of *Tobacco mosaic virus* for seedling inoculation of tomato crops. Neth J Plant Path 78:110–112

Rast ATB (1975) Variability of tobacco mosaic virus in relation to control of tomato mosaic in glass house tomato crops by resistance breeding and cross protection. Agricultural Research Report 834, Pudoc, Wageningen, 76 pp

Rast ATB (1982) Resistance of capsicum species to tobacco, tomato and pepper strains of tobacco mosaic virus. Neth J Plant Pathol 88:163–170

Ratcliff F, Harrison BD, Baulcombe DC (1997) A similarity between viral defense and gene silencing in plants. Science 276:1558–1560

Ratcliff F, MacFarlane SA, Baulcombe DC (1999) Gene silencing without DNA: RNA-mediated cross-protection between viruses. Plant Cell 11:1207–1215

Ratcliff F, Martin-Hernandez AM, Baulcombe DC (2001) Tobacco rattle virus as a vector for analysis of gene functions by silencing. Plant J 25:237–245

Rathi YPS (1979) Temik treatment of pigeonpea seeds for prevention of sterility mosaic. Acta Botanica Indica 7:90–91

Rattan RS, Arvind Bindal (2000) Development of okra hybrids resistant to yellow vein mosaic virus. Veg Sci 27:121–125

Raupach GS, Liu L, Murphy JF, Tuzun S, Kloepper JW (1996) Induced systemic resistance in cucumber and tomato against cucumber mosaic cucumovirus using plant growth promoting rhizobacteria (PGPR). Plant Dis 80:891–894

Raviv M, Antignus Y (2004) UV radiation effects on pathogens and insect pests of greenhouse-grown crops. Photochem Photobiol 79:219–226

Reddy DVR (1991) Groundnut viruses and virus diseases: distribution, identification and control. Rev Plant Pathol 70:665–678

Reddy DVR, Amin PW, Mc Donald D, Ghanekar AM (1983) Epidemiology and control of groundnut bud necrosis and other diseases of legume crops in India caused by *Tomato spotted wilt virus*. In: Plumb RT, Thresh JM (eds) Plant virus epidemiology. Blackwell Scientific Publications, Oxford, pp 93–102

Reddy DVR, Nolt BL, Hobbs HA, Reddy AS, Rajeswari R, Rao AS, Reddy DDR, McDonald D (1988) Clump virus in India: Isolates host range, transmission and management. Association of Applied Biologists, Wellesbourne pp 239–246

Reddy MV, Sharma SB, Nene YL (1990) Pigeon pea: disease management. In: Nene YL, Hall SD, Sheila VK (eds) The pigeonpea. CAB International, Wallingford, pp 303–347

Reddy AS, Reddy LJ, Mallikarjuna N, Abdurahman MD, Reddy YV, Bramel PJ, Reddy DVR (2000) Identification of resistance to *Peanut bud necrosis virus* (PBNV) in wild Arachis germplasm. Ann Appl Biol 137:135–139

Reddy DVR, Sudarshana MR, Fuchs M, Rao NC, Thottappilly G (2009) Genetically engineered virus-resistant plants in developing countries: current status and future prospects. Adv Virus Res 75:185–220

Reestman AJ (1960) The cultivation of potatoes with the use of ‘seed’ in a moderate state of health. Landbonwoorlichting 17:64–70

Regner F, da Camara-Machado A, Laimer de Camara-Machado M, Steinkellner H, Mattanovich D, Hanzer V (1992) Coat protein mediated resistance to *Plum pox virus* in *Nicotiana clevelandii* and *N. benthamiana*. Plant Cell Rep 11:30–33

Reichert I (1959) Rapport sur les resultants d’une e’tude faite en Espagne sur les virus et les autres affections des agrumes sous les auspices de I E.P.P.O. Fruits Prim Afr N 29:96–107

Reichert I, Bental A (1960) Citrus varieties in Israel, Infected with tristeza. Ktavim Res Stn Rehovot Israel 10:53–58

Reimann-Phillipp U (1998) Mechanism of resistance: expression of coat protein. In: Foster GD, Taylor SC (eds) Methods in molecular biology, plant virology protocols: from virus isolation to transgenic resistance, vol 2. Humana Press, Totowa, pp 521–532

Reitz SR, Yearby EL, Funderburk JE, Stavisky J, Momol MT, Olson SM (2003) Integrated management tactics for Frankliniella thrips (Thysanoptera: Thripidae) in field-grown pepper. J Econ Entomol 96:1201–1214

Rich S (1960) Terraclor controls Olpidium on Lettuce. Inîectivity differences between Olpidium from roots of Spinach and Lettuce. Plant Dis Rep 44:352–353

Richardson DE (1958) Some observations on the tobacco veinal necrosis strain of *Potato virus Y*. Plant Path 7:133–135

Richardson LJ, Racicot HN (1951) The comparative efficacy of tuber indexing and tuber unit planting in the elimination of virus diseases from seed potatoes. Am Potato J 28:765–775

Riley DG, Joseph SV, Srinivasan R (2012) Reflective Mulch and Acibenzolar-*S*-methyl treatments relative to Thrips (Thysanoptera: Thripidae) and tomato spotted wilt virus incidence in tomato. J Econ Entomol 105:1302–1310

Ritzenthaler C (2005) Resistance to plant viruses; old issue, new answers. Curr Opin Biotechnol 16:118–122

Rivera CT, Ling KC, Ou SH (1969) Suspect host range of *Rice tungro virus* Phill. Phytopathol 5:16–17

Robert U, Zel J, Ravnikar M (1998) Thermotherapy in virus elimination from garlic. Influences on shoot multiplication from meristem and bulb formation In vitro. Science Hort 73:193–202

Rodier A, Assie J, Marchand JL, Herve Y (1995) Breeding maize lines for complete and partial resistance to *Maize streak virus* (MSV). Euphytica 81:57–70

Robbins WW (1921) Mosaic of sugar beets. Phytopathology 48:349–365

Robertson DG (1966) Seed-borne viruses of cowpea in Nigeria. B.Sc. Thesis, University of Oxford, Oxford, England, 111 pp

Robertson LAD (1987) The role of *Bemisia tabaci* Gennadius in the epidemiology of ACMV in East Africa: Biology, population dynamics and interaction with cassava varieties. In: Proceedings of the International Seminar on *African cassava mosaic disease* and its control, Yamoussoukro, Ivory Coast, pp 57–63

Robinson RA (1980) New concepts in breeding for disease resistance. Annu Rev Phytopathol 18:189–210

Roca WM, Bryan JE, Roca MR (1979) Tissue culture for the international transfer of potato genetic resources. Am Potato J 56:l–10

Rodoni B (2009) The role of biosecurity in preventing and controlling emerging plant virus disease epidemics. Virus Res 141:150–157

Rodriguez O, Salibe AA, Pomben J Jr (1974) Reaction of nucellar Hamlin orange on Rangpur lime to several exocortis strains. In: Weathers LG, Cohen M (eds) Proceedings of the 6th Conference of IOCV, Berkeley, pp 114–116

Roggero P, Pennazio S, Masenga V, Tavella L (2001) Resistance to tospoviruses in pepper. In: Marullo R, Mound L (eds.) Thrips and Tospoviruses: Proceedings of the 7th international symposium on Thysanoptera, Reggio Calabria, Italy, 2–7 July 2001, pp 105–110

Roistacher CN, Nauer EM (1968) Frequency of virus infection in citrus budwood introduced into the United States. In: Proceedings of the 4th conference of the IOCV. University of Florida Press, Gainesville, pp 386–391

Roistacher CN, Calavan EC, Blue RL (1969) *Citrus exocortis virus*—chemical inactivation on tools, tolerance to heat and separation of isolates. Plant Dis Rep 53:333–336

Roistacher CN, Navarro L, Murashige T (1976) Recovery of citrus selections free of several viruses, exocortis viroid and Spiroplasma citri by shoot tip grafting in vitro. In: Proceedings of the 7th conference of the IOCV, Riverside, CA, pp 186–193

Rose DJW (1974) The epidemiology of maize streak disease in relation to population densities of Cicadulina spp. Ann Appl Biol 76:199–207

Ross JP (1977) Effect of aphid transmitted *Soybean mosaic virus* on yields of closely related resistant and susceptible soybean lines. Crop Sci 17:869–872

Rotham PG (1967) Aluminum foil fails to protect winter oats from aphid vectors of *Barley yellow dwarf virus*. Plant Dis Rep 51:354–355

Rowhani A, Uyemoto JK, Golino DA, Martelli GP (2005) Pathogen testing and certification of Vitis and Prunus species. Annu Rev Phytopathol 43:261–278

Ruiz MT, Voinnet O, Baulcombe DC (1998) Initiation and maintenance of virus-induced gene silencing. Plant Cell 10:937–946

Russell GE (1970) Effects of mineral oil on *Myzus persicae* (Sulz) and its transmission of *Beet yellows virus*. Bull Entomol Res 59:691–694

Ryan HJ (1969) Plant quarantine in California. Division of Agricultural Science, University of California, Berkeley, CA, 251 pp

Ryder EJ (1970) Screening for resistance to lettuce mosaic. Hortic Sci 5:47–48

Ryder EJ (1976) The nature of resistance to lettuce mosaic. In: Proceedings of the Eucarpia meeting on leafy vegetables, Wageningen, The Netherlands, pp 110–118

Rykova LM (1977) An experiment with growing virus free potato. Opvt vozdelyvaniya bez virus no go Kartofelya. Zaschita Rastenii 10:22–23

Sabhiki HS, Sekhon PS, Gupta VK, Sohu RS (2004) Identification of cotton leaf curl immune genotypes in *Gossypium hirsutum*. Crop Improv 31:66–70

Sacchettini JC, Poulter CD (1997) Creating isoprenoid diversity. Science 277:1788–1789

Saikia AK, Muniyappa V (1989) Epidemiology and control of *Tomato leaf curl virus* in Southern India. Trop Agric (Trinidad) 66:350–354

Salaudeen MT, Menkir A, Atiri GI, Lava Kumar P (2013) Genetic resistance and gene action of maize germplasm to *Maize streak virus*. In: International symposium on plant virus epidemiology, 28th Jan–1st Feb 2013, Arusha, Tanzania (Abstract, pp 039, p 140)

Salibe AA, Cortez RE (1966) Studies on the leaf mottling disease of citrus. Philipp J Plant Ind 33:187–194

Salibe AA, Moreira S (1965) Tahiti lime bark disease is caused by exocortis virus. In: Conference of IOCV, pp 143–147

Salzmann R (1953) Uber das Totspritzen der kartoffelstanden als Massnahme zur verhinderung der Virusansfreitung. Landw Jb Schweiz NF 2:707–738

Sama S, Hasanuddin A, Manwan I, Cabunagan RC, Hibino H (1991) Integrated management of rice tungro disease in South Sulawesi. Indonesia Crop Prot 10:34–40

Sanderfoot AA, Lazarowitz SG (1996) Getting it together in plant virus movement: cooperative interactions between bipartite geminivirus movement proteins. Trends Cell Biol 6:353–358

Sanders PR, Sammons B, Kaniewski W, Haley L, Layton J, Lavallee BJ, Delannay X, Tumer NE (1992) Field resistance of transgenic tomatoes expressing the tobacco mosaic virus or tomato mosaic virus coat protein genes. Phytopathology 82:683–690

Sanford JC, Johnston SA (1985) The concept of parasite-derived resistance. Deriving resistance genes from the parasites our genome. J Theor Biol 113:395–405

Saraswathi T, Rajamani K, Indra N, Kempuchetty N (2002) Control measures for Indian cassava mosaic disease. J Root Crops 28:76–78

Sastry KS (1984) Management of plant virus diseases by oil sprays. In: Misra A, Polosa H (eds) Virus ecology. South Asian Publishers, New Delhi, pp 31–57

Sastry KS (1989) Tomato leaf curl virus management by carbofuran plus oil combination. J Turkish Phytopathol 18:11–16

Sastry KS (2013a) Seed-transmitted plant virus diseases. Springer, New York, 327 pp

Sastry KS (2013b) Plant virus and viroid diseases in the tropics. Volume-1: Introduction of plant viruses and sub-viral agents, classification, assessment of loss, transmission and diagnosis. Springer, New York (In Press)

Sastry KSM, Singh SJ (1973a) Field evaluation of insecticides for control of whitefly (*Bemisia tabaci*) in relation to the incidence of yellow vein mosaic of okra (*Abelmoschus esculentus*). Indian Phytopathol 26:129–138

Sastry KSM, Singh SJ (1973b) Restriction of yellow vein mosaic virus spread in okra through the control of vector, whitefly Bemisia tabaci. Indian J Mycol Plant Pathol 3:76–80

Sastry KS, Singh SJ (1982) Effect of plant extracts of some medicinal plants on the infectivity of tobacco mosaic virus. Herba Hungarica 21:101–105

Sastry KS, Sastry KSM, Singh SJ (1974) Influence of different insecticides on *Tomato leaf curl virus* incidence in the field. Pesticides 8:41–42

Sastry KSM, Singh SJ, Sastry KS (1976) Density of tomato plant population and application of insecticides in relation to incidence of *Tomato leaf curl virus*. Indian J Mycol Plant Pathol 6:8–13

Sastry KSM, Singh SJ, Sastry KS (1977) Effect of border cropping and the use of insecticides in relation to the incidence of tomato leaf curl virus (TLCV). Indian J Hortic 34:319–322

Sastry KSM, Singh SJ, Sastry KS (1978) Studies on epidemiology of *Tomato leafcurl virus*. Indian J Hortic 35:269–277

Sastry KS, Satyanarayana A, Singh SJ, Rajendran R (1981) Screening of different bean cultivars against *Common bean mosaic virus*. Pulse Crops Newsl 1:30

Satapathy MK, Anjaneyulu A (1984) Use of Cypermethrin, a synthetic pyrethroid in the control of rice tungro virus disease and its vector. Trop Pest Manag 30:170–178

Sathiamoorthy S, Uma S, Selvarajan R, Shyam B, Singh HP (1998) Multiplication of virus-free banana plants through shoot tip culture. NRCB, Tiruchirapalli, 60 pp

Saxena RC, Waldbauer GP, Liquido NJ, Puma BC (1981) Effect of neem seed oil on rice leaffolder, Cnaphalocrocis medinalis. In: Proceedings of the 2nd international Neem conference, Rottach-Engern, pp 189–204

Sayama H, Sato T, Komimato M, Natsuaki T, Kaper JM (1993) Field testing of a satellite containing attenuated strains of *Cucumber mosaic virus* for tomato production in Japan. Phytopathology 83:405–410

Schaad NW (1979) Serological identification of plant pathogenic bacteria. Annu Rev Phytopathol 17:123–147

Schaad NW, Frederick RD, Shaw J, Schneider WL, Hickson R, Petrillo MD, Luster DG (2003) Advances in molecular-based diagnostics in meeting crop biosecurity and phytosanitary issues. Annu Rev Phytopathol 41:305–324

Schalk JM, Creighton CS, Fery RL, Sitterly WR, Davis BW, Mc.Fadden TL, Augustine D (1979) Reflective film mulches influences insect control and yield in vegetables. J Am Soc Hortic Sci 104:759–762

Schenk MF, Hamelink R, der Vlugi Van, Vermunt AMW, Kaarsenmaker RC, Stigger ICCMM (2010) The use of attenuated isolates of Pepino mosaic virus for cross protection. Eur J Plant Pathol 127:240–261

Schepers A, Bus CB (1979) Control of potato virus YN with mineral oil. Bedrijfsontwikkeling 10:308–312

Schepers A, Bus CB, De Bokx, Cuperus C (1977) De verspreiding Van YN-virus in aardappelen. Landbonwkundig Tijdschrift, 89–95

Schlosser E (1988) Epidemiology and management of *Polymyxa betae* and *Beet necrotic yellow vein virus*. In: Cooper JI, Asher MJC (eds) Viruses with fungal vectors. Association of Applied Biologists, Wellesbourne, pp 281–292

Schmelzer K, Wolf P (1975) Possibilities of control of virus diseases on cucumber and tomato under glasshourse and plastics'. Moglichkeiten der Bekampfung von virus erk rank unogen bei Gurke und Tomato under glas and platen. Nachrichtenblatt fur den Pflanzenschutz in der DDR 29:21–26

Schneider IR (1954) The effect of purines, purine analogs and related compounds on the multiplication of tobacco mosaic virus. Phytopathology 44:243–247

Scholthof KBG, Scholthof HB, Jackson AO (1993) Control of plant virus diseases by pathogen-derived resistance in transgenic plants. Plant Physiol 102:7–12

Schuster DJ (2004) Squash as a trap crop to protect tomato from whitefly-vectored *Tomato yellow leaf curl*. Int J Pest Manag 50:281–284

Schutz S, Weißbecker B, Klein A, Hummel HE (1997) Host plant selection of the Colorado potato beetle as influenced by damage induced volatiles of the potato plant. Naturwissenschaften 84:212–217

Schwartz HF, Gent DH, Fichtner SM, Hammon R, Cranshaw WS, Mahaffey L, Camper M, Otto K, McMillan M (2009) Straw mulch and reduced-risk pesticide impacts on Thrips and *IrisYellow Spot Virus* on Western-grown onions. Southwestern Entomologist 34(1):13–29

Schwinn FJ (1994) Seed treatment—a pancea for plant protection? In: Seed treatment: progress and prospects Mono. 57, BCPC, Thornton Health, UK, pp 3–14

Scofield SR, Huang L, Brandt AS, Gill BS (2005) Development of a virus-induced gene silencing system for hexaploid wheat and its use in functional analysis of the Lr21-mediated leaf rust resistance pathway. Plant Physiol 138:2165–2173

Seaton KA, Cook DF, Hardie DC (1997) The effectiveness of a range of insecticides against western flower thrips (*Frankliniella occidentalis*) (Thysanoptera: Thripidae) on cut flowers. Aust J Agric Res 48:781–788

Sediva J, Novak P, Laxa J, Kanka J (2006) Micropropagation, detection and elimination of DMV in the Czech collection of Dahlia. Acta Hortic 725:495–498

Seetharam P, Venugopala Rao N, Hari Satya Narayana N, Chenga Reddy V (2011) Evaluation of insecticides to control whitefly, *Bemisia tabaci* vector of yellow vein mosaic disease on Mesta. Indian J Plant Protect 39:1–3

Selvarajan R, Balasubramanian V (2008) Banana viruses. In: Rao GP, Mytra A, Ling KS (eds) Characterization, diagnosis and management of plant viruses. Studium press LLC, Houston, pp 109–124

Semancik JS (2003) Consideration for the introduction of viroids for economic advantage. In: Hadidi A, Flores R, Randles JW, Semancik JS (eds) Viroids. CSIRO Publishing, Collingwood, pp 357–362

Semancik JS, Rakowski AG, Bash JA, Gumpf DJ (1997) Application of selected viroids for dwarfing and enhancement of production of 'Valencia' orange. J Hortic Sci 72:563–570

Semancik JS, Bash JA, Gumpf DJ (2002) Induced dwarfing of citrus by transmissible small nuclear RNA (TsnRNA). In: Duran-Vila N, Milne RG, da Graca JV (eds) Proceedings of the 15th Conference of the International Organization of Citrus Virologists. IOCV, Riverside, pp 390–394

Setti B, Bencheikh M, Henni JE, Neema C (2011) Advances of molecular markers application in plant pathology research. Eur J Sci Res 50:110–123

Shabir Hussain Wani and Gulzar S. Sanghera (2010) Genetic engineering for viral disease management in plants. Not Sci Biol 2:20–28

Shahid Ali, Kadian MS, Akhtar M, Ortiz O (2013) Epidemiology of aphid vectors of potato viruses in North-Eastern hills of India. In: 12th International symposium on plant virus epidemiology, 28th Jan–1st Feb 2013, Arusha, Tanzania (Abstract, pp 062, p 163)

Shamloul AM, Minafra A, Hadidi A, Giunchedi L, Waterworth HE, Allam EK (1995) Peach latent mosaic viroid: nucleotide sequence of an Italian isolate, sensitive detection using RT-PCR and geographic distribution. Acta Hortic 386:522–530

Shands WA (1977) Control of aphid borne potato virus Y in potatoes with oil emulsions. Am Potato J 54:179–187

Shands WA, Webb RE, Schultz ES (1962) Tests with milk and rice polish to prevent infection in Irish potato with virus Y transmitted by aphids. Am Potato J 39:36–39

Shapovalov M, Blood HL, Christiansen RM (1941) Tomato plant populations in relation to curly top control. Phytopathology 31:864

Sharma SR (1975) Further studies on cowpea viruses. Ph.D. thesis, P.G. School, IARI, New Delhi, 500 pp

Sharma Indu, Dubey GS (1984) Control of *Urdbean leaf crinkle virus* through heat treatment, chemotherapy and resistance. Indian Phytopathol 37:26–30

Sharma SR (1988) *Banana bunchy top virus*. Int J Trop Plant Dis 6:19–41

Sharma SR (1989) Performance of a cross protection trial in Acid Lime (Citrus aurantifolia) against severe strain of Tristeza virus. Int J Trop Plant Dis 7:203–214

Sharma RD, Swarup (1982) Hitherto unrecorded plant parasite nematodes of *Andropogon gayanus* Kunth. var. *bisquanmulatus* Stapf. from Cerrado region of Brazil. In: Lordello LGE (ed) Trabalhos Apresentados na VI Reuniao Brasileira de Nematologia. Publicacao No. 6. Fortaleza, Brazil, pp 99–102

Sharma SR, Varma A (1982) Control of virus diseases by oil sprays. Zbl Mycrobiol 137:329–347

Sharma SR, Varma A (1984) Effect of some cultural practices on virus infection in cowpea. Z Acker Pflazenbau 153:23–31

Sharma BR, Arora SK, Dhanju KC, Ghai TR (1993) Performance of okra cultivars in relation to yellow vein mosaic and yield. Indian J Virol 9:139–142

Sharma U, Chowfla SC, Thakur PD, Handa A, Gupta B (2005) Management of radish mosaic disease in Himachal Pradesh. In: Sharma RC, Sharma JN (eds) Integrated plant disease management. Scientific Publishers, Jodhpur, pp 195–198

Sharma KK, Lavakumar P, Waliyar F, Nigam S, Lavanya M, Reddy AS, Swamykrishnan T (2006) Development and evaluation of transgenic peanuts for induced resistance to Indian peanut clump virus. Indian J Virol 17:12

Sharma S, Singh B, Rani G, Zaidi AA, Hallan VK, Nagpal AK, Virk GS (2008) In vitro production of Indian citrus ringspot virus (ICRSV) free Kinnow plants employing thermotherapy coupled with shoot tip grafting. Plant Cell Tiss Org 92:85–92

Sharp GL, Martin JM, Lanning SP, Blake NK, Brey CW, Sivamani E, Qu R, Talbert LE (2002) Field evaluation of transgenic and classical sources of wheat streak mosaic virus resistance. Crop Sci 42:105–110

Shea KN (1964) Apple virus diseases in Queensland. Qd Agric J 90:682–685

Sheffield FMI (1968) Closed quarantine procedures. Ann Appl Mycol 47:1–8

Shelton AM, Badenes-Perez FR (2006) Concepts and applications of trap cropping in pest management. Annu Rev Entomol 51:285–309

Shepherd RJ, Hills FJ (1970) Dispersal of beet yellows and beet mosaic viruses in the inland valleys of California. Phytopathology 60:798–804

Sherwood JL (1987) Comparision of filter paper immunobinding assay, Western blotting and enzyme linked immunosorbent assay for the detection of wheat streak mosaic virus. Phytopathol Z 118:68–75

Shi AN, Chen PY, Li DX, Zheng CM, Zhang B, Hou AF (2009) Pyramiding multiple genes for resistance to *Soybean mosaic virus* in soybean using molecular markers. Mol Breeding 23:113–124

Shi AN, Vierling R, Grazzini R, Chen PY, Caton H, Panthee D (2011) Identification of molecular markers for Sw-5 gene of tomato spotted wilt virus resistance. Am J Biotechnol Mol Sci 1:8–16

Shiboleth YM, Gal-on A, Koch M, Rabinowitch HD, Salomon R (2001) Molecular characterization of *Onion yellow dwarf virus* (OYDV) infecting garlic (*Allium sativum* L.) in Israel: thermotherapy inhibits virus elimination by meristem tip culture. Ann Appl Biol 138:187–195

Shirshikar SP (2003) Influence of different sowing dates on the incidence of sunflower necrosis disease. HELIA 26:109–116

Shirshikar SP (2008) Integrated management of sunflower necrosis disease. HELIA 31:27–34

Shukla VD, Anjaneyulu A (1980) Evaluation of systemic insecticides for control of rice tungro. Plant Dis 64:79–792

Shukla DD, Teakle DS (1989) *Johnsongrass mosaic virus*. AAB descriptions, No. 340

Shukla DD, Gough KH, Ward CW (1987) Coat protein of potyviruses. III: Comparison of amino acid sequences of the coat proteins of four Australian strains of sugarcane mosaic virus. Arch Virol 96:59–74

Silberschmidt K (1957) Cross protection (pre immunity) tests with two strains of *Potato virus Y*. Archos Inst Biol St Paulo 23:125–150

Silva Rosales L, Becerra L, Leor N, Ruiz Castra S, Teliz Oritz D, Noacarrazana JC (2000) Coat protein sequence comparison of three Mexican isolates of *Papaya ring spot virus* with other geographical isolates reveal a close relationship to American and Australian isolates. Arch Virol 145:835–843

Simmonds JH (1934) Bunchy top of the banana and its control. Qld Agric J 41:241–244

Simmonds JH (1959) Mild strain protection as a means of reducing losses from the Queensland woodiness virus in the passion vine. Qld J Agric Sci 16:371–380

Simons JN (1956) The *Pepper vein banding mosaic virus* in the Everglades area of South Florida. Phytopathology 46:53–57

Simons JN (1957) Three strains of *Cucumber mosaic virus* affecting bell pepper in the everglades area of South Florida. Phytopathology 47:145–150

Simons JN (1958) Viruses affecting vegetable crops in the Everglades and adjacent areas of South Florida. Report of Florida Agricultural Experimental Station, 1957, p 251

Simons JN (1960) Factors affecting field spread of *Potato virus Y* in South Florida. Phytopathology 50:424–428

Simons JN, Zitter TA (1980) Use of oil to control aphid borne viruses. Plant Dis 64:542–546

Simons JN, Conover RA, Walter JM (1956) Correlation of occurrence of *Potato virus Y* with areas of potato production in Florida. Plant Dis Rep 40:531–533

Simons JN, McLean DL, Kinsey MG (1977) Effect of mineral oil on probing behaviour and transmission of stylet-borne virus by Myzus persicae. J Econ Entomol 70:309–315

Singh A (1977) Practical plant physiology. Kalyani Publishers, Ludhaiana, p 66

Singh SJ (1981) The effect of different oils on the inhibition of transmission of *Pumpkin mosaic virus* by aphids. Z Pfl Krankh Pfl Schutz 87:86–98

Singh RP (1985) Clones of *Solanum berthaultii* resistant to *Potato spindle tuber viroid*. Phytopathology 75:1432–1434

Singh SJ (1992) Management of viral diseases of horticultural crops by use of oils. In: Sokhi SS, Chandel SS, Singh PP (eds) Progress in plant pathological research. Indian Society of Plant Pathologists, Ludhiana, pp 223–240

Singh S, Awasthi LP (2006) Protection of mungbean and urdbean crops against vector-borned *Mungbean yellow mosaic virus* through botanicals. Abstract, OP 8/07. In: XVI Annual convention of IVS and International symposium on management of vector-borne viruses, held at ICRISAT, Patancheru, Hyderabad (India)

Singh RA, Nagaich BB (1976) Effect of power oil on aphid transmission of *Potato virus Y*. JIPA 3:21–23

Singh SJ, Sastry KSM (1974) Wilt of pineapple—a new virus disease in India. Indian Phytopathol 27:298–303

Singh SJ, Sharma SR (1989) Immunodiagnosis of virus diseases of horticultural crops. Indian J Virol 5:99–114

Singh JP, Varma JP (1977) Effect of some insecticides and oils on incidence of yellow mosaic and grain yield of mung bean cv. Varsha. Proc Natl Acad Sci India 47(B):219–225

Singh SJ, Sastry KS, Sastry KSM (1979a) Efficacy of different insecticides and oil in the control of leaf curl virus disease of Chillies. Z Pflkrankh Pflschuntz 86:253–256

Singh SJ, Sastry KSM, Sastry KS (1979b) Effect of barrier cropping combined with insecticidal sprays on the control of yellow vein mosaic virus of okra. In: Proceedings of the symposium retrospect and prospects of research in Botany in India, 24–26 Feb, University of Gorakhpur, India

Singh D, Garg HR, Rathore PK (2006) Elite sources of resistance against *Cotton leaf curl disease* in upland cotton. J Cotton Res Dev 20:304–305

Singh BR, Dubey VK, Aminuddin (2007) Inhibition of mosaic disease of gladiolus caused by Bean yellow mosaic- and Cucumber mosaic viruses by virazole. Sci Hortic 114:54–58

Singh HP, Uma S, Selvarajan R, Karihaloo JL (2011) Micro propagation for production of quality banana planting material in Asia-pacific. In: Asia-pacific consortium on Agricultural Biotechnology (APCoAB, New Delhi, India), p 92

Singh MK, Chandel V, Hallan V, Ram R, Zaidi AA (2009) Occurrence of *Peanut stripe virus* on patchauli plants by meristem tip culture. J Plant Dis Prot 116:226

Sittiyos P, Poehlman JM, Sehgal OP (1979) Inheritance of resistance to *Cucumber mosaic virus* in mungbean. Crop Sci 19:51–53

Sivamani E, Huet H, Shen P, Ong CA, de Kochko A, Fauquet CM, Beachy RN (1999) Rice plant *Oryzasativa* L. containing *Rice tungro spherical virus* (RTSV) coat protein transgenes are resistant to virus infection. Mol Breed 5:177–185

Sivamani E, Brey CW, Dyer WE, Talbert LE, Qu R (2000) Resistance to *Wheat streak mosaic virus* in transgenic wheat express-ing the viral replicase (Nib) gene. Mol Breeding 6:469–477

Sivamani E, Brey CW, Talbert LE, Young MA, Dyer WE, Kaniewski WK, Qu R (2002) Resistance to *Wheat streak mosaic virus* in transgenic wheat engineered with the viral coat protein gene. Transgenic Res 11(1):31–41

Slack SA, Singh RP (1998) Control of viruses affecting potatoes through seed potato certification programs. Breeding for resistance to plant viruses. In: Hadidi A, Khetarpal RK, Koganezawa H (eds) Plant virus disease control. APS Press, St. Paul, pp 249–260

Slightom JL, Chee PP, Gonsalves D (1990) Field testing of cucumber plants which express the CMV coat protein gene: field plot design to test natural infection pressures. In: Nijkamp HJJ, van der Plass LPHW, van Aartriik J (eds) Progress in plant cell and molecular biology. Kluwer Academic Publisher, Dordrecht, 201 pp

Slusarek S (1973) Effect of spraying broad bean with oil emulsion on the spread of *Bean common mosaic virus*. Zeszyty Problemowe Postepow Nank Rolniczych 142:137–142

Slykhuis JT (1955) Aceria tulipae Kiefer (Acarnia: Eriophyidae) in relation to the spread of wheat streak mosaic. Phytopathology 45:116–128

Slykhuis JT (1970) Factors determining the development of wheat spindle streak mosaic caused by a soil-borne virus in Ontario. Phytopathology 60:319–331

Slykhuis JT, Andrews JE, Pittmann UJ (1957) Relation of date of seeding winter wheat in southern Alberta to losses from wheat streak mosaic, root rot and rust. Can J Plant Sci 37:113–127

Slykhuis JT, Zillinsky FJ, Young M, Richards WR (1959) Notes on the epidemiology of *Barley yellow dwarf virus* in eastern Ontario. Plant Dis Rep 262(Suppl):317–322

Smartt J (1961) The diseases of groundnuts in Northern Rhodesia. Empire J Exp Agric 29:79–87

Smith FF, Webb RE (1969) Repelling aphids by reflective surfaces—a new approach to the control of insect transmitted viruses. In: Maramorosch K (ed) Viruses, vectors and vegetation. Inter Science Publishers, New York, pp 631–639

Smith FF, Johnson GV, Kahn RP, Bing A (1964) Repelling of aluminum to transient aphid virus vectors. Phytopathology 54:748 (Abstr.)

Smith FF, Boswell AL, Webb RE (1972) Repellent mulches for control of the gladiolus thrips. Environ Entomol 1:672–678

Smith GR, Joyce PA, Handley JA, Sithisarn P, Maugeri MM, Bernard MJ, Berding N, Dale JL, Harding RM, Wilson JR (eds) (1996) Genetically engineering to sugarcane mosaic and Fiji disease viruses in sugarcane, in Sugar-2000 Symposium: Sugarcane: Research towards efficient and sustainable production. CSIRO division of Tropical Crops and Pastures: Brisbane, Australia, pp 138–140

Smith BJ, Keen NT, Becker JO (2003) Acibenzolar S methyl induces resistance to *colletotrichum lagenarium* and *Cucumber mosaic virus* in cantaloupe. Crop Prot 22:769–774

Smykal P, Safarova D, Navratil M, Dostalova R (2010) Marker assisted pea breeding: eIF4E allele specific markers to pea seed-borne mosaic virus (PSbMV) resistance. Mol Breed 26:425–438

Soler-Aleixandre S, Lopez C, Cebolla-Cornejo J, Nuez F (2007) Sources of resistance to *Pepino mosaic virus* (PepMV) in tomato. Hort Science 42:40–45

Somani AK, Kushwah VS, Khurana SMP, Garg ID (2007) Integrated management of stem necrosis disease of potato in India. Potato J 34:231–234

Sowell G Jr (1982) Resistance to *Tobacco mosaic virus* in pepper introductions. Plant Dis 66:1062–1064

Spiegel S (1998) Virus certification of strawberries. In: Hadidi A, Khetarpal RK, Koganezawa H (eds) Plant virus disease control. APS Press, St Paul, pp 320–325

Spiegel S, Frison EA, Converse RH (1993) Recent development in therapy and virus-detection procedures for international movements of clonal plant germ plasm. Plant Dis 77:176–1180

Spiegel S, Stein A, Tam Y (1995) In vitro thermotherapy of Rosaceous fruit trees. Acta Hort 386:419–420

Sreekanth M, Sreeramulu M, Rao RDVJP, Babu BS, Babu TR (2002) Effect of sowing date on *Thrips palmi* Karny population and *Peanut bud necrosis virus* incidence in greengram (*Vigna radiata* L. Wilczek). Indian J Plant Prot 30:16–21

Sreekanth M, Sreenivasulu M, Rao RDVJP, Babu BS, Babu TR (2003) Relative efficacy and economics of different imidacloprid schedules against *Thrips palmi*, the vector of *Peanut bud necrosis virus* on mungbean. Indian J Plant Prot 31:43–47

Sreekanth M, Sreeramulu M, Rao RDVJP, Babu BS, Babu TR (2004) Effect of inter cropping on *Thrips palmi* population and *Peanut bud necrosis virus* incidence in mungbean. Indian J Plant Prot 32:45–48

Sreenivas T, Reddy MV, Jain KC, Reddy MSS (1997) Studies on inheritance and allelic relationships for strain 2 of *Pigeonpea sterility mosaic* pathogen. Ann Appl Biol 130:105–110

Sreenivasulu P, Subba Reddy CHV (2006) Managing plant viruses through genomic manipulations. In: Gadevar AV, Singh BP (eds) Plant protection in new millennium, vol I. Satish Serial Publishing House, New Delhi, pp 463–514

Sreenivasulu P, Subba Reddy CHV, Ramesh B, Kumar PL (2008) Important virus diseases of groundnut. In: Rao GP, Paul Khurana SM, Lenardon SL (eds) Characterization diagnosis and management of plant viruses, vol-1. Industrial crops. Studium Press LLC, Houstan, pp 47–98

Srinivasan K, Mathivanan N (2011) Plant growth promoting microbial consortia mediated classical biocontrol of *Sunflower necrosis virus disease*. J Biopesticides 4:65–72

Srinivasulu B, Jeyarajan R (1989) Resistance of rice varieties to *Rice tungro virus* (RTV) and its green leafhopper (GLH) vector in Tamilnadu, India. IRRN 14:14

Srivastava A, Srivastava S, Verma HN (2006) Induction of antiviral resistance in *Carica papaya* by systemic resistance inducing protein from *Clerodendrum aculeatum*. Abstract, OP 8/05. In: XVI Annual convention of IVS and International symposium on management of vector-borne viruses, held at ICRISAT, Patancheru, Hyderabad (India)

Stace-Smith R, Mellor FC (1970) Eradication of *Potato spindle tuber virus* by thermotherapy and axillary bud culture. Phytopathology 60:1857–1858

Stanley PA, Calvo FJ, Urbaneja A (2005) Augumentive biological control of *Bemisia tabaci* biotype Q in Spanish greenhouse pepper production using *Eretmoceus mundus*. Crop Prot 24:829–835

Stannard MC, Evans JC, Long JK (1975) Effect of transmission of exocortis dwarfing factors into Washington navel orange trees. Aust J Exp Agric 15:136–141

Staples R, Allington WB (1956) Streak mosaic of wheat in Nebraska and its control. Res Bull Neb Agric Exp Stn 178:1–40

Stapleton JJ, Summers CG (2002) Reflective mulches for management of aphids and aphid-borne virus diseases in late-season cantaloupe (*Cucumis melo* L. var. *cantalupensis*). Crop Prot 21:891–898

Stark DM, Beachy RN (1989) Protection against potyvirus infection in transgenic plants: evidence for broad spectrum resistance. Bio/Technology 7:1257–1262

Stark DM, Register JC, Nejidat A, Beachy RN (1990) Toward a better understanding of coat protein mediated protection. In: Lamb CJ, Beachy RN (eds) Plant gene transfer. Alan R. Lis, New York, pp 275–287

Stevens MR, Scott SJ, Gergerich RC (1992) Inheritance of a gene for resistance to *Tomato spotted wilt virus* (TSWV) from Lycopersicon peruvianum Mill. Euphytica 59:9–17

Stevens MR, Scott SJ, Gergerich RC (1994) Evaluation of seven Lycopersicon species for resistance to *Tomato spotted wilt virus* (TSWV). Euphytica 80:79–84

Stone OM, Hollings M (1976) Report of the Glasshouse Crops Research Institute, 1975, 119 pp

Storey HH (1935) Virus diseases of East African plants III. Rosette disease of groundnuts. East Afr Agric J 1:206–211

Stover RH (1997) Banana (Musa spp.). In: Hewitt WB, Chiarappa L (eds) Plant health and quarantine in international transfer of genetic resources. CRC Press, Boca Raton, pp 71–79

Strickland AH (1951a) The entomology of swollen shoot of cacao. I. The insect species involved with notes on their biology. Bull Entomol Res 41:725–748

Strickland AH (1951b) The entomology of swollen shoot of Cocoa. II. The binomics and ecology of the species involved. Bull Entomol Res 42:65–103

Stubbs LL (1948) A new virus disease of carrots: its transmission, host range and control. Aust J Sci Res Ser B 1:303–332

Stubbs GC (1961) Some island races of butterflies and their conservation. In: Nature conservation: in Western Malaysia, Malay. National Society, Kuala Lumpur, pp 240–243

Stubbs LL (1964) Transmission and protective inoculation studies with viruses of the citrus tristeza complex. Aust J Agric Res 15:752–770

Stubbs LL, Guy JAD, Stubbs KL (1963) Control of *Lettuce necrotic yellows virus* disease by the destruction of common sowthistle (*Sonchus oleraceus*). Aust J Exp Agric Animal Husb 3:215–218

Stuchi ES, da Silva SR, Dondio LC, Sempionato OR, Reiff ET (2007) Field performance of "marsh seedles' grapefruit on trifoliate orange inoculated with viroids in Brazil. Scientia Agriculturae (Piradicaba, Brazil) 64:582–588

Subba Reddy ChV, Sreenivasulu P, Sekhar G (2011) Duplex-immunocapture RT-PCR for detection and discrimination of two distinct potyviruses naturally infecting sugarcane (*Saccharum* spp. hybrid). Indian J Exp Biol 49:68–73

Subrahmanyam P, Hildebrand GL, Naidu RA, Reddy JL, Singh AK (1998) Sources of resistance to groundnut rosette disease in global groundnut germplasm. Ann Appl Biol 138:473–485

Subrahmanyam P, van der Merwe PJA, Chiyembekeza AJ, Chandra S (2002) Integrated management of groundnut rosette disease. Afr Crop Sci J 10:99–110

Sudarshana MR, Roy G, Falk BW (2007) Methods for engineering resistance to plant viruses. Methods Mol Biol 354:183–196

Sudhakar N, Balu PM, Krishna Kumar S, Periyar Selvam S, Murugesan K (2006) Induction of resistance in *Capsicum annum* against *Cucumber mosaic virus* by plant growth promoting rhizobacteria. Abstract OP 9/25. In: XVI Annual convention of IVS and International symposium on management of vector-borne viruses, held at ICRISAT, Patancheru, Hyderabad (India)

Sudhir Kumar S, Ashok Kumar CT, Shivaraju C, Thippaiah M (2012) Efficacy of biochemical agents in the management of onion thrips, *thrips tabaci* (L.) on onion. J Biol Control 26:173–178

Sujatha M, Vijay S, Vasavi S, Veera Reddy P, Chander Rao S (2012) Agro bacterium mediated transformation of cotyledons of mature seeds of multiple genotypes of sunflower (Helianthus annuus L.). Plant Cell Tissue and Organ Culture 110:275–287

Summers CG, Stapleton JJ (1999) Management of aphids, silverleaf whiteflies, and corn stunt leafhoppers using reflective plastic mulch and insecticides: 1998 season review. U.C. Plant Prot Q 9:2–7

Summers CG, Stapleton JJ (2002) Use of UV reflective mulch to delay the colonization and reduce the severity of *Bemisia argentifolii* (Homoptera: Aleyrodidae) infestations in cucurbits. Crop Prot 21:921–928

Summers CG, Stapleton JJ, Newton AS, Duncan RA, Hart D (1995) Comparison of sprayable and film mulches in delaying the onset of aphid-transmitted virus diseases in zucchini squash. Plant Dis 79:1126–1131

Summers CG, Mitchell JP, Stapleton JJ (2004) Management of aphid borne viruses and *Bemisia argentifolii* (Homoptera: Aleyrodidae) in zucchini squash by using UV reflective plastic and wheat straw mulches. Environ Entomol 33:1447–1457

Summers CG, Stapleton JJ, Michell JP (2007) Aphid and aphid-borne virus management: Squash disease control. In: Encyclopedia of pest managenent. Taylor and Francis. CABI publications, London, pp 17–20

Suteri BD (1986) Resistance in soybean cultivars to two strains of *Soybean mosaic virus*. Indian Phytopathol 39:578–580

Tachibana Y (1981) Control of aphid-borne viruses in faba bean by mulching with silver polyethylene film. FABIS Newsl 3:56

Tacke E, Salamini F, Rohde W (1996) Genetic engineering of potato for broad-spectrum protection against virus infection. Nat Biotech 14:1597–1601

Tan RR, Wang LP, Hong N, Wang GP (2010) Enhanced efficiency of virus elimination following thermotherapy of shoot-tip cultures of pear. Plant Cell Tiss Org 101:229–235

Tanaka H, Yamada S, Kishi K (1968) Cross protection tests on the tristeza virus strains carried in Satsuma mandarin and Hassaku trees. Bull Hortic Res Stn Okitsu B 8:79–90

Tao X, Zhou X (2004) A modified viral satellite DNA that suppresses gene expression in plants. Plant J 38:850–860

Tarafder P, Mukhopadhyay S (1979) Potential of weeds to spread rice tungro in West Bengal, India. Int Rice Res Newsl 4:11

Tarafder P, Mukhopadhyay S (1980) Further studies on the potential of weeds to spread tungro in west Bengal, India. Int Rice ResNewsl 5:10

Tarafdar J, Biswas S, Chowdhury AK (2001) Reaction of *Rice tungro virus* and its vector *Nephotettix virescens* (Distant) with some traditional and high yielding varieties of rice. J Int Acad 5:142–149

Tarr SAJ (1951) Leaf curl disease of cotton. Commonwealth Mycological Institute, Kew 55 pp

TCP NCS (2008) National Certification System for Tissue Culture Raised Plants: standard operating procedures for recognized tissue culture production facility. Department of Biotechnology, Government of India

Teakle DS, Pritchard AJ (1971) Resistance of Krish sorghum to four strains of *Sugarcane mosaic virus* in Queensland. Plant Dis Rep 55:596–598

Temple SR, Morales FJ (1986) Linkage of dominant hypersensitive resistance to *Bean common mosaic virus* to seed color in *Phaseolus Vulgaris* L. Euphytica 35:331–333

Tennant PF, Gonsalves C, Ling KS, Fitch M, Manshardt R, Slightom JL, Gonsalves D (1994) Differential protection against *Papaya ringspot virus* isolates in coat protein gene transgenic papaya and classically cross-protected papaya. Phytopathology 84:1359–1366

Tepfer M (2002) Risk assessment of virus-resistant transgenic plants. Annu Rev Phytopathol 40:467–491

Thakur MR (1986) Breeding for disease resistance in okra. Veg Sci 13:310–315

Thakur PD, Thakur VS, Bharadwaj SV (1995) Disease reaction of pea germplasm to *Pea seed-borne mosaic virus*. Plant Dis Res 10:140–143

Theuri JM, Bock KR, Woods RD (1987) Distribution, host range and some properties of a virus disease of sunflower. Trop Pest Manag 33:202–206

Thira SK, Cheema SS, Kang SS (2004) Pattern of bud necrosis disease development in groundnut crop in relation to different dates of sowing. Plant Dis Res (Ludhiana)19:125–129

Thomas W (1973) Seed transmitted *Squash mosaic virus*. N Z J Agric Res 16:561–567

Thomas JE, Geering ADW, Gambley CF, Kessling AF, White M (1997) Purification, properties and diagnosis of *Banana bract mosaic potyvirus* and its distinction from abaca mosaic potyvirus. Phytopathology 87:698–705

Thomsen A (1975) Crop reduction in pear trees caused by virus infection. Acta Hortic 44:119–122

Thorne G, Allen MW (1950) *Paratylenchus hamatus* n. sp. and *Xiphinema index* n. sp., two nematodes associated with fig roots, with a note on *Paratylenchus anceps* Cobb. Proc Helminth Soc Wash 17:27–25

Thottappilly G, Hamilton RI, Huguenot C, Rossel HW, Furneaux MT, Gumedzoe MY, Shoyinka SA, Naik DM, Konate G, Atcham-Agneroh T, Haciwa HC, Anno-Nyako FO, Saifodine N, Wangai A, Lamptey P, Gubba A, Mbwaga AM, Neya J, Offei SK (1993) Identification of cowpea viruses and their strains in tropical Africa—an international pilot project. FAO Plant Prot Bull 41:65–72

Thouvenel J-C, Fauquet C, Fargette D, Fishpool LDC (1988) *Peanut clump virus* in West Africa. In: Cooper JI, Asher MJC (eds) Developments in applied biology II. Viruses with fungal vectors. Association of Applied Biologists, Wellesbourne, pp 247–254

Thresh JM (1959) The control of *Cacao swollen shoot disease* in Nigeria. Trop Agric Trin 36:35–45

Thresh JM (1964) Black current revision disease. Jubilee Annu Rep East Malling Res Stn Kent 1963:184–189

Thresh JM (1976) Gradients of plant virus diseases. Ann Appl Biol 82:381–406

Thresh JM (1988) Eradication as a virus disease control measure. In: Clifford BC, Lester E (eds) Control of plant diseases. Costs and benefits. Blackwell Scientific Publications, Oxford, pp 155–194

Thresh JM (1989) Insect-borne viruses of rice and the green revolution. Trop. Pest Manag 35:264–272

Thresh JM (2003) Control of plant virus diseases in sub-saharan Africa: the possibility and feasibility of an integrated approach. Afr Crop Sci J 11:199–223

Thresh JM (2006) Control of tropical plant virus diseases. Adv Virus Res 67:245–295

Thresh JM, Lister RM (1960) Copping experiments on the spread and control of cacao swollen-shoot disease in Nigeria. Ann Appl Biol 48:65–74

Thresh JM, Otim-Nape GW, Jennings DL (1994) Exploiting resistance to *African cassava mosaic virus*. Ann Appl Biol 39:51–60

Thresh JM, Otim-Nape GW, Fargette D (1998) The control of *African cassava mosaic virus* disease : Phytosanitation and/or resistance. In: Hadidi A, Khetarpal RK, Kaganezauva H (eds) Plant virus disease control. APS Press, St Paul, pp 670–677

Timian RG (1971) *Barley stripe mosaic virus* in North Dakota. N D Agric Exp Stn Farm Res 28:3–6

Timian RG, Franckowiak JD (1987) Location of a factor for *Barley stripe mosaic virus* reaction on chromosome. 1. Barley Genetic Newsl 17:79–82

Timian RG, Sisler WW (1955) Prevalence, sources of resistance, and inheritance of resistance to *Barley stripe mosaic virus* (false stripe). Plant Dis Rep 39:550–552

Timmerman GM, Pither-Joyce MD, Cooper PA, Russell AC, Goulden DS, Butler R, Grant JE (2001) Partial resistance of transgenic peas to *Alfalfa mosaic virus* under greenhouse and field conditions. Crop Sci 41:846–853

Tingey WM, Laubengayer JE (1981) Defence against the green peach aphid and potato leafhopper by glandular trichomes of Solanum berthaultii. J Econ Entomol 74:721–725

Tiongco ER, Chancellor TCB, Villareal S, Magbanus M, Teng PS (1998) Roguing as a tactical control for the *Rice tungro virus disease*. J Plant Prot Trop 11:45–52

Tiongco ER, Truong HX and Cabunagan RC (2008) Designing appropriate tungro management strategies. In: Tiongco ER, Angeles ER, Sebastian LS (ed) *Rice tungro virus disease*: a paradigm in disease management. Philippine Rice Research Institute, Science City of Munoz, pp 213–251

Toba HH, Kishaba AN, Bohn GW, Hield H (1977) Protecting muskmelons against aphidborne viruses. Phytopathology 67:1418–1423

Todd JM (1961) The incidence and control of aphid-borne potato virus diseases in Scotland. Eur Potato J 4:316–329

Tomassoli L, Kaniewski V, Ilardi T, Mitsky J Layton, Barba M (1995) Transgenic tomato plants resistant to CMV: two years of field trials. In: Proceedings of the 8th conference on virus diseases of vegetables, Prague, Czech Republic

Tomenius K, Clapham D, Meshi T (1987) Localization by immunogold cytochemistry of the virus coded 30 K protein in plasmodesmata of leaves infected with tobacco mosaic virus. Virology 160:363–371

Tomlinson JA (1962) Control of lettuce mosaic by the use of healthy seeds. Plant Pathol 11:61–64

Tomlinson JA, Carter AL (1970) Seed transmission of *Cucumber mosaic virus* in chickweed. Plant Dis Rep 54:150–151

Tomlinson JA, Faithfull EM (1979) Effects of fungicides and surfactants on the zoospores of Olpidium brassicae. Ann Appl Biol 93:13–19

Tomlinson JA, Faithfull EM (1980) Studies on the control of lettuce big-vein disease in recirculated nutrient solutions. Acta Hortic 98:325–332

Tomlinson JA, Thomas BJ (1986) Studies on *Melon necrotic spot virus* disease of cucumber and on the control of the fungus vector (*Olpidium radicale*). Ann Appl Biol 108:71–80

Tomlinson JA, Walker VM (1973) Further studies on seed-transmission in the ecology of some aphid-transmitted virus. Ann Appl Biol 73:292–298

Torrance L (1995) Use of monoclonal antibodies in plant pathology. Eur J Plant Pathol 101:351–363

Torres AC, Teixeira DMC, Moita AW, Campos M (1996) Recovery of virus-free plants of sweet potato through direct regeneration of shoot tip. Revista Brasileira de Fisiologia Vegetal 8(3):209–213

Toscano NC, Wyman TJ, Kido K, Johnson H Jr, Mayberry K (1979) Reflective mulches foil insects. Calif Agric 33:17–19

Townsend GR (1947) Celery mosaic in the Everglades. Plant Dis Rep 31:118–119

Traore O, Traore MD, Fargette D, Konate G (2006) Rice seed beds as a source of primary infection by *Rice yellow mottle virus*. Eur J Plant Pathol 115:181–186

Tricoli DM, Carney KJ, Russell PF, Mc Master JR, Groff DW, Hadden KC, Himmerl PT, Hubbard JP, Boeshore ML, Quemada HD (1995) Field evaluation of transgenic squash containing single or multiple virus coat protein gene constructs for resistance to *Cucumber mosaic virus*, *Watermelon mosaic virus 2* and *Zucchini yellow mosaic virus*. Biotechnology 13:1458–1465

Tripathi L, Tripathi JN, Tushemereirwe WK (2008) Rapid and efficient production of transgenic East African Highland Banana (Musa spp.) using intercalary meristematic tissues. Afr J Biotechnol 7:1438–1445

Turnage MA, Muangsan N, Peele CG, Robertson D (2002) Geminivirus-based vectors for gene silencing in *Arabidopsis*. Plant J 30:107–117

Turner NE, O'Connell KMO, Nelson RS, Sanders PR, Beachy RN, Fraley RT, Shah DM (1987) Expression of *Alfalfa mosaic virus* coat protein gene confers cross-protection in transgenic tobacco and tomato plants. EMBO J 6:1181–1188

Turyamureeba G, Mwanga ROM, Odongo B, Ocitti p'Obwoya, Carey EE (1998) Application for inclusion of a crop/variety in the National Cultivar List. Submission to the variety release committee for release of sweetpotato varieties. The Ugandan Ministry of Agriculture, Animal Industries and Fisheries, Uganda

Tyagi H, Uma G, Rajasubramaniam S, Mathur S, Dutt N, Paul S, Tandon V, Suri SS, Rajam VV, Dasgupta I (2006) Engineering transgenic resistance against tungro viruses in rice using multiple approaches. 16th annual convention of IVS and international symposium on management of vector-borne viruses, held at ICRISAT, Patancheru, Hyderabad (India). Abstr.: OP 8/04

Tyagi H, Rajasubramaniam S, Rajam MV, Dasgupta I (2008) RNA-interference in rice against *Rice tungro bacilliform virus* results in its decreased accumulation in inoculated rice plants. Transgenic Res 17:897–904

Uchanski M, Skirvin RM, Norton MA (2002) The use of in vitro thermotherapy to obtain Turnip mosaic virus-free horseradish plants. Acta Hortic 631:175–179

Ullman DE, Qualset CO, McLean DL (1988) Feeding Response of *Rhopalosiphum padi* to *Barley Yellow Dwarf Virus* in resistant and susceptible barley varieties. Environ Entomol 17:988–991

Ulrychova M, Blattny C (1961) L'action Synergique de la simazine et des viroses de plantes en taut que methode de la detection eventuelle des maladies a virus. Biol Plant Praha 3:122–125

Upadhyaya NM, Li Z, Wang MB, Chen S, Gong ZX, Waterhouse PM (2001) Engineering for virus resistance in rice. In: Khush GS, Brar DS, Hardy B (eds) Rice genetics. Proceedings of the 4th international rice genetics symposium, pp 405–421

Upstone ME (1974) Effects of inoculation with the Dutch mutant strain of *Tobacco mosaic virus* on the cropping of commercial glasshouse tomatoes. Report of the Agricultural Development and Adv Service (Science Arm) for 1972, pp 162–165

USDA (1933–1978) Export certification Manuel, vols 1 and 2. USDA, Washington, DC

Vanachter A (1995) Development of Olpidium and Pythium in the nutrient solutions of NFT grown lettuce, and possible control methods. Acta Hortic 382:187–196

Van Den Bosch R (1976) Informe sobre el control biologico de los afldos de los cereales en Chile. Agric Tech (Chile) 36:141–145

Van der Merwe PJA, Subrahmanyan P (1997) Screening of rosette-resistance early groundnut breeding lines for yield and other characteristics. Int Arachis Newsl 17:23–24

Vander plank JE (1947) The relation between size of plant and the spread of systemic diseases. 1. A discussion of ideal cases and a new approach to the problems of control. Ann Appl Biol 34:376–387

Vander plank JE (1949) The relation between the size of the fields and the spread of plant-disease into them. Part III. Examples and discussion. Empire J Expt Agric 17(67):141–147

Vander plank JE (1963) Plant disease: epidemics and control. Academic Press, New York, 349 pp

Vander plank JE (1968) Disease resistance in plants. Academic Press, New York 206 pp

Vander plank JE, Anderssen EE (1944) Kromneck disease of tobacco. Farming S Afr 19:391–394

Vanderschuren H, Moreno I, Anjanappa RB, Zainuddin IM, Gruissem W (2012) Exploiting the combination of natural and genetically engineered resistance to *Cassava mosaic* and *Cassava brown streak viruses* infecting cassava production in Africa. Online: http://dx/plos.org.10.1371/journal/pone.0045277

Vanderveken J (1968) Effects of mineral oils and lipids on aphid transmission of *Beet mosaic* and *Beet yellows viruses*. Virology 34:807–809

Vanderveken J (1972) Inhibition de la transmission du virus de la mosaique de la Luzerne par Myzus persicae l'aide, d'huile. Parasitica 28:39–45

Vanderveken J (1973) Recherché du mecanisme de l'inhibition de la transmission aphidienne des phyto-virus par des substances hilleuses. Parasitica 29:1–15

Vanderveken J (1977) Oils and other inhibitors on non-persistent virus transmission. In: Harris KF, Maramorosch K (eds) Aphids as virus vectors. Academic Press, New York, pp 435–454

Vanderveken J, Coutisse S (1975) Essai de lutte contre le virus de la mosaique du tabacchez la tomate par la premonition. Meded Fac Landbouwwet Rijksuniv Genet 40:791–797

Vanderveken J, Dutrecq A (1970) Contribution a l'etude de l'action inhibitrice d'une huile minerale sur la transmission aphidienne des phytovirus. Annls Phytopathol 2:387–402

Vanderveken J, Ohn JC (1968) Inhibition de la transmission du virus de la jannisse de la betterave par *Aphis fabae*, scop. a l'aide de pulverizations d'emulsions huileuses. Meded Rijksfakulteit Land bouwweten-schappen Gent 33:1215–1221

Vanderveken J, Semal J (1966) Aphid transmission of *Beet yellow virus* inhibited by mineral oil. Phytopathology 56:1210–1211

Vanderveken J, Vilain N (1967) Inhibition de la transmission aphidienne de quelques phytovirus an moyen de pulverizations d'huile. Annls Epiphyt 18:125–132

Vanderveken J, Bourge JJ, Semal J (1966) Effects d'huiles minerales sur la transmission de phytovirus par pucerons. Meded Rijksfakulteit Landbouwwe-tenschappen Gent 31:974–980

Vanderveken J, Semal J, Vanderwalle R (1968) Perspectives d'utilisation des huiles dans la lutte contre la transmission des phytovirus. Annls Gembloux 74:47–52

Vander Want JPM (1954) Onderzoeking over virusziekten van de boon (Phaseolus vulgaris L.). Ph.D. Thesis, Wageningen Agricultural University, Wageningen, The Netherlands

Vander Wilk F, Willink DP-L, Huisman MJ, Huttinga H, Goldbach R (1991) Expression of *Potato leafroll luteovirus* coat protein gene in transgenic potato plants inhibits viral infection. Plant Mol Biol 17:431–439

Van Dun CMP, Bol JF (1988) Transgenic tobacco plants accumulating *Tobacco rattle virus* coat protein resist infection with *Tobacco rattle virus* and *Pea early browning virus*. Virology 167:649–650

Van Dun CMP, Bol JF, Van Vloten-Doting L (1987) Expression of *Alfalfa mosaic virus* and *Tobacco rattle virus* coat protein genes in transgenic plants. Virology 159:299–311

Van Dun CMP, van Vloten-Doting L, Bol JF (1988a) Expression of *Alfalfa mosaic virus* cDNA 1 and 2 in transgenic tobacco plants. Virology 163:572–578

Van Dun CMP, Verduin B, van Vloten-Doting L, Bol JF (1988b) Transgenic tobacco expressing *Tobacco streak virus* or mutated *Alfalfa mosaic virus* coat protein does not cross protect against *Alfalfa mosaic virus* infection. Virology 164:383–389

Van Emden HF (1982) Principles of implimentation of IPM. In: Cameron P, Wearing CH, Kain WM (eds) Proceedings of the Australian workshop on development and implementation of IPM. Government Printer, Auckland, p 9

Van Os EA (1980) Complete mechanism of the growing of cut chrysanthemums in nutrient film. ISOSC Proceedings 5th Congress, pp 187–196

Vani S, Varma A, More TA, Srivastava KP (1989) Use of mulches for the management of mosaic disease in muskmelon. Indian Phytopathol 42:227–235

Van Regenmortel MHV (1982) Serology and immunochemistry of plant viruses. Academic Press, New York 268 pp

Van Rheenen HA, Hasselbach OE, Muigai SGS (1981) The effect of growing beans together with maize on the incidence of bean diseases and pests. Neth J Plant Pathol 87:193–199

Varma A (1993) Integrated management of plant viral diseases. Ciba Found Symp 177:140–155

Varma A (2006) Ecological impact of GM crops resistant to viruses. In: The 9th international symposium on the biosafety of genetically modified organisms, Jeju Island, Korea, pp 54–58

Varma A, Mitter N (2001) Durable host-plant resistant, a desirable trait for integrated disease management. In: Peng S, Hardy B (eds) Rice research for food security and poverty alleviation. IRRI, Manila, pp 324–344

Varma A, Khetarpak RK, More TA (1991) Genetical resistance in plants to viruses. In: Proceedings of the Golden Jubilee symposium on genetics research and education: current trends and the next 50 years held at IRI, New Delhi, vol 1, pp 127–128

Varma A, Jain RK, Bhat AI (2002) Virus resistant transgenic plants for environmentally safe management of viral diseases. Indian J Biotech 1:73–86

Vasane R, Shailesh RM, Kothari (2006) Optimization of secondary hardening process of banana plantlets (*Musa paradisiaca* L. var. Grand Naine). Indian J Biotechnol 5:394–399

Vasanthi VJ, Shanmugam V, Ramiah M (2001) Elimination of Indian cassava mosaic virus in cassava (*Manihot esculenta* Crantz) through meristem tip culture and specific detection by serological means. Orissa J Hortic 29:69–74

Vasudeva RS (1959) Plant virus research in India. Indian Phytopathol 12:1–7

Vasudevan A, Oh TK, Park JS, Lakshmi SV, Choi BK, Kim SH, Lee HJ, Ji J, Kim JH, Ganapathi A, Kim SC, Choi CW (2008) Characterization of resistance mechanism in transgenic Nicotiana benthamiana containing Turnip crinkle virus coat protein. Plant Cell Rep 27:1731–1740

Veelar BM, Ferreira DI, Niederwieser JG (1992) Elimination of ornithogalum mosaic virus in the ornithogalum rogel cv. through meristem tip culture and chemotherapy. Plant Cell Tissue Organ Cult 29:51–55

Venkata Ramana C, Prasada Rao RDVJ, Reddy IP, Venkata Rao P, Reddy YN (2006) Screening of tomato germplasm and wild relatives against *Peanut bud necrosis virus* (PBNV) disease. Indian J Plant Prot 34:59–61

Venkatesh HM (2000) Studies on tomato leaf curl geminivirus and *Bemisia tabaci* (Gennadius): molecular detection farmers perception and sustainable management. Ph.D. Thesis, University of Agricultural Sciences, Bangalore

Verhoyen M (1973) Premises resultats des essais de pulverisation d'huile pour lutter contre la dissemination du virus de la mosaique du Celeri (Celery mosaic virus) an champ. Phytopath Z 78:289–300

Verma N, Ram R, Hallan V, Kumar K, Zaidi AA (2004) Production of *Cucumber mosaic virus*-free chrysanthemums by meristem tip culture. Crop Prot 23:469–473

Verma N, Ram R, Zaidi AA (2005) In vitro production of Prunus necrotic ringspot virus-free begonias trough chemo- and thermotherapy. Sci Hortic 103:239–247

Verniere C, Botella L, Dubois A, Chabrier C, Duran-Vila N (2002) Properties of citrus viroids: symptom expression and dwarfing. In: Duran-Vila N, Milne RG, da Graca JV (eds) Proceedings of the 15th conference of International Organization of Citrus Virologists. IOCV, Riverside, pp 240–248

Verniere C, Perrier X, Dubois C, Dubois A, Botella L, Chabrier C, Bove JM, Duran-Vila N (2004) Citrus viroids: symptom expression and effect on vegetative growth and yield of Clementine trees grafted on trifoliate orange. Plant Dis 88:1189–1197

Vidalakis G, Bash JA, Semancik JS (2007) Transmissible small nuclear RNA dwarfing of commercial citrus on Carrizo citrange rootstock. In: Proceedings of the 17th conference of International Organization of Citrus Virologists. IOCV, Riverside

Vidalakis G, Pagliaccia D, Bash JA, Semancik JS (2010) Effects of mixtures of citrus viroids as transmissible small nuclear RNA (TsnRNA) on tree dwarfing and commercial scion performance on Carrizo citrange rootstock. Ann Appl Biol 157:415–423

Vidavsky F, Czosnek H (1998) Tomato breeding lines resistant and tolerant to tomato yellow leaf curl virus issued from *Lycopersicon hirsutum*. Phytopathology 88:910–914

Vidavsky F, Leviatov S, Millo J, Rabinowitch HD, Kedar N, Czosnek H (1998) Response of tolerant breeding lines of tomato, *Lycopersicon esculentum,* originating from three different sources (*L. peruvianum, L. pimpinellifolium* and *L. chilense*) to early controlled inoculation by tomato yellow leaf curl virus (TYLCV). Plant Breeding 117:165–169

Vilardebo A (1955) La cochenille de l'ananas Pseudococcus brevipes CKL et le wilt qu'elle provoque. Fruits 10:59

Voinnet O (2005) Induction and suppression of RNA silencing: insights from viral infections. Nat Rev Genet 6:206–220

Vuittenez A (1957) Lutte preventive contre le court-noué de la vigne par desinfection chimique du sol avant plantation (Preventive control of grapevine fanleaf by soil chemical disinfection before planting). C R Acad Agric France 43:185

Vuittenez A (1958) Transmission par graffage dune virose du type enroulement folaire commune dons les vignobles de lest et du centrest de la France. C R Acad Agric France 44:313–316

Vuittenez A (1960) Nouvelles observations sur l'activite des traitements chimiques du sol pour l'eradication des virus de la degenerescence infectieuse (New observations on the activity of soil treatments for eradicating the viruses of infectious degeneration). C R Acad Agric France 46:89–96

Vuittenez A (1961) Les Nematodes vecteurs de virus et le probleme de la degenerescence infectieuse de la vigne. Les Nematodes C.N.R.A. Versailles 55:78

Vuittenez A (1970) Fanleaf of grapevines. In: Frazier NW et al (eds) Virus diseases of small fruits and grapevines. University of California Press, Berkeley, pp 217–228

Vuylsteke D, De Langhe E (1985) Feasibility of *In vitro* propagation of bananas and plantains. Trop Agric (Trinidad) 62:323–328

Walkey DGA (1968) The production of virus-free Rhabarber (*Rheum rhaponticum*) by apical culture. J Hortic Sci 43:283–287

Walkey DGA (1978) *In vitro* methods for virus elimination. In: Thorpe TA (ed) Frontiers in plant tissue culture. University Calgary Press, Calgary, pp 245–254

Walkey DGA, Cooper VC (1971) Effect of western celery mosaic on celery crops in Britain and occurrence of the virus in umbelliferous weeds. Plant Dis Rep 55:268–271

Walkey DGA, Dance MC (1979) The effect of oil sprays on aphid transmission of turnip mosaic, beet yellows, *Bean common mosaic*, and *Bean yellow mosaic viruses*. Plant Dis Rep 63:877–881

Walkey DGA, Innes NL (1979) Resistance to *Bean common mosaic virus* in dwarf beans (*Phaseolus vulgaris* L.). J Agric Sci 92:101–108

Walkey DGA, Pink DAC (1984) Resistance in vegetable marrow and other Cucurbita spp. to two British strains of *Cucumber mosaic virus*. J Agric Sci 102:197–205

Walkey DGA, Ward CM, Phelps K (1985) Studies on *Lettuce mosaic virus* resistance in commercial lettuce cultivars. Plant Phytopathol 34:545–551

Walkey DGA, Cooper VC, Crisp P (1974) The production of virus-free cauliflowers by tissue culture. J Hortic Sci 49:273–275

Walkey DGA, Webb MJW, Bolland CJ, Miller A (1987) Production of virus-free garlic (*Allium sativum* L.) and shallot (*A. ascalonicum* L.) by meristem-tip culture. J Hortic Sci 62:211–220

Wallace JL, Murphy AL (1938) Studies on the epidemiology of curly top in southern Idaho with special reference to sugarbeets and weed hosts of the vector *Eutettix tenellus*. USDA Tech Bull 624:1–47

Wallis RL (1967) Some host plants of the green peach aphid and beet western yellows virus in pacific Nothwest. Ann Appl Biol 73:293–298

Wallis RL, Turner JE (1969) Burning weeds in drainage ditches to suppress population of green peach aphids and incidence of beet Western yellows disease in sugarbeets. J Econ Entomol 62:307–312

Walsh JA (1992) Resistant watercress. Grower 118:18–21

Walsh JA (1998) Chemical control of fungal vectors of plant viruses. In: Hadidi A, Khetarpal RK, Koganezawa H (eds) Plant virus disease control. APS Press, St Paul, pp 196–207

Walsh JA (2000) Transgenic approaches to disease resistant plants as exemplified by viruses. In: Dickinson M, Beynon J (eds) Molecular plant pathology. Annual plant reviews, vol 4. Sheffield Academic Press, Sheffield, pp 218–252

Walsh KB, Guthrie JN, White DT (2006) Control of phytoplasma disease of papaya in Australia using netting. Aust Plant Pathol 25:49–54

Wang PJ, Hu NY (1980) Regeneration of virus-free plants through *in vitro* culture. In: Frechter A (ed) Advances in biochemical engineering. Plant Cell Culture II. Springer, Berlin, pp 61–99
Wang HL, Yeh SD, Chiu RI, Gonsalves D (1987) Effectiveness of cross protection by mild mutants of *Papaya ringspot virus* for control of ringspot disease in Taiwan. Plant Dis 71:491–497
Wang HL, Gonsalves D, Provvidenti R, Lecoq HL (1991) Effectiveness of cross protection by a mild strain of zucchini yellow mosaic virus in cucumber, melon, and squash. Plant Dis 75:203–207
Wang JiXuan, Liu Zhi, Xie XiuHua, Wu Bin, Tong ZhaoGuo (2000) The reaction of virus-free apple tree growth and fruit production. Acta Horticulturae Sinica 27:157–160
Wang L, Wang G, Hong N, Tang R, Deng X, Zhang H (2006a) Effect of thermotherapy on elimination of *Apple stem grooving virus* and *Apple chlorotic leafspot virus* for *in vitro* cultured pear shoot tips. Hortic Sci 41:729–732
Wang Q, Liu Y, Xie Y, You M (2006b) Cryotherapy of potato shoot tips for efficient elimination of *Potato leafroll virus* (PLRV) and *Potato virus Y* (PVY). Potato Res 49:119–129
Wang Q, Cuellar WJ, Rajamaki ML, Hirata Y, Valkonen JP (2008) Combined thermotherapy and cryotherapy for efficient virus eradication: relation of virus distribution, subcellular changes, cell survival and viral RNA degradation in shoot tips. Mol Plant Pathol 9(2):237–250
Wang K, Hu Y, Liu Y, Mi N, Fan Z, Liu Y, Wang Q (2010a) Design, synthesis, and antiviral evaluation of Phenanthrene-based tylophorine derivatives as potential antiviral agents. J Agric Food Chem 58(23):12337–12342
Wang K, Su B, Wang Z, Wu M, Li Z, Hu Y, Fan Z, Mi N, Wang Q (2010b) Synthesis and antiviral activities of Phenanthroindolizidine alkaloids and their derivatives. J Agric Food Chem 58(5):2703–2709
Wang QC, Panis B, Engelmann F, Lambardi M, Valkonen JPT (2009) Cryotherapy of shoot tips: a technique for pathogen eradication to produce healthy planting materials and prepare healthy plant genetic resources for cryopreservation. Ann Appl Biol 154:351–363
Wani SH, Sanghera GS (2010) Genetic engineering for viral disease management in plants. Not Sci Biol 2:20–28
Wanitchakorn R, Harding RM, Dale JL (1997) *Banana bunchy top virus* DNA-3 encodes the viral coat protein. Arch Virol 142:1673–1680
WARDA (1999) Integrated management of *Rice yellow mottle virus* (RYMV) in low land and irrigated eco-systems in West Africa. In: Annual report 1998/1999, submitted to DFID, WARDA, Bouake, Cote d'Ivoire, pp 11–18
Warwick D, Demski JW (1998) Susceptibility and resistance of soybeans to *Peanut stripe virus*. Plant Dis 72:19–21
Wasswa P, Alicai T, Mukasa SB (2010) Optimisation of *in vitro* techniques for cassava brown streak virus elimination from infected cassava clones. Afr Crop Sci J 18:235–241
Watanabe KN (1994) Molecular genetics. In: Bradshaw JE, Mackay GR (eds) Potato genetics. CAB International, Wallingford
Waterhouse PM, Graham MW, Wang MB (1998) Virus resistance and gene silencing in plants can be induced by simultaneous expression of sense and antisense RNA. Proc Natl Acad Sci U S A 95:13959–13964
Waterworth P, Kahn RP (1978) Thermotherapy and aseptic bud culture of sugarcane to facilitate the exchange of germ plasm and passage through quarantine. Plant Dis Rep 62:72–776
Wathankul L (1964) A study on the host range of tungro and orange leaf viruses of rice. M.S. Thesis, University of Philippines
Watson MA (1959) Cereal virus diseases in Britain. NAASQ Rev 43:93–102
Watson MA, Plumb RT (1972) Transmission of plant-pathogenic viruses by aphids. Annu Rev Entomol 17:425–452
Watson MA, Hull R, Blencowe JW, Hamlyn BGM (1951) The spread of beet yellows and beet mosaic viruses in the sugar beet root crop. I. Field observations on the virus diseases of sugar beet and their vectors *Myzus persicae* Sulz. and *Aphis fabae* Koch. Ann Appl Biol 38:743–764

Way MJ, Heathcote GD (1966) Interactions of crop density of field beans, abundance of Aphis fabae scop, virus incidence and aphid control by chemicals. Ann Appl Biol 57:409–423

Webb RE, Bohn GW (1962) Resistance to cucurbit viruses in *Cucumis melo*. Phytopathology 52:1221

Webb RE, Larson RN, Walker JC (1952) Relationships of *Potato leaf roll virus* strains. Res Bull Agric Exp Stn Coll Agric Univ Wisconsin 178:1–38

Webley DP, Stone LEW (1972) Field experiments on potato aphids and virus spread in South Wales 1966/9. Ann Appl Biol 72:197–203

Wei G, Kloepper JW, Tuzun S (1991) Induction of systemic resistance of cucumber to *Colletotrichum orbiculare* by select strains of plant growth-promoting rhizobacteria. Phytopathology 81:1508–1512

Wellman FL (1932) Control of celery mosaic by eradicating wild hosts. Phytopathology 22:30

Wellman FL (1935) The host range of *Southern celery-mosaic virus*. Phytopathology 25:377–404

Wellman FL (1937) Control of celery mosaic by eradicating wild hosts. Tech Bull USDA 548:1–16

Welsh MF (1964) Evidence for suppression of apple virus disease symptoms by virus interference. Proc Can Phytopathol Soc 31:15

Wenzl H (1970) Die Bekampfung des Y-virus der kartoffel durch Olspritzungen und deren Auswirking auf die Phytopathora–krantfaule. Pflanzenschutz–Berichte 41:25–36

Wenzl H, Foschum H (1973) Die Bekamp fung des Y-virus der Kartoffel durch olsprit zungen. Z Pflanzenkrankh und Pflanzenschutz 80:341–345

Welz HG, Schechert A, Pernet A, Pixley KV, Geiger HH (1998) A gene for resistance to the *Maize streak virus* in the African CIMMYT maize in bred line CML 202. Mol Breed 4:147–154

Werner K, Friedt W, Ordon F (2005) Strategies for pyramiding resistance genes against the *Barley yellow mosaic virus* complex (BaMMV, BaYMV, BaYMV-2). Mol Breed 16:45–55

Wheeler GFC (1961) Glasshouse crops in soilless composts. Comm Grower 3436:920–923

White JG (1983) The use of methylbromide and carbendazim for the control of lettuce big-vein disease. Plant Pathol 32:151–157

Whitney JD, Wheaton TA, Castle WS, Tucker DPH (1995) Tree height, fruit size, and fruit yield affect manual orange harvesting rates. Proc Fla State Hortic Soc 108:112–118

Wilcoxson RD, Peterson AG (1960) Resistance of Dollard red clover to the pea aphid, *Macrosiphum pisi*. J Econ Entomol 53:863–865

Wilson CR (2001) Resistance to infection and translocation of *Tomato spotted wilt virus* in potatoes. Plant Pathol 50:402–410

Wintermantel WM, Zaitlin M (2000) Transgene translatability increases effectiveness of replicase-mediated resistance to *Cucumber mosaic virus*. J Gen Virol 81:587–595

Winter-Nielsen P (1972) Traktorredskaber som spredere of virus x i kartofler. Tidsskrift for Planteavl 76:297–307

Wisler GC, Duffus JE (2000) A century of plant virus management in the Salinas Valley of California, 'East of Eden'. Virus Res 71:161–169

Witcher W, Smith FH (1980) Control of *Watermelon mosaic virus* on summer squash with oil. Phytopathology 70:572

Wolf FA (1933) Roguing as a means of control of tobacco mosaic. Phytopathology 23:831–833

Wolfenbarger DO (1966) Incidence–distance and incidence–time relationships of papaya virus infections. Plant Dis Rep 50:908–909

Wolfenbarger DO, Adlerz WC (1971) Will aluminum control some aphid-transmitted virus diseases? Sunshine State Agric Res Rep 16:4–5

Wolfenbarger DO, Moore WD (1967) Mulch treatments of squash and tomatoes with respect to virus infections and yields. Proc Fla State Hortic Soc 80:217–221

Woodford JAT, Gordon SC, Foster GN (1988) Side-band application of systemic granular pesticides for the control of aphids and potato leafroll virus. Crop Prot 7:96–105

Wright NS, Hughes EC (1964) Effect of defoliation date on yield and leaf roll incidence in potato. Am Potato J 41:83–91

WTO (1995) Agreement on the application of phytosanitary measures. In: Results of the Uruguay Round of multilateral trade negotiations: the legal tests. World Trade Organization, Geneva

Wyman JA, Tocano NC, Kido K, Johnson H, Mayberry KS (1979) Effects of mulching on the spread of aphid-transmitted *Watermelon mosaic virus* to summer squash. J Econ Entomol 72:139–143

Xia Y, Fan Z, Yao J, Liao Q, Li W, Qua F, Peng L (2006) Discovery of bitriazolyl compounds as novel antiviral candidates for combating the Tobacco mosaic virus. Bioorg Med Chem Lett 16:2693–2698

Xiansong Y (2010) Rapid production of virus-free plantlets by shoot tip culture in vitro of purple-coloured sweet potato (*Ipomoea batatas* (L.) Lam.). Pak J Biol 42:2069–2075

Xiuhua Z, Shuzhen Z, Wengi C, Bo T (1976) The inhibitory effects of oil emulsion on the aphid transmission of non-persistent plant viruses. Acta Microbiologia Sinica 16:142–147

Xu P, Zhang Y, Kang L, Roossinck MJ, Mysore KS (2006) Computational estimation and experimental verification of off-target silencing during post transcriptional gene silencing in plants. Plant Physiol 142:429–440

Yadav JS, Ogwok E, Wagaba H, Patil BL, Bagewadi B, Alicai T, Gaitan-Solis E, Taylor NJ, Fauquet CM (2011) RNAi-mediated resistance to *Cassava brown streak uganda virus* in transgenic cassava. Mol Plant Pathol 12:677–687

Yamaga H, Munakata T (1991) Production of virus free apple planting stock by meristem culture. Tech Bull Food Fertilizer Technol Center 126:10–17

Yamasaki S, Sakai J, Kamisoyama S, Goto H, Okuda M, Hanada K (2009) Control of russet disease in sweet potato plants using a protective mild strain of *Sweet potato feathery mottle virus*. Plant Dis 93:190–194

Yang H, Ozias-Akins P, Gulbreath AK, Gorbet DW, Weeks JR, Mandal B, Pappu HR (2004a) Field evaluation of *Tomato spotted wilt virus* resistance in transgenic peanuts. Plant Dis 88:259–264

Yang Y, Sherwood TA, Patte CP, Hiebert E, Polston JE (2004b) Use of *Tomato yellow leafcurl virus* (TYLCV) rep gene sequences to engineer TYLCV resistance in tomato. Phytopathology 94:490–496

Yang WZ, Hsiao CH, Chang WN (1986) Screening cucumbers from resistance to viruses and inheritance to *Zucchini yellow mosaic virus*. J Agric Res China 35:192–201

Yang X, Yie Y, Zhu F, Liu Y, Kang L, Wang X, Tien P (1997) Ribozyme mediated high resistance against *Potato spindle tuber viroid* in transgenic potatoes. Proc Natl Acad Sci U S A 94:4861–4865

Yeh S-D, Gonsalves D (1994) Practices and perspective of control of *Papaya ringspot virus* by cross protection. In: Harris KF (ed) Advances in disease vector research, vol 10. Springer, New York, pp 237–257

Yeh S-D, Gonsalves D, Wang H-L, Namba R, Chiu R-J (1988) Control of *Papaya ringspot virus* by cross protection. Plant Dis 72:375–380

Yeh SD, Chao CH, Cheng YH, Chen CC (1997) Serological comparison of four distinct tospo viruses by polyclonal antibodies to purified nucleocapsid proteins. Acta Hortic 431:123–134

Yeh SD, Bau HJ, Cheng YH, Yu TA, Yang JS, Drew RA (1998) Greenhouse and field evaluations of coat protein transgenic papaya resistant to *Papaya ringspot virus*. Acta Hortic 461:321–328

Yencho G, Cohen M, Byrne P (2000) Application of tagging and mapping insect resistance loci in plants. Annu Rev Entomol 45:393–422

Young ND, Tanksley SD (1989) RFLP analysis of the size of chromosomal segments retained around the Tm-2 locus of tomato during backcross breeding. Theor Appl Genet 77:353–359

Youtsey CO (1978) A method for virus-free propagation of citrus-shoot-tip grafting. Citrus Ind 59:39

Yu H, Kumar PP (2003) Post-transcriptional gene silencing in plants by RNA. Plant Cell Rep 22:167–174

Zaitlin M, Anderson JM, Perry KL, Zhang L, Palukaitis P (1994) Specificity of replicase-mediated resistance to cucumber mosaic virus. Virology 201:200–205

Zakay Y, Navot N, Zeidan M, Kedar N, Rabinowitch H, Czosnek H, Zamir D (1991) Screening of *Lycopersicon* accessions for resistance to Tomato yellow leafcurl virus: presence of viral DNA and symptom development. Plant Dis 75:279–281

Zaman SM, Quraishi A, Hassan G, Raziuddin S, Ali A, Khabir A, Gul N (2001) Meristem culture of potato (*Solanum tuberosum* L.) for production of virus-free plantlets. Online J Biol Sci 1:898–899

Zanek MC, Reyes CA, Cervera M, Peña EJ, Velazquez K, Costa N, Plata MI, Grau O, Pena L, Garcia ML (2008) Genetic transformation of sweet orange with the coat protein gene of Citrus psorosis virus and evaluation of resistance against the virus. Plant Cell Rep 27:57–66

Zaumeyer WJ, Meiners JP (1975) Diseases resistance in beans. Annu Rev Phytopathol 13:313–334

Zerbini FM, Michelmore RW, Gilbertson RL (1995) Resistance to LMV infection in lettuce due to the expression of different forms of the LMV coat protein gene. Phytopathology 85:1138

Zettler FW, Christie RG, Edwardson JR (1967) Aphid transmission of virus from leaf sectors correlated with intracellular inclusions. Virology 33:549–552

Zettler FW, Elliott MS, Purcifull DE, Mink GI, Gorbet DW, Knauft DA (1993) Production of peanut seed free of *Peanut stripe* and *Peanut mottle viruses* in Florida. Plant Dis 77:747–749

Zhang ZH, Xiao M, Yang HY, Li H, Gao XY, Du GD (2006) Evaluation and comparison on methods of virus elimination from the strawberry plants. J Fruit Sci 23:720–723

Zhao WG, Wang JG, Li ZM, Yang Z (2006) Synthesis and antiviral activity against *Tobacco mosaic virus* and 3D-QSAR of alpha-substituted-1,2,3-thiadiazoleacetamides. Bioorg Med Chem Lett. 16(23):6107–6111

Zheng L, Wayper PJ, Gibbs AJ, Fourment M, Rodoni BC, Gibbs MJ (2008) Accumulating variation at conserved sites in potyvirus genomes in driven by species discovery and affects degenerate primer design. PLOS ONE 3. P.e. 1586. doi:10.137/Journal.Pone.0001586

Zhou C, Zhou Y (2012) Strategies for viral cross protection in plants. Methods Mol Biol 894:69–81

Ziebell H, Carr JP (2010) Cross protection: a century of mystery. Adv Virus Res 76:212–264

Zilka S, Faingersh G, Rotbaum A, Malca N, Tam Y, Spiegel S, Malca N (2002) In vitro production of virus-free pear plants. Acta Hortic 596:477–479

Zimmerman GS (1979) Reducing the spread of *Potato leaf roll virus*, *Alfalfa mosaic virus* and *Potato virus Y* in seed potatoes by trapping aphids on sticky yellow polythene sheets. Potato Res 22:123–131

Zimmerman GS, Pilowsky M (1975) Experiments for protecting tomatoes, from *Tobacco mosaic virus* (TMV) by prior infection with the virulent strain M II-16. Phytopathology 3:75 (Abstr.)

Zink FW, Grogan RG, Welch JE (1956) The effect of the percentage of seed transmission upon subsequent spread of *Lettuce mosaic virus*. Phytopathology 46:662–664

Zink FW, Grogan RG, Bardin R (1957) The comparative effect of mosaic-free seed and roguing as a control for common lettuce mosaic. Proc Am Soc Hortic Sci 70:277–280

Zinnen TM, Vachris JW (1990) Insecticidal soap reduces infection by two mechanically transmitted plant viruses. Plant Dis 74:201–202

Zitter TA (1977) Epidemiology of aphid-borne viruses. In: Harris KF, Maramorasch K (eds) Aphids as virus vectors. Academic Press, London, pp 385–412

Zitter TA (1978) Virus control with oil sprays. Am Veg Grower 26:12–13

Zitter TA (1979) Methods for controlling the most common vegetable viruses in South Florida. Belle Glade AREC Report EC-1979-1980

Zitter TA (1991) Tomato mosaic and tobacco mosaic. In: Jones JB, Jones JP, Stall RE, Zitter TA (eds) Compendium of tomato diseases. APS Press, St. Paul, p 39

Zitter TA, Everett PH (1979) Use of mineral oil sprays to reduce the spread of tomato yellows virus disease in Florida. University of Florida Immokalee ARC Research Report S.F. 79–1, 7 pp

Zitter TA, Ozaki HT (1978) Aphid borne vegetable viruses controlled with oil sprays. Proc Fla State Hortic Soc 91:287–289

Zitter TA, Simons JN (1980) Management of viruses by alteration of vector efficiency and cultural practices. Annu Rev Phytopathol 18:289–310

Zitter TA, Pink DAC, Walkey DGA (1991) Reaction of *Cucumber pepo* L. cv. Cindrella to strain of cucumber mosaic virus. Cucurbit Genet Coop Rep 14:125–128

Zschiegner HJ, Kramer W, Sasso O, Fritzsche R, Dubnik H (1971) Successful restriction of the spread of nonpersistent viruses by methods of virus vector control. In: Proceedings of the 6th British insecticide fungicide conference, pp 319–323

Zschiegner HJ, Eisenbrandt K, Fritzsche R (1974) Beitrag Zum Wirkings mechanismus Von Mineralolen bei der Verhinderung der virusuber tragung durch Vektoren. Archiv fur Phytopathologie und Pflanzen Schutz 10:371–382

Zulfeghar Ahmed, Patil MS (2004) Screening of okra varieties against *Okra yellow vein mosaic virus*. Karnataka J Agric Sci 17:613–614

Zwick RW, Westigard PH (1978) Prebloom petroleum oil applications for delaying pear psylla (Homoptera-psyllidae) oviposition. Can Entomol 110:225–236

Index

K. S. Sastry and T. A. Zitter, *Plant Virus and Viroid Diseases in the Tropics*,
DOI: 10.1007/978-94-007-7820-7, © Springer Science+Business Media B.V. 2014

C

D

T

U

Printed by Printforce, the Netherlands